P9-CQG-452

Angular velocity and linear velocity

$$v_T = \omega r \qquad (6.8)$$

kg/ms^2

Average angular acceleration

$$\bar{\alpha} = \frac{\Delta \omega}{\Delta t} = \frac{\omega_f - \omega_i}{\Delta t} \qquad (6.9)$$

Tangential acceleration

$$a_T = \alpha r \qquad (6.10)$$

Centripetal acceleration

$$a_r = \frac{v_T^2}{r} \qquad (6.11)$$

$$a_r = \omega^2 r \qquad (6.12)$$

ANGULAR KINETICS

Torque

$$T = F \times r \qquad (5.1)$$

Static equilibrium

$$\Sigma T = 0 \qquad (5.2)$$

Center of gravity

$$\Sigma(W \times r) = (\Sigma W) \times r_{cg} \qquad (5.3)$$

Moment of inertia

$$I_a = \Sigma m_i r_i^2 \qquad (7.1)$$

inertia cog

$$I_a = mk_a^2 \qquad (7.2)$$

Moment of inertia: parallel axis theorem

$$I_b = I_{cg} + mr^2 \qquad (7.3)$$

Angular momentum

$$H_a = I_a \omega_a \qquad (7.4)$$

Angular momentum of the human body

$$H_a = \Sigma(I_i \omega_i + m_i r^2_{i/cg} \omega_{i/cg}) \qquad (7.5)$$

Conservation of angular momentum

$$H_i = I_i\omega_i = I_f\omega_f = H_f = \text{constant if } \Sigma T = 0 \qquad (7.7)$$

Angular version of Newton's 2nd law

$$\Sigma T_a = I_a \alpha_a \qquad (7.9)$$

$$\Sigma \overline{T}_a = \frac{\Delta H_a}{\Delta t} = \frac{(H_f - H_i)}{\Delta t} \qquad (7.10)$$

Angular impulse-momentum

$$\Sigma \overline{T}_a \Delta t = (H_f - H_i)_a \qquad (7.11)$$

FLUID MECHANICS

Pressure

$$P = \frac{F}{A}$$

Density

$$\rho = \frac{m}{V} \qquad (8.3)$$

Drag force

$$F_D = \frac{1}{2} C_D \rho A v^2 \qquad (8.5)$$

Lift force

$$F_L = \frac{1}{2} C_L \rho A v^2 \qquad (8.6)$$

MECHANICS OF MATERIALS

Stress

$$\sigma = \frac{F}{A} \qquad (9.1)$$

Shear stress

$$\tau = \frac{F}{A} \qquad (9.2)$$

Strain

$$\varepsilon = \frac{\ell - \ell_o}{\ell_o} \qquad (9.4)$$

Elastic modulus

$$E = \frac{\Delta \sigma}{\Delta \varepsilon} \qquad (9.5)$$

Abbreviations for Variables and Subscripts Used in Equations

Variables

a = instantaneous linear acceleration
$\bar{a}$ = average linear acceleration
A = area
C_D = coefficient of drag
C_L = coefficient of lift
d = displacement
e = coefficient of restitution
E = energy
E = elastic modulus or Young's modulus
F = force
$\bar{F}$ = average force
F_d = dynamic friction force
F_s = static friction force
ΣF = net force = sum of forces
g = acceleration due to gravity
G = gravitational constant
h = height
H = angular momentum
I = moment of inertia
k = radius of gyration
k = stiffness or spring constant
KE = kinetic energy
ℓ = distance traveled or length

L = linear momentum
m = mass
P = power
P = pressure
P = force
PE = gravitational potential energy
r = radius
r = moment arm
R = normal contact force
s = instantaneous linear speed
$\bar{s}$ = average linear speed
t = time
T = torque
u = pre-impact velocity
U = work done
v = instantaneous linear velocity
v = post-impact velocity
$\bar{v}$ = average linear velocity
V = volume
W = weight
x = horizontal position
y = vertical position
α = instantaneous angular acceleration
$\bar{\alpha}$ = average angular acceleration
Δ = change in … = final − initial

ε = strain
μ = coefficient of friction
ρ = density
σ = stress
Σ = sum of …
τ = shear stress
θ = angular position
ω = instantaneous angular velocity
$\bar{\omega}$ = average angular velocity

Subscripts

a = axis
b = axis
d = dynamic
cg = center of gravity
D = drag
f = final or ending
i = initial or starting
i = one of a number of parts
L = lift
o = original or undeformed
r = radial
s = static
T = tangential
x = horizontal
y = vertical

HUMAN KINETICS WEB RESOURCE

How to access the supplemental web resource

We are pleased to provide access to a web resource that supplements your textbook, *Biomechanics of Sport and Exercise, Third Edition.* This resource offers all of the problems from the book for easy reference, plus select questions worked out step-by-step in a format that provides you with hints as you work toward the solution.

Accessing the web resource is easy!
Follow these steps if you purchased a new book:

1. Visit **www.HumanKinetics.com/BiomechanicsOfSportAndExercise**.

2. Click the <u>third edition</u> link next to the corresponding third edition book cover.

3. Click the Sign In link on the left or top of the page. If you do not have an account with Human Kinetics, you will be prompted to create one.

4. If the online product you purchased does not appear in the Ancillary Items box on the left of the page, click the Enter Key Code option in that box. Enter the key code that is printed at the right, including all hyphens. Click the Submit button to unlock your online product.

5. After you have entered your key code the first time, you will never have to enter it again to access this product. Once unlocked, a link to your product will permanently appear in the menu on the left. For future visits, all you need to do is sign in to the textbook's website and follow the link that appears in the left menu!

→ Click the Need Help? button on the textbook's website if you need assistance along the way.

How to access the web resource if you purchased a used book:

You may purchase access to the web resource by visiting the product's website, **www.HumanKinetics.com/BiomechanicsOfSportAndExercise**, or by calling the following:

800-747-4457 . . . . U.S. customers
800-465-7301 . . . . .Canadian customers
+44 (0) 113 255 5665 . . . . European customers
08 8372 0999 . . . . Australian customers
0800 222 062 . . . . .New Zealand customers
217-351-5076 . . . . .International customers

For technical support, send an e-mail to:
support@hkusa.com . . . . U.S. and international customers
info@hkcanada.com . . . . Canadian customers
academic@hkeurope.com . . . . European customers
keycodesupport@hkaustralia.com . . . . . Australian and New Zealand customers

HUMAN KINETICS
The Information Leader in Physical Activity & Health

01–2013

Product: Biomechanics of Sport and Exercise, Third Edition, web resource

Key code: MCGINNIS-V32VIH-OSG

This unique code allows you access to the web resource.

Access is provided if you have purchased a new book. Once submitted, the code may not be entered for any other user.

How to Access MaxTRAQ Educational 2D Software

We are pleased to provide access to MaxTRAQ Educational 2D motion analysis software tailored to this text. MaxTRAQ Educational 2D is a downloadable software module that allows you to track and analyze human movement, distance, and angles in video clips. See the section How to Use MaxTRAQ on page xi in the book for more information.

Follow these steps to access the MaxTRAQ Educational 2D software:

1. Go to **www.motionanalysisproducts.com/Books/PM-BSE-R3.html**. Here you will find the download page for MaxTRAQ Educational. If you have purchased a new book, click on the New Books link under Book Downloads-MaxTRAQ Educational to download the software installer. If you have purchased a used book, you can click on the Used Books link to purchase a copy of the software.
2. After you have saved the MaxTRAQ installer to a folder on your computer, double-click it to install the software. See How to Install MaxTRAQ Educational at **www.motionanalysisproducts.com/Books/PM-BSE-R3.html** for more information.
3. After the software has installed, open the program. When the License window appears, click on Enter Key at the lower left. Enter the MaxTRAQ Educational key code printed on this sheet. If you purchased a Used Books key code, it will be sent via e-mail.
4. Take the time to view the MaxTRAQ Tutorial Video and How to Use the Video Files document at **www.motionanalysisproducts.com/Books/PM-BSE-R3.html**. The video clips you will use to complete the exercises in the text will be downloaded with the software and saved to your local disk under Program Files\Motion Analysis\MaxTRAQP\VideoFiles.
5. Send any questions or inquiries via e-mail to **support@motionanalysisproducts.com**. If you need assistance, please provide information on the platform and operating system you use, the type of question you have, the exact text of the error message received, where in the program the error was received, and any steps you may have already taken to fix the problem. Also include the key code for the book. For many questions, it may be best to begin by working with your professor or a fellow student who has successfully used the software.

MaxTRAQ Educational operates in Windows 2000, XP, Vista, or Windows 7 Pentium Class, 512 MB, True Color only, 32 bit recommended. Final files will be delivered in Windows format only. MAP supports only Windows 32 bit. If you are using a Mac, you can try running Virtual PC or a similar program on your Mac.

Questions? Difficulty activating your account?
Contact **support@motionanalysisproducts.com**

MAP
Motion Analysis Products

Product: MaxTRAQ Educational 2D Software

Key code: 00002X-KA58KH-7-J4D3D-G73BPJ-8F32G5-R6429K

This unique code allows you to activate the MaxTRAQ Educational 2D software for a period of one year:

Access is provided if you have purchased a new book. This code can only be used on one computer and, once submitted, the code may not be entered for any other user.

Biomechanics of Sport and Exercise

Third Edition

Peter M. McGinnis

State University of New York, College at Cortland

Human Kinetics

Library of Congress Cataloging-in-Publication Data

McGinnis, Peter Merton, 1954-
  Biomechanics of sport and exercise / Peter M. McGinnis. -- 3rd ed.
    p. ; cm.
  Includes bibliographical references and index.
  I. Title.
  [DNLM: 1. Biomechanics. 2. Exercise--physiology. 3. Sports--physiology. WE 103]
  612.7'6--dc23

                              2012034730

ISBN-10: 0-7360-7966-1 (print)
ISBN-13: 978-0-7360-7966-2 (print)

Copyright © 2013, 2005, 1999 by Peter M. McGinnis

All rights reserved. Except for use in a review, the reproduction or utilization of this work in any form or by any electronic, mechanical, or other means, now known or hereafter invented, including xerography, photocopying, and recording, and in any information storage and retrieval system, is forbidden without the written permission of the publisher.

The web addresses cited in this text were current as of August 2012, unless otherwise noted.

Acquisitions Editors: Amy N. Tocco and Loarn D. Robertson, PhD; **Developmental Editor:** Katherine Maurer; **Assistant Editors:** Brendan Shea and Susan Huls; **Copyeditor:** Joyce Sexton; **Indexer:** Nancy Ball; **Permissions Manager:** Dalene Reeder; **Graphic Designer:** Nancy Rasmus; **Graphic Artist:** Dawn Sills; **Cover Designer:** Keith Blomberg; **Photograph (cover):** © Mike Kemp/Tetra Images/age fotostock; **Photographs (interior):** Courtesy Peter M. McGinnis unless otherwise noted; photos on pages 5, 22, 51, 87, 115, 125, 150, 207, 239, 277, 315, and 361 © Human Kinetics; **Photo Asset Manager:** Laura Fitch; **Photo Production Manager:** Jason Allen; **Art Manager:** Kelly Hendren; **Associate Art Manager:** Alan L. Wilborn; **Art Style Development:** Joanne Brummett; **Illustrations:** © Human Kinetics; **Printer:** Thomson-Shore, Inc.

Printed in the United States of America   10 9 8 7 6 5 4 3 2 1

The paper in this book is certified under a sustainable forestry program.

Human Kinetics
Website: www.HumanKinetics.com

United States: Human Kinetics
P.O. Box 5076
Champaign, IL 61825-5076
800-747-4457
e-mail: humank@hkusa.com

Canada: Human Kinetics
475 Devonshire Road Unit 100
Windsor, ON N8Y 2L5
800-465-7301 (in Canada only)
e-mail: info@hkcanada.com

Europe: Human Kinetics
107 Bradford Road
Stanningley
Leeds LS28 6AT, United Kingdom
+44 (0) 113 255 5665
e-mail: hk@hkeurope.com

Australia: Human Kinetics
57A Price Avenue
Lower Mitcham, South Australia 5062
08 8372 0999
e-mail: info@hkaustralia.com

New Zealand: Human Kinetics
P.O. Box 80
Torrens Park, South Australia 5062
0800 222 062
e-mail: info@hknewzealand.com

E4696

This third edition is dedicated to the memory of two strong, outgoing, and adventurous women whose presence raised the spirits of those around them—my mother, Doris Joye McGinnis (1925-2009), and my friend, colleague, and former student, Julianne Abendroth (1962-2011).

contents

This textbook was written to introduce undergraduate students to the field of sport and exercise biomechanics. The text is primarily intended for undergraduate students majoring in kinesiology, exercise science, or physical education, but it is suitable for students in other human movement fields as well. Most of the examples and applications appearing in the book are from sport or exercise, but examples from clinical and everyday human movement activities have also been included. No matter what human movement field readers are interested in, knowledge of mechanics will be valuable to them in their work. Many human movement professionals' only formal instruction in the mechanics of human movement occurs during a single undergraduate course in kinesiology or biomechanics. This book was developed with this constraint in mind. The goal of the book and its ancillary materials is to present an introduction to the biomechanics of human movement in a clear, concise, user-friendly manner.

This third edition is an improvement over the previous edition in a number of ways. Photos and select figures have been updated. New material has been added, including new sport examples, new sample problems, and discussion of new technologies used by researchers in quantitative biomechanical analysis. Review questions and problems have been added to many chapters, and many problems have been enhanced with new diagrams to help students visualize the mechanics of real-world scenarios.

Also new to this edition is the inclusion of video motion analysis software along with the book. With the purchase of a new textbook, students receive a license and instructions to download an educational version of the motion analysis software MaxTRAQ, along with number of video clips. Chapters 2, 3, 4, and 6 include exercises that require use of the MaxTRAQ software to measure kinematic variables in the various video clips that accompany the software. The software can also be used to analyze human movement in video clips recorded and supplied by the user, and chapter 16

available at
your campus bookstore
or HumanKinetics.com

includes an exercise in which students can record and analyze their own video clip.

The organization of the book remains intact. The introductory chapter provides an introduction to biomechanics, which includes justifications for the study of biomechanics. It also includes an overview of the organization of mechanics and an introduction to measurement systems. The rest of the book is divided into three parts.

Part I concerns external biomechanics, or external forces and their effects on the body and its movement. Rigid-body mechanics with applications to human movement is the primary topic of this part of the text. Mechanics is one of the most difficult topics for undergraduate students of human movement to understand, so this is the most important and largest part of the book. The order of presentation of topics in this part differs from that in most other biomechanics texts. Chapter 1 presents the concepts of force and static equilibrium. With forces as the example, this chapter also introduces vector addition and resolution. The trigonometry used to add and resolve forces is also explained. Chapter 2 discusses linear motion and its description. This chapter includes equations that describe the motion of an object undergoing constant acceleration and their application to describing projectile motion. Chapter 3 presents the causes of linear motion and introduces Newton's three laws of motions, as well as the conservation of momentum principle. Chapter 4 discusses mechanical work and energy principles and also the mechanical work done by muscles. Torque, moment of force, and center of gravity are introduced in chapter 5 preceding a discussion of angular kinematics in chapter 6. The causes of angular motion are presented in chapter 7, along with the angular analogs to Newton's three laws of motion. Part I concludes with a discussion of fluid mechanics in chapter 8.

Part II concerns internal biomechanics, or internal forces and their effects on the body and its movement. This part begins with a discussion of the mechanics of biological materials in chapter 9. Stress and strain are introduced in this chapter, along with various concepts of material strength. Overviews of the skeletal system, muscular system, and nervous system control are then presented in chapters 10, 11, and 12.

Part III deals with the application of biomechanics. Commonsense methods of applying biomechanics to the analysis of sport or human movement skills are presented in the first three chapters of this part. The first of these chapters, chapter 13, presents procedures for completing qualitative biomechanical analyses to improve technique. Chapter 14 presents a method of qualitative biomechanical analysis to improve training. A qualitative procedure for identifying active muscle groups in phases or parts of movements is emphasized in this chapter. Chapter 15 is an examination of how qualitative biomechanical analysis can be used to help understand the causes of injury. This chapter was written by Steven McCaw. Chapter 16 is the concluding chapter of the book. It gives an overview of the technology used in conducting quantitative biomechanical analyses.

Throughout the book, and especially in part I, the objective has been to allow students to discover the principles of mechanics for themselves. Common activities are observed, and explanations for these activities are then developed. The resulting explanations reveal the underlying mechanical concepts. This discovery process requires more active participation by the reader, but it results in a better understanding of the subject matter.

What makes this book unique among biomechanics textbooks is its order of presentation. In most undergraduate biomechanics textbooks, functional anatomy is presented before mechanics. This textbook presents mechanics first. Bones and ligaments are the structural elements that support the human body. Muscles are the motors that move this structure. Understanding how the forces exerted by bones and ligaments support the body, and how the forces and torques produced by muscles do work to move the body's limbs, requires a knowledge of forces and their effects. Mechanics is the study of forces and their effects. Thus mechanics should precede the study of the musculoskeletal system.

This book is also unique in its order of presentation of mechanical topics. The mechanics section of most biomechanics textbooks begins with linear kinematics and then deals with linear kinetics, angular kinematics, and finally angular kinetics. This book introduces forces before presenting linear kinematics. Because forces are the causes of changes in motion and forces are in equilibrium if no changes in motion occur, it makes sense to define and understand forces before discussing motion.

This is especially true if projectile motion and the equations for projectile motion are discussed under linear kinematics. Since projectile motion is influenced by the force of gravity, an understanding of this force should precede discussion of the effects of the force. Similarly, torques are introduced before the discussion of angular kinematics.

Since mechanics uses equations to describe relationships or to define quantities, some knowledge of mathematics (primarily algebra) is necessary. I have tried to write the book in such a way that even students with weak math skills can succeed in learning biomechanics. However, success in learning the material will come more easily to those who are better prepared mathematically.

Appendix A lists the principal units used for mechanical quantities in the International System of Units, as well as prefixes and the conversions to customary units used in the United States.

Each chapter of this book includes elements intended to help the reader learn the material. Each chapter begins with a list of objectives and an opening scenario leading to questions that readers can answer after reading and understanding the material presented in the chapter. Practical examples of concepts are integrated into the text throughout each chapter. Sample problems are presented, and the step-by-step procedures for solving them are illustrated. Problems and review questions appear at the end of each chapter to test the reader's understanding of the material presented. More problems and review questions have been added to this edition. Answers to problems and most of the review questions appear in appendix B. Finally, the educational version of the video motion analysis software MaxTRAQ and accompanying video clips are included with this new edition. Exercises that require the use of this software appear at the ends of chapters 2, 3, 4, 6, and 16, and the solutions to these exercises appear in appendix B. These video analysis problems may more clearly illustrate principles presented in each chapter.

Throughout the text, I have tried to explain and illustrate the concepts as clearly as possible and in such a way that you, the reader, are actively involved in discovering them. Still, you might find that some of the material is challenging. Occasionally, you might find yourself distracted or confused while reading. Don't give up! Your effort will be worthwhile.

acknowledgments

A book is not produced by one individual—a team of people is involved. I thank the members of that team: the folks at Human Kinetics, including Amy Tocco, Kate Maurer, and especially Loarn Robertson, whose patience I tested; Victoria Berger of Motion Analysis Products, who was responsible for the inclusion of the MaxTRAQ software; Steve McCaw of Illinois State University, who was responsible for chapter 15; my students and colleagues who provided suggestions for improvement; and my extended family, whose support and encouragement is further acknowledged by the appearance of their names in problems and questions throughout the book. Special thanks to my wife, Boodie, for her love and support.

student and instructor resources

Student Resources

Students, visit the free student web resource at www.HumanKinetics.com/BiomechanicsOfSportAndExercise. The web resource takes select end-chapter problems from the textbook and provides a sequence of hints to guide you through the problems, helping you to develop your skills and gain confidence working through biomechanical problems. It also provides downloadable PDF copies of all end-chapter questions and problems. Problems that are worked through in the web resource are marked with the web resource icon:

In addition, with the purchase of a new textbook you will receive access to the MaxTRAQ motion analysis software. To access and download this software, visit www.motionanalysisproducts.com/Books/PM-BSE-R3.html. If you have purchased a used book or e-book, you may purchase the software separately at the MaxTRAQ site. The MaxTRAQ software is compatible with Windows operating systems only. For more on MaxTRAQ, see the section "How to Use MaxTRAQ."

Instructor Resources

The instructor guide, test package, and image bank are free to course adopters and are accessed at www.HumanKinetics.com/BiomechanicsOfSportAndExercise.

Instructor Guide

Specifically developed for instructors who have adopted *Biomechanics of Sport and Exercise, Third Edition,* the instructor guide includes chapter summaries, objectives, and outlines, plus lecture ideas and sample outlines, student activities, and teaching tips. The instructor guide also includes PDFs that show the mathematical and graphic work done to arrive at the correct answer for each of the end-chapter review problems. These can be used as teaching aids or for evaluating student work.

Test Package

The test package includes a bank of over 380 questions created especially for *Biomechanics of Sport and Exercise, Third Edition.* Various question types are included, such as true-or-false, fill-in-the-blank, essay and short answer, and multiple choice. The test package is available for download in three different formats: Rich Text (.rtf), Respondus, and Learning Management System (LMS).

Image Bank

New for the third edition, the image bank includes most of the illustrations, photos, and tables from the text, sorted by chapter. These are provided as separate files for easy insertion into lecture presentations and other course materials, providing instructors with the flexibility to create customized resources.

Several chapters in this book

include activities that use MaxTRAQ Educational 2D software. MaxTRAQ Educational (ME) is a downloadable software module that allows you to track and analyze human movement, distance, and angles in video clips. ME operates only under Windows operating systems (Windows XP, Windows Vista, and Windows 7).

To use ME to complete the activities in this text, you will need to download the software, install it on your computer, and enter a key code to activate it. If you have purchased a new print book, you may enter the code provided on the blue letter bound into the front of the text. If you have purchased a used book or e-book, you will need to purchase the MaxTRAQ Educational 2D software directly from the MaxTRAQ site and your key code will be sent to you via e-mail.

To download the software, go to **www.motionanalysis products.com/Books/PM-BSE-R3.html**. If you have purchased a new book, click on the New Books link under Book Downloads—MaxTRAQ Educational to download the software installer. If you have purchased a used book or e-book, you can click on the Used Books link to purchase a copy of the software.

After you have saved the MaxTRAQ installer to a folder on your computer, double-click it to install the software. See How to Install MaxTRAQ Educational at www.motionanalysisproducts.com/Books/PM-BSE-R3. html for more information. The first time you open the program, a License window will appear. Click on Enter Key at the lower left. Carefully enter the key code that came with this book or the code you were sent via e-mail (cut and paste the code) if you purchased the software access separately. Once you have entered your key code, access is provided for one year.

Tips for Using MaxTRAQ Educational 2D

- The software download for this book includes 12 video clips created by the author. When you download the software, these video clips are saved to your local disk at Program Files\Motion Analysis\ MaxTRAQP\VideoFiles. You may also use the MaxTRAQ software to analyze other video clips that are saved in .avi format.
- A tutorial video and guidance on using the video files are available at www.motionanalysis products.com/Books/PM-BSE-R3.html. See the links in the lower-right corner under Short Tutorials and Tours. The video shows several versions of MaxTRAQ. Follow along with the manual tracking version.

- The video clips provided with the software require different deinterlacing settings. Follow the instructions at the beginning of each set of exercises to set the deinterlacing options for the video clip used.

- To view a video clip, open it using File—Open, then use the player buttons in the center bottom of the screen. The buttons to the left and right of the center Stop button advance or rewind the video one frame at a time. You will use these a lot as you complete the exercises. Also note that you can adjust the brightness and contrast of a video if needed to improve visibility.

- Note that the frame rate of the video clips varies. The frame rate is shown to the lower right of the MaxTRAQ screen in hertz (Hz). Hertz are cycles per second, so a frame rate of 30 Hz equals 30 frames per second (fps). For some of the video clips, MaxTRAQ does not recognize the true frame rate and the exercise will instruct you in doing a conversion.

- The software functions you will use the most to complete the exercises in this text are setting the scale, digitizing points on the subject, and calculating the changes in the x and y coordinates of the points as the subject moves.

  - To set the scale, open a video, then go to View—Tools and select Show Scale. Next, click on Tools in the top menu bar and Select Scale. In the window that pops up, enter the length of a known reference in the video (for the video clips included with the software, see the exercise instructions for the length to enter). Click OK and then use the mouse to left-click each end point of the known distance in the frame (for example, two points known to be 500 cm apart on a wall). When you click the first point, nothing will appear on the screen, but when you click the second, a scale will appear. To hide the scale, go to View—Tools and uncheck Show Scale.

- To digitize points on the subject, use the arrows for Number of Points on the right side of the MaxTRAQ screen to select the number of points you will need to digitize in any single frame of video (often this will be only one). Click the Digitize button in the right-hand column, then position the cursor over the spot on the video that you want to digitize and left-click the mouse button. To digitize points in subsequent frames of video, just advance the video, position the cursor over the spot on the video, and left-click again. Repeat this in each frame and for each point that you want to digitize.

- To improve the accuracy of your digitizing, you can enlarge the video image around the point you wish to digitize. Position the cursor near the point and right-click the mouse. Select Zoom and then Point Zoom in the windows that appear. To zoom back out to regular view, right-click, select Zoom, and Point Zoom again. To adjust the magnification of the Point Zoom, select Tools—Options on the top menu bar. In the window that appears, select General from the choices on the left,

then set the Point Zoom Magnification under the first option in the right side of the window. It is recommended that you use Point Zoom when digitizing the ends of the reference measure.

- For each digitized point, the coordinates along the x (horizontal) and y (vertical) axes will be displayed next to the point in that order (x, y). To show or hide these coordinates, go to View—Point Markers and check or uncheck Show Coordinates.

- If you make a mistake when creating a scale or digitizing a point, you can right-click the element and select Delete Scale or Delete Point to remove it.

- If you are digitizing more than one point in a frame, make sure that the point you want to digitize next (point 1, point 2, and so on) is highlighted in the right column before you click to set the location of that point.

- For many questions, it may be best to begin by working with your professor or a fellow student who has successfully used the software. If you need further advice, send questions or inquiries via e-mail to **support@motionanalysisproducts. com.**

Why Study Biomechanics?

objectives

When you finish reading this introduction, you should be able to do the following:

- Define biomechanics

- Define sport and exercise biomechanics

- Identify the goals of sport and exercise biomechanics

- Describe the methods used to achieve the goals of sport and exercise biomechanics

- Be somewhat familiar with the history and development of sport and exercise biomechanics

- Define mechanics

- Outline the organization of mechanics

- Define length and the units of measurement for length

- Define time and the units of measurement for time

- Define mass and the units of measurement for mass

© Lukas Blazek | Dreamstime.com

You are watching the Olympic Games on television when you see a high jumper successfully jump over a crossbar set more than a foot above his head. The technique he uses looks very awkward. He approaches the bar from the side, and when he jumps off the ground he turns his back toward the bar. His head and arms clear the bar first; he then arches his back and finally kicks his legs out and up to get them over the bar. He lands in the pit in an ungainly position: on his shoulders and back with his legs extended in the air above him. You think to yourself, *How can he jump so high using such an odd-looking technique?* Certainly, there must be another technique that is both more effective and more graceful looking. Biomechanics might provide you with some insights to answer this and other questions you have about human movement.

What is your motivation for studying

biomechanics? What can you gain from learning about biomechanics? How will a working knowledge of biomechanics assist you in your future endeavors? Will the time you spend learning about biomechanics be worthwhile? You should consider these questions before investing a lot of time learning about biomechanics.

If you are like most readers of this book, you are probably an undergraduate student majoring in kinesiology, physical education, exercise science, or another human movement science. If so, your answer to the question "Why study biomechanics?" may be that you are enrolled in a required course in biomechanics for which this is the required text. You are studying biomechanics to earn a required credit so that you can graduate. If this is your answer, and for the majority of readers this may be true, you are probably unable to answer the other questions because you do not have enough prior knowledge of biomechanics to know how it can benefit you. So let me give you some reasons for studying biomechanics. This may provide you with some intrinsic motivation to get started on the task of learning about biomechanics.

You are probably planning a career as a physical education teacher, coach, or some other physical activity specialist; and you probably are or have been active as a participant in one or more sports or fitness activities. Suppose a student or athlete asks you, "Why do I have to do this skill this way?" or "Why isn't this technique better?" (see figure I.1). Perhaps you have even asked such questions yourself. Was the coach or teacher able to answer your questions? Were these questions asked of you? Could you answer them? Traditional teaching and coaching methods tell you what techniques to teach

or coach, whereas biomechanics tells you why those techniques are best to teach or coach. A good knowledge of biomechanics will enable you to evaluate techniques used in unfamiliar sport skills as well as to better evaluate new techniques in sports familiar to you.

⟩ Perhaps the best outcome of studying and using biomechanics will be improved performances by your athletes or the accelerated learning of new skills by your students.

Figure I.1 Studying biomechanics will help you understand why some sport techniques work and some don't.

Athletic training and physical therapy students, as well as other students of sports medicine, will also benefit from a knowledge of biomechanics. A good knowledge of biomechanics will help in diagnosing the causes of an injury. It can provide the mechanical basis for taping techniques, braces, and orthotic devices. An understanding of biomechanics may also guide therapists in their prescriptions for rehabilitation and indicate to exercise specialists what exercises may be dangerous for certain individuals.

What Is Biomechanics?

What is this science that promises so much? Before going any further, we should agree on a definition of the word *biomechanics*. What is biomechanics? How have you heard this word used by others? How have you used this word? Your immediate response may be that biomechanics has something to do with determining the best techniques used by athletes in various sport skills. Indeed, some biomechanists are involved in such work, and we just highlighted technique evaluation as a primary reason for studying biomechanics. But biomechanics encompasses much more than that.

Let's look in the library for clues to the definition. Several journals have the word "biomechanics" or a derivation of it in their title. These include the *Journal of Biomechanics*, the *Journal of Biomechanical Engineering*, the *Journal of Applied Biomechanics*, *Sports Biomechanics*, *Clinical Biomechanics*, *Applied Bionics and Biomechanics*, and *Computer Methods in Biomechanics and Biomedical Engineering*. Looking through the tables of contents of these journals, we find that the *Journal of Applied Biomechanics* and *Sports Biomechanics* contain numerous articles about sport biomechanics, such as "Upper-limb kinematics and coordination of short grip and classic drives in field hockey," "Hydrodynamic drag during gliding in swimming," "Effects of bat grip on baseball hitting kinematics," "The influence of the vaulting table on the handspring front somersault," and "Biomechanics of skateboarding: Kinetics of the Ollie." These articles support our sport-related definition of biomechanics. However, a look through the other journals reveals a broad range of topics that at first may appear unrelated, such as "Regional stiffening of the mitral valve anterior leaflet in the beating ovine heart," "Biomechanical model of human cornea based on stromal microstructure," "Simulation of pulmonary air flow with a subject-specific boundary condition," and "Strain-energy function and three-dimensional stress distribution in esophageal biomechanics." There's even an article titled "Biomechanics of fruits and vegetables." From these titles we can deduce that biomechanics is not limited to sport;

it is not even limited to human activities; indeed, it is not even limited to animal activities! The range of titles indicates that biomechanics could include not only the study of the movement of an athlete, but also the study of the airflow in the athlete's lungs and the strength of the athlete's tissues.

Let's return to the word itself and examine it directly for clues about its definition. The word *biomechanics* can be divided into two parts: the prefix *bio-* and the root word *mechanics*. The prefix *bio-* indicates that biomechanics has something to do with living or biological systems. The root word *mechanics* indicates that biomechanics has something to do with the analysis of forces and their effects. So it appears that **biomechanics** is the study of forces and their effects on living systems. This comes very close to the definition of biomechanics presented by Herbert Hatze in 1974: "Biomechanics is the study of the structure and function of biological systems by means of the methods of mechanics" (p. 189). This is a much broader field of study than you may have at first thought. The study of the structure and function of plants as well as animals is encompassed by the definition of biomechanics.

> Biomechanics is the study of forces and their effects on living systems.

What Are the Goals of Sport and Exercise Biomechanics?

Now let's focus on our specific topic of interest in biomechanics. Biomechanics includes the study of all living things, plant and animal; animal biomechanics includes only animals as subjects of study; human biomechanics includes only humans; and sport and exercise biomechanics includes only humans involved in exercise and sport. We might define **sport and exercise biomechanics** as the study of forces and their effects on humans in exercise and sport.

Performance Improvement

The ultimate goal of sport and exercise biomechanics is performance improvement in exercise or sport. A secondary goal is injury prevention and rehabilitation. This secondary goal is closely related to the first and could almost be considered part of the primary goal, because an uninjured athlete will perform better than an injured athlete. Well, how do biomechanists work toward achieving these goals?

⊃ The ultimate goal of sport and exercise biomechanics is performance improvement in exercise or sport.

Technique Improvement

The most common method for improving performance in many sports is to improve an athlete's technique. This is highlighted here as one motivation for studying biomechanics, and it is probably what you thought of when asked how a biomechanist goes about trying to improve an athlete's performance.

The application of biomechanics to improve technique may occur in two ways: Teachers and coaches may use their knowledge of mechanics to correct actions of a student or athlete in order to improve the execution of a skill, or a biomechanics researcher may discover a new and more effective technique for performing a sport skill. In the first instance, teachers and coaches use qualitative biomechanical analysis methods in their everyday teaching and coaching to effect changes in technique. In the second instance, a biomechanics researcher uses quantitative biomechanical analysis methods to discover new techniques, which then must be communicated to the teachers and coaches who will implement them.

Let's look at a simple example of the first case. As a coach, suppose you observe that your gymnast is having difficulty completing a double somersault in the floor exercise. You might suggest three things to the gymnast to help her successfully complete the stunt: (1) jump higher, (2) tuck tighter, and (3) swing her arms more vigorously before takeoff. These suggestions may all result in improved performance and are based on biomechanical principles. Jumping higher will give the gymnast more time in the air to complete the somersault. Tucking tighter will cause the gymnast to rotate faster due to conservation of angular momentum. Swinging the arms more vigorously before takeoff will generate more angular momentum, thus also causing the gymnast to rotate faster. In general, this is the most common type of situation in which biomechanics has an effect on the outcome of a skill. Coaches and teachers use biomechanics to determine what actions may improve performance.

⊃ Coaches and teachers use biomechanics to determine what actions may improve performance.

The second general situation in which biomechanics contributes to improved performance through improved technique occurs when biomechanics researchers develop new and more effective techniques. Despite the common belief that new and revolutionary techniques are regularly developed by biomechanists, such developments are rare. Perhaps the reason is that biomechanics as a discipline is a relatively new science. The much more common outcome of biomechanics research is the discovery of small refinements in technique. One example of biomechanics research that did greatly affect the technique and performances in a sport occurred in swimming in the late '60s and early '70s. Research done by Ronald Brown and James "Doc" Counsilman (1971) indicated that the lift forces acting on the hand as it moved through the water were much more important in propelling a swimmer through the water than previously thought. This research indicated that rather than pulling the hand in a straight line backward through the water to produce a propulsive drag force, the swimmer should move the hand back and forth in a sweeping action as it is pulled backward to produce propulsive lift forces as well as propulsive drag forces (see figure I.2). This technique is now taught by swimming teachers and coaches throughout the world.

Other examples of sports in which dramatic changes in technique produced dramatic improvement in performance include javelin throwing, high jumping, and cross-country skiing. In 1956, before the Summer Olympic Games in Melbourne, Felix Erasquin, a 48-year-old retired discus thrower from the Basque region of Spain, experimented with an unconventional way of throwing the javelin. Erasquin had experience in barra vasca, a traditional Basque sport that involved throwing an iron bar called a palanka. A turn was used to throw the palanka, and Erasquin incorporated this turn in his innovative javelin throwing technique. Rather than throwing using the conventional technique—over the shoulder with one hand from a run—Erasquin held the javelin with his right hand just behind the grip. The tip of the javelin pointed down to his right, and the tail was behind his back and pointed upward. During the run-up, Erasquin spun around like a discus thrower and slung the javelin from his right hand, which guided the implement. To reduce the frictional forces acting on the javelin as it slid through his hand, it had been dunked in soapy water to make it slippery. The outstanding results achieved by Erasquin and others with this technique attracted international attention. Several throwers using this "revolutionary" technique recorded throws that were more than 10 m beyond the existing javelin world record. Officials at the International Amateur Athletic Federation (IAAF), the governing body for track and field, became so alarmed that they altered the rules for the event, and this unconventional technique became illegal (see figure I.3). None of the records set with the Spanish technique were recognized as official world records.

© Human Kinetics/J. Wiseman, reefpix.org

Figure I.2   Swimming techniques have been influenced by biomechanics.

Figure I.3   Current IAAF rules require athletes to throw the javelin over the shoulder with one hand.

In 1968, most world-class high jumpers used the straddle technique (figure I.4*a*). But at the Olympics in Mexico City, the gold medalist in the high jump used a technique few had ever seen. Dick Fosbury, an American from Oregon State University, used a back layout technique

to jump 7 ft 4 1/4 in. (2.24 m). The technique became known as the Fosbury Flop (figure I.4*b*). Its advantages over the straddle technique were its faster approach run and its ease of learning. No biomechanics researcher had developed this technique. Fosbury achieved success

Figure I.4 Before 1968, most high jumpers used the straddle technique *(a)*; but after 1968, many switched to the Fosbury Flop technique *(b)*, the technique used by almost all elite high jumpers today.

© Zuma Press/Icon SMI

with it in high school and continued using and jumping higher with it despite its dramatic differences from the conventional straddle technique. His successes led others to adopt it, and now all world-class high jumpers use the Fosbury Flop.

In the late '70s, Bill Koch, an American cross-country skier, began experimenting with a new skating technique he had observed marathon skiers using in Europe. The technique he experimented with was much different from the traditional diagonal stride skiing technique in which cross-country skiers moved their skis parallel to each other in set tracks. In the 1976 Olympic Games in Innsbruck, Austria, Koch surprised the world by winning a silver medal in the 30K cross-country skiing event. More surprising were his performances in the 1982 to 1983 season, when he became the first American ever to win the World Cup. Koch used the skating technique in achieving this title. By the mid-1980s, the skating technique was used by virtually all elite Nordic ski racers. Beginning with the 1992 Winter Olympics, there were separate competitions for traditional (diagonal stride) and freestyle (skating) cross-country skiing.

With the exception of the swimming example, these examples of new and dramatically different techniques leading to improved performances all happened without the apparent assistance of biomechanics. Maybe this is evidence of the skill of teachers, coaches, and athletes. Through repeated observation, trial and error, and possibly some application of mechanical principles, they have successfully developed excellent techniques for performing skills in most sports without the assistance of biomechanical researchers. But perhaps these improved techniques would have been developed sooner if more teachers and coaches had a working knowledge of biomechanics.

Equipment Improvement

How else can biomechanics contribute to performance improvement? What about improved designs for the equipment used in various sports? Shoes and apparel constitute the equipment used in almost every sport. The equipment worn may have an effect on the performance, either directly or through injury prevention. Can you think of any sports in which improvements in apparel or shoes have changed performances? What about swimming, ski jumping, and speed skating?

Let's look at swimming to see how swimsuit design has changed performances in that event. A hundred years ago swimmers competed in woolen swimsuits, and women's suits had skirts. The wool was replaced by silk and then by synthetic fibers, and the skirt disappeared as swimsuit manufacturers made their swimsuits slicker and more hydrodynamic. Perhaps the most dramatic advance in swimsuit design occurred in February 2008 when Speedo introduced its LZR Racer swimsuit. The Speedo LZR Racer was designed by Speedo scientists and engineers to minimize muscle vibration and reduce drag with compression panels that streamlined the shape of the swimmer. The LZR swimsuits had polyurethane panels and no stitched seams. Within six weeks of its introduction, 13 world records were set by swimmers wearing the Speedo LZR Racer. At the 2008 Beijing Olympic Games, swimmers wearing the Speedo suits set 23 world records and won more than 90% of all the gold medals in swimming.

Besides shoes and apparel, many sports require the use of some sort of implement as well. Think of sports in which an implement is used. How have changes in implements changed performances in these sports? What about bicycling, skiing, tennis, golf, pole vaulting, javelin throwing? Lighter and better-designed implements have not only contributed to improved performances by elite athletes in these sports; they have contributed to improved performances by recreational participants as well.

Let's examine javelin throwing as an example of a sport in which a basic application of mechanics to the equipment design changed the event dramatically. In 1952, Frank "Bud" Held made the United States Olympic team in the javelin. At the 1952 Olympics in Helsinki, he placed ninth behind his American teammates, who won the gold and silver medals. Upon returning to the United States, Bud met with his brother, Dick Held, who had some engineering expertise, and together they designed and built a more aerodynamic javelin. The increased surface area of their new javelin gave it more lift, causing it to "fly" farther. In 1953, Bud Held used one of his javelins to break the existing world record for the javelin throw. The Held brothers were not biomechanists, but their knowledge of mechanics enabled them to improve the design of the javelin. The records continued to be broken as others began using the Held javelin. In 1955, the IAAF implemented rules that limited the size of the javelin so that further increases in its surface area and lift were constrained. Before 1953, the world record in the javelin was 258 ft 2 3/8 in. (78.70 m), set in 1938. With the use of modern aerodynamic javelins based on the Held design, the world record in the event eventually progressed to 343 ft 10 in. (104.80 m), in 1984. In 1986, the IAAF effectively reduced the distance of the men's javelin throw by again changing the rules governing construction of the javelin. The new specifications prevented the javelin from "sailing" so far. Despite this attempt to limit performances, by 1990 the world record with the new javelin rules exceeded 300 ft (91.44 m); and by the turn of the century, the record was 323 ft 1 in. (98.48 m). In 1999, the IAAF implemented similar changes to the rules governing the construction of the women's javelin. These are examples of application of mechanics to *limit* performance in a sport.

The rules makers in many sports, including such popular sports as golf, tennis, cycling, and baseball, regulate the designs of the equipment used in their sports to keep the sports challenging. In spite of these efforts, recent innovations in equipment design have had major impacts on the record books in recent Olympic Games. A number of world records were set in speed skating at the 1998 Winter Olympics in Sapporo, Japan, when the "klap" skate made its first widespread appearance. During that year, world records were set in all but one of the 10 long-track speed skating events (men and women). The Speedo LZR swimsuit had a similar effect on the swimming events in the 2008 Olympic Games in Beijing as described previously; however, FINA (Fédération Internationale de Natation), the international governing body for the sport of swimming, revised its rules regarding swimsuits in 2009 and again in 2010, and the Speedo

LZR Racer may no longer be worn in FINA-sanctioned events (see figure I.5).

Training Improvement

How else can biomechanics contribute to improved performance in sports and physical activities? What about training? Biomechanics has the potential to lead to modifications in training and thus improvements in performance. This application of biomechanics can occur in several ways. An analysis of the technique deficiencies of an athlete can assist the coach or teacher in identifying the type of training the athlete requires to improve. The athlete may be limited by the strength or endurance of certain muscle groups, by speed of movement, or by one specific aspect of his technique. Sometimes the limitation is obvious. For example, a gymnast attempting an iron cross maneuver requires tremendous strength in the adductor muscles of the shoulder (see figure I.6). A mechanical analysis of the maneuver would reveal

© Faugere/DPPI-SIPA/Icon SMI

Figure I.5 Use of the Speedo LZR Racer swimsuit contributed to the numerous world records set by swimmers in the 2008 Summer Olympic Games.

© Stephen Bartholomew/Action Plus/Icon SMI

Figure I.6 A male gymnast's ability to perform an iron cross maneuver may be limited by the strength of his shoulder adductor muscles.

this, but it is already obvious to gymnastics coaches and observers. In other sport skills, the strength requirements may not be so obvious.

> An analysis of the technique deficiencies of an athlete can assist the coach or teacher in identifying the type of training the athlete requires to improve.

Consider the sport of pole vaulting. High school–age boys and girls learning this event often reach a plateau in their performances. A common technique deficiency among beginning pole-vaulters involves the athlete's movement on the pole. The vaulter does not get the hips above the head or hands during the later stages of the vault. This prevents the vaulter from achieving an inverted position, and the heights the vaulter clears are modest. This flaw is easily identified by the coach or teacher, but despite repeated instructions to the vaulter about getting the hips up, the vaulter is still unable to correct this flaw in technique. Why? Many young vaulters do not have the shoulder extensor strength required to swing upside down into an inverted position. A biomechanical analysis of this flaw in the vaulter's technique would reveal that greater shoulder extensor strength is required. The coach or teacher could then devise a training program that would strengthen the vaulter's shoulder extensor muscles enough that she could complete this aspect of the vault successfully.

Figure skating provides another example of how biomechanical analysis can lead to changes in training and ultimately performance improvement. Training camps for junior female skaters held at the U.S. Olympic Training Center in Colorado Springs in the mid-1980s included biomechanical analyses of the skaters attempting double and some triple jumps. Many of the skaters who attempted triple twisting jumps were unsuccessful. An initial analysis revealed that some were unsuccessful in the triples because they were not bringing their arms in tight enough to cause them to spin faster while they were in the air (see figure I.7). Further biomechanical analysis revealed that they were unable to bring their arms in tight enough or quickly enough due to inadequate strength in their arm and shoulder musculature. After their training programs were modified to include upper body strength training to increase arm and shoulder strength, several of the skaters were able to complete triple jumps successfully in subsequent training camps.

One final example of a simple biomechanical analysis that revealed deficiencies in training occurred in cross-country skiing in the late 1970s. An analysis of an international competition included timing skiers over

Figure I.7 Successfully completing triple jumps in figure skating requires strong arms and shoulders.

specific sections of the course. The results of this analysis indicated that the American skiers were just as good as the race leaders on the flat and downhill sections of the course, but they were losing on the uphills. This result encouraged the coaches to allocate more training time to uphills and place more emphasis on perfecting the uphill skiing techniques of the skiers.

Injury Prevention and Rehabilitation

Some believe that injury prevention and rehabilitation should be the primary goal of sport and exercise biomechanics. Biomechanics is useful to sports medicine professionals in identifying what forces may have caused an injury, how to prevent the injury from recurring (or occurring in the first place), and what exercises may assist with rehabilitation from the injury. Biomechanics can be used to provide the basis for alterations in technique, equipment, or training to prevent or rehabilitate injuries.

⊃ Some believe that injury prevention and rehabilitation should be the primary goal of sport and exercise biomechanics.

Techniques to Reduce Injury

Gymnastics provides an example of how biomechanics may aid in reducing injury. Research funded in part by the United States Olympic Committee and the United States Gymnastics Association is concerned with the impact forces that gymnasts experience when landing from stunts and the strategies they use to reduce these forces (McNitt-Gray 1991; McNitt-Gray, Yokoi, and Millward 1993; McNitt-Gray, Yokoi, and Millward 1994). Judges award higher points to gymnasts who "stick" their landings. But such landings may involve greater and more dangerous impact forces. These impact forces are the cause of overuse injuries in many gymnasts. A landing in which the gymnast flexes at the knees, hips, and ankles may reduce the impact forces, but it also results in a lower score. One outcome of this research was a rule change allowing for landing strategies that reduced these impact forces without penalty to the gymnast's score.

Tennis elbow (lateral epicondylitis) is an overuse type of injury that afflicts many novice tennis players. Biomechanics research has revealed that one cause of tennis elbow is overexertion of the extensor carpi radialis brevis muscle (Morris, Jobe, and Perry 1989). Several biomechanists (Blackwell and Cole 1994; Riek, Chapman, and Milner 1999) have implicated faulty technique during backhand strokes as a possible reason for the overexertion. Tennis players who maintain a neutral wrist position through ball impact during a backhand stroke are less likely to develop tennis elbow than those with flexed wrists.

Equipment Designs to Reduce Injury

An example of biomechanics affecting the design of sport equipment to reduce injury involves the running shoe industry. Following Frank Shorter's marathon gold medal in the 1972 Olympics, the United States experienced a running boom. Unfortunately, this boom in participation was accompanied by a boom in running-related injuries. The increase in injuries led runners to become more sophisticated in their selection of running shoes. Thus a boom in biomechanics research on running and running shoes began in the 1970s. An annual shoe ranking published in *Runner's World* magazine included results of biomechanical tests conducted on the shoes at a university biomechanics laboratory. Some of the shoe companies hired biomechanists as consultants, and some funded biomechanics research in other university biome-

chanics laboratories. In 1980, Nike established the Nike Sport Research Laboratory to further the development of athletics and athletic shoes by means of studies in biomechanics, exercise physiology, and functional anatomy.

Running shoes available in the early 1970s were too stiff for many inexperienced runners, and impact injuries such as shinsplints and stress fractures became common. Shoe manufacturers responded to this by producing softer shoes. However, the softer shoes did not provide as much stability or control as the harder shoes; and as a result, ankle, knee, and hip joint injuries increased among runners. Biomechanics research sponsored by various shoe companies has led to many of the features offered in modern running shoes, which provide stability as well as cushioning. These improvements have resulted in fewer running injuries.

Sport and exercise biomechanics can lead to performance improvement and may aid in injury prevention and rehabilitation through improvements in technique, equipment design, and training. Most of the examples in the previous sections illustrate how biomechanics can play a role in performance improvement. A few of the examples, such as the technique changes in the javelin throw, high jump, and cross-country skiing, demonstrate that radically different and improved techniques occur in sport without the apparent assistance of biomechanics researchers. In fact, there are very few examples of biomechanics contributing to the development of new techniques or equipment to improve performance. Why? The answer may be that the people who can most affect sport techniques—the teachers, coaches, and athletes—for the most part are not knowledgeable in biomechanics; but their constant trial-and-error motivation allows them to stumble upon improved techniques. As more teachers, coaches, and athletes become exposed to biomechanics, improvements in techniques may occur more rapidly. Sport and exercise biomechanics is still a relatively young field, though. The population of practicing sport biomechanists is too small to effect changes in many sports. When the technique changes described earlier for the javelin and high jump occurred, the number of people with a knowledge of biomechanics was very small. In fact, before 1960, the word *biomechanics* was used by only a handful of people. A brief review of the history of sport biomechanics can give us more insight into why biomechanics has not had the impact on sport that it seems capable of.

The History of Sport Biomechanics

The history of sport biomechanics is partly the history of **kinesiology**. The word *kinesiology* was first used in the late 19th century and became popular during the 20th century, whereas the word *biomechanics* did not become popular until the 1960s. The roots of the word *kinesiology* give its definition as the study of movement, but in its present-day usage, kinesiology is defined as the study of human movement.

An individual course in kinesiology was a required part of the undergraduate curriculum in physical education at many American schools for most of the 20th century. The major content of such a course was usually applied anatomy with some mechanics and possibly physiology thrown in. Usually this was the only course in which a future coach or physical education teacher received any exposure to mechanics. In many instances, with the emphasis on applied anatomy, the amount of mechanics the future practitioner was exposed to was not enough to be of much practical use.

Researchers concerned with the biomechanics of human movement were active throughout the 20th century, although the mechanics of human and animal motion have intrigued scientists since at least the time of Aristotle (see *De motu animalium* [in Smith and Ross 1912]). In the last decades of the 19th century, Etienne Jules Marey wrote *Le Mouvement* ([1895] 1972), in which he described the use of a variety of devices, including cameras and pressure-sensitive instruments, to measure and record forces and motions produced by man (and animals) in a variety of activities. His well-instrumented "biomechanics" laboratory was the precursor to modern biomechanics and exercise physiology laboratories.

> ## The mechanics of human and animal motion have intrigued scientists since at least the time of Aristotle.

An early example of sport and exercise biomechanics research appeared in *The Baseball Magazine* in 1912. The publishers commissioned a study to determine the speed of a baseball pitched by Walter Johnson. At that time, Johnson was the "king of speed pitchers." Although biomechanics is now thought of as a relatively new field of study, the prelude to the article in *The Baseball Magazine* stated, "*The Baseball Magazine* makes an excursion into an absolutely new field of scientific investigation in its study of the National Game" (Lane 1912, p. 25).

Archibald V. Hill conducted studies of the mechanics and energetics of sprinting in the 1920s (cited in Braun 1941). This work was continued by Wallace Fenn in the 1930s (cited in Cureton 1939). Although he was more well known as an exercise physiologist, Thomas Cureton also wrote about the mechanics of swimming (1930) and various track and field skills in the 1930s (Cureton 1939). He also described techniques for analyzing movements

in sport using motion picture cameras (Cureton 1939). During this time, Arthur Steindler wrote one of the first "biomechanics" textbooks (1935). The 1940s saw World War II, and research in the biomechanics of sport was not a priority. In 1955, the book *Scientific Principles of Coaching* by John Bunn was published. This was one of the first texts to emphasize the mechanics rather than the anatomical aspects of human movement in sports.

By the 1960s, use of the term *biomechanics* became popular, and more people were involved in sport and exercise biomechanics research. In 1967, the First International Seminar on Biomechanics was held in Zurich, Switzerland. The majority of papers presented at this conference dealt with the mechanics of human movement. This seminar was a success, and international conferences on biomechanics have been held every two years since then. In 1968, the *Journal of Biomechanics* was first published. Several papers in the first volume of this journal dealt with sport biomechanics. During the 1960s, several graduate programs in biomechanics were established within physical education departments, and a few of these programs offered doctoral degrees.

In 1973, the International Society of Biomechanics was formed, followed closely by formation of the American Society of Biomechanics in 1977. Sport and exercise biomechanists were involved in the formation of each of these organizations, although the membership of the societies included scientists with a variety of interests. In the early 1980s, the International Society of Biomechanics in Sport was formed to represent the interests of sport biomechanists. In 1985, the *International Journal of Sports Biomechanics* began publication, and in 1992 it changed its name to the *Journal of Applied Biomechanics.* The most recent journal to exclusively feature sport biomechanics articles is *Sport Biomechanics,* whose first issue appeared in 2002.

Research in sport and exercise biomechanics increased steadily throughout the last few decades of the 20th century and into the 21st century. The number of people involved in sport and exercise biomechanics also increased tremendously during this time. One reason for this boom has been the advent of the modern digital computer, which allows for easier data collection and analysis from the high-speed film or video cameras and electronic force-measuring platforms used in biomechanics research. Without a computer, the time required to compute measurements accurately from film data and do quantitative biomechanical research was grossly excessive, accounting for the dearth of research in sport and exercise biomechanics before the 1960s. Anatomical research was not as difficult to complete, and thus courses in kinesiology were weighted toward applied anatomy. With the increase in sport and exercise biomechanics research in the last three decades, the content of many kinesiology courses has been reexamined, and mechanics is covered more thoroughly now. Many kinesiology courses have been renamed as biomechanics courses.

We began this section on the history of sport biomechanics by asking why biomechanics has not had the impact on sport that it seems capable of. The answer seems clearer now. Biomechanics and education in biomechanics have not been in existence long enough to have had a great impact yet. But as more sport and exercise professionals (including you) learn and understand biomechanics, its impact will be more noticeable. Before learning more about biomechanics, you should become familiar with mechanics and the measurement system used in biomechanics.

The Organization of Mechanics

Our study of sport and exercise biomechanics necessitates a knowledge of mechanics. Within our definition of biomechanics, we briefly defined **mechanics** as the analysis of forces and their effects. A more complete definition may be that mechanics is the science concerned with the effects of forces acting on objects. The objects we are concerned with in sport and exercise biomechanics are humans and the implements they may be manipulating in sport and exercise.

⟳ Mechanics is the science concerned with the effects of forces acting on objects.

Mechanics may be divided into several branches (see figure I.8): rigid-body mechanics, deformable-body mechanics, fluid mechanics, relativistic mechanics, and quantum mechanics. In rigid-body mechanics, objects are assumed to be perfectly rigid. This simplifies the analyses. In deformable-body mechanics, the deformation of objects is considered. These deformations complicate the analyses. Fluid mechanics is concerned with the mechanics of liquids and gases. Relativistic mechanics is concerned with Einstein's theory of relativity, and quantum mechanics is concerned with quantum theory. Each branch of mechanics is best suited for describing and explaining specific features of our physical world. **Rigid-body mechanics** is best suited for describing and explaining the gross movements of humans and implements in sport and exercise, so the concepts of rigid-body mechanics will be important in our study of sport and exercise biomechanics. Because some sport and exercise activities occur in fluid environments, we will also learn some concepts of fluid mechanics. Most of part I of this book (chapters 1 through 8) deals with concepts of rigid-body mechanics, with chapter 8 giving a brief overview of fluid mechanics.

In rigid-body mechanics, the objects being investigated are assumed to be perfectly rigid; that is, they do not deform by bending, stretching, or compressing. In describing and explaining the gross movements of the human body and any implements in sport and exercise, we will consider the segments of the human body as rigid bodies that are linked together at joints. In reality, the segments of the body do deform under the actions of forces. These deformations are usually small and don't appreciably affect the gross movements of the limbs or the body itself, so we can get away with considering the body as a system of linked rigid bodies. Repeated small deformations may lead to overuse injuries, however, so we will discuss the deformations of the tissues of the human body in chapter 9, which covers one part of deformable-body mechanics, the mechanics of biological materials.

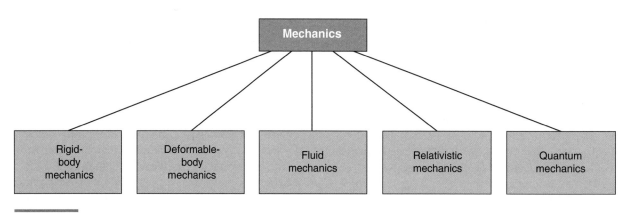

Figure I.8 The branches of mechanics.

Rigid-body mechanics is subdivided into **statics**, or the mechanics of objects at rest or moving at constant velocity, and **dynamics**, or the mechanics of objects in accelerated motion (see figure I.9). Dynamics is further subdivided into **kinematics** and **kinetics**. Kinematics deals with the description of motion, whereas kinetics deals with the forces that cause or tend to cause changes in motion. Our first exploration into biomechanics is concerned with statics. The discussion of forces in the next chapter covers many of the principles of statics. Before we can proceed, though, you must first understand some fundamental concepts of mechanics.

Basic Dimensions and Units of Measurement Used in Mechanics

Mechanics is a quantitative science, and therefore so is biomechanics. We want to describe human movement and its causes in a quantitative manner. If something is quantifiable, certain aspects of it are measurable and can be expressed in numbers. To measure something, we must have some common units of measurement. But first, let's think about what we want to measure in biomechanics.

Suppose we are observing a football player (in American football) returning a kickoff. What terms might we use in describing his runback? We may talk about his position on the field—where did he catch the ball, and where was he tackled? How far did he run? How long did it take him to run that far? How fast was he? In what directions did he run? If you were an opposing player, you might also be interested in how big he is and what it will feel like when you tackle him. Some of the mechanical terms we might use in our description are speed, inertia, power, momentum, force, mass, weight, distance, velocity, and acceleration. (You may not be familiar with some of these terms. If you don't know what they mean, don't worry about it now; they'll all be defined in this or future chapters.) Now let's sort through our descriptions and try to figure out what basic dimensions we need to describe the parameters we've listed.

Length

One basic dimension we may want to measure is length. We need some measurement of length to describe the position of the player on the field and how far he ran. So length is used to describe the space in which movement occurs. Length is also an important dimension in many other sports. It is the most important dimension in sports such as shot putting or high jumping, where how far or how high is the actual measure of performance. In other sports, length may not be the actual measure of performance, but it may be a critical component. How far a golfer can drive the ball off the tee is one way of determining success in golf. How far a batter can hit the ball is one determinant of success in baseball. Stride length in running is one component that determines a runner's velocity.

Length is also an important dimension when we consider the anthropometry of athletes. An athlete's height may be one determinant of success in sports such as basketball or high jumping. Similarly, the length of the implements used may affect performance in sports such as golf, baseball, lacrosse, and pole vaulting. Finally, length may be a dimension of the sport specified by the rules of the activity, such as the length and width of a football field or playing court, the height of a basket, or the length of a race.

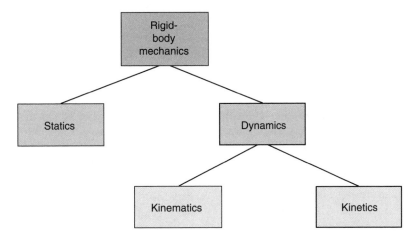

Figure I.9   The branches of rigid-body mechanics.

Length is an important dimension, but how do we measure it? You are probably most familiar with the length measures of inches, feet, yards, or miles if you are an American student. But as scientists, we should be using the Système International d'Unités (International System of Units) or SI units. There are many systems of measurement, but the International System is by far the most widely used and is internationally recognized. This measurement system is based on the metric system. The SI unit of measurement for length is the meter, abbreviated as m, which is about 3.28 ft or 39 in. One foot is about 0.3048 m. Other units for length are the centimeter (cm) and inch (in). There are 2.54 cm in an inch and 100 cm in a meter. For longer distances, lengths are often measured in kilometers (km) or miles. There are 1000 m in a kilometer and 1.609 km in a mile.

Time

Time is another basic dimension we may have used to describe aspects of the ball carrier's run in our football example. Time is an important dimension of performance in almost all sports. In sports that involve races, time is the performance measure. The athlete with the shortest time from start to finish wins the race. In other sports, time is an important determinant of success. A goalie's reaction and movement times in large part determine her success in hockey, lacrosse, team handball, and water polo. Similar situations exist for athletes in tennis, racquetball, squash, and handball when they are attempting to return the ball. Time may also be a dimension specified by the rules of the activity, such as game time, time to begin a play in American football, time-outs in basketball, time in the penalty box in hockey, and time to initiate a throw or jump in track and field.

Time can be measured in seconds, minutes, hours, days, weeks, months, years, and so on. All of these units of measure are based on the second, which is the SI unit of measure for time, so we will use the second as our unit of measure for time. We will abbreviate second as s.

We now have the basic dimensions required for motion to occur: space and time—space to move in and time during which to move. With measurements of time and length, we can describe motion completely. If we consider some of the descriptors we used to characterize the motion of the running back in the football example, many of them involved some measure of time and length. Speed and velocity, for instance, are derived from length and time and are expressed as some unit of length per unit of time. Acceleration is also derived from length and time measures. It is expressed as some unit of length per unit of time multiplied by itself, or squared. (If you don't know the definition of speed, velocity, or

acceleration, don't worry; these terms will be clearly defined in chapter 2.)

Some of the descriptors used in our football example were not really descriptors of the motion of the running back, but rather of some aspects of the running back himself. These included the terms *momentum, power, force, inertia, mass,* and *weight.* What are the basic dimensions used in these descriptors?

Mass and Inertia

What quality of the football running back makes it difficult for him to start running and difficult for a tackler to stop him from running? Something about the size of the football player affects how difficult it is to start or stop his motion. In mechanics, we refer to the property of an object to resist changes in its motion as **inertia**. You may have heard this word before or maybe even used it. Think about how you used it. You probably used "inertia" to indicate that something had a resistance or a reluctance to change what it was doing. This is close to the mechanical definition of inertia, but we must specify that the change being resisted is a change in motion.

How is inertia measured? Let's look at another example. Who is more difficult to start or stop moving, an Olympic shot-putter or an Olympic gymnast? Obviously, the shot-putter is more difficult to move. That athlete has more inertia. So, apparently, a larger object has more inertia than a smaller object. But what do we mean by larger? Well, maybe it's the weight of the object that determines its inertia. Not exactly—it is true that heavier objects have more inertia, but weight is not the measure of inertia. **Mass** is the measure of inertia, whereas **weight** is the measure of the force of gravity acting on an object. On the moon, the Olympic shot-putter would weigh less than he does on earth, but he would still be just as difficult to start or stop moving horizontally, so his mass is the same as on the earth.

⟩ **Mass is the measure of inertia, whereas weight is the measure of the force of gravity acting on an object.**

How important is mass in sport and exercise? Let's think about sports in which the athlete may have to change the motion of something. The athlete may have to change the motion of an implement such as a ball or racket; change the motion of another athlete, as in football or judo; or change the motion of his own body or body parts. So the mass of the implement or athlete or body part has a great effect on the execution of the performance, because the mass of the object to be moved or stopped determines how much effort is required to get the object to move or to stop the object from moving.

The kilogram is the SI unit of measurement for mass. You may have used the kilogram as a measure for weight as well. Mechanically speaking, describing something as weighing a certain number of kilograms is incorrect. In the SI system, the kilogram is a unit of measure for mass, not weight. Because weight is the force of gravity acting on an object, the unit of measure for weight is the unit of measure for force. Force and its units of measurement are further described and defined in the next chapter, but its units of measurement can be derived from the dimensions of length, time, and mass. The kilogram is the SI unit of measure for mass and is abbreviated as kg. One kilogram weighs 2.2 pounds. One kilogram is 1000 grams.

We have now defined the three basic dimensions used in the SI system of measurement as length, time, and mass. Correspondingly, the three basic SI units of measurement are the meter, the second, and the kilogram. All other mechanical quantities and all other SI units used in mechanics can be derived from these three dimensions and their units of measurement. Tables showing the basic dimensions and their SI units, as well as prefixes used in the SI system, are included in appendix A. In the United States, the three basic dimensions commonly used are length, time, and force (rather than mass). The units of measurement for these are the foot, the second, and the pound. Conversions between the SI units of measurement and other units of measurement, including those commonly used in the United States, are presented in appendix A.

Summary

Biomechanics is the study of forces and their effects on living systems, whereas sport and exercise biomechanics is the study of forces and their effects on humans in exercise and sport. Biomechanics may be a useful tool for physical educators, coaches, exercise scientists, athletic trainers, physical therapists, and others involved in human movement. Application of biomechanics may lead to performance improvement or the reduction and rehabilitation of injury through improved techniques, equipment, or training.

Biomechanics is a relatively new term that became popular only in the 1960s. Biomechanical studies of human movements were reported throughout the 20th century. In the late 1960s and early 1970s, professional societies devoted to biomechanics began meeting, and journals devoted to biomechanics appeared. The advent and widespread use of electronic digital computers made biomechanical research more feasible throughout the 1970s and 1980s.

Sport and exercise biomechanics is concerned primarily with that branch of mechanics called *rigid-body mechanics*. Statics and dynamics are the subdivisions of rigid-body mechanics. Kinematics and kinetics are the further subdivisions of dynamics.

The fundamental dimensions used in mechanics are length, time, and mass. The SI units of measurement for these dimensions are the meter (m) for length, the second (s) for time, and the kilogram (kg) for mass. All the other dimensions we will be using in biomechanics are derived from these three fundamental units. We now have knowledge of the basic dimensions used in a mechanical analysis of human movement. Carrying out such an analysis may be difficult at this point, however. In the chapters that follow, you will learn techniques that will make the mechanical analysis of human movements easier. Before you can do that, though, you need to have a better understanding of forces—the topic of the next chapter.

KEY TERMS

biomechanics (p. 3)
dynamics (p. 13)
inertia (p. 14)
kinematics (p. 13)
kinesiology (p. 10)

kinetics (p. 13)
mass (p. 14)
mechanics (p. 12)
rigid-body mechanics (p. 12)

sport and exercise
   biomechanics (p. 3)
statics (p. 13)
weight (p. 14)

REVIEW QUESTIONS

1. Explain the difference between biomechanics and kinesiology.
2. In what ways will biomechanics be useful to you in your career?
3. Give an example of how a physical education teacher might use biomechanics.
4. Give an example of how a coach might use biomechanics.
5. Give an example of how an athletic trainer might use biomechanics.
6. Give an example of how a physical therapist might use biomechanics.
7. Give an example of how a personal trainer might use biomechanics.
8. What are the goals of sport and exercise biomechanics?
9. What are the methods for achieving the goals of sport and exercise biomechanics?
10. What professional societies are specifically concerned with biomechanics?
11. What technology contributed to the growth of biomechanics in the 1970s?
12. What are the three basic dimensions in mechanics?
13. What are the SI units for the three basic dimensions used in mechanics?

PROBLEMS

1. What is your height in meters?
2. What is your mass in kilograms?
3. The largest athlete at the 2012 London Olympics was judo athlete Ricardo Blas Jr. from Guam. He weighed 480.5 pounds. What was his mass in kilograms?
4. In the Winter Olympic sport of curling, the rock's mass is 18 kg. What is the rock's weight in pounds?
5. If you run a mile, how many meters have you run?
6. How many yards are there in a 400 m track?
7. How much longer is a 25 m pool than a 25 yd pool?
8. In 1993, Javier Sotomayor of Cuba set the high jump world record by clearing a height of 2.45 m. How high is that in feet and inches?
9. How much does a 100 kg mass weigh in pounds?
10. A marathon is 26 miles 385 yd. How many kilometers is this?
11. How many miles in 100 km?
12. How many centimeters in 1 yd?
13. The distance from goal line to goal line in an American football field is 100 yards. How many meters is this?
14. The mass of an official basketball for women is 567 grams. How much does a women's basketball weigh in pounds?
15. A FIFA-approved size 5 soccer ball must have a circumference of 68.5 to 69.5 cm and a mass of 420 to 445 grams.
    a. What is the circumference in inches of a FIFA-approved size 5 soccer ball?
    b. How much does a FIFA-approved size 5 soccer ball weigh in pounds?

External Biomechanics

External Forces and Their Effects
on the Body and Its Movement

The movements of the body as a whole are determined by the external forces that act on the body. Part I concerns these forces and the effects of these forces on the body and its motion. Almost all of this part of the book is devoted to rigid-body mechanics and its applications to human movement. Part I includes eight chapters. The first four chapters deal with linear kinematics and kinetics. Chapters 5 through 7 cover angular kinematics and kinetics. Part I concludes with a discussion of fluid mechanics in chapter 8. ■

chapter 1

Forces

Maintaining Equilibrium or Changing Motion

objectives

When you finish this chapter, you should be able to do the following:

- Define force
- Classify forces
- Define friction force
- Define weight
- Determine the resultant of two or more forces
- Resolve a force into component forces acting at right angles to each other
- Determine whether an object is in static equilibrium, if the forces acting on the object are known
- Determine an unknown force acting on an object, if all the other forces acting on the object are known and the object is in static equilibrium

© Brian Cahn/Zuma Press/Icon SMI

A gymnast mainains a precarious position on one foot during a balance beam routine. A rock climber clings by his fingertips to the face of a cliff. A cyclist is motionless on her bicycle at the start of a race. A diver is supported only by his toes on the edge of the diving board before executing a back dive. What are the forces that act on each of these athletes? How do the athletes manipulate these forces in order to maintain balance? The information presented in this chapter provides you with the knowledge you need to answer these questions.

At every instant throughout our lives, our bodies are subjected to forces. Forces are important for motion because they enable us to start moving, stop moving, and change directions. Forces are also important even if we aren't moving. We manipulate the forces acting on us to maintain our balance in stationary positions. To complete a biomechanical analysis of a human movement, we need a basic understanding of forces: how to add them to produce a resultant force, how to resolve forces into component forces, and how forces must act to maintain equilibrium.

What Are Forces?

Simply defined, a **force** is a push or a pull. Forces are exerted by objects on other objects. Forces come in pairs: The force exerted by one object on another is matched by an equal but oppositely directed force exerted by the second object on the first—action and reaction. A force is something that accelerates or deforms an object. In rigid-body mechanics, we ignore deformations and assume that the objects we analyze do not change shape. So, in rigid-body mechanics, forces do not deform objects, but they do accelerate objects if the force is unopposed. Mechanically speaking, something accelerates when it starts, stops, speeds up, slows down, or changes direction. So a force is something that can cause an object to start, stop, speed up, slow down, or change direction.

⟩ Simply defined, a force is a push or a pull.

Our most familiar unit of measurement for force is the pound, but the SI unit of measurement for force is the newton, named in honor of the English scientist and mathematician Isaac Newton (we'll learn more about him in chapter 3). The newton is abbreviated as N. One newton of force is defined as the force required to accelerate a 1 kg mass 1 m/s², or algebraically as follows:

$$1.0 \text{ N} = (1.0 \text{ kg})(1.0 \text{ m/s}^2) \qquad (1.1)$$

One newton of force is equal to 0.225 lb of force, or 1 lb equals 4.448 N. You may remember the story of Isaac Newton's discovery of gravity when an apple fell on his head. This story is probably not true, but it provides a good way to remember the size of a newton. A typical ripe apple weighs about 1 N.

Think about how to describe a force. For instance, suppose you want to describe the force a shot-putter exerted on a shot at the instant shown in figure 1.1. Would describing the size of the force provide enough information about it to predict its effect? What else would we want to know about the force? Some other important characteristics of a force are its point of application, its direction (line of action), and its sense (whether it pushes or pulls along this line). A force is what is known as a vector quantity. A **vector** is a mathematical representation of anything that is defined by its size or magnitude (a number) and its direction (its orientation). To fully describe a force, you must describe its size and direction.

If we want to represent a force (or any other vector) graphically, an arrow makes a good representation. The length of the arrow indicates the size of the force, the shaft of the arrow indicates its line of application, the arrowhead indicates its sense or direction along that line of application, and one of the arrow's ends indicates the point of application of the force. This is a good time to emphasize that the point of application of the force also defines which object the force is acting on (and thus defines which of the pair of forces—action or reaction—we are examining).

In this chapter and the next three chapters, we'll simplify rigid body mechanics even further by assuming that the rigid bodies we analyze are point masses or particles. The objects we'll examine are not really point masses or particles—they have a size and occupy space—but in analyzing the objects as particles, we'll assume that all the forces acting on these objects have the same point of application. Given these assumptions, the dimensions and shape of the objects do not change the effect of the forces acting on the object.

Classifying Forces

Now let's consider the different types of forces and how they are classified. Forces can be classified as internal or external.

Internal Forces

Internal forces are forces that act within the object or system whose motion is being investigated. Remember, forces come in pairs—action and reaction. With internal forces, the action and reaction forces act on different parts of the system (or body). Each of these forces may affect the part of the body it acts on, but the two forces do not affect the motion of the whole body because the forces act in opposition.

> Internal forces are forces that act within the object or system whose motion is being investigated.

Force acting on shot-putter

Force acting on shot

Figure 1.1 The forces acting on a shot-putter and a shot at the instant before release.

In sport biomechanics, the objects whose motion we are curious about are the athlete's body and the implements manipulated by the athlete. The human body is a system of structures—organs, bones, muscles, tendons, ligaments, cartilage, and other tissues. These structures exert forces on one another. Muscles pull on tendons, which pull on bones. At joints, bones push on cartilage, which pushes on other cartilage and bones. If the forces acting at the ends of an internal structure are pulling forces are referred to as tensile forces, and the structure is under tension. If tensile forces act at the ends of an internal structure, the forces pulling on it are referred to as compressive forces, and the structure is under compression. Internal forces hold things together when the structure is under tension or compression. Sometimes the tensile or compressive forces acting on a structure are greater than the internal forces that keep it from withstanding them. When this happens, the structure fails and breaks. This type of failure in the body occurs when muscles pull too hard, tendons or ligaments tear, and bones break.

We think of these forces as the structures that produce the forces that cause us to change our motion. Actually, because these forces are, by definition, only internal forces, they are incapable of producing changes in the motion of the body's center of mass. It seems that internal forces can produce motions of the body's body, but these motions will not produce any change in motion of the body's center of mass unless external forces are acting on the system. The body is able to change its motion only if it can push or pull against some external object. Imagine a defensive player in basketball jumping up to block a shot (see figure 1.2). If she has been fooled by the shooter and jumps too early, she can't stop herself in midair to wait for the shooter to shoot. The only external force acting on her in this case is gravity. She needs to touch something to create another external force to counteract the force of gravity. So she has to get her feet back on the ground. Then she can push against the ground and create an external reaction force that causes her to jump up again. The ground provides the external force that causes the change in motion of the basketball player.

Internal forces may be important in the study of exercise and sport biomechanics if we are concerned about the nature and causes of injury, but they cannot produce any changes in the motion of the body's center of mass. External forces are solely responsible for that.

External Forces

External forces are those forces that act on an object as a result of its interaction with the environment surrounding it. We can classify external forces as contact forces or noncontact forces. Most of the forces we think about are contact forces. These occur when objects are

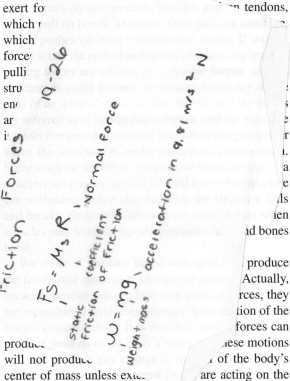

Figure 1.2 A basketball player cannot change her motion once she has jumped into the air.

touching each other. Noncontact forces are forces that occur even if the objects are not touching each other. The gravitational attraction of the earth is a noncontact force. Other noncontact forces include magnetic forces and electrical forces.

⟳ **External forces are those forces that act on an object as a result of its interaction with the environment surrounding it.**

In sports and exercise, the only noncontact force we will concern ourselves with is the force of gravity. The force of gravity acting on an object is defined as the weight of the object. Remember that we defined 1 N of force as the force that would accelerate a 1 kg mass 1 m/s². If the only force acting on an object is the force of the earth's gravity, then the force of gravity will accelerate the object. This is the case when we drop something (if the force of air resistance can be ignored). Scientists have

precisely measured this acceleration for various masses at various locations around the earth. It turns out to be about 9.81 m/s² (or 32.2 ft/s²) downward no matter how large or small the object is. This acceleration is called **gravitational acceleration** or the **acceleration due to gravity** and is abbreviated as g.

⟳ **Weight is the force of gravity acting on an object.**

Now let's see if we can figure out the weight of something if we know its mass. If a 1 N force accelerates a 1 kg mass 1 m/s², then how large is the force that would accelerate a 1 kg mass 9.81 m/s²? Another way of asking this question is, How much does 1 kg weigh?

? N = (1 kg)(9.81 m/s²) = Weight of 1 kg = Force of gravity acting on 1 kg

If we solve this equation, we find that 1 kg weighs 9.81 N. On the earth, mass (measured in kilograms) and weight (measured in newtons) are proportional to each other by a factor of 9.81. The weight of an object (in newtons) is its mass (in kilograms) times the acceleration due to gravity (9.81 m/s²), or,

⟳ $$W = mg \qquad (1.2)$$

where

W = weight (measured in newtons),
m = mass (measured in kilograms), and
g = acceleration due to gravity = 9.81 m/s².

To estimate the weight of something, multiplying its mass by 9.81 m/s² may be difficult to do in your head. For quick approximations, let's round 9.81 m/s² to 10 m/s² and use that as our estimate of the acceleration due to gravity. This will make things easier, and our approximation won't be too far off because our estimate of g is only 2% in error. If more accuracy is required, the more precise value of 9.81 m/s² should be used for g.

Contact forces are forces that occur between objects in contact with each other. The objects in contact can be solid or fluid. Air resistance and water resistance are examples of fluid contact forces, which are further discussed in chapter 8. The most important contact forces in sport occur between solid objects, such as the athlete and some other object. For a shot-putter to put the shot, the athlete must apply a force to it, and the only way the athlete can apply a force to the shot is to touch it. To jump up in the air, you must be in contact with the ground and push down on it. The reaction force from the ground pushes up on you and accelerates you up into the air. To accelerate yourself forward and upward as you take a running step, you must be in contact with the ground and push backward and downward against it. The reac-

Forces

23

tion force from the ground pushes forward and upward against you and accelerates you forward and upward.

Contact forces can be resolved into parts or components—the component of force that acts perpendicular to the surfaces of the objects in contact and the component of force that acts parallel to the surfaces in contact. We call the first component of contact force a normal contact force (or normal reaction force), where normal refers to the fact that the line of action of this force is perpendicular to the surfaces in contact. During a running step, when the runner pushes down and backward on the ground, the normal contact force is the component of force that acts upward on the runner and downward on the ground. The second component of the contact force is called **friction**. The line of action of friction is parallel to the two surfaces in contact and opposes motion or sliding between the surfaces. So when the runner pushes down and backward on the ground during a running step, the frictional force is the component of force that acts forward on the runner and backward on the ground (see figure 1.3). The frictional force is the component of the contact force responsible for changes in the runner's horizontal motion. Frictional forces are primarily responsible for human locomotion, so an understanding of friction is important.

Friction

The frictional force just described is dry friction, which is also referred to as Coulomb friction. Another type of friction is fluid friction, which develops between two layers of fluid and occurs when dry surfaces are lubricated. The behavior of fluid friction is complicated; and because fluid friction occurs less frequently in sport, we will limit our discussion to dry friction. Dry friction acts between the nonlubricated surfaces of solid objects

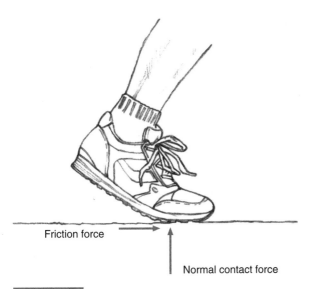

Friction force

Normal contact force

Figure 1.3 Normal contact force and friction force acting on a runner's foot during push-off.

or rigid bodies in contact and acts parallel to the contact surfaces. Friction arises as a result of interactions between the molecules of the surfaces in contact. When dry friction acts between two surfaces that are not moving relative to each other, it is referred to as **static friction**. Static friction is also referred to as **limiting friction** when we describe the maximum amount of friction that develops just before two surfaces begin to slide. When dry friction acts between two surfaces that are moving relative to each other, it is referred to as **dynamic friction**. Other terms for dynamic friction are *sliding friction* and *kinetic friction*.

Friction and Normal Contact Force

Try self-experiment 1.1 to see how friction is affected by normal contact force.

Self-Experiment 1.1

Let's do some experimentation to learn more about friction. Place a book on a flat horizontal surface such as a desk or tabletop. Now push sideways against the book and feel how much force you can exert before the book begins to move. What force resists the force that you exert on the book and prevents the book from sliding? The resisting force is static friction, which is exerted on the book by the table or desk. If the book doesn't slide, then the static friction force acting on the book is the same size as the force you exert on the book, but in the opposite direction. So, the effects of these forces are canceled, and the net force acting on the book is zero. Put another book on top of the original book and push again (see figure 1.4). Can you push with a greater force before the books begin to move? Add another book and push again. Can you push with an even greater force now? As you add books to the pile, the magnitude (size) of the force you exert before the books slide becomes bigger, and so does the static friction force.

How did adding books to the pile cause the static friction force to increase? We increased the inertia of the pile by increasing its mass. This shouldn't affect the static friction force, though, because there is no apparent way an increase in mass could affect the interactions of the molecules of the contacting surfaces. It is these interactions that are responsible for friction. We also increased the weight of the pile as we added books to it. Could this affect the static friction force? Well, increasing the weight would increase the normal contact force acting between the two surfaces. This would increase the interactions of the molecules of the contacting surfaces, because they would be pushed together harder. So it is not the weight of the books that caused the increased static friction force, but the increase in the normal contact force. If

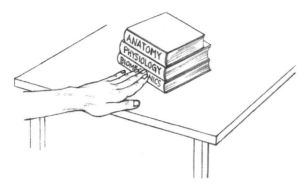

Figure 1.4 Adding books to the stack increases static friction between the bottom book and the table.

we measured this normal contact force and the friction force, we would find that the friction force is proportional to the normal contact force. As one increases, the other increases proportionally. This is true for both static and dynamic friction.

In self-experiment 1.1, the friction force was horizontal and the normal contact force was a vertical force influenced by the weight of the books. Is friction force only a horizontal force? Is the normal contact force always vertical and related to the weight of the object that friction acts on? Try self-experiment 1.2 to answer these questions.

Self-Experiment 1.2

Now try holding the book against a vertical surface, such as a wall (see figure 1.5). Can you do this if you push against the book only with a horizontal force? How hard must you push against the book to keep it from sliding down the wall? What force opposes the weight of the book and prevents the book from falling? The force of your hand pressing against the book is acting horizontally, so it can't oppose the vertical force of gravity pulling down on the book. The force acting upward on the book is friction between the book and the wall (and possibly between the book and your hand). The force you exert against the book affects friction since the book will slide and fall if you don't push hard enough. Again we see that friction is affected by the normal contact force—the contact force acting perpendicular to the friction force and the contact surfaces.

> Friction force is proportional to the normal contact force and acts perpendicular to it.

Friction and Surface Area

What else affects friction? What about surface area? Let's try another experiment (self-experiment 1.3) to see if

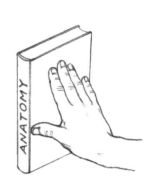

Figure 1.5 Friction force between the book and the wall and between the book and your hand is enough to hold the book up.

increasing or decreasing surface area in contact affects friction force.

Self-Experiment 1.3

Does surface area of contact affect friction? Take a hardback book and lay it on a table or desk. (It's important that you use a hardback book.) Push the book back and forth across the table and get a feeling for how large the dynamic and static friction forces are. Try to exert only horizontal forces on the book. Now, try the same thing with the book standing on its end (as in figure 1.6). Use a rubber band to hold the book closed, but don't let the rubber band touch the table as you're sliding the book. Are there any noticeable differences between the frictional forces you feel with the book in its different orientations? Try it with another hardback book.

With the different orientations in self-experiment 1.3, the surface areas in contact between the book and table varied dramatically, but friction did not change noticeably. In fact, dry friction, both static and dynamic, is not affected by the size of the surface area in contact. This statement is probably not in agreement with your intuitions about friction, but you have just demonstrated it to yourself. If that isn't enough to convince you that dry friction is unaffected by surface area, let's try to explain it.

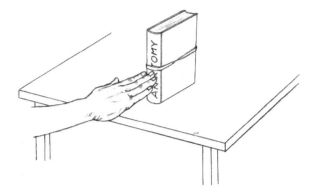

Figure 1.6 A book on its end has a smaller area of contact with the table. Does this reduced area of contact affect the frictional force between the book and the table?

Dry friction arises due to the interaction of the molecules at the surface areas in contact. We have seen that if we press these surfaces together with greater force, the interactions of the molecules will be greater and friction will increase. It makes sense to say that if we increase the area of the surfaces in contact, we also increase the number of molecules that can interact with each other, and thus we create more friction. But if the force pushing the surfaces together remains the same, with the greater surface area in contact, this force is spread over a greater area, and the pressure between the surfaces will be less (pressure is force divided by area). So the individual forces pushing each of the molecules together at the contact surfaces will be smaller, thus decreasing the interactions between the molecules and decreasing the friction. This looks like a trade-off. The increase in surface area increases the number of molecular interactions, but the decrease in pressure decreases the magnitude of these interactions. So the net effect of increasing surface area is zero, and friction is unchanged.

> Dry friction is not affected by the size of the surface area in contact.

Friction and Contacting Materials

Friction is affected by the size of the normal contact force, but it is unaffected by the area in contact. What about the nature of the materials that are in contact? Is the friction force on rubber-soled shoes different than the friction force on leather-soled shoes? Let's try one more experiment (self-experiment 1.4) to investigate how the nature of the materials in contact affects the friction force between them.

Self-Experiment 1.4

Let's observe the difference between the frictions of a book on the table and a shoe on the table. Place the book on the table and put an athletic shoe on top of it. Push the book back and forth across the table and get a feeling for how large the dynamic and static friction forces are. Now, put the shoe on the table, sole down, and place the book on top of it. Push the shoe back and forth across the table and get a feeling for how large the dynamic and static friction forces are. Which produced larger frictional forces with the table, the book or the shoe? What changed between the two conditions? In the two conditions, the weight and mass of the objects being moved (the shoe and book) stayed the same. The surface area of contact changed, but we have determined that friction is unaffected by that. The variable that must be responsible for the changes in the observed frictional force is the difference in the type of material that was in contact with the table. Greater friction existed between the table and the softer and rougher sole of the shoe than between the table and the smoother and harder book cover.

One more observation about friction must be made. When you moved the book back and forth across the table in the self-experiments, was it easier to get the book started or to keep the book moving? In other words, was static friction larger or smaller than dynamic friction? It was easier to keep the book moving than to get it started moving, so static friction is larger than dynamic friction.

Let's summarize what we now know about dry friction. Friction is a contact force that acts between and parallel to the two surfaces in contact. Friction opposes relative motion (or impending relative motion) between the surfaces in contact. Friction is proportional to the normal contact force pushing the two surfaces together. This means that as the normal contact force increases, the frictional force increases as well. If the normal contact force doubles in size, the frictional force will double in size also. Friction is affected by the characteristics of the surfaces in contact. Greater friction can be developed between softer and rougher surfaces than between harder and smoother surfaces. Finally, static friction is greater than dynamic friction. Mathematically, we can express static and dynamic friction as

$$F_s = \mu_s R \qquad (1.3)$$

$$F_d = \mu_d R \qquad (1.4)$$

where

F_s = static friction force,
F_d = dynamic friction force,
μ_s = coefficient of static friction,
μ_d = coefficient of dynamic friction, and
R = normal contact force.

The coefficient of friction is a number that accounts for the different effects that materials have on friction. Mathematically, the coefficient of friction, abbreviated with the Greek letter mu, is just the ratio of friction force to normal contact force.

> Mathematically, the coefficient of friction is the ratio of friction force over normal contact...

Friction in Sport and Human Movement

Friction is an important... movement. L... shoes we wea... forces betwe... most athletic... the materials... friction. In s... sliding is des... activities ha... skiing, we a... bottoms of... In racket sports... large frictional forces are desirable so that... hold of the implement. The grips are made of material such as leather or rubber, which have large coefficients of friction. We may even alter the grips to increase their coefficients of friction by wrapping athletic tape on them, spraying them with tacky substances, or using chalk on our hands. Think about the variety of sports you have been involved in and how friction affects performance in them. In everyday activities, the friction between footwear and floors is important in preventing slips and falls.

We now know about several of the various external forces that can act on us in sport activities; gravity, friction, and contact forces are the major ones. In most sport and exercise situations, more than one of these external forces will act on the individual. How do we add up these forces to determine their effect on the person? What is a net force or a resultant force?

Addition of Forces: Force Composition

The net force acting on an object is the sum of all the external forces acting on it. This sum is not an algebraic sum; that is, we can't just add up the numbers that represent the sizes of the forces. The net force is the vector sum of all the external forces. Remember that we define a force as a push or pull, and that forces are vector quantities. This means that the full description of a force includes its magnitude (how large is it?) and its direction (which way

does it act?). Visually, we can think of forces as arrows, with the length of the shaft representing the magnitude of the force, the orientation of the arrow representing its line of application, and the arrowhead indicating its direction of action along that line. When we add vectors such as forces, we can't just add up the numbers representing their sizes. We must also consider the directions of the forces. Forces are added using the process of vector addition. The result of vector addition of two or more forces is called resultant force. The vector addition of all the external forces acting on an object is the **net force**. It is also referred to as the resultant force, because it results from the addition of all the external forces. Now we will learn how to carry out vector addition of forces.

> The vector addition of all the external forces acting on an object is the net force.

Colinear Forces

To begin our discussion of vector addition, let's start with ... case that involves colinear forces. If you look ...y at the word *colinear*, you may notice that the word ...ne appears in it. **Colinear forces** are forces that have the same line of action. The forces may act in the same direction or in opposite directions along that line. Now here's the situation. You are on a tug-of-war team with two others. You pull on the rope with a force of 100 N, and your teammates pull with forces of 200 N and 400 N. You are all pulling along the same line—the line of the rope. To find out the resultant of these three forces, we begin by graphically representing each force as an arrow, with the length of each arrow scaled to the size of the force. First, draw the 100 N force that you exerted on the rope. If you were pulling to the right, the force you exerted on the rope might be represented like this:

100 N
—————▶

Now draw an arrow representing the 200 N force. Put the tail of this force at the arrowhead of the 100 N force. If it is scaled correctly, this arrow should be twice as long as the arrow representing the 100 N force.

100 N   200 N
————▶—————————▶

Now draw the arrow representing the 400 N force. Put the tail of this force at the arrowhead of the 200 N force. This arrow should be four times as long as the arrow representing the 100 N force and twice as long as the arrow representing the 200 N force. Your drawing should look something like this:

[Handwritten margin notes:]
Adding Forces 26–31
- Locomotion
- Resultant Force = Addition of vectors or missing vector
- Net force = all external forces
- Colinear = same line, diff direction ex. ——→←—
- Concurrent = same point line
- Sample Prob 1.1 = Net

100 N 200 N 400 N

An arrow drawn from the tail of the 100 N force to the tip of the 400 N force represents the resultant force, or the vector sum of the 100 N, 200 N, and 400 N forces, if we put the arrowhead on the end where it meets the tip of the 400 N force.

700 N

If we measure the length of this arrow, it turns out to be seven times as long as the 100 N force. The resultant force must be a 700 N force acting to the right. But this is what we would have found if we added the magnitudes of the three forces algebraically:

100 N + 200 N + 400 N = 700 N

Does that mean that vector addition and algebraic addition are the same? No! This is true only when the forces all act along the same line and in the same direction.

> When forces act along the same line and in the same direction, they can be added using regular algebraic addition.

Now let's consider the forces the opposing team exerts on the rope. That team also consists of three members. The forces they exert on the rope are to the left and are 200 N, 200 N, and 200 N, respectively. What is the resultant of these three forces?

200 N 200 N 200 N

600 N

We can determine the resultant by graphically representing the three forces as arrows and connecting the tail of the first arrow with the tip of the last arrow, as we did previously. We could also have added the magnitudes of the forces algebraically, because all three forces acted along the same line in the same direction.

200 N + 200 N + 200 N = 600 N

Now what is the resultant force acting on the rope as a result of your team pulling to the right and the opposing team pulling to the left? In this case, we have the three forces from your team pulling to the right

100 N 200 N 400 N

and the three forces from the opposing team pulling to the left:

200 N 200 N 200 N

These forces are all still colinear because they act along the same line, the line of the rope in this case. If we follow the procedure we used before, we add the forces graphically by lining the vectors up tip to tail. We have done this for the three forces from your team. The tail of the 200 N force is lined up with the tip of the 100 N force, and the tail of the 400 N force is lined up with the tip of the 200 N force. We have done this for the opposing team's forces as well. Now, to add up all of these forces, we line up the tail of the 200 N force of the opposing team with the tip of the 400 N force of your team (we also could have lined up the tail of your 100 N force with the tip of the 200 N force from the opposing team):

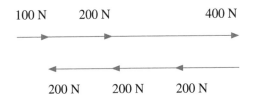
100 N 200 N 400 N

200 N 200 N 200 N

We find the resultant force by drawing an arrow from the tail of the 100 N force to the tip of the last 200 N force, with the tip of the arrow at the end lined up with the tip of the 200 N force and the tail at the end lined up with the tail of the 100 N force:

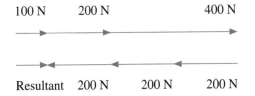
100 N 200 N 400 N

Resultant 200 N 200 N 200 N

If we measure the length of this resultant vector, we see that it is the same length as the 100 N force. The resultant force is 100 N to the right.

We could have arrived at the same resultant if we replaced the forces exerted by your team with its 700 N resultant force to the right and the forces by the opposing team with its 600 N resultant force to the left.

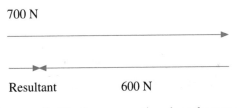
700 N

Resultant 600 N

Because all of the forces are acting along the same line, the resultant force could also be found through algebraic means. Now, rather than just adding up the forces as we did for each team, we must also consider the direction in which the forces act. Let's arbitrarily say that forces acting to the right are positive. Then forces acting to the left must be considered negative. So the resultant force acting on the rope as a result of your team pulling on it to the right and the opposing team pulling on it to the left can now be determined algebraically by adding up the positive forces from your team and the negative forces from the opposing team.

$$100 \text{ N} + 200 \text{ N} + 400 \text{ N} + (-200 \text{ N}) + (-200 \text{ N}) + (-200 \text{ N}) = +100 \text{ N}$$

Adding a negative number is just like subtracting it, so we could also write this as

$$100 \text{ N} + 200 \text{ N} + 400 \text{ N} - 200 \text{ N} - 200 \text{ N} - 200 \text{ N} = +100 \text{ N}.$$

The positive sign associated with our answer of 100 N indicates that the resultant force acts in the positive direction. We set up our positive direction to the right, so the resultant force is a force of 100 N to the right.

If forces are colinear, we may add them using vector addition by graphically representing each force as an arrow and arranging the force arrows tip to tail. We determine the resultant force by drawing an arrow from the tail of the first force to the tip of the last force. This arrow has its tip at the tip of the last force, and it represents the resultant force. We may also add colinear forces algebraically if we take into account the senses of the forces on the line along which they act by assigning positive or negative signs to the magnitudes of the forces. Positive forces act in one direction along the line, and negative forces act in the opposite direction along the line.

Concurrent Forces

If forces do not act along the same line but do act through the same point, the forces are **concurrent forces**. As long as we model objects as point masses, the forces acting on these objects will be considered colinear forces if they act along the same line, and concurrent forces if they do not act along the same line. It is not until chapter 5, when we begin modeling objects as true rigid bodies and not point masses, that the forces we consider can be nonconcurrent forces.

Now let's consider a situation in which the external forces are not colinear but are concurrent. A gymnast is about to begin his routine on the high bar. He jumps up and grasps the bar, and his coach stops his swinging by exerting forces on the front and back of the gymnast's torso. The external forces acting on the gymnast are the force of gravity acting on the mass of the gymnast, a horizontal force of 20 N exerted by the coach pushing on the front of the gymnast, a horizontal force of 30 N exerted by the coach pushing on the back of the gymnast, and an upward vertical reaction force of 550 N exerted by the bar on the gymnast's hands. The gymnast's mass is 50 kg. What is the net external force acting on the gymnast?

First, how large is the force of gravity that acts on the gymnast? If you remember, earlier in this chapter we said that the force of gravity acting on an object is the object's weight. What is the gymnast's weight? Weight is defined with equation 1.2

$$W = mg$$

where W represents weight in newtons, m represents mass in kilograms, and g represents the acceleration due to gravity, or 9.81 m/s². For a good approximation, we can round 9.81 m/s² to 10 m/s² and make our computations easier. If we want more accuracy, we should use 9.81 m/s² rather than 10 m/s². So, using the rougher approximation for g, the gymnast weighs

$$W = mg = (50 \text{ kg})(10 \text{ m/s}^2) = 500 \text{ kg m/s}^2 = 500 \text{ N}.$$

This weight is a downward force of 500 N. We now have all the external forces that act on the gymnast. A drawing of the gymnast and all the external forces that act on him is shown in figure 1.7.

SAMPLE PROBLEM 1.1

A spotter assists a weightlifter who is attempting to lift a 1000 N barbell. The spotter exerts an 80 N upward force on the barbell, while the weightlifter exerts a 980 N upward force on the barbell. What is the net vertical force exerted on the barbell?

Solution

Assume that upward is the positive direction. The 80 N force and the 980 N force are positive, and the 1000 N weight of the barbell is negative. Adding these up gives us the following:

$$\Sigma F = (+80 \text{ N}) + (+980 \text{ N}) + (-1000 \text{ N}) = 80 \text{ N} + 980 \text{ N} - 1000 \text{ N} = +60 \text{ N}$$

The symbol, Σ, that appears before the F in the above equation is the Greek letter sigma. In mathematics it is the summation symbol. It means to sum or add up all items indicated by the variable following the Σ. In this case, ΣF means sum all of the forces or add up all of the forces.

The net vertical force acting on the barbell is a 60 N force acting upward.

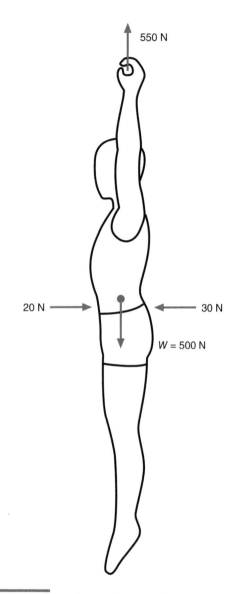

Figure 1.7 Free-body diagram showing the external forces acting on a gymnast hanging from the horizontal bar.

Now we can begin the process of determining the resultant of these forces. Just as we did with the colinear forces, we can represent each force graphically with an arrow, scaling the length of the arrow to represent the magnitude of the force, orienting the arrow to show its line of application, and using an arrowhead to show its sense or direction. As with the colinear forces, if we line up the forces tip to tail, we can find the resultant. Let's do that. First, draw the 20 N horizontal force acting to the right. Now draw the 550 N upward force so that its tail begins at the head of the 20 N force. Draw the 30 N horizontal force to the left so that the tail of this force begins at the head of the 550 N force. Draw the 500 N downward force of gravity so that the tail of this force begins at the head of the 30 N force. You should now have a drawing that looks something like figure 1.8.

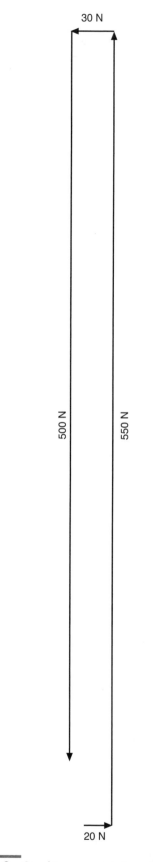

Figure 1.8 Graphic representation of all forces acting on the gymnast.

The head of the 500 N downward force and the tail of the 20 N horizontal force do not connect. The resultant of the four forces can be represented by an arrow connecting the tail of the 20 N horizontal force (the first force in our drawing) with the head of the 500 N downward force (the last force in our drawing). Figure 1.9 shows the construction of the resultant force. This resultant force is directed from the tail of the 20 N horizontal force to the head of the 500 N downward force. The resultant is thus directed upward and slightly to the left. The size of the resultant force is indicated by the length of its arrow. Using the same scale that was used to construct the other forces in figure 1.9, we can estimate that the magnitude of the resultant force is about 51 N.

If we describe the direction of the resultant force as "upward and slightly to the left," we haven't provided a very precise description. Can the direction of the force be described with more precision than that? We could describe the angle that the force makes with a vertical line or a horizontal line. Measuring clockwise from a vertical line, this force is about 11° from vertical. This angular description is much more precise than the description of the force as "upward and slightly to the left."

If vertical and horizontal forces act on a body, we can add forces graphically, as we did to determine the resultant force. Is there any way we can determine the resultant force without using graphical means? Is there a mathematical technique we can use? Let's again consider the four forces acting on the gymnast. Horizontally, there are two forces acting: a 20 N force to the right and a 30 N force to the left. Vertically, there are also two forces acting: a 500 N force downward and a 550 N force upward. Can we just add up all of these forces algebraically? If we did, we would have

20 N + 30 N + 500 N + 550 N = 1100 N.

This is much different than what we determined graphically. Maybe we need to consider the downward forces as negative and the forces to the left as negative. Using this method, we have

20 N + (−30 N) + (−500 N) + 550 N =
20 N − 30 N − 500 N + 550 N = 40 N.

This is much closer to the graphical result, but it still is not correct. We also don't know in which direction the resultant acts. Let's try one more method. Consider the horizontal and vertical forces separately and determine what the horizontal resultant force is and what the vertical resultant force is. Now the problem is similar to the colinear force problems we solved earlier.

Horizontally, we have a 20 N force acting to the right and a 30 N force acting to the left. Previously, we arbitrarily decided that forces to the right were positive, and we assigned a negative value to forces that acted to the left, so the resultant horizontal force is

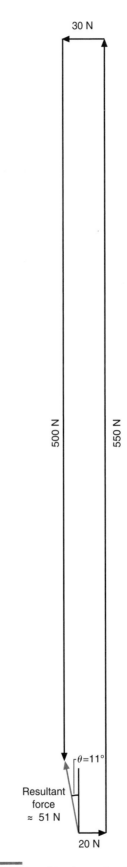

Figure 1.9 Graphic determination of resultant force acting on the gymnast.

20 N + (−30 N) = 20 N − 30 N = −10 N.

The negative sign associated with this force indicates that it acts to the left. The resultant horizontal force is 10 N acting to the left.

Vertically, we have a 500 N force acting downward and a 550 N force acting upward. Let's call upward our positive direction and assign a negative value to the downward force. The resultant vertical force is

(−500 N) + 550 N = +50 N.

The positive sign associated with this force indicates that it acts in an upward direction. The resultant vertical force is 50 N acting upward.

Using this method, the resultant force can be expressed as a 10 N horizontal force acting to the left and a 50 N vertical force acting upward. Is this equivalent to the 51 N resultant force that acts upward and slightly to the left at 11° from vertical? How can a 51 N force be equivalent to a 50 N force and a 10 N force? Add the horizontal resultant force of 10 N and the vertical force of 50 N graphically to determine their resultant force. Draw the forces tip to tail, as shown in figure 1.10.

Now draw the resultant by connecting the tail of the 10 N horizontal force with the tip of the 50 N vertical force. How does this force compare to the resultant shown in figure 1.9? They look identical. Measure the resultant in figure 1.10, and measure the angle it makes with the vertical. The resultant force is about 51 N and makes an angle of 11° with vertical. It is identical to the resultant force shown in figure 1.9. Apparently, a 50 N force and a 10 N force can be equivalent to a 51 N force.

Trigonometric Technique

Take a closer look at the shape created by the three forces in figure 1.10. It's a triangle. In fact, it's a right triangle—one of the angles in the triangle is a 90° angle. The 90° angle is formed between the sides of the triangle representing the horizontal resultant force and the vertical resultant force. There are special relationships among the sides of a right triangle. One of these relates the lengths of the two sides that make the right angle to the length of the side opposite the right angle. If A and B represent the two sides that make up the right angle and C represents the hypotenuse (the side opposite the right angle), then

$$A^2 + B^2 = C^2. \tag{1.5}$$

This relationship is called the Pythagorean theorem. For our force triangle, then, we can substitute 10 N for A and 50 N for B and then solve for C, which represents the resultant force.

$$(10\ N)^2 + (50\ N)^2 = C^2$$

$$100\ N^2 + 2500\ N^2 = C^2$$

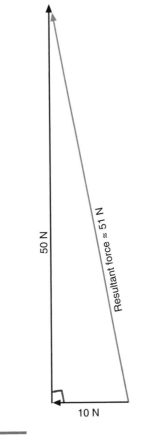

Figure 1.10 Vector sum of the net horizontal force and net vertical force acting on the gymnast.

resultant = $A^2 + B^2 = C^2$

$$2600\ N^2 = C^2$$

$$C = 51\ N$$

This gives us an answer identical to what we got when we actually measured the graphical representation of the force.

Let's take another look at the right triangle we ended up with in figure 1.10. Besides the Pythagorean theorem, there are other relationships between the sides and the angles of a right triangle. If we know the lengths of any two sides of a right triangle, we can determine the length of the other side and the size of the angle between the sides as well. Conversely, if we know the length of one side of a right triangle and the measurement of one of the angles other than the right angle, we can determine the lengths of the other sides and the measurement of the other angle using trigonometry. Trigonometry was not a prerequisite for using this book, and the intent of this book is not to teach you trigonometry, but a knowledge of some of the tools of trigonometry will assist you in the study of biomechanics.

Basically, what trigonometry tells us is that a ratio exists among the lengths of the sides of right triangles that have similar angles. Look at the right triangles in figure 1.11. They are all different sizes, but the angles are all the same, and the sides all change proportionally. If you lengthened one side of any of these triangles, you would have to lengthen the other sides as well to keep the angles of the triangle unchanged. So relationships exist between the lengths of the sides of a right triangle and the angles in a right triangle.

These relationships can be expressed as ratios of one side to another for each size of angle that may exist between two sides of a right triangle. Here are the relationships that may be helpful:

$$\sin\theta = \frac{\text{opposite side}}{\text{hypotenuse}} \tag{1.6}$$

$$\cos\theta = \frac{\text{adjacent side}}{\text{hypotenuse}} \tag{1.7}$$

$$\tan\theta = \frac{\text{opposite side}}{\text{adjacent side}} \tag{1.8}$$

In these equations, θ, which is pronounced "theta," represents the angle; opposite refers to the length of the side of the triangle opposite the angle theta; adjacent refers to the length of the side of the triangle adjacent to the angle theta; and hypotenuse refers to the length of the side of the triangle opposite the right angle. The term *sin* refers to the word *sine;* *cos* refers to the word *cosine;* and *tan* refers to the word *tangent.* Any modern scientific calculator includes the functions for sine, cosine and tangent. The right triangle in figure 1.12 has these three sides labeled for you.

An easy technique for remembering these trigonometric relationships is the following sentence:

Some Of His	$\sin\theta = \dfrac{\textbf{o}\text{pposite side}}{\textbf{h}\text{ypotenuse}}$
Children Are Having	$\cos\theta = \dfrac{\textbf{a}\text{djacent side}}{\textbf{h}\text{ypotenuse}}$
Trouble Over Algebra.	$\tan\theta = \dfrac{\textbf{o}\text{pposite side}}{\textbf{a}\text{djacent side}}$

The first letter of each of these words matches the first letter in each of the trigonometric variables listed in the equations. You may know of other mnemonic devices for memorizing these relationships.

Equations 1.6, 1.7, and 1.8 can be used to determine the length of an unknown side of a right triangle if the length of another side is known and one of the two angles other than the 90° angle is known. If the angle and the hypotenuse are known, the opposite side could be determined using equation 1.6, and the adjacent side could be determined using equation 1.7.

If the sides of the right triangle are known, then the inverse of the trigonometric function is used to compute the angle:

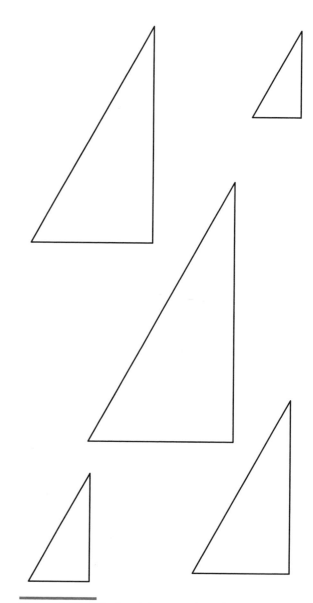

Figure 1.11 Similar right triangles. The triangles are different sizes, but the corresponding angles of each triangle are the same.

$$\theta = \arcsin\left(\frac{\text{opposite side}}{\text{hypotenuse}}\right) \tag{1.9}$$

$$\theta = \arccos\left(\frac{\text{adjacent side}}{\text{hypotenuse}}\right) \tag{1.10}$$

$$\theta = \arctan\left(\frac{\text{opposite side}}{\text{adjacent side}}\right) \tag{1.11}$$

The arcsine, arccosine, and arctangent functions are used to compute one of the angles in a right triangle if the lengths of any two sides are known.

Now let's go back to the resultant forces acting on the gymnast in figure 1.10. We used the Pythagorean theorem

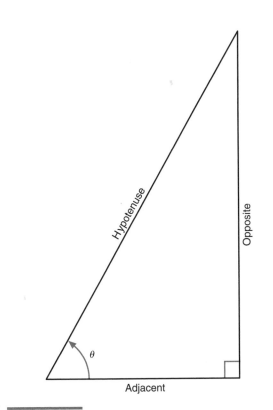

Figure 1.12 Parts of a right triangle.

to compute the size of the resultant force, 51 N. But in what direction is it acting? Let's determine the angle between the 51 N resultant force (the hypotenuse of the triangle) and the 10 N horizontal force (the adjacent side). The 50 N vertical force is the side opposite the angle. Using equation 1.11 gives the following:

$$\theta = \arctan\left(\frac{\text{opposite side}}{\text{adjacent side}}\right)$$

$$\theta = \arctan\left(\frac{50\text{ N}}{10\text{ N}}\right) = \arctan(5)$$

To determine the angle θ, we use inverse of the tangent function or the arctangent. On most scientific calculators, the arctangent function is the second function for the tangent key and is usually abbreviated as $\tan^{-1}$ or atan. Using a calculator (make sure its angle measure is programmed for degrees rather than radians), we find that

$$\theta = \arctan(5) = 78.7°.$$

The angles in a triangle add up to 180°. In a right triangle, one angle is 90°, so the sum of the other two angles is 90°. The other angle in this case is thus 11.3° (i.e., 90° − 78.7°). This is pretty close to the value we arrived at earlier using the graphical method when we measured the angle directly with a protractor.

If forces are concurrent but not colinear, we can add the forces to determine their resultant by graphically

representing the forces as arrows and arranging them tip to tail. The resultant force will be represented by an arrow drawn from the tail of the first force to the tip of the last force represented. Alternatively, if the forces are directed only horizontally or vertically, we can algebraically add up all the horizontal forces to determine the resultant horizontal force, then add up all the vertical forces and determine the resultant vertical force. The size of the resultant of these two forces can be determined using the Pythagorean theorem, and its direction can be determined using trigonometry.

Resolution of Forces

What if the external forces acting on the object are not colinear and do not act in a vertical or horizontal direction? Look back at figure 1.1 and consider the forces acting on a shot during the putting action. Imagine that at the instant shown, the athlete exerts a 100 N force on the shot at an angle of 60° above horizontal. The mass of the shot is 4 kg. What is the net force acting on the shot? First, we need to determine the weight of the shot. Using the rough approximation for g, the shot weighs

$$W = mg = (4\text{ kg})(10\text{ m/s}^2) = 40\text{ N}.$$

Now we can determine the net external force by graphically adding the 40 N weight of the shot to the 100 N force exerted by the athlete. Try doing this. Your graphic solution should be similar to figure 1.13. If we measure the resultant force, it appears to be about 68 N. It acts upward and to the right at an angle a little less than 45°.

Is there another method we could use to determine this resultant, as we did with the gymnast problem earlier? Recall that the external forces acting on the gymnast were all horizontal or vertical forces. In that problem, we could just sum the horizontal forces and the vertical forces algebraically to find the resultant horizontal and vertical forces. In the shot-putting problem, we have one vertical force, the shot's weight, but the force from the athlete is acting both horizontally and vertically. It is pushing upward and forward on the shot. Because this 100 N force acts to push the shot both horizontally and vertically, perhaps it can be represented by two different forces: a horizontal force and a vertical force.

Graphical Technique

Let's start by looking at the problem graphically. We want to represent the 100 N force that acts forward and upward at 60° above horizontal as a pair of forces. The pair of forces we are trying to find are called the horizontal and vertical components of this 100 N force. You are probably familiar with the word *component*. Components are the parts that make up a system. The horizontal and vertical force components of the 100 N force are the parts that

The vertical ground reaction force (normal contact force) acting under a runner's foot is 2000 N, while the frictional force is 600 N acting forward. What is the resultant of these two forces?

Solution:

Step 1: Draw the forces.

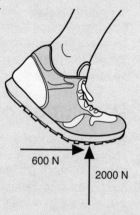

600 N

2000 N

Step 2: Draw the resultant force. Let the two known forces represent two sides of a box. Draw the other two sides of the box. The resultant force is the diagonal of this box, with one end at the point of application of the other two forces.

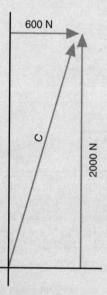

600 N

C

2000 N

Step 3: Use the Pythagorean theorem (equation 1.5) to compute the size of the resultant force:

$$A^2 + B^2 = C^2$$

$$(2000\ N)^2 + (600\ N)^2 = C^2$$

$$4{,}000{,}000\ N^2 + 360{,}000\ N^2 = C^2$$

$$4{,}360{,}000\ N^2 = C^2$$

$$2088\ N = C$$

Step 4: Use the arctangent function (equation 1.11) to determine the angle of the resultant force with horizontal:

$$\theta = \arctan\left(\frac{\text{opposite side}}{\text{adjacent side}}\right)$$

$$\theta = \arctan\left(\frac{2000\ N}{600\ N}\right) = \arctan$$

$$\theta = 73.3°$$

— Inverse tan to find angle of resultant force with the horizontal force

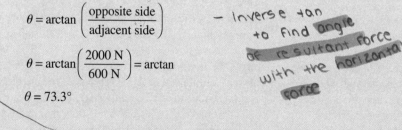

2088 N

2000 N

θ

600 N

34

make up or have the same effect as the 100 N force. We can think of the 100 N force as the resultant of adding the horizontal and vertical components of this force. Let's draw the 100 N force as a vector, as shown in figure 1.14a.

Think about how we graphically determined the resultant of two forces—we lined up the arrows representing these forces end to end and then drew an arrow from the tail of the first force arrow to the tip of the last force arrow in the sequence. This last force arrow we drew was the resultant. Now we want to work that process in reverse. We know what the resultant force is, but we want to know what horizontal and vertical forces can be added together to produce this resultant.

Draw a box around the 100 N force so that the sides of the box align vertically or horizontally and so that the 100 N force runs diagonally through the box from corner to corner (see figure 1.14b). Notice that the box is actually two triangles with the 100 N force as the common side. In each triangle, the other two sides represent the horizontal and vertical components of the 100 N force. In the upper triangle, the 100 N resultant force is the outcome when we start with a vertical force and add a horizontal force to it. The tail of the vertical force is the point of application of forces, and we add the horizontal force to it by aligning the tail of the horizontal force to the tip of the vertical force. In the lower triangle, the 100 N force is the outcome when we start with a horizontal force and add a vertical force to it. The tail of the horizontal force is the point of application of the forces, and we add the vertical force to it by aligning the tail of the vertical force to the tip of the horizontal force. The triangles are identical, so we can use either one. Let's choose the lower triangle. Put arrowheads on the horizontal and vertical force components in this triangle (see figure 1.14c). Now measure the lengths of these force vectors. The horizontal force component is about 50 N, and the vertical force component is about 87 N.

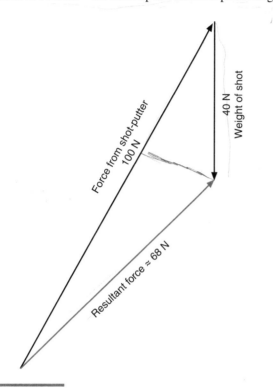

Figure 1.13 Graphic determination of resultant force acting on the shot.

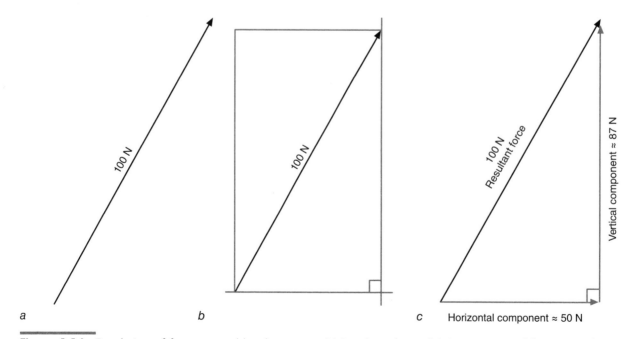

Figure 1.14 Resolution of force exerted by shot-putter. *(a)* Resultant force. *(b)* Construction of force triangle. *(c)* Resolution into component forces.

Since we are working with a right triangle, the Pythagorean theorem (equation 1.5) must apply.

$$A^2 + B^2 = C^2$$

For our force triangle, then, we can substitute 50 N for *A*, 87 N for *B*, and 100 N for *C*. Let's check to see if it works.

$$(50 \text{ N})^2 + (87 \text{ N})^2 = (100 \text{ N})^2$$

$$2500 \text{ N}^2 + 7569 \text{ N}^2 = 10,000 \text{ N}^2$$

$$10,069 \text{ N}^2 \approx 10,000 \text{ N}^2$$

Although 10,069 doesn't exactly equal 10,000, the difference is less than 1%. That's pretty close, especially considering our accuracy in measuring the length of the force arrows. If the measurement accuracy were increased, the difference between the two numbers would become closer to zero.

To complete the original problem, we would include the 40 N weight of the shot as a downward force. This would be subtracted algebraically from the 87 N upward component of the force exerted by the athlete. The resulting vertical force acting on the shot would be

$$(-40 \text{ N}) + 87 \text{ N} = +47 \text{ N}.$$

A 47 N force acts upward on the shot. We still have the 50 N horizontal component of the force exerted by the athlete. If we add this to the 47 N vertical force, using the Pythagorean theorem (equation 1.5), we get

$$A^2 + B^2 = C^2$$

$$(50 \text{ N})^2 + (47 \text{ N})^2 = C^2$$

$$2500 \text{ N}^2 + 2209 \text{ N}^2 = C^2$$

$$4709 \text{ N}^2 = C^2$$

$$C = 68.6 \text{ N}$$

This is close to the answer we got when we used the graphical technique to solve for the resultant force acting on the shot in figure 1.13. In this problem, we actually resolved a force into components, added these components to other forces along the same lines, and then added the resultant component forces back together to find the net resultant force.

The process of determining what two force components add together to make a resultant force is called force resolution. We resolved a force into its components. The word *resolve* sounds like *re-solve*, which is what we did. We had the resultant force, and we solved the problem backward—we *re-solved* it—to determine the forces that added together to yield this resultant. But it was all done graphically. We want a nongraphical technique for doing this.

Trigonometric Technique

The force triangle we ended up with in figure 1.14*c* is a right triangle. Besides the Pythagorean theorem, there are other relationships between the sides and the angles of a right triangle. Some of these relationships can be described by the sine, cosine, and tangent functions, which were defined by equations 1.6, 1.7, and 1.8. Let's see if we can use any of these relationships to resolve the 100 N force that the shot-putter exerts on the shot into horizontal and vertical components.

First, draw the 100 N force as an arrow acting upward and to the right 60° above horizontal, as we did in figure 1.14*a*. Now, just as we did in figure 1.14*b*, draw a box around this force so that the sides of the box are horizontal or vertical and the 100 N force runs diagonally through the box from corner to corner. Let's consider the lower of the two triangles formed by the box and the 100 N diagonal of the box (see figure 1.15). The 100 N force is the hypotenuse of this right triangle. The horizontal side of the triangle is the side adjacent to the 60° angle. The length of this side can be found using the cosine function defined by equation 1.7:

$$\cos \theta = \frac{\text{adjacent side}}{\text{hypotenuse}}$$

$$\cos 60° = \frac{\text{adjacent side}}{100 \text{ N}}$$

$$(100 \text{ N}) \cos 60° = \text{adjacent side}$$
$$= \text{horizontal force component}$$

Using a scientific calculator (make sure its angle measure is programmed for degrees rather than radians), we find that the cosine of 60° is 0.500. Substitute this number for cos 60° in the previous equation:

$$(100 \text{ N}) \cos 60° = (100 \text{ N})(0.500)$$
$$= \text{adjacent side} = 50 \text{ N}$$

The horizontal component of the 100 N force is 50 N. Now find the vertical component of the 100 N force. The side of the triangle opposite the 60° angle represents the vertical component of the 100 N force. We can find the length of this side by using the sine function defined by equation 1.6:

$$\sin \theta = \frac{\text{opposite side}}{\text{hypotenuse}}$$

$$\sin 60° \frac{\text{opposite side}}{100 \text{ N}}$$

$$(100 \text{ N}) \sin 60° = \text{opposite side}$$
$$= \text{vertical force component}$$

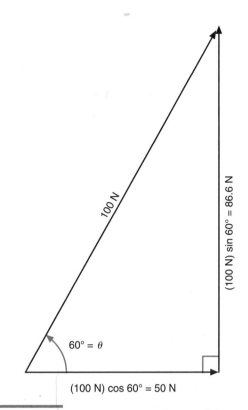

Figure 1.15 Trigonometric resolution of force into horizontal and vertical components.

If we compute the sine of 60° with our scientific calculator, we find that it equals 0.866. Substitute this number for sin 60° in the last equation:

$$(100 \text{ N}) \sin 60° = (100 \text{ N})(0.866)$$
$$= \text{opposite side} = 86.6 \text{ N}$$

The vertical component of the 100 N force is 86.6 N.

We can now determine the net force acting on the shot by adding up all the horizontal forces to get the resultant horizontal force and then adding up all the vertical forces to get the resultant vertical force. The only horizontal force acting on the shot is the horizontal component of the 100 N force. This resultant horizontal force is thus 50 N. Two vertical forces act on the shot: the weight of the shot acting downward (40 N) and the 86.6 N upward vertical component of the 100 N force. If we add these two forces, we get a resultant vertical force of 46.6 N acting upward. Using the Pythagorean theorem, we can then find the net force acting on the shot:

$$(50 \text{ N})^2 + (46.6 \text{ N})^2 = C^2$$

$$2500 \text{ N}^2 + 2172 \text{ N}^2 = C^2$$

$$4672 \text{ N}^2 = C^2$$

$$C = 68.4 \text{ N}$$

The net external force acting on the shot is 68.4 N.

To complete the problem we need to know the direction of this net external force. We can use a trigonometric relationship to determine the angle this net force (68.4 N) makes with horizontal. The force triangle is made up of the 46.6 N upward force (the opposite side); the 50 N horizontal force (the adjacent side); and the resultant of these two forces, the net force of 68.4 N (the hypotenuse). The angle we are trying to determine is between the net force and the horizontal force. Let's use the arctangent function (equation 1.11) to determine the angle:

$$\theta = \arctan \left(\frac{\text{opposite side}}{\text{adjacent side}} \right)$$

$$\theta = \arctan \left(\frac{46.6 \text{ N}}{50 \text{ N}} \right) = \arctan (0.932)$$

$$\theta = 43°$$

The net external force acting on the shot is a force of 68.4 N acting forward and upward at an angle of 43° above horizontal. This is almost identical to the answer we got using the graphical techniques.

We now have two ways to resolve a force into its components. We can do it graphically or through the use of trigonometric relationships. Resolving forces into components makes force addition easier, because we just add up the horizontal forces algebraically to determine the resultant horizontal force and add up the vertical forces algebraically to determine the resultant vertical force. In some instances, we may want to express the net force acting on an object as a pair of forces: the resultant horizontal force and the resultant vertical force. Then, if we were interested in the horizontal motion of the object, we would be interested only in the horizontal component of the net force. Likewise, if we were interested in the vertical motion of an object, we would be interested only in the vertical component of the net force.

Static Equilibrium

You should now have an understanding of what a net external force is and how to determine the net external force acting on an object if all the external forces that act on the object are known. The techniques you've learned will be useful in our analysis of an object that is at rest or moving with constant velocity (zero acceleration). In each of these cases, the external forces acting on the object are in equilibrium, and they result in a net force of zero—if the net force wasn't zero, the object would be accelerating as a result of the forces. If the object is at rest, the forces are in equilibrium and the object is described as being in a state of **static equilibrium**. The branch of mechanics dealing with the study of objects in static equilibrium is called statics.

The biceps muscle exerts a pulling force of 800 N on the radius bone of the forearm. The force acts at an angle of 30° to the radius in an anterior and superior direction. How large is the component of this force that pulls the radius toward the elbow joint, and how large is the component of this force that pulls perpendicular to the radius?

Solution:

Step 1: Draw the force and show the angle it makes with the radius.

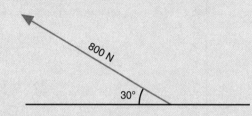

Step 2: Draw a box around the force with sides parallel to and perpendicular to the radius. The 800 N force is the diagonal of the box. Two triangles are formed by the box and the 800 N diagonal. Choose the one with the side along the radius. This is the force triangle.

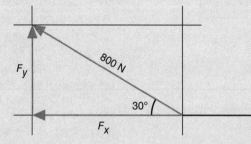

Step 3: Use the cosine function to compute the force component pulling toward the elbow joint (the side of the triangle parallel to the radius).

$$\cos \theta = \frac{\text{adjacent side}}{\text{hypotenuse}}$$

$$\cos 30° = \frac{F_x}{800 \text{ N}}$$

$$F_x = (800 \text{ N}) \cos 30°$$

$$F_x = 693 \text{ N}$$

Step 4: Use the sine function to compute the force component pulling perpendicular to the radius (the side of the triangle perpendicular to the radius).

$$\sin \theta = \frac{\text{opposite side}}{\text{hypotenuse}}$$

$$\sin 30° = \frac{F_y}{800 \text{ N}}$$

$$F_y = (800 \text{ N}) \sin 30°$$

$$F_y = 400 \text{ N}$$

Statics enables us to analyze two types of situations. In the first situation, we may know that an object, such as a person, is not moving, but we may want to know what external forces are acting on the person. An example of this could be seen in gymnastics. A gymnast's coach may want to know how strong a gymnast must be to hold a certain position, such as an iron cross. A static analysis of the position would give the coach knowledge of the forces the athlete would have to withstand to hold that position. The coach could then test the athlete to determine if he was strong enough to react to the external forces whose magnitudes were determined by the static analysis. In the second situation, we may know what forces act on the athlete but want to know if these forces will result in static equilibrium. Let's use the gymnast example again. The coach may know how strong the gymnast is and thus know how large the external forces are to which the gymnast can react. A static analysis of the position in question could then be done to determine whether or not the gymnast's strength is great enough to hold that position.

Free-Body Diagrams

Let's start with a simple situation to test our understanding. A woman in skates is standing still on the ice. The woman's mass is 50 kg. What external forces act on her? Perhaps a picture of the situation would help us visualize the problem (see figure 1.16a).

One force that we know acts on the woman is the force of gravity, which is the woman's weight. Let's indicate that in the drawing with an arrow pointing downward through the woman's center of gravity. You probably have some notion of what center of gravity is. We'll define it in more specific terms later, in chapter 5. For now, we'll say that the center of gravity of an object is the imaginary point in space through which the force of gravity acts on that object.

Is weight the only force acting on the woman? If it is, then the woman should be accelerating downward 9.81 m/s² as a result of this force. Because she isn't, another force must be acting that cancels out the effect of the weight. If you remember our discussion of types of forces earlier in this chapter, we said that when two objects touch each other, they may exert forces on each other. The woman's skates are touching the ice and exerting a force on the ice. The ice is exerting a force on the woman as well. This is the reaction force from the ice. If we drew an arrow representing this force, it would be an upward arrow whose arrowhead is just contacting the skates. But does this arrow represent a force pushing up on the skates or a force pulling up on the ice? This is confusing. Because we are concerned only with the forces acting on the skater, why don't we just make a drawing of the skater alone, isolated from the rest of the environment? Any points where the skater touches something external

to her are places where external forces may be acting. This is shown in figure 1.16b. In this drawing, it is clear that the reaction force from the ice is acting only on the skater and not on the ice. This type of drawing is called a **free-body diagram**. It is a mechanical representation, in this case, of the skater. The free-body diagram is a valuable tool for doing mechanical analyses, so we'll be using free-body diagrams frequently.

⮑ In a free-body diagram, only the object in question is drawn along with all of the external forces that act on it.

The free-body diagram of the woman on skates now shows the two vertical forces that act on the skater: her weight and the reaction force from the ice. Do any horizontal forces act on her? What about friction? On ice, friction is very small; and because the skater is not accelerating forward or backward, the frictional force under her skates must be zero. The free-body diagram thus shows all of the external forces that act on the skater. To draw a free-body diagram of any situation, the system in question (in this case, the skater) is isolated from the environment (in this case, anything that is not the skater), and reaction forces are drawn as arrows at the points of contact of the system with the environment. The noncontact force of gravity is also drawn as an arrow acting downward from the center of gravity of the system.

Static Analysis

Now that we have shown all the external forces acting on the skater, we can finish our analysis. If an object is not moving, it is in static equilibrium. In static equilibrium, acceleration is zero, and the sum of all external forces acting on the object is zero. Mathematically, the situation can be described with an equation:

⮑ $$\Sigma F = 0 \qquad\qquad (1.12)$$

This is the equation of static equilibrium. The term ΣF represents the net external force (the resultant of the external forces) or the vector sum of the external forces. If the only external forces acting are colinear forces, as in the situation with the ice-skater, we find the vector sum of the external forces by simply adding them algebraically. But the senses of the forces must be taken into account. The sense of a force or any vector is the direction in which the arrow of the force or vector points along its line of action. For vertical forces—forces acting up or down—let's give upward forces positive signs and downward forces negative signs.

Getting back to the analysis, the upward reaction force that the ice exerts on the skater has a positive sign, but

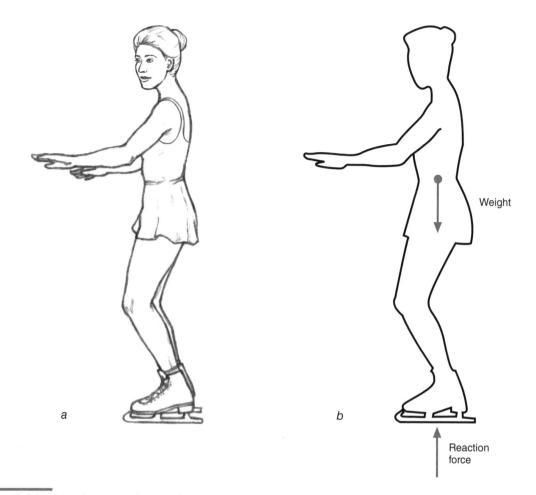

a   *b*   Weight

Reaction force

Figure 1.16   *(a)* A skater standing on the ice; *(b)* free-body diagram of the skater.

we don't know how large the force is yet. That's what we want to find out. The downward force of gravity that acts on the skater has a negative sign because it's downward. We said that the skater's mass was 50 kg. What is her weight? Remember we defined weight, W, with the equation $W = mg$, where m represents mass in kilograms and g represents acceleration due to gravity, or about 10 m/s² downward. So the skater's weight is

$$W = mg = (50 \text{ kg})(-10 \text{ m/s}^2)$$
$$= -500 \text{ kg m/s}^2 = -500 \text{ N}.$$

We can now write the equilibrium equation for the ice-skater:

$$\Sigma F = R + (-500 \text{ N}) = R - 500 \text{ N} = 0$$

where R represents the reaction force from the ice. Solving this equation for R gives us the following:

$$R - 500 \text{ N} = 0$$

$$R - 500 \text{ N} + 500 \text{ N} = 0 + 500 \text{ N}$$

$$R = +500 \text{ N}$$

The reaction force from the ice is +500 N, exactly equal to the weight of the skater but acting in the upward direction. This is true in all situations of static equilibrium where only two external forces act. The external forces must be equal in size but act in opposite directions.

Here's another example. An 80 kg weightlifter has lifted 100 kg over his head and is holding it still. He and the barbell are in static equilibrium. What is the reaction force from the floor that must act on the weightlifter's feet? Draw the free-body diagram first. Should the barbell be included in the free-body diagram of the weightlifter? Because the problem concerns the reaction force between the athlete's feet and the floor, the system can be considered to be the weightlifter alone or the weightlifter and the barbell. In both cases, the reaction force occurs at the boundary between the environment and the system.

First, let's try this analysis using a free-body diagram of the weightlifter alone (see figure 1.17). Three external forces act on the weightlifter: (1) the reaction force from the floor acting upward on the athlete's feet, (2) the weight of the athlete acting downward through the athlete's center of gravity, and (3) the reaction force from the barbell acting downward on the athlete's hands.

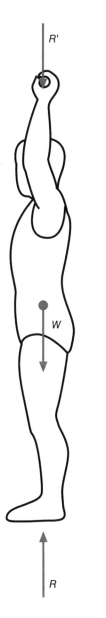

R'

W

R

Figure 1.17 Free-body diagram of a weightlifter alone.

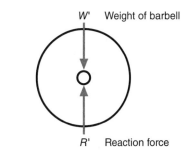

W'   Weight of barbell

R'   Reaction force

Figure 1.18 Free-body diagram of a barbell.

The weight of the barbell is

$$W' = mg = (100 \text{ kg})(-10 \text{ m/s}^2) = -1000 \text{ N}.$$

Now use the equilibrium equation to determine the reaction force exerted by the hands on the barbell:

$$\Sigma F = R' + (-1000 \text{ N}) = R' - 1000 \text{ N} = 0$$

$$R' - 1000 \text{ N} + 1000 \text{ N} = 0 + 1000 \text{ N}$$

$$R' = +1000 \text{ N}$$

The reaction force from the hands is a 1000 N force acting upward on the barbell. The force exerted by the hands on the barbell is equal in magnitude but opposite in direction to the force exerted by the barbell on the hands. Since the force exerted by the hands on the barbell is 1000 N acting upward, the reaction force exerted by the barbell on the hands must be a force of 1000 N acting downward. We can now solve for the reaction force from the floor. The weight of the athlete is

$$W = mg = (80 \text{ kg})(-10 \text{ m/s}^2) = -800 \text{ N}.$$

Now use the equilibrium equation to determine the reaction force from the floor:

$$\Sigma F = R + (-1000 \text{ N}) + (-800 \text{ N}) = R - 1800 \text{ N} = 0$$

$$R - 1800 \text{ N} + 1800 \text{ N} = 0 + 1800 \text{ N}$$

$$R = +1800 \text{ N}$$

The reaction force from the floor is an 1800 N force acting upward on the weightlifter's feet.

Now let's consider the athlete and the barbell as a system. In the free-body diagram, include the barbell. The free-body diagram is shown in figure 1.19.

What external forces are shown in the free-body diagram as acting on this system? The vertical reaction force from the floor acting upward on the athlete's feet is one. This is the force whose magnitude we want to determine. The other forces are the noncontact forces of gravity that act vertically downward through the center of gravity of the athlete and through the center of gravity of

With only one equilibrium equation, only one unknown force can be determined, and in this situation we have two unknown forces: the reaction force from the floor and the reaction force from the barbell. The magnitude of the athlete's weight can be determined because we know the athlete's mass. The free-body diagram of the weightlifter and its equilibrium equation must be used to determine the reaction force from the floor. So another equation must be used to determine the other unknown force, the reaction force from the barbell.

To determine the reaction force from the barbell, a free-body diagram of the barbell would be useful. The external forces acting on the barbell are the weight of the barbell and the reaction force from the hands acting upward on the barbell. The free-body diagram of the barbell is shown in figure 1.18.

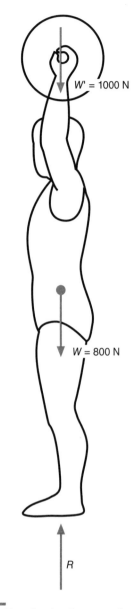

$W' = 1000$ N

$W = 800$ N

R

Figure 1.19 Free-body diagram of weightlifter with barbell.

the barbell. We have already computed these gravitational forces as the weight of the athlete, 800 N, and the weight of the barbell, 1000 N. We can now use the equilibrium equation to determine the reaction force from the floor:

$$\Sigma F = R + (-800 \text{ N}) + (-1000 \text{ N}) = R - 1800 \text{ N} = 0$$

$$R - 1800 \text{ N} + 1800 \text{ N} = 0 + 1800 \text{ N}$$

$$R = 1800 \text{ N}$$

The reaction force from the floor is a force of 1800 N acting upward on the weightlifter. This force is exactly equal to the combined weight of the athlete and the barbell. Representing the athlete and the barbell as the system

in a free-body diagram was a less complicated way of solving this problem, because the reaction force between the hands and the barbell did not have to be computed.

Let's try one more problem. Brian, a 200 kg strongman competitor, is competing in the truck pull event in a strongman competition. Brian wears a harness attached to a cable, and the cable is attached to a truck and trailer. Brian is also using his hands to pull on a rope that is attached to an immovable object in front of him. Brian is attempting to move the truck, but it hasn't yet started moving and neither has Brian. Brian is in a state of static equilibrium. Brian pulls forward and slightly upward on the cable with a force of 2200 N acting at an angle of 14° above horizontal. Brian also pulls on the rope with a horizontal force of 650 N. How much force does the ground exert against Brian's feet?

First, draw a free-body diagram of Brian and show the external forces acting on him (see figure 1.20). Show the force of gravity, or weight, acting on Brian. How large is this force? Let's be more precise now and use 9.81 m/s² as our value for g.

$$W = mg = (200 \text{ kg})(-9.81 \text{ m/s}^2) = -1962 \text{ N}$$

Now show the force that the cable from the truck exerts against Brian. We know that the force Brian exerts on the cable is 2200 N, but what is the force exerted by

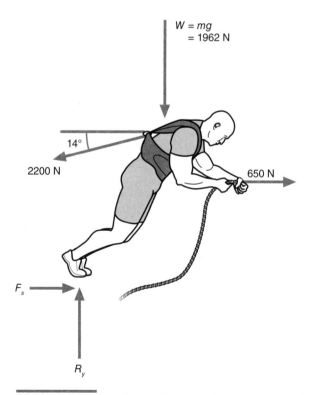

$W = mg$
$= 1962$ N

14°

2200 N

650 N

F_s

R_y

Figure 1.20 Free-body diagram showing external forces acting on a strongman attempting to pull a truck.

the cable on Brian? Remember that forces come in pairs, and the forces are equal in size but opposite in direction. The force the cable exerts against Brian is 2200 N, but it pulls backward and slightly downward at angle of 14° below horizontal.

Now show the force that the rope exerts against Brian. We know that Brian pulls back on the rope with a force of 650 N, but what is the force exerted by the rope on Brian? Again, remember that forces come in pairs, and the forces are equal is size but opposite in direction. The force the rope exerts against Brian is also 650 N, but it pulls forward on Brian in the horizontal direction.

Are any other forces acting on Brian? What about the ground reaction force? This force should be pushing upward and forward on Brian. We don't know the exact direction of this force, only its general direction. Rather than drawing this force as one force on the free-body diagram, let's represent the force with its horizontal and vertical components—friction force (F_s) and normal contact force (R_y), respectively. Are any other external forces acting on Brian? No. In this situation, horizontal and vertical forces are acting as well as the force from the cable, which doesn't pull straight horizontally or vertically but at an angle.

A neat trick to simplify a problem like this is to use two equilibrium equations: one to represent the horizontal component of the net external forces and another for the vertical component of the net external forces. In other words, if there is no horizontal acceleration, this means that the sum of all the horizontal forces must be zero; and if there is no vertical acceleration, this means that the sum of all the vertical forces must be zero. The equilibrium equation (equation 1.12)

$$\Sigma F = 0$$

becomes two equilibrium equations—one for horizontal forces,

$$\Sigma F_x = 0, \tag{1.13}$$

and one for vertical forces,

$$\Sigma F_y = 0, \tag{1.14}$$

where

ΣF_x = Net horizontal force = Horizontal component of the net external forces and

ΣF_y = Net vertical force = Vertical component of the net external forces.

But, before we can make use of these equations in our strongman problem, we have to make sure all of the forces in the free-body diagram have been resolved into horizontal and vertical components. All of the forces in

the free-body diagram shown in figure 1.20 are horizontal or vertical with the exception of the 2200 N force from the cable. Before we go any further, we have to resolve this 2200 N force into its vertical and horizontal components. Do this by drawing a box around the 2200 N force with the sides of the box horizontal or vertical. Two triangles are formed by the box and the 2200 N diagonal. Choose one of these triangles and use the sine and cosine functions to compute the horizontal (P_x) and vertical (P_y) components of the 2200 N force. The subscript x refers to horizontal or the horizontal component of a variable, while the subscript y refers to vertical or the vertical component of a variable.

$$P_x = (2200 \text{ N})\cos 14°$$

$$P_x = 2135 \text{ N}$$

$$P_y = (2200 \text{ N})\sin 14°$$

$$P_y = 532 \text{ N}$$

Now redraw the free-body diagram of Brian and replace the 2200 N force from the rope with its horizontal and vertical components, P_x and P_y (see figure 1.21). We are now ready to use the static equilibrium equations to solve for the friction force (F_s) and normal contact force (R_y) acting on Brian's feet. The equilibrium equations become

$$\Sigma F_x = F_s + 650 \text{ N} + (-2135 \text{ N}) = 0$$

$$\Sigma F_x = F_s + 650 \text{ N} - 2135 \text{ N} = 0$$

$$F_s = +1485 \text{ N}$$

and

$$\Sigma F_y = R_y + (-532 \text{ N}) + (-1962 \text{ N}) = 0$$

$$\Sigma F_y = R_y - 532 \text{ N} - 1962 \text{ N} = 0$$

$$R_y = +2494 \text{ N}$$

where F_s represents the static friction force exerted by the ground and R_y represents the normal contact force. A positive sign in the horizontal direction indicates that the force is to the right or forward, and a negative sign indicates that it is to the left or backward. A positive sign in the vertical direction indicates that the force is upward, and a negative sign indicates that it is downward.

So the reaction force that the ground exerts on Brian can be represented by its component forces of 1485 N forward and 2494 N upward. To complete our analysis, we should determine the resultant of these two forces and its direction. The magnitude of the resultant force can be determined using the Pythagorean theorem (equation 1.5):

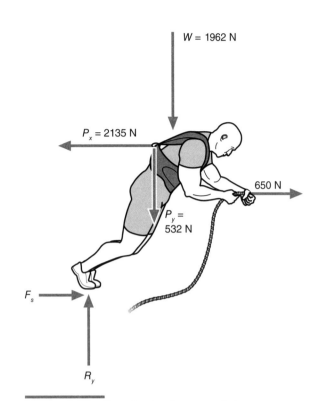

W = 1962 N

P_x = 2135 N

650 N

P_y = 532 N

F_s

R_y

Figure 1.21 Free-body diagram of the strongman with all forces resolved into horizontal and vertical components.

$$A^2 + B^2 = C^2$$

$$(1485 \text{ N})^2 + (2494 \text{ N})^2 = C^2$$

$$2{,}205{,}225 \text{ N}^2 + 6{,}220{,}036 \text{ N}^2 = C^2$$

$$8{,}425{,}261 \text{ N}^2 = C^2$$

$$C = 2903 \text{ N}$$

The resultant reaction force exerted by the ground on Brian is 2903 N. To fully describe this force, we must describe its direction—the angle the force makes with horizontal. When we resolved a force into its horizontal and vertical components, we knew the angle and used trigonometry to compute the components. Now, we know the components and want to find the angle. To determine the angle θ we use the inverse of the tangent function or the arctangent. On most scientific calculators, the arctangent function is the second function for the tangent key and is usually abbreviated as $\tan^{-1}$. We know that

$$\tan \theta = \frac{\text{opposite side}}{\text{adjacent side}}$$

$$\tan \theta = \frac{2494 \text{ N}}{1485 \text{ N}}$$

$$\theta = \arctan\left(\frac{2494 \text{ N}}{1485 \text{ N}}\right)$$

$$\theta = \arctan(2.0162) = 59°$$

The resultant force exerted by the ground on Brian is a force of 2903 N acting forward and upward on Brian at an angle of 59° from horizontal.

Let's summarize the steps involved in determining reaction forces in a static situation. First, isolate the system in question from the environment and draw a free-body diagram. Show all the external forces that act on the system. Don't forget the gravitational forces (weights). If the external forces that act on a system are all colinear, only one equilibrium equation needs to be used; otherwise, use the equilibrium equation for the horizontal direction and the equilibrium equation for the vertical direction. Resolve any forces that are not vertical or horizontal into horizontal and vertical components. Positive and negative signs associated with forces are determined by the direction of the forces. Solve the equilibrium equations using algebra. Use the Pythagorean theorem to compute the vector sum of the resultant horizontal force and the resultant vertical force. Use the arctangent function to determine the direction of this force (the angle it makes with horizontal). Alternatively, the reaction forces can be solved using the graphical methods discussed early in the chapter to demonstrate force addition and force resolution. Either method is acceptable, and the two methods should give the same result.

Summary

Forces are pushes or pulls. They are vector quantities, so they are described by a size and a direction of action. Use arrows to represent forces graphically. Internal forces hold things together and cannot cause changes in the motion of a system. External forces may cause changes in the motion of the system. The most commonly felt external forces are gravity and contact forces. Friction and the normal reaction force are the two components of a contact force. Forces are added with the use of vector addition. We can accomplish this using graphical techniques, or using algebra if the forces are resolved into horizontal and vertical components. If the net external force acting on an object is zero, the object is standing still or moving in a straight line at constant velocity. If the object is standing still, it is in a state of static equilibrium, and the external forces acting on it are balanced and sum to a net force of zero. If the external forces acting on an object do not balance (i.e., they do not sum to zero), the object is not in equilibrium and will change its state of motion. In the next chapter, you will learn about motion and how it is described. Then you will be able to analyze objects that are not in equilibrium.

KEY TERMS

acceleration due to gravity
  (g) (p. 22)
colinear forces (p. 26)
compressive forces (p. 21)
concurrent forces (p. 28)
contact forces (p. 22)
dynamic friction (p. 23)

external forces (p. 21)
force (p. 20)
free-body diagram (p. 39)
friction (p. 23)
gravitational acceleration (g) (p. 22)
internal forces (p. 21)
limiting friction (p. 23)

net force (p. 26)
resultant force (p. 26)
static equilibrium (p. 37)
static friction (p. 23)
tensile forces (p. 21)
tension (p. 21)
vector (p. 20)

REVIEW QUESTIONS

1. The SI unit of measurement for force is named after what English mathematician and scientist?
2. What is the difference between mass and weight?
3. How does the total weight of a team affect the result of a tug-of-war contest?
4. One technique for initiating a turn in downhill skiing involves flexing at the knees and hips so that the skier "drops" towards the skis. Explain how this "unweighting" action affects the frictional force between the skier's skis and the snow.
5. For three different activities, describe the ways in which friction force is either enhanced or diminished by the design of the implement used or by the actions of the person.
6. Which technique is better for sliding a heavy object across the floor, pushing forward and upward on the object or pushing forward and downward on the object? Why?

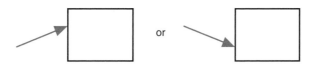

7. In cars without antilock braking systems (ABS), it is possible to stomp on the brake pedal and cause the wheels to lock. This causes the tires to skid on the pavement. If you want to stop a moving car (without ABS) in the shortest distance possible, which technique is better—stomping on the brakes and skidding the tires or pumping the brakes and slowing the wheels down but not skidding the tires? Why?
8. A simple way to determine the coefficient of limiting friction between two materials is to place an object made from one of the materials on a flat surface made of the other material. The flat surface is then slowly tilted until the object on it begins to slide. The angle the flat surface makes with horizontal is measured at the instant the sliding begins. This angle is sometimes called the angle of friction. Using trigonometric techniques, how is this angle algebraically related to the coefficient of limiting friction?

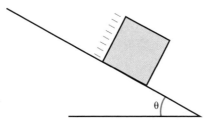

9. Can the vector sum (the resultant) of two forces ever be equal to the algebraic sum of the two forces? Explain.
10. Can the vector sum (the resultant) of two forces ever be larger than the algebraic sum of the two forces? Explain.

11. Draw a free-body diagram of a runner at the instant of foot touchdown.

12. Draw a free-body diagram of a baseball catcher at the instant the ball strikes his glove.

PROBLEMS

1. When offensive lineman Aaron Gibson reported to the Buffalo Bills in 2006, he was the largest NFL player ever with a mass of 186 kg. What was Aaron's weight? 186 Kg × 9.81 ε^{-5}

2. At the 2008 Beijing Olympics, Matthias Steiner of Germany won the gold medal in weightlifting in the unlimited weight class (>105 kg). He lifted 203 kg in the snatch and 258 kg in the clean and jerk for a total of 461 kg. How much does 461 kg weigh? −2N

3. Fitness instructor Kim weighs 491 N. What is her mass?

4. Mairin is a 55 kg road cyclist. Her bicycle weighs 100 N. What is the combined weight of Mairin and her bicycle?

5. The coefficient of static friction between Jimmy's hand and his tennis racket is 0.45. How hard must he squeeze the racket (what is R?) if he wants to exert a total force of 200 N along the longitudinal axis of the racket?

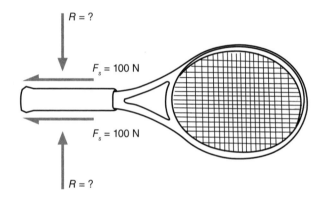

6. The coefficient of static friction between the sole of Ken's shoes and the basketball court floor is 0.67. If Ken exerts a normal contact force of 1400 N when he pushes off the floor to run down the court, how large is the friction force exerted by Ken's shoes on the floor? 1400   1.8N

7. Emma is working in a shoe test lab measuring the coefficient of friction for tennis shoes on a variety of surfaces. The shoes are pushed against the surface with a force of 400 N, and a sample of the surface material is then pulled out from under the shoe by a machine. The machine pulls with a force of 300 N before the material begins to slide. When the material is sliding, the machine has to pull with a force of only 200 N to keep the material moving.

    a. What is the coefficient of static friction between the shoe and the material?

    b. What is the coefficient of dynamic friction between the shoe and the material?

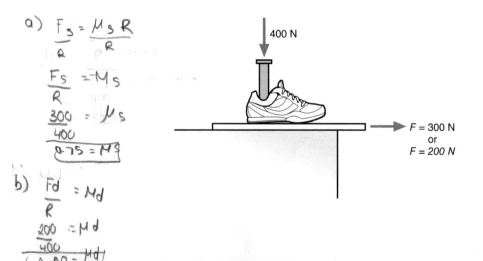

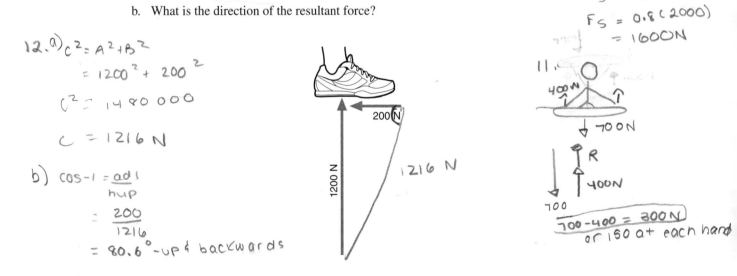

8. $F_s = M_s R$

$\mu_s = 0.55$

$80 kg \times 9.81 m/s^2 = 784.8$

$F_s = 0.55 \times 784.8 \quad = 432 N$

8. Billy is trying to slide an 80 kg box of equipment across the floor. The coefficient of static friction between the box and the floor is 0.55. If Billy pushes only sideways (horizontally) against the box, how much force must he push with to initiate movement of the box?

9. A sprinter is just coming out of the starting block, and only one foot is touching the block. The sprinter pushes back (horizontally) against the block with a force of 800 N and pushes down (vertically) against the block with a force of 1000 N.

   a. How large is the resultant of these forces?

   b. What is the direction of the resultant force?

a) $c^2 = A^2 + B^2$

$c^2 = 1000^2 + 800^2$

$= 1640,000$

$= 1281 N$

1281 N

1000 N

800 N

b) $\cos{-1} = \dfrac{adj}{hyp}$

$= \dfrac{800}{1281}$

$= 51.3°$

10. The sprinter from problem 9 is now out of the blocks and running. If the coefficient of static friction between the track shoe and the track is 0.80, and the sprinter exerts a vertical force of 2000 N downward on the track, what is the maximum horizontal force he can generate under his shoe?

11. Katie exerts a 400 N upward force on a 700 N barbell that is resting on the floor. The barbell does not move. How large is the normal contact force exerted by the floor on the barbell?

12. Daisy walks across a force platform, and the forces exerted by her foot during a step are recorded. The peak vertical reaction force is 1200 N (this force acts upward on Daisy). At the same instant, the braking frictional force is 200 N (this force acts backward on Daisy).

   a. How large is the resultant of these forces?

   b. What is the direction of the resultant force?

10. $F_s = M_s R$

$F_s = 0.8(2000)$

$= 1600 N$

11.

400 N

700 N

R

400N

700

$700 - 400 = 300 N$

or 150 at each hand

12. a) $c^2 = A^2 + B^2$

$= 1200^2 + 200^2$

$c^2 = 1480000$

$c = 1216 N$

b) $\cos{-1} = \dfrac{adj}{hyp}$

$= \dfrac{200}{1216}$

$= 80.6° - up \& backwards$

1200 N

200 N

1216 N

13. The quadriceps pulls on the patella with a force of 1000 N while the patellar tendon pulls on the patella with a force of 1000 N also. The knee is in a flexed position, so the angle between these two forces is 120°. A compressive force from the femoral condyles is the only other significant force acting on the patella. If the patella is in static equilibrium, how large is the compressive force exerted by the femoral condyles on the patella?

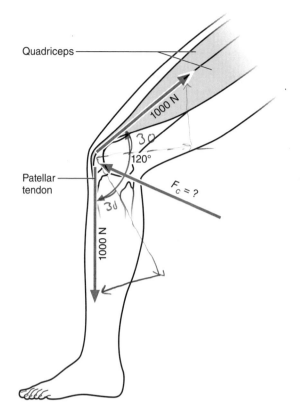

Quadriceps

1000 N

30

120°

Patellar
tendon

$F_c = ?$

30

1000 N

14. The ground reaction force acting on Carter during his long jump is 4500 N acting forward and
    upward at an angle of 78° from horizontal. Carter's mass is 70 kg. Other than gravity, this is the
    only external force acting on Carter.

    686.7 N

    a. What is the vertical component of this ground reaction force?

    b. What is the horizontal component of this ground reaction force?

    c. How big is the net force, the sum of all the external forces, acting on Carter?

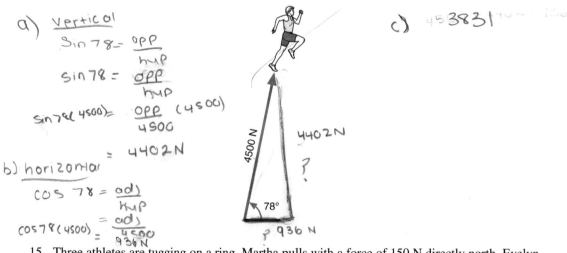

a) verticol

$\sin 78 = \dfrac{opp}{hyp}$

$\sin 78 = \dfrac{opp}{hyp}$

$\sin 78 (4500) = \dfrac{opp}{4500} (4500)$

$= 4402 N$

b) horizontal

$\cos 78 = \dfrac{adj}{hyp}$

$\cos 78 (4500) = \dfrac{adj}{4500}$

$= 936 N$

c) 4538.31

4500 N   4402 N

?

78°

936 N

15. Three athletes are tugging on a ring. Martha pulls with a force of 150 N directly north. Evelyn
    pulls with a force of 100 N directly east. Sara pulls with a force of 200 N directly southwest
    (225° clockwise from north).

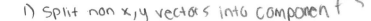

1) split non x,y vectors into components

a. How big is the resultant of the three forces?
b. What is the direction of the resultant force?

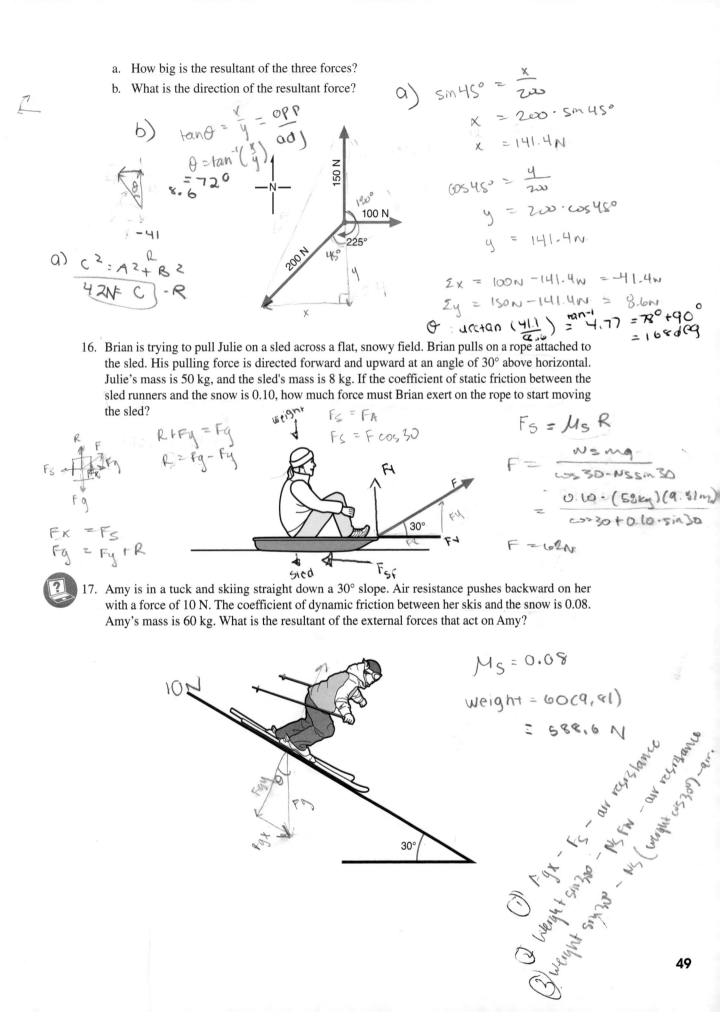

a) $\sin 45° = \dfrac{x}{200}$

$x = 200 \cdot \sin 45°$

$x = 141.4 N$

$\cos 45° = \dfrac{y}{200}$

$y = 200 \cdot \cos 45°$

$y = 141.4 N$

$\Sigma x = 100N - 141.4N = -41.4N$

$\Sigma y = 150N - 141.4N = 8.6N$

$\theta = \arctan\left(\dfrac{41.1}{8.6}\right) = 4.77 = 78° + 90°$
$= 168 deg$

b) $\tan\theta = \dfrac{x}{y} = \dfrac{opp}{adj}$

$\theta = \tan^{-1}\left(\dfrac{x}{y}\right)$

$= 72°$

8.6

-41

a) $c^2 = A^2 + B^2$

$42N = C = -R$

16. Brian is trying to pull Julie on a sled across a flat, snowy field. Brian pulls on a rope attached to the sled. His pulling force is directed forward and upward at an angle of 30° above horizontal. Julie's mass is 50 kg, and the sled's mass is 8 kg. If the coefficient of static friction between the sled runners and the snow is 0.10, how much force must Brian exert on the rope to start moving the sled?

$R + F_y = F_g$

$R = F_g - F_y$

$F_x = F_s$

$F_g = F_y + R$

$F_s = F_A$

$F_s = F \cos 30$

$F_s = \mu_s R$

$F = \dfrac{\mu_s \, mg}{\cos 30 - \mu_s \sin 30}$

$= \dfrac{0.10 \cdot (58kg)(9.81 m/s)}{\cos 30 + 0.10 \cdot \sin 30}$

$F = 62N$

17. Amy is in a tuck and skiing straight down a 30° slope. Air resistance pushes backward on her with a force of 10 N. The coefficient of dynamic friction between her skis and the snow is 0.08. Amy's mass is 60 kg. What is the resultant of the external forces that act on Amy?

10 N

$\mu_s = 0.08$

weight = $60(9.81)$

$= 588.6 N$

① $F_{gx} - F_s - $ air resistance
② weight $\sin 30 - \mu_s F_N - $ air resistance
③ weight $\sin 30 - \mu_s$ (weight $\cos 30$) $- $ air.

49

18. A golf club makes impact with a 0.045 kg golf ball during a drive off the tee. The loft angle of the club (the angle between the club face and vertical) is 10°. The normal contact force exerted by the club on the ball is a 7500 N force directed forward and upward 10° above horizontal. The friction force exerted by the club on the ball is 700 N directed forward and downward 80° below horizontal.

   a. What is the magnitude of the vector sum of the two club head forces acting on the golf ball?

   b. What is the direction of this resultant force?

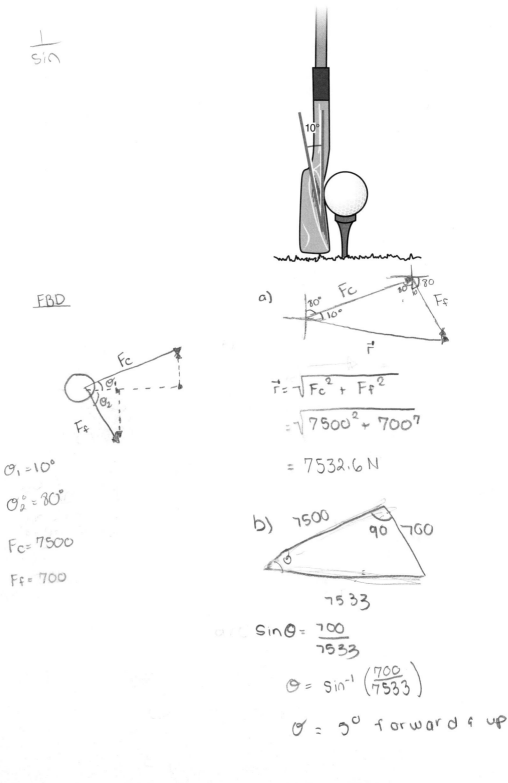

$\frac{1}{\sin}$

FBD

$\theta_1 = 10°$

$\theta_2° = 80°$

$F_c = 7500$

$F_f = 700$

a)

$\vec{r} = \sqrt{F_c^2 + F_f^2}$

$= \sqrt{7500^2 + 700^2}$

$= 7532.6 \text{ N}$

b) 7500   90 700

7533

$\sin\theta = \frac{700}{7533}$

$\theta = \sin^{-1}\left(\frac{700}{7533}\right)$

$\theta = 5° \text{ forward \& up}$

chapter 2

Linear Kinematics

Describing Objects in Linear Motion

objectives

When you finish this chapter, you should be able to do the following:

- Distinguish between linear, angular, and general motion
- Define distance traveled and displacement and distinguish between the two
- Define average speed and average velocity and distinguish between the two
- Define instantaneous speed and instantaneous velocity
- Define average acceleration
- Define instantaneous acceleration
- Name the units of measurement for distance traveled and displacement, speed and velocity, and acceleration
- Use the equations of projectile motion to determine the vertical or horizontal position of a projectile given the initial velocities and time

The world's best female sprinters are lined up at the starting line in the 100 m dash finals at the Olympic Games. The winner will earn the title of world's fastest woman. The starter's pistol goes off, and Shelly-Ann jumps into an early lead. At 50 m she has a 1 m lead on the other runners. But during the last 40 m of the race, Carmelita slowly reduces that lead. At the finish, Shelly-Ann finishes less than 1 m ahead of Carmelita and wins the race. Shelly-Ann wins the title of world's fastest woman, but was her top speed really faster than Carmelita's? Whose acceleration was greater? Were both athletes accelerating during the entire race? Did either athlete decelerate? What performance parameters could be used to account for the last 40 m of the race? These questions concern the kinematic measures of performance covered in this chapter.

This chapter is about the subbranch of mechanics called kinematics. Dynamics is the branch of rigid-body mechanics concerned with the mechanics of moving objects. Kinematics, the topic of this chapter, is the branch of dynamics concerned with the description of motion. The outcomes of many sporting events are kinematic measures, so an understanding of these measures is important. Some of the kinematic terminology introduced in this chapter may sound familiar to you (speed, velocity, acceleration, and so forth). You may believe that you already know all about these terms, but we will be using them in specific ways. The precise mechanical definitions may not agree with the meanings you associate with the terms, and there will be misunderstandings unless our definitions agree. With that in mind, let's begin.

⟳ Kinematics is the branch of dynamics concerned with the description of motion.

Motion

What is motion? Can you define it? We might define motion as the action or process of a change in position. Movement is a change in position. Moving involves a change in position from one point to another. Two things are necessary for motion to occur: space and time—space to move in and time during which to move. To make the study of movement easier, we classify movements as linear, angular, or both (general).

Linear Motion

Linear motion is also referred to as translation. It occurs when all points on a body or object move the same distance, in the same direction, and at the same time. This can happen in two ways: rectilinear translation or curvilinear translation.

Rectilinear translation is the motion you probably would think of as linear motion. Rectilinear translation occurs when all points on a body or object move in a straight line so that the direction of motion does not change, the orientation of the object does not change, and all points on the object move the same distance.

Curvilinear translation is very similar to rectilinear translation. Curvilinear translation occurs when all points on a body or object move so that the orientation of the object does not change and all points on the object move the same distance. The difference between rectilinear and curvilinear translation is that the paths followed by the points on an object in curvilinear translation are curved, so the direction of motion of the object is constantly changing, even though the orientation of the object does not change.

Try to think of some examples of linear motion in sports or human movement. What about a figure skater gliding across the ice in a static position? Is her motion rectilinear or curvilinear? What about a sailboarder zipping across the lake in a steady breeze? Is it possible for the sailboarder's motion to be rectilinear? What about a bicyclist coasting along a flat section of the road? (In each of these examples, it is possible for the athletes to achieve rectilinear motion.) Can you think of any examples of curvilinear motion? Can a gymnast on a trampoline experience linear motion? How? What about a diver? A ski jumper? A skateboarder rolling along a flat section of concrete? An in-line skater? (It's possible for the gymnast, diver, and ski jumper to achieve curvilinear motion. The gymnast, diver, skateboarder, and in-line skater can achieve both rectilinear and curvilinear motion. The ski jumper can achieve rectilinear motion during the in-run to the jump, and curvilinear motion during the flight phase of the jump.)

To determine whether a motion is linear, imagine two points on the object in question. Now imagine a straight line connecting these two points. As the object moves, does the line keep its same orientation; that is, does the line point in the same direction throughout the movement? Does the line stay the same length during the movement? If both of these conditions are true throughout the movement, the motion is linear. If both points on the imaginary line move in parallel straight lines during the motion, the motion is rectilinear. If both points on the imaginary line move in parallel lines that are not straight, the motion is curvilinear. Now try to think of more examples of linear motion in sport. Would you classify the motions you thought of as rectilinear or curvilinear?

Angular Motion

Angular motion is also referred to as rotary motion or rotation. It occurs when all points on a body or object move in circles (or parts of circles) about the same fixed central line or axis. Angular motion can occur about an axis within the body or outside of the body. A child on a swing is an example of angular motion about an axis of rotation external to the body. An ice-skater in a spin is an example of angular motion about an axis of rotation within the body. To determine whether or not a motion is angular, imagine any two points on the object in question. As the object moves, are the paths that each of these points follow circular? Do these two circular paths have the same center or axis? If you imagine a line connecting the two imaginary points, does this line continuously change orientation as the object moves? Does the line continuously change the direction in which it points? If these conditions are true, the object is rotating.

Examples of angular motion in sports and human movement are more numerous than examples of linear motion. What about a giant swing on the horizontal bar? Are parts of this motion rotary? What about individual movements of our limbs? Almost all of our limb movements (if they are isolated) are examples of angular motion. Hold your right arm at your side. Keeping your upper arm still, flex and extend your forearm at the elbow joint. This is an example of angular motion. Your forearm rotated about a fixed axis (your elbow joint). During the flexing and extending, your wrist moved in a circular path about your elbow joint. Every point on your forearm and wrist moved in a circular path about your elbow joint. Consider each limb and the movements it can make when movement about only one joint is involved. Are these movements rotary—that is, do all the points on the limb move in circular paths about the joint?

Let's consider motion about more than one joint. Is the limb's motion still angular? Extend your knee and hip at the same time. Was the movement of your foot angular? Did your foot move in a circular path? Was the motion of your foot linear?

General Motion

Combining the angular motions of our limbs can produce linear motions of one or more body parts. When both the knee and hip joints extend, you can produce a linear motion of your foot. Similarly, extension at the elbow and horizontal adduction at the shoulder can produce a linear motion of the hand. **General motion** is a combination of linear and angular motions. Try self-experiment 2.1.

Self-Experiment 2.1

Grab hold of a pencil that is lying flat on a desk or a table. While keeping the pencil flat on the table, try to move the pencil rectilinearly across the table. Can you do it? You produced that motion by combining angular motions of your hand, forearm, and upper arm. The total motion of our limbs is called general or mixed motion.

General motion is the most common type of motion exhibited in sports and human movement. Running and walking are good examples of general motion. In these activities, the trunk often moves linearly as a result of the angular motions of the legs and arms. Bicycling is another example of general motion. Think of various human movements in sports and consider how you would classify them.

Classifying motion as linear, angular, or general motion makes the mechanical analysis of movements easier. If a motion can be broken down into linear components and angular components, the linear components can be analyzed using the mechanical laws that govern linear motion. Similarly, the angular components can be analyzed using the mechanical laws that govern angular motion. The linear and angular analyses can then be combined to understand the general motion of the object.

⊃ **Classifying motion as linear, angular, or general motion makes the mechanical analysis of movements easier.**

Linear Kinematics

Now let's examine linear motion in more detail. Linear kinematics is concerned with the description of linear motion. Questions about speed, distance, and direction are all inquiries about the linear kinematics of an object.

Try self-experiment 2.2 to identify some of the characteristics of linear motion.

Self-Experiment 2.2

How would you describe something that is moving? Roll a ball across the floor. Describe its movement. What words do you use? You might describe how fast or slow it is going, mention whether it is speeding up or slowing down, and note that it is rolling and not sliding. You also might say something about where it started and where it might end up. Or you might describe its direction: "It's moving diagonally across the room," or "It's moving toward the wall or toward the door." After it stops, you might say how far it traveled and how long it took to get to where it went. All of the terms you used to describe the motion of the ball are words that concern the kinematics of linear motion.

Position

The first kinematic characteristic we might describe about an object is its position. Our definition of motion—the action or process of change in position—refers to position. Mechanically, **position** is defined as location in space. Where is an object in space at the beginning of its movement or at the end of its movement or at some time during its movement? This may not seem like such an important characteristic at first, but consider the importance of the positions of players on the field or court in sports such as football, tennis, racquetball, squash, soccer, field hockey, ice hockey, and rugby. The strategies employed often depend on where the players on each team are positioned.

Let's start with a simple example. Consider a runner competing in a 100 m dash (see figure 2.1). How would you go about describing the runner's position during the race? You might describe the runner's position relative to the starting line: "She's 40 m from the start." Or you might describe the runner's position relative to the finish line: "She's 60 m from the finish." In both cases, you have used a measure of length to identify the runner's position relative to some fixed, nonmoving reference.* The references were the starting line and the finish line. Some concept of direction was also implied by your description and the event itself. When you say the runner is 40 m from the start, this is usually interpreted to mean

---

*Are these really fixed, nonmoving references? Relative to the surface of the earth they are, but the earth itself is moving around the sun in the solar system. And the solar system is moving in the galaxy. And the galaxy is moving in the universe. So, it is difficult to define a position in terms of an absolute nonmoving reference frame. For our purposes, however, we will consider anything that doesn't move relative to the earth's surface a fixed reference.

that the runner is 40 m in front of the start and toward the finish line. Mechanically, if we used the starting line as our reference, we would say that the runner is at +40 m. If the runner was on the other side of the starting line, we would describe the runner's position as −40 m. We use the positive and negative signs to indicate which side of the starting line the runner is on.

This example of the 100 m dash is only one-dimensional. We were concerned about only one dimension—the dimension from the starting line to the finish line. Only one number was required to identify the position of the runner in the race. Now let's consider a two-dimensional situation. Imagine you are watching a game of American football. A running back has broken out of the backfield and is running toward the goal line. He is on the opposing team's 20 yd line. To describe his position, you would say he is 20 yd from the goal line. But to fully describe his position, you would also have to give information about his location relative to the sidelines. Using the left sideline as a reference, you could then describe his position as 20 yd from the goal line and 15 yd from the left sideline. This is shown in figure 2.2a.

In this situation, it might be helpful to set up a **Cartesian coordinate system** to help identify the location of the runner. Cartesian coordinates are named after René Descartes (1596-1650), a French philosopher and mathematician who is credited with inventing analytic geometry. You may remember this type of coordinate system from high school mathematics. First, we would need to locate a fixed reference point for our coordinate system. This fixed point is called the origin, because all our position measurements originate from it. Let's put the origin for this system at the intersection of the left sideline and the running back's goal line. We could put the origin at any fixed point; we chose the intersection of the goal line and the sideline because it was convenient. Imagine the x-axis lying along the goal line with zero at the origin and positive numbers to the right on the playing field. Imagine the y-axis lying along the left sideline with zero at the origin and positive numbers increasing as you move toward the opposite goal. With this system, we could identify the running back's position with two numbers corresponding to his x- and y-coordinates in yards as follows: (15, 80). This situation is shown in figure 2.2b. The x-coordinate of 15 indicates that he is 15 yd from the left sideline on the field, and the y-coordinate of 80 indicates that he is 80 yd from his goal line or 20 yd from scoring, because we know that the goal lines are 100 yd apart.

In three dimensions, we would need three numbers to describe the position of an object in space. For example, how would you describe the position of the ball during a game of racquetball? We might set up a three-dimensional Cartesian coordinate with one axis in the vertical direction and two axes in the horizontal plane. If we put the

Faugere/DPPI-SIPA/Icon SMI

Figure 2.1 How would you describe a runner's position in a 100 m dash?

point of reference or origin in the lower left front corner of the court (where the front wall, left side wall, and floor intersect), the x-axis would be the line along the intersection of the front wall and floor. The y-axis would be the line along the intersection of the front wall and the left side wall, and the z-axis would be the line along the intersection of the left side wall and the floor. This is shown in figure 2.3. If the ball were 3 m to the right of the left side wall, 2 m above the floor, and 4 m away from the front wall, its x-, y-, and z-coordinates in meters would be (3, 2, 4).

> In three dimensions, we would need three numbers to describe the position of an object in space.

To describe the position of something in space, we need to identify a fixed reference point to serve as the origin of our coordinate system. For our purposes, any point fixed relative to the earth will do. Then we set up a Cartesian coordinate system. If we are describing the position of objects in only one dimension, only one axis is needed; for two dimensions, two axes are needed; and for three dimensions, three axes are needed. The axes of

this system may point in any direction that is convenient, as long as they are at right angles to each other if we are describing the position of something in two or three dimensions. Typically, one axis will be oriented vertically (the y-axis), and the other axis (the x-axis) or axes (the x- and z-axes) will be oriented horizontally. Each of these axes will have a positive and negative direction along them. The x-coordinate of an object is the distance the object is away from the plane formed by the y- and z-axes. The y-coordinate of an object is the distance the object is away from the plane formed by the x- and z-axes, and the z-coordinate of an object is the distance the object is away from the plane formed by the x- and y-axes. Units of length are used to describe position.

Distance Traveled and Displacement

Now we have a method of describing and locating the position of an object in space. This is our first task in describing motion. If we remember how we defined motion—the action or process of change in position—our next task will be discovering a way to describe or measure change in position. How would you do this?

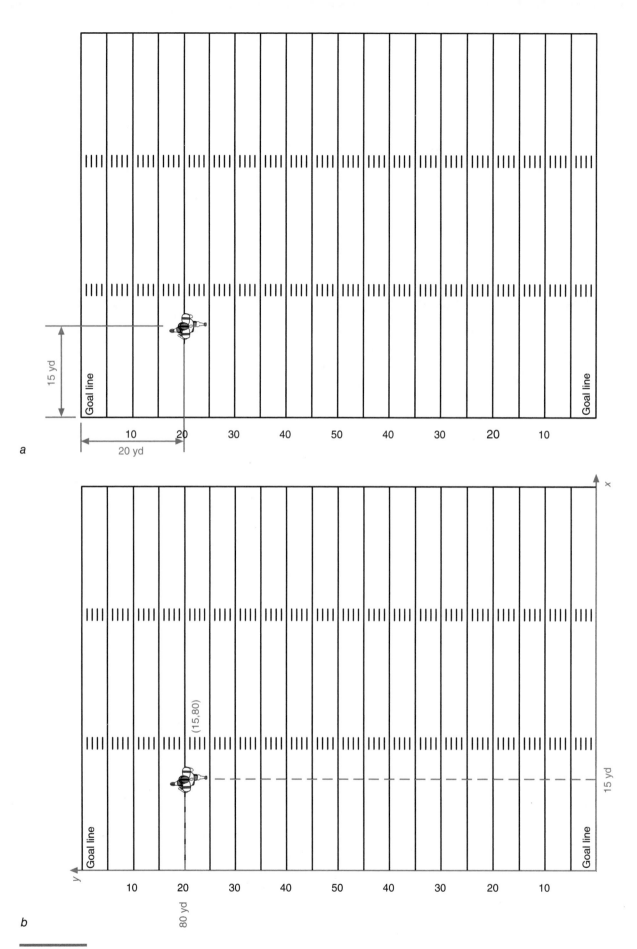

Figure 2.2 The location of a running back on a football field using the sideline and the opponent's goal line as references (a) or using a Cartesian coordinate system (b).

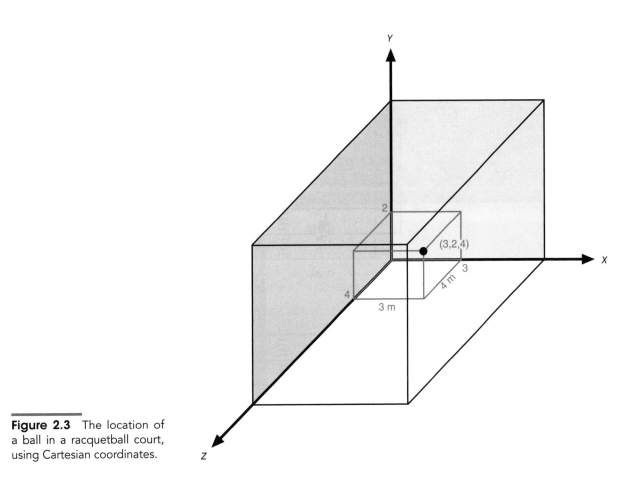

Figure 2.3 The location of a ball in a racquetball court, using Cartesian coordinates.

Distance Traveled

Let's use a football example again. Suppose a football player has received the kickoff at his 5 yd line, 15 yd from the left sideline. His position on the field (using the Cartesian coordinate system we established in the previous section) is (15, 5) when he catches the ball. He runs the ball back following the path shown in figure 2.4a. He is finally tackled on his 35 yd line, 5 yd from the left sideline. His position on the field at the end of the play is (5, 35). If we measure the length of the path of his run with the ball, it turns out to be 48 yd. So we might describe this run as a run of 48 yd to gain 30 yd.

Another way of saying this is to say that the runner's displacement was +30 yd in the y-direction and −10 yd in the x-direction, or a resultant displacement of 31.6 yd toward the left sideline and goal. The distance traveled by the runner was 48 yd. We've used two different terms to describe the runner's progress: **displacement** and **distance traveled**. Distance traveled is easily defined—it's simply a measure of the length of the path followed by the object whose motion is being described, from its starting (initial) position to its ending (final) position. Distance traveled doesn't mean a whole lot in a football game, though, because the direction of travel isn't considered. Displacement does take into account the direction of travel.

Displacement

Displacement is the straight-line distance in a specific direction from initial (starting) position to final (ending) position. The **resultant displacement** is the distance measured in a straight line from the initial position to the final position. Displacement is a vector quantity. If you recall from chapter 1, we said force was also a vector quantity. A vector has a size associated with it as well as a direction. It can be represented graphically as an arrow whose length represents the size of the vector and whose orientation and arrowhead represent the direction of the vector. Representation of displacement with an arrow is appropriate and communicates what displacement means as well. Figure 2.4b shows the path of the player in the kick return example. The arrow from the initial position of the player to where he was tackled represents the displacement of the back.

⊃ Displacement is the straight-line distance in a specific direction from starting (initial) position to ending (final) position.

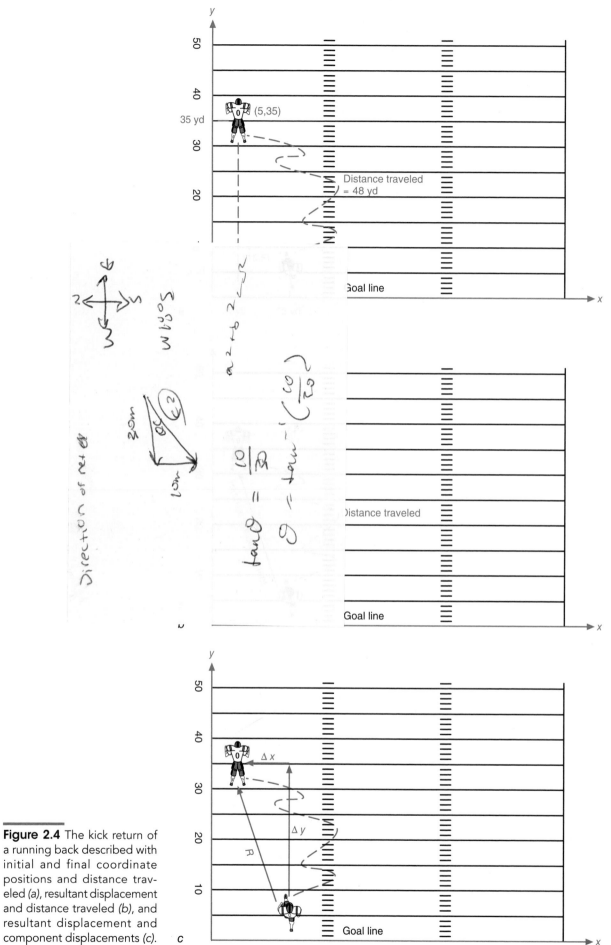

Figure 2.4 The kick return of a running back described with initial and final coordinate positions and distance traveled (a), resultant displacement and distance traveled (b), and resultant displacement and component displacements (c).

If you also recall from chapter 1, vectors can be resolved into components. In the football example, the resultant displacement of the running back does not indicate directly how many yards the back gained. But if we resolve this resultant displacement into components in the x-direction (across the field) and y-direction (down the field toward the goal), we then have a measure of how effective the run was. In this case, the y-displacement of the running back is the measure of importance. His initial y-position was 5 yd and his final y-position was 35 yd. We can find his y-displacement by subtracting his initial position from his final position:

$$d_y = \Delta y = y_f - y_i \qquad (2.1)$$

where

d_y = displacement in the y-direction,

Δ = change, so Δy = change in y-position,

y_f = final y-position, and

y_i = initial y-position.

If we put in the initial (5 yd) and final (35 yd) values for y-position, we get the runner's y-displacement:

$$d_y = \Delta y = y_f - y_i = 35 \text{ yd} - 5 \text{ yd}$$

$$d_y = +30 \text{ yd}$$

The runner's y-displacement or displacement down the field was +30 yd. The positive sign indicates that the displacement was in the positive y-direction or toward the goal (a gain in field position in this case). This measure is probably the most important measurement to the coaches, players, and fans because it indicates the effectiveness of the kick return.

We may also be curious about the player's displacement across the field (in the x-direction). We can use the same equation to determine the x-displacement:

$$d_x = \Delta x = x_f - x_i \qquad (2.2)$$

where

d_x = displacement in the x-direction,

Δx = change in x-position,

x_f = final x-position, and

x_i = initial x-position.

If we put in the initial (15 yd) and final (5 yd) values for x-position, we get the runner's x-displacement:

$$d_x = \Delta x = x_f - x_i = 5 \text{ yd} - 15 \text{ yd}$$

$$d_x = -10 \text{ yd}$$

The runner's x-displacement or displacement across the field was −10 yd. The negative sign indicates that the displacement was in the negative x-direction or toward the left sideline.

We could find the resultant displacement of the runner similarly to the way we found a resultant force. Graphically, we could do this by drawing the arrows representing the component displacements of the runner in the x- and y-directions. Look at figure 2.4c. Put the tail of the x-displacement vector at the tip of the y-displacement vector, and then draw an arrow from the tail of the y-displacement vector to the tip of the x-displacement vector. This arrow represents the resultant displacement.

We could also determine this resultant displacement by starting with the x-displacement vector and then putting the tail of the y-displacement vector at the tip of the x-displacement vector. We would determine the resultant by drawing an arrow from the tail of the x-displacement vector to the tip of the y-displacement vector. We should get the same resultant as determined using the method shown in figure 2.4c.

We could determine this resultant displacement in still another way, using trigonometric relationships. The displacement vectors arranged as shown in figure 2.4c form a triangle, specifically, a right triangle with the hypotenuse represented by the resultant displacement. As explained in chapter 1, the size of the hypotenuse can be determined as follows. If A and B represent the two sides that make up the right angle and C represents the hypotenuse, then

$$A^2 + B^2 = C^2 \qquad (2.3)$$

$$(\Delta x)^2 + (\Delta y)^2 = R^2.$$

For our displacements, then, we can substitute −10 yd for Δx and +30 yd for Δy and then solve for R, which represents the resultant displacement.

$$(-10 \text{ yd})^2 + (30 \text{ yd})^2 = R^2$$

$$100 \text{ yd}^2 + 900 \text{ yd}^2 = R^2$$

$$1000 \text{ yd}^2 = R^2$$

$$R = \sqrt{1000 \text{ yd}^2} = 31.6 \text{ yd}$$

To find the direction of this resultant displacement, we can use the relationship between the two sides of the displacement triangle.

$$\tan \theta = \frac{\text{opposite side}}{\text{adjacent side}} \qquad (2.4)$$

$$\theta = \arctan \left(\frac{\text{opposite side}}{\text{adjacent side}} \right)$$

$$\theta = \arctan \left(\frac{\Delta x}{\Delta y} \right)$$

In these equations, θ, which is pronounced "theta," represents the angle between the resultant displacement vector and the y-displacement vector. To find the value of θ, substitute −10 yd for Δx and +30 yd for Δy.

$$\theta = \arctan\left(\frac{-10 \text{ yd}}{30 \text{ yd}}\right)$$

To determine the angle θ, we use the inverse of the tangent function or the arctangent. On most scientific calculators, the arctangent function is the second function for the tangent key and is usually abbreviated as $\tan^{-1}$.

$$\theta = \arctan(-.333)$$

$$\theta = -18.4°$$

We can now describe several qualities of movement—initial and final positions, distance traveled, and displacement. Distance traveled can be described by a single number that represents the length of the path followed by the object during its motion. Displacement, however, is a vector quantity, so it is expressed with a length measurement and a direction. The resultant displacement is the length of a straight line from the initial position to the final position in the direction of motion from the initial position to the final position. Components of the resultant displacement may also be used to describe displacement of the object in specific directions. In some situations (such as our football examples), a component displacement is more important than the resultant displacement.

Now let's see if we understand the concept of displacement. Imagine two downhill ski racers, Tamara and Cindy, competing on the same course. They start at the same starting position and finish the race at the same finish point. Tamara takes wider turns than Cindy does, so the length of the path Tamara follows is longer. Who has the greater resultant displacement from start to finish? Because they start at the same spot and finish at the same spot, their resultant displacements are the same. Now consider a 100 m swimming race in a 50 m pool. Which measure (displacement or distance traveled) is more meaningful? In a 100 m swimming race in a 50 m pool, you have to start and finish in the same place, so your displacement is zero! Distance traveled is the more meaningful measure. What about a 400 m running race around a 400 m oval track? Or a 100 m running race on a straight section of track?

Speed and Velocity

We can now describe an object's position, and we have measures (distance traveled and displacement) for describing its change in position, but how do we describe how quickly something changes its position? When we speak of how fast or slow something moves, we are describing its speed or velocity. Both are used to refer to the rate of motion. You have probably used both of these terms, perhaps interchangeably.

Speed

Are speed and velocity the same thing? Mechanically, speed and velocity are different. **Speed** is just rate of motion. More specifically, it is the rate of distance traveled. It is described by a single number only. **Velocity** is rate of motion in a specific direction. More specifically it is the rate of displacement. Since displacement is a vector quantity, so is velocity. Velocity has a magnitude (number) and a direction associated with it.

⮌ Speed is rate of motion; velocity is rate of motion in a specific direction.

Average speed of an object is distance traveled divided by the time it took to travel that distance. Mathematically, this can be expressed as

$$\bar{s} = \frac{\ell}{\Delta t} \tag{2.5}$$

where

$\bar{s}$ = average speed,

ℓ = distance traveled, and

Δt = time taken or change in time.

The units for describing speed are a unit of length divided by a unit of time. The SI unit for describing speed is meters per second. You have probably used other units of measurement for speed. If you have driven a car, you are probably more familiar with miles per hour or kilometers per hour. These are also units of measurement for speed.

Average speed is an important descriptor of performance in many sport activities. In some activities, average speed is in fact the measure of success. Consider almost any type of racing event (swimming, running, cycling, and so on). The winner is the person who completes the specified distance in the shortest time. The average speed of the winner is the distance of the race divided by the time. The winner's average speed over the race distance will always be the fastest among all the competitors if everyone raced the same distance.

This one number, average speed, doesn't tell us much about what went on during the race itself, though. It doesn't tell us how fast the racer was moving at any specific instant in the race. It doesn't tell us the maximum speed reached by the racer during the race. It doesn't indicate when the racer was slowing down or speeding up. Average speed for the whole race is just a number indicating that, on average, the competitor was moving that fast. To find out more about the speed of a competitor in a race, a coach or athlete may want to measure more than one average speed.

Let's look at a 100 m dash as an example. At the 12th IAAF World Championships in Athletics in Berlin in 2009, the men's 100 m dash was won by Usain Bolt of Jamaica in an astounding world-record time of 9.58 s. The second-place finisher, Tyson Gay of the United States, finished in 9.71 s, the fastest time ever for a second-place finish. Comparing the average speeds of these two sprinters over the entire 100 m using equation 2.5, we find the following:

$$\text{Average speed} = \bar{s} = \frac{\ell}{\Delta t}$$

Bolt: Gay:

$$\bar{s} = \frac{100 \text{ m}}{9.58 \text{ s}} \qquad \bar{s} = \frac{100 \text{ m}}{9.71 \text{ s}}$$

$$\bar{s} = 10.44 \text{ m/s} \qquad \bar{s} = 10.30 \text{ m/s}$$

To find out more about how the two sprinters ran this race, we might have timed them for the first 50 m of the 100 m as well. Bolt's time for the first 50 m was 5.47 s. Gay's time for the first 50 m was 5.55 s. Their average speeds for the first 50 m of the race were

$$\text{Average speed} = \bar{s}_{0\text{-}50m} = \frac{\ell}{\Delta t} .$$

Bolt: Gay:

$$\bar{s}_{0\text{-}50m} = \frac{50 \text{ m}}{5.47 \text{ s}} \qquad \bar{s}_{0\text{-}50m} = \frac{50 \text{ m}}{5.55 \text{ s}}$$

$$\bar{s}_{0\text{-}50m} = 9.14 \text{ m/s} \qquad \bar{s}_{0\text{-}50m} = 9.01 \text{ m/s}$$

Their average speeds from 50 to 100 m also could be determined:

$$\text{Average speed} = \bar{s}_{50\text{-}100m} = \frac{\ell}{\Delta t} = \frac{100 \text{ m} - 50 \text{ m}}{\Delta t}$$

Bolt: Gay:

$$\bar{s}_{50\text{-}100m} = \frac{100 \text{ m} - 50 \text{ m}}{9.58 \text{ s} - 5.47 \text{ s}} \qquad \bar{s}_{50\text{-}100m} = \frac{100 \text{ m} - 50 \text{ m}}{9.71 \text{ s} - 5.55 \text{ s}}$$

$$\bar{s}_{50\text{-}100m} = \frac{50 \text{ m}}{4.11 \text{ s}} \qquad \bar{s}_{50\text{-}100m} = \frac{50 \text{ m}}{4.16 \text{ s}}$$

$$\bar{s}_{50\text{-}100m} = 12.17 \text{ m/s} \qquad \bar{s}_{50\text{-}100m} = 12.02 \text{ m/s}$$

With two numbers to describe each runner's speed during the race, we know much more about how each runner ran the race. Usain Bolt took the lead in the first 50 m. His average speed was 0.13 m/s faster than Gay's over this portion of the race. Both sprinters were even faster over the second 50 m, but Bolt's average speed over the second 50 m was 0.15 m/s faster than Gay's.

If we wanted to know which athlete had the fastest top speed in the 100 m, we would have to record split times at more frequent intervals in the race. This would give us even more information about the performance of each sprinter. Sport scientists at the 12th IAAF World Championships in Athletics in Berlin recorded the split times at the 20, 40, 60, and 80 m marks for the finalists in the men's 100 m dashes (IAAF 2009). The split times the scientists recorded were used to estimate the 10 m split times for Usain Bolt and Tyson Gay shown in table 2.1.

These 10 m split times can be used to determine the average speed of each sprinter during each 10 m interval. To do this we divide the distance covered in each interval, 10 m in this case, by the time taken to run that distance, the interval time. Table 2.2 shows the values of each runner's average speed over each 10 m interval.

Now we have much more information about each sprinter's performance. From table 2.2, we can tell that

Table 2.1 Elapsed and Interval Times for Each 10 m Interval for Usain Bolt and Tyson Gay in the Men's 100 m Dash Final at the 12th IAAF World Championships in Athletics in Berlin, 2009

Position (m)	Usain Bolt		Tyson Gay	
	Elapsed time (s)	Interval time (s)	Elapsed time (s)	Interval time (s)
0	0		0	
10	1.89	1.89	1.91	1.91
20	2.88	.99	2.92	1.01
30	3.78	.90	3.83	.91
40	4.64	.86	4.70	.87
50	5.47	.83	5.55	.85
60	6.29	.82	6.39	.84
70	7.10	.81	7.20	.81
80	7.92	.82	8.02	.82
90	8.75	.83	8.86	.84
100	9.58	.83	9.71	.85

Table 2.2 Interval Times and Average Speeds for Each 10 m Interval for Usain Bolt and Tyson Gay in the Men's 100 m Dash Final at the 12th IAAF World Championships in Athletics in Berlin, 2009

	Usain Bolt		Tyson Gay	
Interval (m)	Interval time (s)	Average speed (m/s)	Interval time (s)	Average speed (m/s)
0-10	1.89	5.29	1.91	5.24
10-20	.99	10.10	1.01	9.90
20-30	.90	11.11	.91	10.99
30-40	.86	11.63	.87	11.49
40-50	.83	12.05	.85	11.76
50-60	.82	12.20	.84	11.90
60-70	.81	12.35	.81	12.35
70-80	.82	12.20	.82	12.20
80-90	.83	12.05	.84	11.90
90-100	.83	12.05	.85	11.76

Bolt was faster than Gay over every interval up to the 60 to 80 m. During the 60 to 70 m interval, both Bolt and Gay reached their maximum speed, and their average speeds during this interval were the same. After 70 m, both runners slowed down, but Gay slowed down more, especially over the last 20 m of the race, from 80 to 100 m.

By taking more split times during the race, we can determine the runners' average speeds for more intervals and shorter intervals. This procedure also gives us a better idea of what each runner's speeds were at specific instants of time during the race. The speed of an object at a specific instant of time is its **instantaneous speed**. The speed of an object may vary with time, especially in an event such as a 100 m dash. The maximum or top speed a runner achieves during a race is an example of an instantaneous speed. An average speed gives us an estimate of how fast something was moving over only an interval of time—not an instant in time. If we are told what a runner's average speed was for an interval of time, we can correctly assume that the runner's instantaneous speed was faster than the average speed during some parts of that interval and slower than the average speed during other parts of that interval.

Think about your car's speedometer. Does it measure average speed or instantaneous speed? Does it indicate how fast you were going during the past hour? During the

past minute? During the past second? The speedometer on your car measures instantaneous speed. It indicates how fast you are going at the instant in time that you are looking at it. Practically speaking, we can think of instantaneous speed as distance traveled divided by the time it took to travel that distance if the time interval used in the measurement is very small. If the word *average* does not precede the word *speed*, you should assume that instantaneous speed is being referred to.

> We can think of instantaneous speed as distance traveled divided by the time it took to travel that distance if the time interval used in the measurement is very small.

Velocity

Now let's turn our attention to velocity. **Average velocity** is displacement of an object divided by the time it took for that displacement. Because displacement is a vector, described by a number (magnitude) and a direction, average velocity is also a vector, described by a number (magnitude) and a direction. Mathematically, this can be expressed as

$$\bar{v} = \frac{d}{\Delta t} \tag{2.6}$$

where

$\bar{v}$ = average velocity,

d = displacement, and

Δt = time taken or change in time.

The units for describing velocity are the same as those for describing speed: a unit of length divided by a unit of time. The SI unit for describing velocity is meters per second. To measure the average velocity of an object, you need to know its displacement and the time taken for that displacement.

Sometimes we are interested in the components of velocity. So, just as we were able to resolve force and displacement vectors into components, we can also resolve velocity vectors into components. To resolve a resultant average velocity into components, we could simply determine the components of the resultant displacement. For the football player returning the kickoff in the example used earlier, the player's displacement from the instant he received the ball until he was tackled was −10 yd in the x-direction (across the field) and +30 yd in the y-direction (down the field). His resultant displacement was 31.6 yd down and across the field (or −71.6° across the field). If this kick return lasted 6 s, his resultant average velocity, using equation 2.6, was

$$\bar{v} = \frac{d}{\Delta t}$$

$$\bar{v} = \frac{31.6 \text{ yd}}{6 \text{ s}}$$

$$\bar{v} = 5.3 \text{ yd/s}$$

This resultant average velocity was in the same direction as the resultant displacement. Similarly, the running back's average velocity across the field (in the x-direction) would be the x-component of his displacement divided by time or

$$\bar{v}_x = \frac{\Delta x}{\Delta t} \qquad (2.7)$$

$$\bar{v}_x = \frac{-10 \text{ yd}}{6 \text{ s}}$$

$$\bar{v}_x = -1.7 \text{ yd/s}$$

The running back's average velocity down the field (in the y-direction), which is the most important of all these velocities, would be the y-component of his displacement divided by time or

$$\bar{v}_y = \frac{\Delta y}{\Delta t} \qquad (2.8)$$

$$\bar{v}_y = \frac{30 \text{ yd}}{6 \text{ s}}$$

$$\bar{v}_y = 5.0 \text{ yd/s}$$

Just as with the displacements, the resultant average velocity is larger than any of its components. And just as with the displacements, the square of the resultant average velocity should equal the sum of the squares of its components. Let's check, starting with equation 2.3.

$$A^2 + B^2 = C^2$$

$$(\bar{v}_x)^2 + (\bar{v}_y)^2 = \bar{v}^2$$

$$(-1.7 \text{ yd/s})^2 + (5.0 \text{ yd/s})^2 = \bar{v}^2$$

$$2.8 \text{ yd}^2/\text{s}^2 + 25.0 \text{ yd}^2/\text{s}^2 = \bar{v}^2$$

$$5.3 \text{ yd/s} = \sqrt{27.8 \text{ yd}^2/\text{s}^2} = \bar{v}$$

This indeed matches the resultant average velocity of 5.3 yd/s we computed from the resultant displacement and elapsed time.

Average velocity and average speed would both be good descriptors to use for the 100 m dash because it is in a straight line. The runner's speed and the magnitude of the velocity toward the finish line would be identical. In such a case, speed and velocity may be used interchangeably with no problem. Generally, if the motion of the object under analysis is in a straight line and rectilinear, with no change in direction, average speed and average velocity will be identical in magnitude. However, if we are speaking of an activity in which the direction of motion changes, speed and the magnitude of velocity are not synonymous. Imagine a 100 m swimming race in a 50 m pool. If the first-place finisher completes the race in 50 s, we can use equation 2.5 to calculate the swimmer's average speed.

$$\bar{s} = \frac{\ell}{\Delta t}$$

$$\bar{s} = \frac{100 \text{ m}}{50 \text{ s}}$$

$$\bar{s} = 2.0 \text{ m/s}$$

What is the swimmer's average velocity? If the swimmer starts and finishes in the same place, the swimmer's displacement is zero, which means the swimmer's average velocity would also have to be zero. In this case, average velocity and average speed do not mean the same thing, and the average speed measurement is a better descriptor.

> **If the motion of the object under analysis is in a straight line and rectilinear, with no change in direction, average speed and average velocity will be identical in magnitude.**

What about instantaneous speed and instantaneous velocity? We haven't discussed instantaneous velocity yet. It is similar to the concept of instantaneous speed except that direction is included. If we measured average velocity over shorter and shorter intervals of time, practically speaking we would soon have a measure of instantaneous velocity. **Instantaneous velocity** is the velocity of an object at an instant in time. When we speak of the magnitude of the resultant instantaneous velocity of an object, that number is the same as the instantaneous speed of the object.

A resultant instantaneous velocity can also be resolved into components in the direction of interest. For the football player running back the kickoff, we could describe his instantaneous resultant velocity, and we could also describe his instantaneous velocity in the x-direction (across the field) or the y-direction (down the field). If we were concerned about how quickly he was gaining yardage, his instantaneous velocity down the field would be important. Similarly, with the downhill ski racers, it is not their instantaneous resultant velocity that is important; it is the component of this velocity in the direction down the hill that will have the greater effect on the result of the race. Try self-experiment 2.3 to illustrate the difference between speed and velocity.

Self-Experiment 2.3

Imagine that you are in a room with four walls. You are facing the north wall. Let's consider north the direction we are interested in, so north is positive. We are interested only in the component of velocity in the north–south directions. As you begin walking forward, toward the north wall, your velocity north is positive. When you stop, your velocity north is zero. As you begin walking backward, toward the south wall, your velocity north is negative (you are moving in the negative direction). If you walk to your right or left, directly east or west, your velocity north is zero, because you are not getting closer to or farther away from the north wall. If you walk forward toward the north wall and begin turning right toward the east wall, your velocity north is positive and then decreases as you turn. If you are walking east and then turn left toward the north wall, your velocity north is zero and then increases as you turn. During all of these turns, your speed may not even be changing, but if your direction of motion changes, then your velocity changes.

Importance of Speed and Velocity

Now let's make sure we realize the importance of speed and velocity in different sport activities. We've already indicated that in racing events, average speed and average velocity are direct indicators of performance. The athlete with the greatest average speed or greatest average velocity will be the winner. In what other sports is speed or velocity important? How about baseball? A good fastball pitch, which moves with a velocity of 90 to 100 mi/h (145 to 160 km/h), is difficult to hit. Why? The faster the ball is pitched, the less time the batter has to react and decide whether or not to swing at the ball. For instance, in 2010, Ardolis Chapman of the Cincinnati Reds threw a fastball pitch that was clocked at 105.1 mi/h. This is equivalent to 154 ft/s or 47 m/s. The distance from the pitching rubber to home plate is 60 ft 6 in., or 60.5 ft (18.4 m). The ball is released about 2 ft 6 in. in front of the rubber, so the horizontal distance it must travel to reach the plate is only 58 ft (60.5 ft − 2.5 ft) or 17.7 m. Another way to say this is that the horizontal displacement of the ball is 58 ft. How much time does a batter have to react to a fastball pitched at 105.1 mi/h? If we assume that this is the average horizontal velocity of the ball during its flight, then, using equation 2.6,

$$\bar{v} = \frac{d}{\Delta t}$$

$$154 \text{ ft/s} = \frac{58 \text{ ft}}{\Delta t}$$

$$\Delta t = \frac{58 \text{ ft}}{154 \text{ ft/s}}$$

$$\Delta t = 0.38 \text{ s}$$

Wow! A batter only has 0.38 s to decide whether or not to swing his bat, and if he does decide to swing it, he has to do so in the time he has left. No wonder hitting a baseball thrown by a major league pitcher is so difficult. The faster the pitcher can pitch the ball, the less time the batter has to react, and the less likely it is that the batter will hit the ball. In 2003, *USA Today* ranked hitting a baseball thrown at more than 90 mi/h as the most difficult thing to do in sports. Speed and velocity are very important in baseball.

Are speed and velocity important in soccer, lacrosse, ice hockey, field hockey, team handball, or any other sport where a goal is guarded by a goalkeeper? The speed of the ball (or puck) when it is shot toward the goal is very important to the goalkeeper. The faster the shot, the less time the goalkeeper has to react and block it.

Are speed and velocity important in the jumping events in track and field? Yes! Faster long jumpers jump farther. Faster pole-vaulters vault higher. Speed is also related to success in the high jump and triple jump.

Can you think of any sports where speed and velocity are not important? There aren't many. Speed and velocity play an important role in almost every sport. Table 2.3 lists the fastest reported speeds for a variety of balls and implements used in sport. The typical speeds of the balls and implements used in these sports are much slower than those reported in table 2.3.

Acceleration

We now have a large repertoire of motion descriptors: position, distance traveled, displacement, speed, and velocity. In addition, we can use component displacements or velocities to describe an object's motion, because displacement and velocity are vector quantities. Did we use any other descriptors at the beginning of this section to describe the motion of a ball rolling across the floor? Let's try another motion of the ball. Throw the ball up in the air and let it fall back into your hand. How would you describe this motion? You might say that the ball moves upward and slows down on the way up, then begins moving downward and speeds up on the way down. Another way to describe how the ball slows down or speeds up would be to say that it decelerates on the way up and accelerates on the way down. Acceleration is a term you are probably somewhat familiar with, but the mechanical definition of acceleration may differ from yours, so we'd better get some agreement.

Table 2.3 Fastest Reported Speeds for Balls and Implements Used in Various Sports

Ball or implement	Mass (g)	Fastest speed (mi/h)	Fastest speed (m/s)
Golf ball	≤45.93	204	91.2
Jai alai pelota	125-140	188	84.0
Squash ball	23-25	172	76.9
Golf club head	–	163	72.9
Tennis ball	56.0-59.4	156	69.7
Baseball (batted)	142-149	120	53.6
Hockey puck	160-170	110	49.2
Baseball (pitched)	142-149	105	46.9
Softball (12 in.)	178.0-198.4	104	46.5
Lacrosse ball	140-149	100	44.7
Cricket ball (bowled)	156-163	100	44.7
Volleyball	260-280	88	39.3
Soccer ball	410-450	82	36.7
Field hockey ball	156-163	78	34.9
Javelin (men)	800	70	31.3
Team handball (men)	425-475	63	28.2
Water polo ball	400-450	60	26.8

SAMPLE PROBLEM 2.1

The average horizontal velocity of a penalty kick in soccer is 22 m/s. The horizontal displacement of the ball from the kicker's foot to the goal is 11 m. How long does it take for the ball to reach the goal after it is kicked?

Solution:

Step 1: Write down the known quantities.

$$v_x = 22 \text{ m/s}$$

$$d_x = 11 \text{ m}$$

Step 2: Identify the variable to solve for.

$$\Delta t = ?$$

Step 3: Review equations and definitions, and identify the appropriate equation with the known quantities and the unknown variable in it.

$$\bar{v} = \frac{d}{\Delta t}$$

Step 4: Substitute values into the equation and solve for the unknown variable. Keep track of the units when doing arithmetic operations.

$$22 \text{ m/s} = \frac{11 \text{ m}}{\Delta t}$$

$$\Delta t = \frac{11 \text{ m}}{22 \text{ m/s}}$$

$$\Delta t = 0.5 \text{ s}$$

Step 5: Check your answer using common sense.
A penalty kick is pretty quick, definitely less than a second. A half second seems reasonable.

Mechanically, **acceleration** is the rate of change in velocity. Because velocity is a vector quantity, with a number and direction associated with it, acceleration is also a vector quantity, with a number and direction associated with it. An object accelerates if the magnitude or direction of its velocity changes.

> When an object speeds up, slows down, starts, stops, or changes direction, it is accelerating.

Average acceleration is defined as the change in velocity divided by the time it took for that velocity change to take place. Mathematically, this is

$$\bar{a} = \frac{\Delta v}{\Delta t}$$

$$\bar{a} = \frac{v_f - v_i}{\Delta t} \qquad (2.9)$$

where

$\bar{a}$ = average acceleration,

Δv = change in velocity,

v_f = instantaneous velocity at the end of an interval, or final velocity,

v_i = instantaneous velocity at the beginning of an interval, or initial velocity, and

Δt = time taken or change in time.

From this mathematical definition of average acceleration, it is apparent that acceleration can be positive or negative. If the final velocity is less (slower) than the initial velocity, the change in velocity is a negative number, and the resulting average acceleration is negative. This happens if an object slows down in the positive direction. You may have thought of this as a deceleration, but we'll call it a negative acceleration. A negative average acceleration will also result if the initial and final velocities are both negative and if the final velocity is a larger negative number than the initial velocity. This occurs if an object is speeding up in the negative direction.

The units for describing acceleration are a unit of length divided by a unit of time divided by a unit of time. The SI units for describing acceleration are meters per second per second or meters per second squared. You may have seen car ads that tout the acceleration capabilities of the car. An ad may say that a car can accelerate from 0 to 60 in 7 s. Using equation 2.9, this would represent an average acceleration for the car of

$$\bar{a} = \frac{v_f - v_i}{\Delta t}$$

$$\bar{a} = \frac{60 \text{ mi/h} - 0 \text{ mi/h}}{7 \text{s}}$$

$$= 8.6 \text{ mi/h/s}$$

This acceleration can be interpreted as follows: In 1 s, the car's velocity increases (the car speeds up) by 8.6 mi/h. If the car is accelerating at 8.6 mi/h/s and moving at 30 mi/h, 1 s later the car will be traveling 8.6 mi/h faster or 38.6 mi/h. Two seconds later the car will be traveling two times 8.6 mi/h faster (17.2 mi/h) or 47.2 mi/h (=30 mi/h + 17.2 mi/h), and so on.

If we measured average acceleration over shorter and shorter time intervals, practically speaking we would soon have a measure of instantaneous acceleration. **Instantaneous acceleration** is the acceleration of an object at an instant in time. Instantaneous acceleration indicates the rate of change of velocity at that instant in time.

Because acceleration is a vector (as are force, displacement, and velocity), it can also be resolved into component accelerations. This is true for both average and instantaneous accelerations. But how is the direction of an acceleration determined? Try self-experiment 2.4 to get a better understanding of acceleration direction. One of the difficulties of understanding acceleration is that it is not directly observed as displacement and velocity are. The direction of motion is not necessarily the same as the direction of the acceleration.

Self-Experiment 2.4

Let's go back to the example of walking around the room with four walls. You are facing the north wall. Again, consider north to be the direction we are interested in, so north is positive. We are interested in describing the motion only in the north–south directions. As you begin walking forward, toward the north wall, your velocity north is positive; and since you speed up in the northerly direction, your acceleration north is positive (your velocity and acceleration are north). When you slow down and stop, your velocity north decreases to zero, and your acceleration north must be negative since you are slowing down in the positive direction. This could also be described as an acceleration in the southerly direction. This is where you may be confused—you were moving north, but your acceleration was south! This is correct, however, since acceleration indicates your change in motion. As you begin walking backward, you speed up toward the south wall; your velocity north is negative and increasing (you are moving in the negative direction), and your acceleration is also negative (or an acceleration in the southerly direction). If you walk to your right or left, directly east or west, your velocity north is zero because you are not getting closer to or farther away from the north wall. Your acceleration is also zero since you are not speeding up or slowing down toward the north wall. If you walk forward toward the north wall and begin turning right toward the east wall, your velocity north is positive and decreases as you turn, so your acceleration north is negative as you turn. If you are walking east and then turn left toward the north wall, your velocity north is zero and then increases as you turn, so your acceleration north is positive as you turn. During all of these turns, your speed may not even be changing, but if your direction of motion changes, then your velocity changes and you are accelerating. Figure 2.5 illustrates the directions of motion and acceleration for various motions in one dimension (along a line).

> ## The direction of motion does not indicate the direction of the acceleration.

Let's summarize some things about acceleration. If you are speeding up, your acceleration is in the direction of your motion. If you are slowing down, your acceleration is in the opposite direction of your motion. If we assign positive and negative signs to the directions along a line, then the acceleration direction along that line is determined as follows. If something speeds up in the positive direction, its acceleration is positive (it accelerates in the positive direction). Think of this as a double positive (+ +), which results in a positive (+). If it slows down in the positive direction, its acceleration is negative (it accelerates in the negative direction). Think of this as a negative positive (− +), which results in a negative (−). If something speeds up in the negative direction, its acceleration is negative (it accelerates in the negative direction). Think of this as a positive negative (+ −), which results in a negative (−). If something slows down in the negative direction, its acceleration is positive (it accelerates in the positive direction). Think of this as a double negative (− −), which results in a positive (+). Remember, though, the algebraic signs + and − are only symbols we use to indicate directions in the real world. Before analyzing a problem, first establish which direction you will identify as +.

— Direction +	v (Direction of motion)	Change in motion (Speeding up +; slowing down −)	a (Direction of acceleration)
Speeding up	+	+	+
Not changing	+	0 (Constant velocity)	0
Slowing down	+	−	−
Speeding up	−	+	−
Not changing	−	0 (Constant velocity)	0
Slowing down	−	−	+

Figure 2.5 The direction of motion and direction of acceleration are the same when the object is speeding up, but opposite to each other when the object is slowing down.

Uniform Acceleration and Projectile Motion

In certain situations, the acceleration of an object is constant—it doesn't change. This is an example of **uniform acceleration**. It occurs when the net external force acting on an object is constant and unchanging. If this is the case, then the acceleration of the object is also constant and unchanging. The motion of such an object can then be described by equations relating time with velocity, position, or acceleration. Using these equations, we can predict the future! If an object undergoes uniform acceleration, its position and velocity at any future instant in time can be predicted. Wow! Can you think of any situations in which the net external force acting on an object is constant and thus the resulting acceleration is uniform? Try self-experiment 2.5 and see if this is an example of uniform acceleration.

Self-Experiment 2.5

Throw a ball straight up into the air and try to describe its motion. Let's use the terms we have learned—displacement, velocity, and acceleration. If we set up a coordinate system with the x-axis oriented horizontally in the direction of the horizontal motion of the ball and the y-axis oriented vertically, how would you describe the vertical motion of the ball? Let's consider the positive direction along the y-axis (vertical axis) as upward. As the ball leaves your hand, it is moving in the upward direction, so its velocity is positive. Is the ball speeding up or slowing down as it goes up? The ball is slowing down in the upward direction, so its acceleration is negative or in the downward direction. When the ball reaches the peak of its flight, its velocity is changing from positive to negative (or from upward to downward), so it is still accelerating downward. After the ball is past its peak, it falls downward, so its velocity is negative (downward). Since the ball speeds up in the downward direction, its acceleration is still negative (downward). Despite the changes in the direction the ball was moving in, its vertical acceleration was always downward while it was in the air. The direction of its acceleration was constant. Was the magnitude of the acceleration constant as well? What forces acted on the ball while it was in the air? If air resistance can be ignored, then the only force acting on the ball was the force of gravity or the weight of the ball. Since the ball's weight does not change while it's in the air, the net external force acting on the ball is constant and equal to the weight of the ball. This means that the acceleration of the ball is constant as well.

Vertical Motion of a Projectile

In self-experiment 2.5, the ball you threw up in the air was a projectile. A **projectile** is an object that has been projected into the air or dropped and is acted on only by the forces of gravity and air resistance. If air resistance is too small to measure, and the only force acting on a projectile is the force of the earth's gravity, then the force of gravity will accelerate the projectile. In the previous chapter we learned that this acceleration, the acceleration due to gravity or g, is 9.81 m/s² downward. This is a uniform acceleration. Now let's see if we can come up with the equations that describe the vertical motion of a projectile such as the ball in self-experiment 2.5.

Since the vertical acceleration of the ball is constant, we already have one equation to describe this kinematic variable. If we define upward as the positive vertical direction, then

$$a = g = -9.81 \text{ m/s}^2. \tag{2.10}$$

The negative sign indicates that the acceleration due to gravity is in the downward direction.

We know what the vertical acceleration of the ball is; perhaps we can use this knowledge to determine its velocity from equation 2.9, which relates acceleration to velocity.

$$\bar{a} = \frac{v_f - v_i}{\Delta t}$$

The acceleration in equation 2.9 is an average acceleration, but in our case we know the acceleration of the ball at any instant in time—it's 9.81 m/s² downward. But since the acceleration is constant, 9.81 m/s² is also the average acceleration. So, we can substitute g for average acceleration, $\bar{a}$, in equation 2.9 and solve for final velocity, v_f:

$$\bar{a} = \frac{v_f - v_i}{\Delta t} = g$$

$$v_f - v_i = g\Delta t$$

$$v_f = v_i + g\Delta t \tag{2.11}$$

Equation 2.9 gives us a means of determining the instantaneous vertical velocity of the ball (v_f) at the end of some time interval (Δt) if we know its initial vertical velocity (v_i) and the length of the time interval. We can predict the future! Look closer at this equation. If you remember your high school algebra, you might recognize this as the equation for a line:

$$y = mx + b \tag{2.12}$$

where

y = dependent variable (plotted on vertical axis),

x = independent variable (plotted on horizontal axis),

m = slope of line = $\dfrac{\Delta y}{\Delta x}$, and

b = intercept.

In equation 2.11,

$v_f = v_i + g\Delta t,$

v_f is the dependent variable, y,

Δt is the independent variable, x,

g is the slope, m, and

v_i is the intercept, b.

The vertical velocity of the ball changes linearly with changes in time—the vertical velocity of the ball is directly proportional to the time that the ball has been in the air.

What about the vertical position of the ball? Perhaps we can use the definition for average velocity from equation 2.8.

$$\overline{v}_y = \frac{\Delta y}{\Delta t}$$

$$\overline{v}_y = \frac{y_f - y_i}{\Delta t}$$

Since velocity is linearly proportional to time (it's defined by a linear equation), the average velocity over a time interval is equal to the velocity midway between the initial and final velocities. This velocity is the average of the initial and final velocities:

$$\overline{v}_y = \frac{v_f + v_i}{2}$$

$$\overline{v}_y = \frac{v_f + v_i}{2} = \frac{y_f - y_i}{\Delta t} \tag{2.13}$$

If we use the expression from equation 2.11,

$v_f = v_i + g\Delta t,$

and substitute it for v_f in equation 2.13,

$$\frac{v_f + v_i}{2} = \frac{y_f - y_i}{\Delta t},$$

we can solve for y_f.

$$\frac{(v_i + g\Delta t) + v_i}{2} = \frac{y_f - y_i}{\Delta t}$$

$$\frac{(2v_i + g\,\Delta t)\Delta t}{2} = \frac{y_f - y_i}{\Delta t}$$

$$\frac{(2v_i + g\Delta t)\Delta t}{2} = y_f - y_i$$

$$\frac{2v_i\Delta t + g(\Delta t)^2}{2} = y_f - y_i$$

$$v_i\Delta t + \frac{1}{2}g(\Delta t)^2 = y_f - y_i$$

$$y_f = y_i + v_i\Delta t + \frac{1}{2}g(\Delta t)^2 \tag{2.14}$$

If you couldn't follow the derivation of equation 2.14, don't worry about it. The result is what is important for our understanding of the motion of the ball. Equation 2.14 gives us a means of determining the vertical position of the ball (y_f) at the end of a time interval (Δt) if we know its initial vertical velocity (v_i) and the length of the time interval.

There is one more equation that describes vertical velocity of the ball as a function of its vertical displacement and initial vertical velocity. The equation is presented here, but we'll have to wait until chapter 4 (see p. 127) for the derivation of this equation.

$$v^2 = v^2 + 2g\Delta y \tag{2.15}$$

Using equations 2.11 and 2.14 (or 2.15), we can now predict not only how fast the ball will be moving vertically, but where it will be as well. We now have four equations to describe the vertical motion of a projectile.

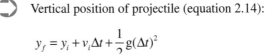 Vertical position of projectile (equation 2.14):

$$y_f = y_i + v_i\Delta t + \frac{1}{2}g(\Delta t)^2$$

Vertical velocity of projectile (equation 2.11 and 2.15):

$$v_f = v_i + g\Delta t$$

$$v^2 = v^2 + 2g\Delta y$$

Vertical acceleration of projectile (equation 2.10):

$$a = g = -9.81 \text{ m/s}^2$$

where

y_i = initial vertical position,

y_f = final vertical position,

$\Delta y = y_f - y_i$ = vertical displacement,

Δt = change in time,

v_i = initial vertical velocity,

v_f = final vertical velocity, and

g = acceleration due to gravity = −9.81 m/s².

If we are analyzing the motion of something that is dropped, the equations are simplified. For a dropped object, $v_f = 0$. If we set the vertical scale to zero at the position the object was dropped from, then $y_i = 0$ as well. For a dropped object, the equations become the following:

Vertical position of falling object:

$$y_f = \frac{1}{2}g(\Delta t)^2 \qquad (2.16)$$

Vertical velocity of falling object:

$$v_f = g\Delta t \qquad (2.17)$$

$$v^2 = 2g\Delta y \qquad (2.18)$$

Imagine that you could safely drop a ball from the top of some tall building and that air resistance is not significant. When you let go of the ball, its vertical velocity is zero. According to equation 2.17, after it has fallen for 1 s, its velocity would be 9.81 m/s downward. According to equation 2.16, its position would be 4.91 m below you. After 2 s, its velocity would be another 9.81 m/s faster, or −19.62 m/s, and its position would be 19.62 m below you. After 3 s its velocity would be another 9.81 m/s faster, or −29.43 m/s, and its position would be 44.15 m below you. Notice that the ball's velocity is just increasing by the same amount (9.81 m/s) during each 1 s time interval, but the ball's position changes by a larger and larger amount during each second it falls (see figure 2.6).

Some other observations about the vertical motion of projectiles may simplify things further. Throw a ball straight up in the air again. How fast is the vertical velocity of the ball at the instant it reaches its peak height? Hmmm. Just before it reached its peak height, it had a small positive velocity (it was going upward slowly). Just after it reached its peak height, it had a small negative velocity (it was going downward slowly). Its vertical velocity went from positive to negative. What number is between positive and negative numbers? How fast is it moving if it's not moving up anymore and hasn't started

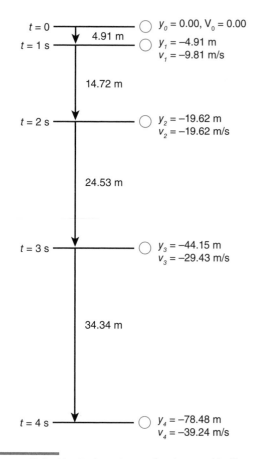

Figure 2.6 Vertical position of a dropped ball at each 1 s interval.

moving downward yet? The ball's vertical velocity at the peak of its flight is zero.

$$v_{peak} = 0 \qquad (2.19)$$

A useful application of this is in the sport of tennis. When you serve a tennis ball, you want to toss it up in the air just high enough that your racket hits it when it is at or near the peak of its flight. Small errors in the timing of your serve won't significantly affect where on the racket the ball hits, because at the peak of its flight the ball's velocity is zero, so it will be near this position for a longer time. If you toss the ball up too high, however, the time during which the ball is in the hitting zone of the racket will be shorter, since the ball is moving much faster as it falls through the hitting zone.

The symmetry of the flight of a projectile is the source of more simplification for our analyses. Toss a ball up again and try to determine which is longer—the time it takes for the ball to reach its peak height or the time it takes for the ball to fall back down from its peak height to its initial height. Wow, those time intervals are close to the same. In fact, they are the same.

$\Delta t_{up} = \Delta t_{down}$ if the initial and final y-positions are the same (2.20)

or,

$\Delta t_{flight} = 2\Delta t_{up}$ if the initial and final y-positions are the same (2.21)

Similarly, the upward velocity of the ball as it passes any height on the way up is the same as the downward velocity of the ball when it passes that same height on the way down. The time it takes for the ball's upward velocity to slow down to zero is the same as the time it takes for the ball's downward velocity to speed up from zero to the same size velocity downward. If you throw a ball upward with an initial vertical velocity of 5 m/s, when you catch it on the way down its velocity will also be 5 m/s but downward.

Horizontal Motion of a Projectile

Now we can describe the vertical motion of a projectile—at least a projectile that is moving only up and down. What about the horizontal motion of a projectile? Try self-experiment 2.6.

SAMPLE PROBLEM 2.2

A volleyball player sets the ball for the spiker. When the ball leaves the setter's fingers, it is 2 m high and has a vertical velocity of 5 m/s upward. How high will the ball go?

Solution:

Step 1: Write down the known quantities and any quantities that can be inferred from the problem

$y_i = 2$ m

$v_i = 5$ m/s

$v_f = v_{peak} = 0$

Step 2: Identify the variable to solve for.

$h = y_f = ?$

Step 3: Review equations and definitions, and identify the appropriate equation with the known quantities and the unknown variable in it (equation 2.15).

$v^2 = v^2 + 2g\Delta y$

Step 4: Substitute values into the equation and solve for the unknown variable. Keep track of the units when doing arithmetic operations.

$v^2 = v^2 + 2g\Delta y$

$0 = (5 \text{ m/s})^2 + 2(-9.81 \text{ m/s}^2)\Delta y$

$\Delta y = \dfrac{(5 \text{ m/s})^2}{2(9.81 \text{ m/s}^2)} = 1.27$ m

$\Delta y = y_f - y_i$

$1.27 \text{ m} = y_f - 2$ m

$y_f = h = 2 \text{ m} + 1.27 \text{ m} = 3.27$ m

Step 5: Check your answer using common sense.
The answer, 3.27 m, is almost 11 feet, which seems about right for a set in volleyball.

Self-Experiment 2.6

Throw a ball in the air from one hand to the other so the ball has both vertical motion and horizontal motion. What forces act on the ball? If we resolve the motion of the ball into horizontal *(x)* and vertical *(y)* components, we know that gravity is an external force that acts in the vertical direction and pulls downward on the ball. What about horizontally (sideways)? Are there any external forces pulling or pushing sideways against the ball to change its horizontal motion once it leaves your hand? The only thing that could exert a horizontal force on the ball is the air through which the ball moves. This force will probably be very small in most cases, and its effect will be too small to notice. If air resistance is negligible, the horizontal velocity of the ball should not change from the time it leaves your hand until it contacts your other hand or another object, since no horizontal forces act on the ball. Try to observe only the horizontal motion of the ball. The ball continues to move in the direction you projected it. It does not swerve right or left. Its horizontal velocity is positive. Does the ball accelerate horizontally while it is in the air? Does it speed up or slow down horizontally? No. Does it change its direction horizontally? No. If the ball doesn't speed up or slow down or change direction, it is not accelerating in the horizontal direction.

SAMPLE PROBLEM 2.3

A punter punts the football. The football leaves the punter's foot with a vertical velocity of 20 m/s and a horizontal velocity of 15 m/s. What is the hang time of the football (how long is it in the air)? (Assume that air resistance has no effect and that the height at landing and at release are the same.)

Solution:

Step 1: Write down the known quantities and any quantities that can be inferred from the problem.

$$y_i = y_f$$

$$v_i = 20 \text{ m/s}$$

$$v_x = 15 \text{ m/s}$$

$$v_{peak} = 0$$

$$\Delta t_{up} = \Delta t_{down}$$

Step 2: Identify the variable to solve for.

$$\Delta t = ?$$

Step 3: Review equations and definitions, and identify the appropriate equation with the known quantities and the unknown variable in it (equation 2.11).

$$\Delta t = \Delta t_{up} + \Delta t_{down} = 2\Delta t_{up}$$

$$v_f = v_i + g\Delta t$$

Step 4: Substitute values into the equation and solve for the unknown variable. Keep track of the units when doing arithmetic operations.

$$v_f = v_i + g\Delta t$$

$$0 = 20 \text{ m/s} + (-9.81 \text{ m/s}^2)(\Delta t_{up})$$

$$\Delta t_{up} = \frac{(-20 \text{ m/s})}{(-9.81 \text{ m/s})} = 2.04 \text{ s}$$

$$\Delta t = 2\Delta t_{up} = 2(2.04 \text{ s}) = 4.08 \text{ s}$$

Step 5: Check your answer using common sense.
Four seconds seems like a reasonable hang time for a punt.

It is difficult to examine or observe the horizontal motion of a projectile separately from its vertical motion, though, because when you observe a projectile, you see the horizontal and vertical motions simultaneously as one motion. How can we view a projectile, such as the ball in self-experiment 2.6, so that we isolate only its horizontal motion? What if we watched the projectile from above? Imagine yourself perched on the catwalk of a gymnasium watching a basketball game. Better yet, imagine watching a football game from the Goodyear blimp. How would the motions of the football or basketball appear to you from these vantage points? If your depth perception was hindered (if you closed one eye), could you see the vertical motion of the football during a kickoff? Could you detect the vertical motion of the basketball during a free throw? The answer is no in both cases. All you see is the horizontal motion of the balls. Does the basketball slow down, speed up, or change direction horizontally as you view it from above? How about the football? If we tried to represent the motion of the basketball as viewed from above in a single picture, it might look something like figure 2.7.

To represent the motion, we show the position of the basketball at four instants in time, each 0.10 s apart. Notice that the images line up along a straight line, so the motion of the ball is in a straight line. Also notice that the displacement of the ball over each interval of time is the same, so the velocity of the ball is constant. The horizontal velocity of a projectile is constant, and its horizontal motion is in a straight line.

> **The horizontal velocity of a projectile is constant and its horizontal motion is in a straight line.**

We derived equations describing the vertical position, velocity, and acceleration of a projectile. Now we can do the same for the horizontal position, velocity, and acceleration of a projectile. We start with the fact that the horizontal velocity of a projectile is constant.

$$v = v_f = v_i = \text{constant} \tag{2.22}$$

If the horizontal velocity is constant, that means there is no change in horizontal velocity. If horizontal velocity doesn't change, then horizontal acceleration must be zero, since acceleration was defined as the rate of change in velocity.

$$a = 0 \tag{2.23}$$

Also, if the horizontal velocity is constant, then the average horizontal velocity of the projectile is the same as its instantaneous horizontal velocity. Since average velocity

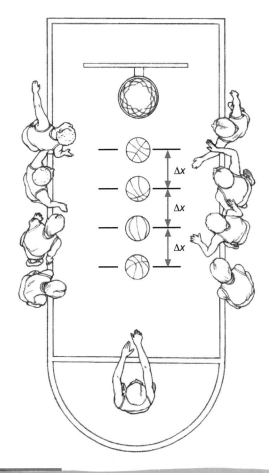

Figure 2.7 An overhead view of a basketball free throw shows that the horizontal displacement, Δx, for each 0.10 s time interval is the same.

is displacement divided by time, displacement is equal to velocity times time (equation 2.6).

$$\bar{v} = \frac{d}{\Delta t}$$

$$\bar{v} = \frac{\Delta x}{\Delta t}$$

$$\Delta x = v\Delta t \tag{2.24}$$

$$x_f - x_i = v\Delta t$$

$$x_f = x_i + v\Delta t \tag{2.25}$$

If our measuring system is set up so that the initial horizontal position (x_i) is zero, then equation 2.25 simplifies to

$$x = v\Delta t. \tag{2.26}$$

Using equations 2.22 and 2.26 (or 2.25), we can now predict not only how fast a projectile will be moving horizontally but where it will be as well. We now have the equations to describe the horizontal motion of a projectile.

➲ Horizontal position of projectile (equations 2.25 and 2.26):

$$x_f = x_i + v\Delta t$$

$$x = v\Delta t \qquad \text{if initial position is zero}$$

➲ Horizontal velocity of projectile (equation 2.22):

$$v = v_f = v_i = \text{constant}$$

➲ Horizontal acceleration of projectile (equation 2.23):

$$a = 0$$

where

x_i = initial horizontal position,

x_f = final horizontal position,

Δt = change in time,

v_i = initial horizontal velocity, and

v_f = final horizontal velocity.

Combined Horizontal and Vertical Motions of a Projectile

We have now developed equations that describe the motion of a projectile in terms of its vertical and horizontal components. Does the vertical motion of a projectile affect its horizontal motion or vice versa? Try self-experiment 2.7.

Self-Experiment 2.7

Put a coin on the edge of a tabletop. Place another same-denomination coin on the end of a ruler or other long, flat object. Place the ruler with the coin on it on the table next to the other coin so that the end of the ruler with the coin on it overhangs the tabletop. Strike the ruler with your hand so that it in turn strikes the coin on the table and knocks the coin off of the table. Simultaneously, the movement of the ruler will dislodge the coin off the end of the ruler. Figure 2.8 shows the setup of the demonstration.

Which coin will strike the floor first? Try it several times to see. The two coins hit the floor at the same time. The coin that is knocked off of the table has a horizontal velocity as it begins to fall, while the coin that slips off of the ruler does not. The two coins fall the same vertical distance, and neither of them has a vertical velocity when it begins to fall. What force pulls the coins toward

the earth? The force of gravity pulls the coins downward and accelerates both downward at the same rate of 9.81 m/s². Does the fact that one coin has a horizontal velocity affect how the force of gravity acts on that coin, and thus affect the vertical acceleration of that coin? No, the force of gravity has the same effect on the coin knocked off the table as it does on the coin that slid off of the ruler.

The vertical and horizontal motions of a projectile are independent of each other. In other words, a projectile continues to accelerate downward at 9.81 m/s² with or without horizontal motion, and the horizontal velocity of a projectile remains constant even though the projectile is accelerating downward at 9.81 m/s². Although the motions of a projectile are independent of each other, an equation can be derived to describe the path of a projectile in two dimensions. Take equation 2.26 and solve for Δt.

$$\Delta t = \frac{x}{v_x}$$

Now substitute this expression for Δt in equation 2.14.

$$y_f = y_i + v_i\Delta t + \frac{1}{2}g(\Delta t)^2$$

$$y_f = y_i + v_{y_i}\left(\frac{x}{v_x}\right) + \frac{1}{2}g\left(\frac{x}{v_x}\right)^2 \qquad (2.27)$$

Equation 2.27 is the equation of a parabola. It describes the vertical (y) and horizontal (x) coordinates of a projectile during its flight based solely on the initial vertical position and vertical and horizontal velocities. Figure 2.9 shows the parabolic path followed by a ball thrown in the air with an initial vertical velocity of 6.95 m/s and an initial horizontal velocity of 4.87 m/s. The ball was

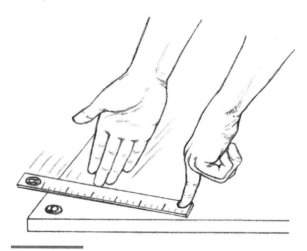

Figure 2.8 The coin experiment demonstrates the independence of the horizontal and vertical components of projectile motion.

Figure 2.9 Stroboscopic photos of a ball in flight taken at equally spaced time intervals. Note the parabolic trajectory.

photographed at a rate of 12 frames per second, so the position of the ball at each 0.0833 s interval is shown in the figure. Notice that the horizontal displacements over each time interval are the same and that the path is symmetrical on either side of the peak. The peak height actually occurs between the ninth and 10th ball images as counted from the left.

Several of the equations that describe projectile motion can be written with only three variables. These equations (2.11, 2.15, and 2.24) are

$$v_f = v_i + g\Delta t$$

$$v_f^2 = v_i^2 + 2g\Delta y$$

$$\Delta x = v\Delta t$$

Equation 2.14 has four variables, but it can be modified by substituting Δy for y_i and y_f as shown to produce equation 2.28 with only three variables.

$$y_f = y_i + v_i\Delta t + \frac{1}{2}g(\Delta t)^2$$

$$y_f - y_i = v_i\Delta t + \frac{1}{2}g(\Delta t)^2$$

$$\Delta y = v_i\Delta t + \frac{1}{2}g(\Delta t)^2 \qquad (2.28)$$

We now have four equations, and each has only three variables. In each of these equations, if two of the variables are known, the equation can be solved for the third variable. Table 2.4 lists these equations and their variables. You can use this table as an aid to help you solve projectile problems by following these steps. First identify the unknown variable that you are trying to determine. Look in the first column, labeled "Unknown variable" in table 2.4, to see if the variable is in the table. If it appears in a row in this column, look to the right in that row to see if you know the values for the two variables listed in the "Known variables" column. If you know the values for those two variables, look to the right in the "Equation" column and plug the values into the equation and solve for the unknown variable. Remember, you might have to solve two or more equations before you get to the equation that has the unknown variable that interests you.

Projectiles in Sport

Examples of projectiles in sports and human movement are numerous. Can you name a few? Here are some examples of projectiles: a shot in flight during a shot put, a basketball in flight, a hammer in flight during a hammer throw, a volleyball in flight, a squash ball in flight, a lacrosse ball in flight, a football in flight, a rugby ball in flight . . . just about any ball used in sport becomes a projectile once it is thrown, released, or hit,

Table 2.4 Solution Guide for Solving Projectile Problems If Two Variables Are Known

Identify the unknown variable in the first column set. Find the two known variables in the second column set that match the row of the unknown variable. Solve for the unknown variable using the equation in the rightmost column of that row.

	Unknown variable *If you want to find this . . .*	Known variables *. . . and you know these,*		Equation *. . . use this equation to find the unknown variable.*
y (vertical)	Δy	v_i	Δt	$\Delta y = v_i \Delta t + \dfrac{1}{2} g (\Delta t)^2$
	v_i	Δt	Δy	
	Δt	Δy	v_i	
	v_i	v_f	Δt	$v_f = v_i + g \Delta t$
	v_f	v_i	Δt	
	Δt	v_f	v_i	
	Δy	v_f	v_i	$v_f^2 = v_i^2 + 2g\Delta y$
	v_i	v_f	Δy	
	v_f	v_i	Δy	
x (horizontal)	Δx	v_x	Δt	$\Delta x = v_x \Delta t$
	v_x	Δt	Δx	
	Δt	Δx	v_x	

Variable definitions:

Δt = time

$\Delta y = y_f - y_i$ = vertical displacement

y_i = initial vertical position

y_f = final vertical position

v_i = initial vertical velocity

v_f = final vertical velocity

g = acceleration due to gravity = -9.81 m/s²

$\Delta x = x_f - x_i$ = horizontal displacement

x_i = initial horizontal position

x_f = final horizontal position

v_x = horizontal velocity

if air resistance is negligible. So in ball sports, the path of the ball cannot be changed in flight if air resistance is negligible. Its path is determined by equation 2.27. Vertically, the ball is constantly accelerating downward, and horizontally it won't slow down or speed up. Once a ball has left our hands and is in flight, our actions and antics cannot change its predetermined course or velocity.

It seems pretty obvious that the balls used in sport are projectiles, but what if we ourselves are the projectiles? Can the human body be a projectile? Are there situations in which the only external force acting on you is the force of gravity? Yes, of course there are! Think of some examples in sport in which the human body is a projectile. How about in running? High jumping? Long jumping? Diving? Pole vaulting? Volleyball? Basketball? Soccer? Football? In each of these sports, there are situations in which the athlete is airborne and the only force acting

on her is gravity. Do the projectile equations govern the motion of an athlete in these situations? Yes! This means that, once an athlete's body has left the ground and become a projectile, the athlete cannot change her path. Once a volleyball player has jumped up to the left to block a shot, the path of her body motion cannot be changed; in other words, once she has jumped up to the left, she won't be able to change direction and block a shot to the right. And once a pole-vaulter releases the pole, he cannot change his motion. Once he has let go of the pole, he no longer has control over where he falls. Once a long jumper leaves the takeoff board and becomes a projectile, her actions while in the air will not affect the velocity of her body. She cannot speed up her horizontal velocity to increase the distance of the jump after leaving the ground. Nor can she turn off gravity to stay in the air longer.

In projectile activities, the initial conditions (the initial position and initial velocity) of the projectile determine the motion that the projectile will have. In sports involving projectiles, the athlete's objective when throwing, kicking, striking, shooting, or hitting the projectile usually concerns one of three things: time of flight, peak height reached by the projectile, or horizontal displacement.

Time of flight of a projectile is dependent on two things: initial vertical velocity and initial vertical position. We can use the equations to mathematically demonstrate this, or we can just make some simple observations. Drop a ball to the floor first from waist height, then from shoulder height, and then from over your head. Which one took the shortest time to reach the floor? Which one took the longest time to reach the floor? The higher the initial height of the projectile, the longer it stays in the air. The shorter the initial height of the projectile, the shorter the time it stays in the air.

Now, rather than dropping the ball, throw it upward. Throw it upward again, but harder this time, and try to release it at the same height. Now throw it down, and again, try to release it at the same height. What should you do if you want the ball to stay in the air longer? The faster the initial upward velocity of the projectile, the longer it stays in the air. The slower the initial upward velocity (or the faster its initial downward velocity), the shorter the time it stays in the air.

Maximizing time in the air is desirable in certain situations in sport such as a football punt or a lob in tennis. Gymnasts and divers also need sufficient time in the air to complete stunts. In these situations, the initial vertical velocity of the projectile is relatively large (compared to the horizontal velocity), and the angle of projection is above 45°. The optimal angle of projection to achieve maximum height and time of flight is 90° or straight up.

In some sport activities, minimizing the projectile's time in the air is important. Examples of these activities include a spike in volleyball, an overhead smash in tennis, throws in baseball, and a penalty kick in soccer. In these situations, the initial upward vertical velocity of the ball is minimized or the ball may even have an initial downward velocity. The projection angle is relatively small—less than 45°—and in some cases even less than zero.

The peak height reached by a projectile is also dependent on its initial height and initial vertical velocity. The higher a projectile is at release and the faster it is moving upward at release, the higher it will go. Maximizing peak height is important in sports such as volleyball and basketball, where the players themselves are the projectiles. Another obvious example of a sport in which maximal peak height is desired is high jumping. Again, the athlete is the projectile. In these activities, the angle of projection is large, above 45°.

Maximizing the horizontal displacement or range of a projectile is the objective of several projectile sports. Examples of these include many of the field events in track and field, including the shot put, hammer throw, discus throw, javelin throw, and long jump. In the discus throw and javelin throw, the effects of air resistance are large enough that our projectile equations may not be accurate in describing the flight of the discus or javelin. For the shot put, hammer throw, and long jump, air resistance is too small to significantly affect things, so our projectile equations are valid. Our analysis of these situations may require the use of equations. If we want to maximize horizontal displacement, then equation 2.24 may be useful.

$$\Delta x = v\Delta t$$

This equation describes horizontal displacement (Δx) as a function of initial horizontal velocity (v) and time (Δt). But time in this case would be total time in the air or the flight time of the projectile. We just saw that the flight time of a projectile is determined by its initial height and its initial vertical velocity. The horizontal displacement of a projectile is thus determined by three things: initial horizontal velocity, initial vertical velocity, and initial height. If the initial height of release is zero (the same as the landing height), then the resultant velocity (the sum of vertical and horizontal velocities) at release determines the horizontal displacement of the projectile. The faster you can throw something, the farther it will go. But what direction should you throw in—more upward (vertical) or more outward (horizontal)?

If the initial velocity of the ball is totally vertical (a projection angle of 90°), the initial horizontal velocity (v in equation 2.24) would be zero, and the horizontal displacement would be zero as well. If the initial velocity of the ball is totally horizontal (a projection angle of zero), the flight time (Δt in equation 2.24) would be zero, and the horizontal displacement would be zero as well. Obviously, a combination of horizontal and vertical initial velocities (and a projection angle somewhere between 0° and 90°) would be better. What combination works best? If the resultant velocity is the same no matter what the angle of projection, then maximum horizontal displacement will occur if the horizontal and vertical components of the initial velocity are equal, or when the projection angle is 45°. If we look at equation 2.24, this makes sense. Horizontal displacement is determined by initial horizontal velocity and time in the air, but time in the air is determined by initial vertical velocity alone (if height of release is zero). It makes sense that these two variables—initial horizontal and vertical velocities—would have equal influence on horizontal displacement.

Let's check to see if this reasoning is confirmed by observations of projection angles in the sport of shot putting. At the 1995 World Track and Field Championships, the average angle of release for the best throw by the six medalists (three men and three women) in the shot put was 35° (Bartonietz and Borgtom 1995). This is much less than the optimal angle of 45°. But wait, does a shot have a height of release? Yes, the shot is released more than 2 m high. Look at figure 2.10, which shows a shot-putter near the instant he releases the shot. The shot is well above the ground. This height is its initial height. The height of release will give the shot more time in the air, so the time in the air does not have to be created by the vertical velocity at release. If the shot-putter doesn't have to give the shot as much vertical velocity at release, he can put more effort into generating horizontal velocity. The optimal projection angle will thus be less than 45°. The higher the height of release, the lower the projection angle.

Is there any other reason why the optimal release angle for shot putting should be less than 45° (other than the fact that shot-putters have a release height of 2 m or more)? Maybe. Our conclusion that 45° was an optimal projection angle for maximizing the horizontal displacement of a projectile relied on two conditions—first, that the release height was zero, and second, that the resultant velocity of the projectile was the same no matter what the projection angle was. For the shot putter, the first assumption was incorrect, so the release angle was less than 45°. What about the second assumption? In shot putting, does the resultant velocity of the shot change if you change the release angle? To answer this question, consider another question: Is it easier to move something faster horizontally or vertically upward? If you have access to a shot, determine whether you can roll it across the floor (move it horizontally) faster than you can throw it straight up. It's more difficult to accelerate objects upward and produce a large upward velocity than it is to accelerate objects horizontally and produce large horizontal velocities. In shot putting (and in most other throwing events), the resultant velocity of the shot increases as the angle of projection decreases below 45°.

If we examine projection angles for the discus throw or the javelin throw, we find that they are even lower than those of the shot put—even though the height of release is lower for the discus and javelin. Why? During the flight of the discus or javelin, the implement is acted on by another force besides gravity—air resistance. If the javelin or discus is thrown correctly, the air resistance force will exert some upward force on the javelin or discus during its flight. This upward force reduces the net downward force acting on the implement and thus causes its downward acceleration to be smaller as well. The result is that the javelin or discus stays airborne longer. Since the lift force gives the javelin or discus more time in the air, the time in the air does not have to be created by the vertical velocity at release. Once again, if the thrower doesn't have to give the javelin or discus as much vertical velocity at release, the athlete can put more effort into generating horizontal velocity. An extreme example of the lift effect of air resistance providing a projectile with more time in the air would be throwing a flying disc or ring such as a Frisbee or an Aerobie for distance. The lift effect of air resistance is so large on these projectiles that the optimal angle of release for maximizing horizontal distance is not much above horizontal.

Let's summarize what we now know about projectiles in sports.

© Alan Stanford/Actionplus/Icon SMI

Figure 2.10 The shot has an initial height at the instant of release.

1. If you want to maximize the time of flight or the height reached by a projectile, the vertical component of release velocity should be maximized, and the projection angle should be above 45°.

2. If you want to minimize the time of flight of a projectile, the upward component of release velocity should be minimized (perhaps so much so that the vertical velocity at release is downward). The projection angle should be much lower than 45° and in some situations may even be below horizontal.

3. If you want to maximize the horizontal displacement of a projectile, release velocity should be maximized and a higher release height is better. The horizontal component of release velocity should be slightly faster than the vertical component so that the projection angle is slightly lower than 45°. The higher the release height and the greater the lift effects of air resistance on the projectile, the farther below 45° the projection angle should be.

The equations governing projectile motion dictate the path that a ball or other thrown object will take once it leaves our hands. Once you release a ball, you no longer have control over it. Likewise, if you yourself become a projectile, the path taken by your body in the air is predetermined by your velocity and position at the instant you leave the ground. Once you have left the ground, if the only force acting on you is the force of gravity, you no longer have control over the path your body will take or your velocity.

Summary

Motion may be classified as linear, angular, or a combination of the two (general motion). Most examples of human movement are general motion, but separating the linear and angular components of the motion makes it easier to analyze the motion. Linear displacement is the straight-line distance from starting point to finish, whereas linear distance traveled represents the length of the path followed from start to finish. Velocity is the rate of change of displacement, whereas speed is the rate of change of distance. Acceleration is the rate of change of velocity. Displacement, velocity, and acceleration are vector quantities and are described by size and direction.

The vertical and horizontal motion of a projectile can be described by a set of simple equations if the only force acting on the projectile is the force of gravity. The horizontal velocity of a projectile is constant, and its vertical velocity is constantly changing at the rate of $9.81 \, \text{m/s}^2$. The path of a projectile and its velocity are set once the projectile is released or is no longer in contact with the ground.

We now have the terms to describe many aspects of the linear motion of an object—distance traveled, displacement, speed, velocity, and acceleration. But what causes linear motion of objects? How do we affect our motion and the motion of things around us? We've gotten some hints in this and the previous chapter. In the next chapter, we will explore the causes of linear motion more thoroughly.

KEY TERMS

acceleration (p. 66)
angular motion (p. 53)
average acceleration (p. 66)
average speed (p. 60)
average velocity (p. 62)
Cartesian coordinate system (p. 54)
curvilinear translation (p. 52)

displacement (p. 57)
distance traveled (p. 57)
general motion (p. 53)
instantaneous acceleration (p. 66)
instantaneous speed (p. 62)
instantaneous velocity (p. 63)
linear motion (p. 52)

position (p. 54)
projectile (p. 68)
rectilinear translation (p. 52)
resultant displacement (p. 57)
speed (p. 60)
uniform acceleration (p. 68)
velocity (p. 60)

REVIEW QUESTIONS

1. Give an example of a human movement involving the whole body that represents curvilinear motion. Do not use the examples given at the beginning of the chapter.

2. Give an example of a human movement involving the whole body that represents rectilinear motion. Do not use the examples given at the beginning of the chapter.

3. Give an example of a human movement involving the whole body that represents angular motion. Do not use the examples given at the beginning of the chapter.

4. Tyler and Jim race each other up a mountain on their bicycles. Tyler rides a road bike on the switchbacks of the twisting and turning mountain road. Jim rides a mountain bike and follows a direct, but steeper, straight-line path up the mountain. They start at the same time and place at the bottom of the mountain and finish at the same time and place at the top of the mountain. From start to finish,

   a. whose distance traveled was longer?

   b. whose displacement was longer?

   c. which rider had the faster average speed?

   d. which rider had the faster average velocity?

   e. who won the race?

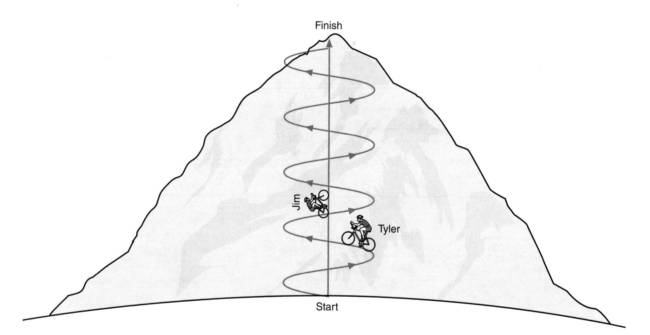

5. Are the sizes of the goals in ice hockey, lacrosse, soccer, field hockey, and team handball related to the speeds of the balls (or puck) used in these games? If so, explain the relationship.

6. Are the sizes of the courts in tennis, volleyball, racquetball, squash, table tennis, and badminton related to the speeds of the balls (or shuttlecock) used in these games? If so, explain the relationship.

7. What factors affect the speeds of the balls and implements listed in table 2.3?

8. When a 100 m sprinter has reached her maximum speed, is her average horizontal velocity faster during the support phase of a step (when her foot is on the ground) or during the flight phase of a step (when she is not in contact with the ground)? Explain.

9. Can a runner moving around a curve at constant speed be accelerating? Explain.

10. If Jim runs around a circle counterclockwise, in which direction (relative to the circle) is his acceleration? Explain.

11. List as many examples as you can of sports or situations in sport in which maximizing a projectile's time in the air is important.

12. List as many examples as you can of sports in which minimizing a projectile's time in the air is important.

13. Elite long jumpers have takeoff angles around 20°. Why do elite long jumpers have takeoff angles so much lower than the theoretically optimal takeoff angle of 45°?

PROBLEMS

1. Sam receives the kicked football on the 3 yd line and runs straight ahead toward the goal line before cutting to the right at the 15 yd line. He then runs 9 yd along the 15 yd line directly toward the right sideline before being tackled.

   a. What was Sam's distance traveled? $12 + 9 = \boxed{21 \text{ yrds}}$

   b. What was Sam's resultant displacement? $c^2 = A^2 + B^2 = 9^2 + 12^2$
   $= \boxed{15 \text{ yrds}}$

   c. How many yards did Sam gain in this play (how far was the ball advanced toward the goal line)?

   $15 - 3 = 12 \text{ yds}$

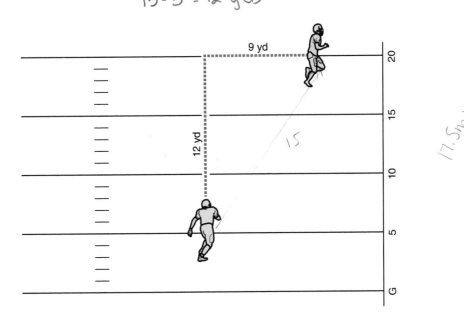

2. During an ice hockey game, Phil had two shots on goal—one shot from 5 m away at 10 m/s and one shot from 10 m away at 40 m/s. Which shot did Brian, the hockey goalie, have a better chance of blocking?   5m at 10m/s - take longer to reach him

3. The horizontal velocity of Bruce's fastball pitch is 40 m/s at the instant it's released from his hand. If the horizontal distance from Bruce's hand to home plate is 17.5 m at the instant of release, how much time does the batter have to react to the pitch and swing the bat?

4. The world-record times for the men's 50 m, 100 m, 200 m, and 400 m sprint races are 5.47 s, 9.58 s, 19.19 s, and 43.18 s, respectively. Which world-record race was run at the fastest average speed?

5. Matt is sailboarding northeast across the river with a velocity of 10 m/s relative to the water. The river current is moving the water north at a velocity of 3 m/s downstream. If the angle between

3. $\Delta x = V x \Delta t$

   $\dfrac{\Delta x}{Vx} = \Delta t$

   $\dfrac{17.5m}{40 m/s} = \Delta t$

   $0.445 = \Delta t$

4. 100 M race at 10.44 m/s²
   Speed = $\dfrac{distance}{time}$
   $= \dfrac{100}{9.58} = 10.44$

5.

81

the relative velocity of the sailboard and the river current is 30°, what is the resultant or true velocity of the sailboard?

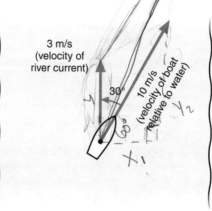

3 m/s
(velocity of
river current)

30°

60°

10 m/s
(velocity of boat
relative to water)

Handwritten annotations in margins:

8

$V_i = 0 m/s$

11.7m

$a = -9.8 m/s$
$V_i = 0 m/s$
$t = 1s$
$\Delta x = ?$

Find y2
add it to y

Use puthagorean
thery to get
res velocity

y
y2
X₁

6. Sean is running a 100 m dash. When the starter's pistol fires, he leaves the starting block and continues speeding up until 6 s into the race, when he reaches his top speed of 11 m/s. He holds this speed for 2 s; then his speed has slowed to 10 m/s by the time he crosses the finish line 11 s after he started the race.

   a. What was Sean's average acceleration during the first 6 s of the race? $g = \frac{v}{t} = \frac{11}{6} = 1.83 m/s$

   b. What was Sean's average acceleration from 6 to 8 s into the race? 0

   c. What was Sean's average velocity for the whole race? average $v = \frac{distance}{time} = \frac{100}{11} = 9.1 m/$

   d. What was Sean's average acceleration from 8 to 11 s into the race? $\frac{\Delta v}{\Delta t} = \frac{-1}{3} = -0.33$

margin: $6 - x = 11 m/s$

7. It is the final seconds of an ice hockey game between the Flyers and the Bruins. The Bruins are down by 1 point. With 20 s left in the game, the Bruins pull the goalie and have him play as a forward in an attempt to tie the game. The Flyers successfully defend their goal for 9 s. With only 1.25 s remaining on the game clock, a Flyer shoots the puck on the ice past the skates and sticks of the other players and toward the Bruins' goal. The puck is 37 m from the goal when it leaves the stick with an initial horizontal velocity of 30 m/s. The shot is perfectly directed toward the empty goal, but the ice slows the puck down at a constant rate of 0.50 m/s² as it slides toward the goal. None of the Bruins can stop the puck before it reaches the goal.

   a. Where is the puck when the game clock reaches zero and the horn sounds to end the game?

   b. Do the Flyers win the game by 1 or 2 points? 2

$7. a) = V(t) + \frac{1}{2} a(t)^2$
$= 30(1.25) + \frac{1}{2}(0.5)(1.25)^2$
$= 37.1$
$37.1 - 37 m from goal$
$= 0.1 meters inside$
$the net$
beside 7: acceleration

8. Mike clears a crossbar while pole vaulting. He releases the pole before he achieves his peak height. It takes him 1 s to fall from his peak height to the landing pit. The landing pit is 1 m high. How high above the ground was Mike at the peak height of his vault?

9. Brian is attempting to high jump over a crossbar set at 2.44 m (8 ft). At the instant of takeoff (when he is no longer in contact with the ground) his vertical velocity is 4.0 m/s, and his center of gravity is 1.25 m high.

   a. What is Brian's vertical acceleration at the instant of takeoff? $-9.81 m/s^2$ -down

   b. How much time elapses after takeoff until Brian reaches his peak height? $\Delta t = \frac{v}{g} = \frac{4 m/s}{9.41 m/s^2} = 0.4$

   c. What peak height does Brian's center of gravity achieve? $\frac{v+v^2}{20} + 1 \frac{H^2}{20(9.81)} + 1.25 = 2.07$

10. Oliver punts a football into the air. The football has an initial vertical velocity of 15 m/s and an initial horizontal velocity of 15 m/s when it leaves the Oliver's foot. The ball experiences a constant vertical acceleration of 9.81 m/s² downward while it is in the air.

    a. What is the ball's horizontal velocity 2 s after it leaves the kicker's foot?

    b. What is the ball's vertical velocity 2 s after it leaves the kicker's foot?

margin bottom:
10. a) 15 m/s forward
b) $v = v_i - g(t)$
$= 15 - 9.91(2)$
$= 15 - 19.62$
$= -4.62 m/s$

c) $\Delta x = V_x \Delta t$
$= 15 \times 2$
$= 30 m$

d) $\Delta y = v_i \Delta t + \frac{1}{2} g(\Delta t)^2 = 15(2) + \frac{1}{2}(9.81)(2)$
$30 - 19.62$
$= 10.38 m$

c. What is the ball's horizontal displacement 2 s after it leaves the kicker's foot?

d. What is the ball's vertical displacement 2 s after it leaves the kicker's foot?

11. Gerri leaves the long jump takeoff board with a vertical velocity of 2.8 m/s and a horizontal velocity of 7.7 m/s.

a. What is Gerri's resultant velocity at takeoff? $2.9^2 + 7.7^2 = c^2$ $= 4.19$

b. What is Gerri's takeoff angle—the angle of her resultant takeoff velocity with horizontal? b) $\tan^{-1}\left(\frac{2.9}{7.7}\right) = 20°$

c. What is Gerri's horizontal velocity just before she lands? 7.7 m/s

d. If Gerri is in the air for 0.71 s, what is her horizontal displacement during this time in the air? $\Delta v = v_x \Delta t$ $= 7.7(0.71)$

e. What is Gerri's vertical velocity at the end of her 0.71 s flight? $v_f = v_i - g(t)$ $= 2.8 - 9.9(0.71)$ $= -4.17$ d) 5.47 m

f. If Gerri's center of gravity was 1.0 m high at the instant of takeoff, how high will it be at the peak of her flight? $\frac{viy^2}{2a} = \frac{2.8^2}{2(9.81)} + 1 = 1.4$

g. How high is Gerri's center of gravity at the end of her flight, when she first hits the pit?

12. Louise spikes a volleyball. At the instant the ball leaves her hand, its height is 2.6 m and its resultant velocity is 20 m/s downward and forward at an angle of 60° below horizontal.

a. How long will it take for the ball to strike the floor if the opposing team does not block it? $x = \cos 60 \times 20 = 10$

b. How far will the ball travel horizontally before it strikes the floor? $y = \sin 60 \times 20 = 17.3$

a) $vi \Delta t + \frac{1}{2} a (t)^2 - \Delta dy = 0$

quadratic formula to find

$\frac{-b \pm \sqrt{b^2 - 4ac}}{2a}$

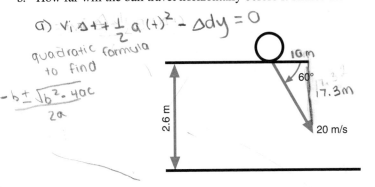

13. Chloe has a vertical velocity of 3 m/s when she leaves the 1 m diving board. At this instant, her center of gravity is 2.5 m above the water.

a. How high will Chloe go? $\frac{viy^2}{2a} + 2.55 = \frac{3^2}{2(9.81)} + 2.5 = 2.96$ m

Find final vel
b. How long will Chloe be in the air before she touches the water? Assume that she first touches the water when her center of gravity is 1 m above the water. ① $v_f^2 = vi^2 + 2g\Delta y$ ② $v_f = vi + gt$

14. Sam fields a baseball hit to him in the left field. He then throws the ball to third to force out the base runner, Mike. Sam releases the ball 1.80 m above the ground with a vertical velocity of 8 m/s and a horizontal velocity of 25 m/s. At the instant Sam releases the ball, he is 41 m from the third baseman, Charlie, and Mike is 13 m from third base and running at 8 m/s toward third. Assume that air resistance does not affect the flight of the ball when answering the following questions.

a. How high in the air does the ball go? $\frac{-vyi^2}{2a} = \frac{-8^2}{2(-9.81)} + 1.81 = 5.06$

b. How much time does it take for the ball to reach the third baseman? $t = \frac{d}{v} = \frac{41}{25} = 1.64$ s

c. How high is the ball when it reaches the third baseman? $df = di + v_y t + \frac{1}{2}a\Delta t^2 = 8(1.64) + \frac{1}{2}(-9.8)(1.64)^2$

d. If Mike maintains a constant velocity of 8 m/s toward third base, does he reach third base before the ball reaches the third baseman? $t = \frac{d}{v} = \frac{13}{8} = 1.625 =$ smaller so yes

15. At the Dallas Cowboys Stadium, the minimum clearance height between the football field and the gigantic video screens that hang over the field is only 90 ft (27.43 m). In the first game played at the Cowboys Stadium on August 21, 2009, a punter's kick hit the video screen.

a. What minimum initial vertical velocity would a football need to have to hit the video screens if it were kicked from a height 1 m above the playing field surface? Assume that air resistance does not affect the flight of the football.

26.43

Vertical Displacement = 27.43 - 1m = 26.43
Vertical vel = ?
Vertical acceleration = -9.81 m/s²

$v_f^2 = vi^2 + 2ad$
$0 = vi^2 + 2(-9.81)(26.43)$
$\sqrt{vi^2} = \sqrt{518.6}$
$vi = 22.77$ m/s

83

b. If the video screens were not in the way, what would be the hang time for a football kicked 27.43 m high? Again, assume that air resistance does not affect the flight of the football and that the ball is 1 m above the playing field when it is kicked.

Motion Analysis Exercises Using MaxTRAQ

If you haven't done so already, review the instructions for downloading and using the educational version of the MaxTRAQ motion analysis software at the beginning of this book, then download and install the software. Once this is done, you are ready to try the following two-dimensional kinematic analyses using MaxTRAQ.

1. Open MaxTRAQ. Select Tools in the menu bar and then open Options under the Tools menu. In the Options submenu, select Video. In the upper right side of the Video window, under Video Aspect Ratio, make sure that Default-Used Preferred Aspect Ratio is selected. In the lower half of the right side of the Video window, under Deinterlace Options, select BOB, select Use Odd Lines First, and select Stretch Image Vertically. Click OK. Close MaxTRAQ and reopen it to have the deinterlace options take effect. Now open the *Run Slow* video from within MaxTRAQ. The video clips will have been downloaded with the software and saved to your local disk under Program Files\Motion Analysis\MaxTRAQP\VideoFiles. In the Open window, make sure that the drop-down menu at the bottom is set to Video Files.

2. When you open the *Run Slow* video, you should see a running track with three yellow balls in the middle lane. These balls are 2.5 m apart, so the yellow ball on the left is 5 m from the yellow ball on the right. Make sure that the scaling/calibration tool is activated by clicking View on the menu bar, then selecting Tools from the drop-down menu, and then making sure that Show Scale is checked. Open the scaling tool by clicking on Tools on the menu bar and selecting Scale. In the Scaling Tool window that opens, set the gauge length to 500 cm, then click OK. Now place the cursor over the left ball and click the left mouse button once (nothing will appear on screen yet); then place the cursor over the right ball and click the left mouse button a second time. The scale should appear in the video window. Hide the scale by selecting View in the menu, then click Tools, and uncheck Show Scale (this appears in the submenu).

   a. What is the stride length of the runner? Advance the video (using the Step Forward button in the video controls panel at the bottom of the screen) until the runner's right foot touches the track; then continue to advance the video to the last instant when the right foot is in contact with the track. Activate the digitizing function by clicking on the Digitize button on the right side of the screen. Place the cursor over the toe of the runner's right foot and click the left mouse button. A mark should appear on the toe along with the horizontal and vertical coordinates of the toe. If the coordinates do not appear, select Point Markers from the View drop-down menu and make sure that Show Coordinates is checked. Record the horizontal coordinate. Advance the video through one full stride to the next instant of takeoff of the right foot. Back up one frame to the last instant of contact of the right foot. Digitize the toe of the runner again and record the horizontal coordinate. The difference between the horizontal coordinates is the runner's stride length in centimeters.

   b. What is the stride rate of the runner in the *Run Slow* video? Use the frame number/time window at the bottom of the MaxTRAQ window to compute stride rate. Determine the time that elapses from takeoff of the right foot until the next takeoff of the right foot (the two points you digitized in part a). To improve accuracy, compute the elapsed time by multiplying the difference in frame numbers by 1/60 second (or just divide by 60). Divide one by this stride duration to calculate the number of strides per second.

   c. What is the average velocity of the runner in the *Run Slow* video during the one full stride you measured in parts a and b?

3. Open the *Run Medium* video in MaxTRAQ. If you did not set the video aspect ratio and deinterlace options in exercise 1, do so now by following those instructions. As you did in exercise 1,

make sure that the scaling/calibration tool is activated by clicking View, then Tools, and making sure that Show Scale is checked. Open the scaling tool by clicking on Tools on the menu bar and selecting Scale. Set the gauge length to 500 cm, then click once over the left ball and once over the right ball to set the scale. Hide the scale by selecting View, then Tools, and unchecking Show Scale.

- a. What is the stride length of the runner in the *Run Medium* video? Measure the stride length from the instant of takeoff of the left foot until the next instant of takeoff of the left foot.
- b. What is the stride rate of the runner in the *Run Medium* video?
- c. What is the average velocity of the runner over one full stride in the *Run Medium* video?

4. Open the *Run Fast* video in MaxTRAQ. If you have not set the video aspect ratio and deinterlace options, do so now by following the instructions in exercise 1. As you did in exercise 1, make sure that Show Scale is checked under View—Tools, then open the scaling tool, set the gauge length to 500 cm, and click once over the left ball and once over the right ball to set the scale.

- a. What is the stride length of the runner in the *Run Fast* video? Measure the stride length from the instant of takeoff of the right foot until the next instant of takeoff of the right foot.
- b. What is the stride rate of the runner in the *Run Fast* video?
- c. What is the average velocity of the runner over one full stride in the *Run Fast* video?

5. Open the *Run Fastest* video in MaxTRAQ. If you have not set the video aspect ratio and deinterlace options, do so now by following the instructions in exercise 1. As you did in exercise 1, make sure that Show Scale is checked under View—Tools, then open the scaling tool, set the gauge length to 500 cm, and click once over the left ball and once over the right ball to set the scale.

- a. What is the stride length of the runner in the *Run Fastest* video? Measure the stride length from the instant of takeoff of the left foot until the next instant of takeoff of the left foot.
- b. What is the stride rate of the runner in the *Run Fastest* video?
- c. What is the average velocity of the runner over one full stride in the *Run Fastest* video?

6. Open the *Run Spring Stilts* video in MaxTRAQ. If you have not set the video aspect ratio and deinterlace options, do so now by following the instructions in exercise 1. As you did in exercise 1, make sure that Show Scale is checked under View—Tools, then open the scaling tool, set the gauge length to 500 cm, and click once over the left ball and once over the right ball to set the scale.

- a. What is the stride length of the athlete in the *Run Spring Stilts* video? Measure the stride length from the instant of takeoff of the left stilt tip until the next instant of takeoff of the left stilt tip.
- b. What is the stride rate of the athlete in the *Run Spring Stilts* video?
- c. What is the average velocity of the athlete over one full stride in the *Run Spring Stilts* video?

7. Open the *Ball Drop* video in MaxTRAQ. If you have not previously set the video aspect ratio and deinterlace options, do so now by following the instructions in exercise 1. You should see the author standing on a ladder against the wall of a building. The bricks of the wall are about 1 ft square, and horizontal white strips of tape are spaced vertically on the wall 1 m apart to the left of the ladder. Make sure the scaling/calibration tool is activated by clicking View on the menu bar, then selecting Tools from the drop down menu, and then making sure that Show Scale is checked. Open the scaling tool by clicking on Tools on the menu bar and selecting Scale. In the Scaling Tool window that opens, set the gauge length to 500 cm, then click OK. Now place the cursor over the left upper corner of the highest strip of tape and click the left mouse button once (nothing will appear on screen yet); then place the cursor over the left upper corner of the lowest strip of tape and click the left mouse button a second time. The scale should appear in the video window. Hide the scale by selecting View in the menu bar, then click Tools, and uncheck Show Scale. What is the vertical displacement of the ball 1 s (60 frames) after it was dropped? The ball was released in the first frame of video. Theoretically, what should the vertical displacement be?

8. Open the *Ball Toss Up* video in MaxTRAQ. If you have not previously set the video aspect ratio and deinterlace options, do so now by following the instructions in exercise 1. Set up the scaling/calibration tool as you did with the *Ball Drop* video using the 500 cm distance between the uppermost and lowermost strips of tape on the wall.

   a. If the ball reaches peak height between frames 47 and 48, or about 0.79 seconds into the video, what is the vertical displacement of the ball during the 0.75 s (45 frames) before peak height (from frame 2 to frame 47)?

   b. The ball reaches peak height between frames 47 and 48 or about 0.79 seconds into the video. What is the vertical displacement of the ball during the 0.75 s (45 frames) after peak height (from frame 48 to frame 93)?

   c. Is the upward displacement of the ball during the 0.75 s before it reaches peak height similar to the downward displacement of the ball during the 0.75 s after it reaches peak height? Should it be?

9. Open the *Ball Toss* video in MaxTRAQ. If you have not previously set the video aspect ratio and deinterlace options, do so now by following the instructions in exercise 1. Set up the scaling/calibration tool as you did with the *Ball Drop* and *Ball Toss Up* videos using the 500 cm distance between the uppermost and lowermost strips of tape on the wall.

   a. Determine the horizontal displacement of the ball over every 9-frame interval starting at frame 5 and ending at frame 86. Are these nine displacements similar to each other? Should they be?

   b. Determine the vertical displacement of the ball over every 9-frame interval starting at frame 5 and ending at frame 86. Are any of the first four displacements similar to any of the last four displacements? Should there be similarities?

   c. The ball reaches peak height between frames 45 and 46 or about 0.76 seconds into the video. What are the horizontal and vertical displacements of the ball during the 0.67 s (40 frames) before peak height (from frame 5 to frame 45)?

   d. The ball reaches peak height between frames 45 and 46 or about 0.76 seconds into the video. What are the horizontal and vertical displacements of the ball during the 0.67 s (40 frames) after peak height (from frame 46 to frame 86)?

   e. Are the answers to question c similar to the answers to question d? Should they be similar?

chapter 3

Linear Kinetics

Explaining the Causes of Linear Motion

objectives

When you finish this chapter, you should be able to do the following:

- Explain Newton's three laws of motion

- Apply Newton's second law of motion to determine the acceleration of an object if the forces acting on the object are known

- Apply Newton's second law of motion to determine the net force acting on an object if the acceleration of the object is known

- Define impulse

- Define momentum

- Explain the relationship between impulse and momentum

- Describe the relationship between mass and weight

You're watching the Olympic weightlifting competition. The weights on the barbell are more than twice the lifter's weight. The athlete approaches the barbell and grips it firmly with both hands. With a sharp grunt, he snatches the bar off the floor and lifts it over his head in one smooth motion. What forces did the athlete have to exert on the barbell to create the movement you just witnessed? Newton's laws of motion—proposed more than 300 years ago—provides us with the basis for analyzing situations like this. This chapter introduces Newton's laws of motion and the application of these laws to the analysis of human movement.

Isaac Newton was an English mathematician. He was born on Christmas day in 1642,* the same year that Galileo died and three months after the death of his father (Westfall 1993, p. 7). He died on March 20, 1727. Newton was a student at Cambridge University and subsequently became a professor there. Many of his ideas about mechanics (and calculus) were conceived during a two-year retreat to his family's estate in Lincolnshire when he was in his early 20s. This retreat was prompted by a plague epidemic in England, which caused temporary closings of the university in Cambridge between 1665 and 1667. Despite the devastation it caused, one benefit of the plague was that it allowed Isaac Newton an uninterrupted period of time to establish the groundwork for his version of mechanics.

More than 20 years passed before Newton shared his work with others in 1686, when his book, *Philosophiae Naturalis Principia Mathematica (Mathematical Principles of Natural Philosophy), or Principia,* as it is commonly referred to, was published. *Principia* was written in Latin, the language used by scientists during that time. In *Principia,* Newton presented his three laws of motion and his law of gravitation. These laws form the basis for modern mechanics. It is these laws that provide the basis for the subbranch of mechanics called kinetics. Dynamics is the branch of mechanics concerned with the mechanics of moving objects. Kinetics is the branch of dynamics concerned with the forces that cause motion. This chapter specifically deals with linear kinetics, or the causes of linear motion. In this chapter, you will learn about Newton's laws of motion and how they can be used to analyze motion. What you learn in this chapter will give you the basic tools to analyze and explain the techniques used in many sport skills.

Newton's First Law of Motion: Law of Inertia

Corpus omne perseverare in statu suo quiescendi vel movendi uniformiter in directum, nisi quatenus a viribus impressis cogitur statum illum mutare. (Newton 1686/1934, p. 644)

This is Newton's first law of motion in Latin as originally presented in *Principia.* It is commonly referred to as the law of inertia. Translated directly, this law states, "Every body continues in its state of rest, or of uniform motion in a straight line, unless it is compelled to change that state by forces impressed upon it" (Newton 1686/1934, p. 13). This law explains what happens to an object if no external forces act on it or if the net external force (the resultant of all the external forces acting on it) is zero. More simply stated, Newton's first law says that if no net external force acts on an object, that object will not move (it will remain in its state of rest) if it wasn't moving to begin with, or it will continue moving at constant speed in a straight line (it will remain in its state of uniform motion in a straight line) if it was already moving.

⟳ **If no net external force acts on an object, that object will not move if it wasn't moving to begin with, or it will continue moving at constant speed in a straight line if it was already moving.**

Let's see how Newton's first law of motion applies to human movement in sport. Can you think of any situa-

---

*Newton's dates of birth (December 25, 1642) and death (March 20, 1727) are dates in the Julian calendar—the calendar in use in England at the time. The calendar in use in much of Europe at that time was the Gregorian calendar, the same calendar used in the United States and most of the world today. The calendars differ by 11 days. In the Gregorian calendar, Newton was born on January 4, 1643, and died on March 31, 1727.

tions in which no external forces act on an object? This is difficult. Gravity is an external force that acts on all objects close to the earth. Apparently, there are no situations in sports and human movement to which Newton's first law of motion applies! Is this true? Perhaps we can find applications of Newton's first law in sport if we consider only motions of an object or body in a specific direction.

We actually already used Newton's first law of motion several times in the previous chapters. In the previous chapter, we analyzed projectile motion. We did this by resolving the motion of the projectile into vertical and horizontal components. Vertically, the velocity of the projectile was constantly changing, and it accelerated downward at 9.81 m/s^2 due to the force of gravity. In the vertical direction, Newton's first law of motion does not apply. Horizontally, however, the velocity of the projectile was constant, and its acceleration was zero because no horizontal forces acted on the projectile. This is a case where Newton's first law of motion does apply! If air resistance is negligible, the net horizontal force acting on a projectile is zero, so the horizontal velocity of the projectile is constant and unchanging. Newton's first law of motion provides the basis for the equations describing the horizontal motion of a projectile that we used in the previous chapter.

Newton's first law of motion also applies if external forces do act on an object, so long as the sum of those forces is zero. So, an object may continue its motion in a straight line or continue in its state of rest if the net external force acting on the object is zero. In chapter 1, we learned about static equilibrium—the sum of all the external forces acting on an object is zero if the object is in static equilibrium. Newton's first law of motion is the basis for static equilibrium. This law also extends to moving objects, however. If an object is moving at constant velocity in a straight line, then the sum of all the external forces acting on the object is zero. Newton's first law of motion basically says that if the net external force acting on an object is zero, then there will be no change in motion of the object. If it is already moving, it will continue to move (in a straight line at constant velocity). If it is at rest, it will stay at rest (not move). Newton's first law can be expressed mathematically as follows:

$$v = \text{constant} \qquad \text{if } \Sigma F = 0 \qquad (3.1a)$$

or

$$\Sigma F = 0 \qquad \text{if } v = \text{constant} \qquad (3.1b)$$

where

$v = $ instantaneous velocity and

$\Sigma F = $ net force.

Since Newton's first law also applies to components of motion, equations 3.1a and 3.1b can be represented by equations for the three dimensions (vertical, horizontal—forward and backward; and horizontal—side to side):

$$v_x = \text{constant} \qquad \text{if } \Sigma F_x = 0 \qquad (3.2a)$$

$$\Sigma F_x = 0 \qquad \text{if } v_x = \text{constant} \qquad (3.2b)$$

$$v_y = \text{constant} \qquad \text{if } \Sigma F_y = 0 \qquad (3.3a)$$

$$\Sigma F_y = 0 \qquad \text{if } v_y = \text{constant} \qquad (3.3b)$$

$$v_z = \text{constant} \qquad \text{if } \Sigma F_z = 0 \qquad (3.4a)$$

$$\Sigma F_z = 0 \qquad \text{if } v_z = \text{constant} \qquad (3.4b)$$

To keep things simple in the problems and examples in this book, we'll mostly confine ourselves to analyses in just two dimensions—vertical *(y)* and horizontal *(x)*.

We have already done analyses based on Newton's first law of motion in chapters 1 and 2, but we didn't know it at the time. Now that we know Newton's first law of motion, let's try another analysis. Imagine holding a 10 lb (4.5 kg) dumbbell in your hand. How large a force must you exert on the dumbbell to hold it still? What external forces act on the dumbbell? Vertically, gravity exerts a force downward equal to the weight of the dumbbell, 10 lb. Your hand exerts a **reaction force** upward against the dumbbell. According to Newton's first law, an object will stay at rest only if there are no external forces acting on the object or if the net external force acting on the object is zero. Since the dumbbell is at rest (not moving), the net external force acting on it must be zero. Figure 3.1 shows a free-body diagram of the dumbbell.

The two external forces acting on the dumbbell are both vertical forces, the force of gravity acting downward and the reaction force from your hand acting upward.

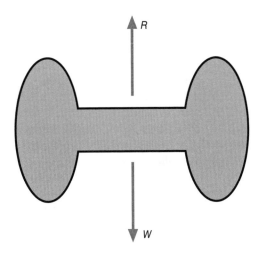

Figure 3.1 Free-body diagram of a dumbbell when held still in the hand. According to Newton's first law, this diagram is also accurate for a dumbbell moving at a constant velocity in a straight line.

Since the dumbbell is not moving (v = constant = 0), we can use equation 3.3b to solve for the reaction force from your hand.

$$\Sigma F_y = 0$$

$$\Sigma F_y = R + (-W) = 0 \qquad (3.5)$$

$$R = W = 10 \text{ lb}$$

where

R = the reaction force from your hand, and

W = the weight of the dumbbell = 10 lb.

When you are holding the 10 lb dumbbell still in your hand, the force you must exert against it is 10 lb upward. The problem was solved with upward considered the positive direction. Since the answer we got was a positive number, it represented an upward force.

Now let's see what happens if the dumbbell is moving. It's best if you actually feel this, so try self-experiment 3.1.

Self-Experiment 3.1

Pick up a dumbbell if you have one—if not, use a book or any other convenient object weighing more than 5 lb (22 N). Let's suppose you do have a dumbbell (if you don't, just imagine that the book or whatever you have is a dumbbell) and that it weighs 10 lb (44.5 N). Hold the dumbbell still in your hand. If you begin lifting the dumbbell, and during the lift it moves at a constant velocity upward, how large a force must you exert against the dumbbell to keep it moving upward at constant velocity? What does it feel like compared to the force required to hold the dumbbell still? Remember, we're trying to find the force you exert when the dumbbell is moving upward at constant velocity, not when it starts upward. What external forces act on the dumbbell? Vertically, gravity still exerts a force downward equal to the weight of the dumbbell, 10 lb, and your hand still exerts a reaction force upward against the dumbbell. According to Newton's first law, an object will move at constant velocity in a straight line only if there are no external forces acting on the object or if the net external force acting on the object is zero. Since the dumbbell is moving at a constant velocity in a straight line, the net external force acting on it must be zero. Refer back to the free-body diagram of the dumbbell in figure 3.1. The two external forces on the dumbbell in this experiment are exactly the same as when the dumbbell was still: the force of gravity acting downward and the reaction force from your hand acting upward. We'll use the same equation, equation 3.5, to solve for the force from your hand. Since the numbers are the same, we get the same reaction force of 10 lb upward.

When you are moving the 10 lb dumbbell upward at a constant velocity, the force you must exert against the dumbbell to keep it moving upward at a constant velocity is a 10 lb upward force. When you are holding the dumbbell still, the force you exert on it is a 10 lb upward force. If you move the dumbbell downward at constant velocity, the force you exert on it would still be a 10 lb upward force.

Newton's first law of motion may be interpreted in several ways:

1. If an object is at rest and the net external force acting on it is zero, the object must remain at rest.

2. If an object is in motion and the net external force acting on it is zero, the object must continue moving at constant velocity in a straight line.

3. If an object is at rest, the net external force acting on it must be zero.

4. If an object is in motion at constant velocity in a straight line, the net external force acting on it must be zero.

Newton's first law of motion applies to the resultant motion of an object and to the components of this resultant motion. Because forces and velocities are vectors, Newton's first law can be applied to any direction of motion. If no external forces act, or if the components of the external forces acting in the specified direction sum to zero, there is no motion of the object in that direction or the velocity in that direction is constant.

> Newton's first law of motion applies to the resultant motion of an object and to the components of this resultant motion.

Conservation of Momentum

Newton's first law of motion provides the basis for the principle of conservation of momentum (if we consider only objects whose mass is constant). Actually, the conservation of momentum principle was first introduced by René Descartes and Christian Huygens (a Dutch mathematician) before Newton published *Principia*, but that's another story. What is momentum? **Linear momentum** is the product of an object's mass and its linear velocity. The faster an object moves, the more momentum it has. The larger a moving object's mass, the more momentum it has. So, momentum is a way of quantifying the motion and inertia of an object together in one measure. Linear momentum is defined mathematically by equation 3.6.

> $$L = mv \qquad (3.6)$$

where

    L = linear momentum,

    m = mass, and

    v = instantaneous velocity.

Newton's first law of motion basically states that the velocity of an object is constant if the net force acting on the object is zero. In sports and human movement, most objects we deal with have a constant mass (at least over the short durations of the activities we may be analyzing). If the velocity of an object is constant, then its momentum is constant as well since mass doesn't change. Momentum is constant if the net external force is zero. This can be expressed mathematically as

    L = constant   if $\Sigma F = 0$ (3.7)

where

    L = linear momentum and

    ΣF = net external force.

Since velocity is a vector quantity (with size and direction), momentum is also a vector quantity. The total momentum of an object can be resolved into components, or if the components of momentum are known, the components can be added together (using vector addition) to determine the resultant momentum. Conservation of momentum applies to the components of momentum, so equation 3.7 can be represented by equations for the three dimensions (vertical, horizontal—forward and backward, and horizontal—side to side).

    L_x = constant   if $\Sigma F_x = 0$ (3.8)

    L_y = constant   if $\Sigma F_y = 0$ (3.9)

    L_z = constant   if $\Sigma F_z = 0$ (3.10)

Using the conservation of momentum principle to analyze a single object is not worthwhile. Why complicate things with mass when all we have to worry about with Newton's first law is velocity? The value of the conservation of momentum principle is more apparent when we are concerned with not a single object, but a group of objects.

The analysis of a system of two or more objects is simplified if all the objects are considered part of one thing, the system. If the objects are considered together as one system, then the forces that the objects exert against each other are internal forces and do not affect the motion of the whole system. We don't have to know what they are! Only external forces—forces exerted by agents external to the system—will change the motion of the system. According to the conservation of momentum principle, the total momentum of a system of objects is constant if

the net external force acting on the system is zero. This principle is represented mathematically in equation 3.11. For a system made up of a number of objects, the sum of the momenta of all the objects at some initial time is equal to the sum of the momenta of all the objects at some later, or final, time if no external forces act on the system. If the system is made up of only one object, this is too simple—velocity and mass are unchanged. But, if the system is made up of two or more objects, the initial and final velocities of an object within the system might not remain unchanged. In the case of a multi-object system, if one object's velocity increases, another object's velocity decreases to keep the total momentum of the system constant.

$L_i = \Sigma(mu) = m_1u_1 + m_2u_2 + m_3u_3 + \ldots = m_1v_1 + m_2v_2 + m_3v_3 + \ldots = \Sigma(mv) = L_f$ = constant (3.11)

where

    L_i = initial linear momentum,

    L_f = final linear momentum,

    m = mass of part of the system,

    u = initial velocity, and

    v = final velocity.

The total momentum of a system of objects is constant if the net external force acting on the system is zero.

The conservation of momentum principle is especially useful for analyzing collisions. Collisions are common in sport: Baseballs collide with bats, tennis balls hit rackets, soccer balls hit feet, defensive linemen collide with offensive linemen in American football, and so on. The outcome of these collisions can be explained with the conservation of momentum principle.

Elastic Collisions

When two objects collide in a head-on collision, their combined momentum is conserved. To see a simple demonstration of this principle, try self-experiment 3.2. We can use this principle to predict the postcollision movements of the objects in certain situations if we know their masses and their precollision velocities.

Self-Experiment 3.2

For this experiment we'll use coins of the same weight, for example two U.S. pennies. Take the two pennies and put them on a table (or other hard, flat surface) about 5 cm (2 in.) apart. Flick one penny (let's call this penny A) into the other penny (let's call this penny B) so that it strikes penny B directly on center and not to one side or the other. Try this a couple of times. What happens? Just after the collision, penny A stops or barely moves, and penny B now moves in the same direction and with about the same velocity as penny A before the collision. Just before the collision, the total momentum of the system was the mass of penny A times its velocity. Since penny B was not moving, it didn't contribute to the total momentum. Just after the collision, the total momentum of the system is the mass of penny B times its velocity. Momentum is conserved. When penny A strikes penny B, it transfers its momentum to penny B.

Let's examine the two-penny collision from self-experiment 3.2. For a system with only two objects, equation 3.11 simplifies to

$$L_i = \Sigma(mu) = m_1 u_1 + m_2 u_2 = m_1 v_1 + m_2 v_2$$
$$= \Sigma(mv) = L_f \qquad (3.12)$$

This equation gives us some information about what occurs after the collision, but not enough to solve for both postcollision velocities unless we know more about what happens after the collision. It is only one equation, but we have two unknown variables (u_1 and u_2). For our penny example, this equation becomes

$$m_A u_A + m_B u_B = m_A v_A + m_B v_B. \qquad (3.13)$$

In the penny example, we do know something about what happens after the collision. We observed that the velocity of penny A was zero (or close to it) after the collision. The precollision velocity of penny B was zero ($u_B = 0$), and the postcollision velocity of penny A was zero ($v_A = 0$), so

$$m_A u_A = m_B v_B. \qquad (3.14)$$

The masses of the pennies are equal, $m_A = m_B$, so

$$u_A = v_B.$$

The velocity of penny A just before the collision equals the velocity of penny B just after the collision. For head on, perfectly elastic collisions of two objects with equal mass, the precollision momentum of each object is totally transferred to the other object postcollision, or more simply stated mathematically,

$$u_1 = v_2 \qquad (3.15)$$

and

$$u_2 = v_1 \qquad (3.16)$$

What if the masses of two colliding objects are not equal? Will momentum totally transfer from one object to the other and vice versa in all perfectly elastic head on collisions?

Self-Experiment 3.3

Let's repeat self-experiment 3.2 but this time substitute a nickel for penny A. Flick the nickel into the penny. What happens after the nickel strikes the penny? In this case the penny rapidly moves in the same direction as the nickel was before the collision, but the nickel also keeps moving the same direction, although much more slowly. If momentum was transferred completely from the nickel to the penny and vice versa, the nickel would not be moving after the collision.

Self experiment 3.3 shows that in a perfectly elastic collision of two objects with unequal masses, the momentum of each object is not exchanged completely with the other object. Let's try to figure out the velocities of the nickel and the penny after the collision in self-experiment 3.3. We know from observation that the postcollision velocity of the nickel is not zero, so, equations 3.15 and 3.16 do not work. Equation 3.13 is still valid, but one equation is not enough if we want to solve for two unknowns – the postcollision velocity of the nickel and the penny. Two equations are needed. In a perfectly elastic collision, not only is momentum conserved, but kinetic energy is conserved as well. We'll learn more about kinetic energy in the next chapter, but conservation of kinetic energy gives us the additional equation we need to determine the postcollision velocities of the nickel and the penny or any two objects in a head on perfectly elastic collision. The equations that describe the postcollision velocities in a head on perfectly elastic collision are

$$v_1 = \frac{2m_2 u_2 + (m_1 - m_2) u_1}{m_1 + m_2} \qquad (3.17)$$

and

$$v_2 = \frac{2m_1 u_1 + (m_2 - m_1) u_2}{m_1 + m_2} \qquad (3.18)$$

Notice that if m_1 equals m_2, then equations 3.17 and 3.18 reduce to equations 3.15 and 3.16 respectively.

Now, let's try to determine the postcollision velocities of the penny and nickel in self-experiment 3.3. Use subscript P to represent the penny and subscript N to represent the nickel. Then, using equation 3.17 and 3.18 we get the postcollision velocities of the nickel and penny as

$$v_N = \frac{2m_P u_P + (m_N - m_P) u_N}{m_N + m_P}$$

and

$$v_P = \frac{2m_N u_N + (m_P - m_N) u_P}{m_N + m_P}$$

The mass of a new U.S. penny minted after 1982 is 2.5 g. The mass of a new U.S. nickel is 5 g. The nickel's mass (m_N) is twice the penny's mass (m_p) or, $m_N = 2m_p$. We also know that the precollision velocity of the penny was zero. Substituting $2m_p$ for m_p and setting u_p to zero in the previous equations gives

$$v_N = \frac{2m_p(0)+\left(2m_p - m_p\right)u_N}{2m_p + m_p} = \frac{\left(m_p u_N\right)}{3m_p} = \frac{1}{3}u_N$$

and

$$v_p = \frac{2\left(2m_p\right)u_N +\left(m_p - m_N\right)(0)}{2m_p + m_p} = \frac{4m_p u_N}{3m_p} = \frac{4}{3}u_N$$

So, the nickel's postcollision velocity (v_N) is one third the nickel's precollision velocity (u_N) and the penny's postcollision velocity is four thirds the nickel's precollision velocity. Suppose the nickel had a velocity of 2 m/s before the collision. Immediately after the collision, the nickel would have a velocity of $^2/_3$ m/s and the penny would have a velocity of $2^2/_3$ m/s. The problem and solution are presented graphically in figure 3.2.

The collisions of the coins in self-experiments 3.2 and 3.3 were examples of perfectly elastic collisions in two dimensions of a moving object with a non-moving object. If the colliding objects have equal masses, as in self-experiment 3.2, equations 3.15 and 3.16 apply. If the colliding objects have different masses, as in self-experiment 3.3, equations 3.17 and 3.18 apply. What happens if both objects in a perfectly elastic collision are moving along the same line but in opposite directions when they collide? Will these equations still apply?

If both objects in a perfectly elastic collision are moving along the same line but in opposite directions, we must remember that momentum is a vector quantity, so the momentum of each object is opposite to the momentum of the other object in the collision. Try self-experiment 3.2 again, but this time flick both pennies toward each other so they strike head-on along the same line. What happens? Immediately after they collide, they

bounce off each other and move in the opposite direction of their precollision velocity. If one penny (let's call this penny A) is moving faster than the other penny (let's call this penny B) before the collision, penny B will be moving at that speed and in that direction after the collision, while penny A will be moving at the slower speed in the opposite direction after the collision. If the collision of the two pennies is perfectly elastic, then the precollision momentum of each is transferred to the other after the collision. Since the collision is perfectly elastic, the precollision momentum of penny A equals the postcollision momentum of penny B, while the precollision momentum of penny B equals the postcollision momentum of penny A. Equations 3.15 and 3.16 apply, and

$$u_A = v_B$$

$$u_B = v_A$$

In our penny example, if penny A was moving to the right with a velocity of 2 m/s, and penny B was moving to the left with a velocity of 1 m/s, then the total momentum of the system would be 2.5 g·m/s to the right (the mass of a penny is 2.5 g). If we set up our positive direction to the right, then

$$u_A = v_B = +2\text{m/s},$$

and

$$u_B = v_A = -1 \text{ m/s}.$$

The postcollision velocity of penny A is 1 m/s to the left, and the postcollision velocity of penny B is 2 m/s to the right. The general illustration of this impact situation is presented in figure 3.3.

Try self-experiment 3.3 again, but this time flick both the penny and the nickel toward each other so they strike head-on along the same line. What happens? Immediately after they collide, they bounce off each other and move in the opposite direction of their precollision velocities. If the precollision velocities of the penny and nickel were equal but opposite in direction, then the postcollision velocity of the penny will be faster than its precollision

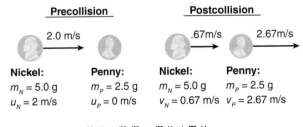

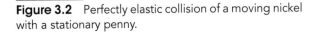

Figure 3.2 Perfectly elastic collision of a moving nickel with a stationary penny.

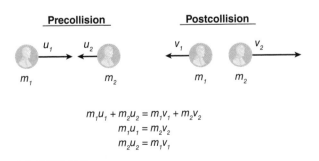

Figure 3.3 Perfectly elastic head-on collision of two pennies moving in opposite directions.

velocity, while the postcollision velocity of the nickel will be slower than its precollision velocity. Let's suppose the penny was moving to the right with a velocity of 2 m/s and the nickel was moving to the left with a velocity of 2 m/s. Remember that the nickel is twice the mass of a penny, 5 g vs. 2.5 g. If we set up our positive direction to the right, then the postcollision velocities of the nickel and penny are still described by equation 3.17 and 3.18.

SAMPLE PROBLEM 3.1

A small (25 g) high-bounce ball is stacked on top of a large (100 g) high-bounce ball, and the two balls are dropped to the floor together from a height of 2 m. The large ball strikes the floor first and rebounds upward to collide with the falling small ball. At the instant of the collision, the large ball has a velocity of 4.4 m/s upward, and the small ball has a velocity of 4.4 m/s downward. If the collision between the two balls is perfectly elastic, how fast is each ball moving immediately after impact?

Solution:

Step 1: List the known quantities.

$$m_{large} = 100 \text{ g}$$

$$m_{small} = 25 \text{ g}$$

$$u_{large} = 4.4 \text{ m/s}$$

$$u_{small} = -4.4 \text{ m/s}$$

Step 2: Identify the variables to solve for.

$$v_{large} = ?$$

$$v_{small} = ?$$

Step 3: Search for equations with the known and unknown variables in them.

$$v_1 = \frac{2m_2u_2 + (m_1 - m_2)u_1}{m_1 + m_2}$$

$$v_2 = \frac{2m_1u_1 + (m_2 - m_1)u_2}{m_1 + m_2}$$

$$v_{large} = \frac{2m_{small}u_{small} + (m_{large} - m_{small})u_{large}}{m_{large} + m_{small}}$$

$$v_{small} = \frac{2m_{large}u_{large} + (m_{small} - m_{large})u_{small}}{m_{large} + m_{small}}$$

Step 4: Now substitute known values and solve for the postimpact velocity of each ball.

$$v_{large} = \frac{2(25 \text{ g})(-4.4 \text{ m/s}) + (100 \text{ g} - 25 \text{ g})(4.4 \text{ m/s})}{(100 \text{ g} + 25 \text{ g})}$$

$$v_{large} = \frac{(-220 \text{ g m/s}) + (330 \text{ g m/s})}{(125 \text{ g})}$$

$$v_{large} = 0.88 \text{ m/s}$$

$$v_{small} = \frac{2(100 \text{ g})(4.4 \text{ m/s}) + (25 \text{ g} - 100 \text{ g})(-4.4 \text{ m/s})}{(100 \text{ g} + 25 \text{ g})}$$

$$v_{small} = \frac{(880 \text{ g m/s}) + (330 \text{ g m/s})}{(125 \text{ g})}$$

$$v_{small} = 9.68 \text{ m/s}$$

Step 5: Common sense check.
Whoa! The postimpact velocity of the large ball sounds reasonable. It should be moving in the downward direction if the downward momentum of the small ball was completely transferred to it. But the 9.68 m/s postimpact velocity of the small ball seems too fast. How can it be moving so much faster than before? Let's make one more check using the original conservation of momentum equation (equation 3.13).

$$m_Au_A + m_Bu_B = m_Av_A + m_Bv_B$$

$$m_{large}u_{large} + m_{small}u_{small} = m_{large}v_{large} + m_{small}v_{small}$$

$$(100 \text{ g})(4.4 \text{ m/s}) + (25 \text{ g})(-4.4 \text{ m/s}) = (100 \text{ g})(0.88 \text{ m/s}) + (25 \text{ g})(9.68 \text{ m/s})$$

$$440 \text{ g·m/s} + (-110 \text{ g·m/s}) = (88 \text{ g·m/s}) + 242 \text{ g·m/s}$$

$$330 \text{ g·m/s} = 330 \text{ g·m/s}$$

Wow, it checks out. This is an example of the "deadly superball trick." If a small ball is stacked on top of a much larger ball and the two balls are dropped together, the small ball will bounce off the large ball with a very fast velocity if the collision is close to perfectly elastic. Warning: This is a sample problem, not a self-experiment. Do not try this trick at home without eye protection and personal protection. It is difficult to drop the balls so that the smaller ball lands perfectly on top center of the larger ball. Off-center impacts will send the smaller ball flying very fast in an unexpected direction!

$$v_N = \frac{2m_p u_p + (m_N - m_p)u_N}{m_N + m_p}$$

$$= \frac{2(2.5 \text{ g})(2 \text{ m/s}) + (5 \text{ g} - 2.5 \text{ g})(-2 \text{ m/s})}{5 \text{ g} + 2.5 \text{ g}}$$

$$v_N = \frac{10 \text{ g m/s} - 5 \text{ g m/s}}{7.5 \text{ g}}$$

$$v_N = +0.67 \text{ m/s}$$

and

$$v_p = \frac{2m_N u_N + (m_p - m_N)u_p}{m_N + m_p}$$

$$= \frac{2(5 \text{ g})(-2 \text{ m/s}) + (2.5 \text{ g} - 5 \text{ g})(2 \text{ m/s})}{5 \text{ g} + 2.5 \text{ g}}$$

$$v_p = \frac{-20 \text{ g m/s} - 5 \text{ g m/s}}{7.5 \text{ g}}$$

$$v_p = -3.33 \text{ g m/s}$$

The postcollision velocity of the penny is 3.33 m/s to the left, and the postcollision velocity of the nickel is 0.33 m/s to the right. This agrees with our observation that the penny rebounded off the nickel with more velocity than it had before the collision while the nickel rebounded off the penny with less velocity than it had before the collision.

We've described two types of linear, perfectly elastic collisions involving two objects. In the first case, a moving object collided with a stationary object. In the second case, two objects moving in opposite directions collided head-on. There is a third type of linear, perfectly elastic collision. In this case, the two objects are moving in the same direction but at different velocities. The faster, overtaking object collides with the slower-moving object. If the two objects have the same mass, the momentum of the faster-moving object is completely transferred to the slower-moving object and equations 3.15 and 3.16 apply. Immediately after the collision, the previously faster-moving object slows to the velocity of the previously slower-moving object, and the previously slower-moving object speeds up to the velocity of the previously faster-moving object. This type of collision and the equations describing it are shown in figure 3.4 for two pennies. If the two objects differ in mass, then equations 3.17 and 3.18 apply.

Inelastic Collisions

The collisions we've examined so far were perfectly elastic collisions. Not all collisions are perfectly elastic. The opposite of a perfectly elastic collision is a per-

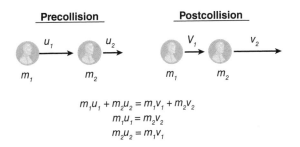

Precollision **Postcollision**

$$m_1 u_1 + m_2 u_2 = m_1 v_1 + m_2 v_2$$
$$m_1 u_1 = m_2 v_2$$
$$m_2 u_2 = m_1 v_1$$

Figure 3.4 Perfectly elastic overtaking collision of two pennies moving in the same direction.

fectly inelastic collision (also called a perfectly plastic collision). In a perfectly inelastic collision, momentum is still conserved, but rather than bouncing off of each other, the objects in the collision stay together after the collision and move together with the same velocity. This postcollision condition gives us the additional information we need to determine the postcollision velocities of objects in an inelastic collision. The equations that follow describe the motion of two objects involved in a linear, perfectly inelastic collision. Whether the collision is elastic or inelastic, we begin with equation 3.12, which describes the conservation of momentum for the collision of two objects.

$$L_i = \Sigma(mu) = m_1 u_1 + m_2 u_2 = m_1 v_1 + m_2 v_2$$
$$= \Sigma(mv) = L_f$$

$$m_1 u_1 + m_2 u_2 = m_1 v_1 + m_2 v_2$$

In a perfectly inelastic collision,

$$v_1 = v_2 = v = \text{final velocity.}$$

Therefore,

$$m_1 u_1 + m_2 u_2 = (m_1 + m_2)v. \tag{3.19}$$

Most collisions occurring in American football are examples of inelastic collisions. Offensive linemen collide with defensive linemen, linebackers tackle running backs, receivers collide with safeties, and so on. In most of these collisions, the two colliding players move together as one following the collision. What happens after these collisions greatly affects the outcome of each play and ultimately the outcome of the game. Can a faster and lighter running back match or exceed the momentum of a slower, more massive defensive player? Football puts a premium on momentum—mass and velocity are equally important. Fast and large players are the most successful. Let's try a problem to illustrate the mechanics of an inelastic collision.

Suppose an 80 kg fullback collides in midair with a 120 kg linebacker at the goal line during a goal line stand.

Just before the collision, the fullback had a velocity of 6 m/s toward the goal line, and the linebacker had a velocity of 5 m/s in the opposite direction. If the collision was perfectly inelastic, would the fullback be moving forward and score a touchdown just after the collision, or would the defense prevail?

To answer this question, we'll begin with equation 3.19:

$$m_1u_1 + m_2u_2 = (m_1 + m_2)v$$

Let's consider the positive direction to be the direction toward the goal line. Substituting the known values for the masses and precollision velocities gives us

$$(80 \text{ kg})(6 \text{ m/s}) + (120 \text{ kg})(-5 \text{ m/s})$$
$$= (80 \text{ kg} + 120 \text{ kg})v.$$

Solving for v (the final velocity of both players) gives us

$$480 \text{ kg·m/s} - 600 \text{ kg·m/s} = (200 \text{ kg})v$$

$$-120 \text{ kg·m/s} = (200 \text{ kg})v$$

$$v = \frac{-120 \text{ kg·m/s}}{200 \text{ kg}} = -0.6 \text{ m/s}$$

The fullback won't score. He and the linebacker will be moving back away from the goal line with a velocity of 0.6 m/s.

Most collisions in sports are neither perfectly elastic nor perfectly inelastic but are somewhere in between. These are elastic collisions, but not *perfectly* elastic collisions. The coefficient of restitution is a means of quantifying how elastic the collisions of an object are.

Coefficient of Restitution

The **coefficient of restitution** is defined as the absolute value of the ratio of the velocity of separation to the velocity of approach. The velocity of separation is the difference between the velocities of the two colliding objects just after the collision. It describes how fast they are moving away from each other. The velocity of approach is the difference between the velocities of the two colliding objects just before the collision. It describes how fast they are moving toward each other. The coefficient of restitution is usually abbreviated with the letter e. Mathematically,

$$e = \left| \frac{v_1 - v_2}{u_1 - u_2} \right| = \left| \frac{v_2 - v_1}{u_1 - u_2} \right| \qquad (3.20)$$

where

   e = coefficient of restitution,

   v_1, v_2 = postimpact velocities of objects one and two, and

   u_1, u_2 = preimpact velocities of objects one and two.

The coefficient of restitution has no units. For perfectly elastic collisions, the coefficient of restitution is 1.0, its maximum value. For perfectly inelastic collisions, the coefficient of restitution is zero, its minimum value. If we know the coefficient of restitution for the two objects in an elastic collision, this along with the conservation of momentum equation gives us the information needed to determine the postcollision velocities of the colliding objects.

The coefficient of restitution is affected by the nature of both objects in the collision. For ball sports, it is most easily measured if the object that the ball collides with is fixed and immovable. Then only the pre- and postimpact velocities of the ball need to be measured. Actually, if the ball is dropped from a specified height onto the fixed impact surface, then the height of the rebound and the height of the drop provide enough information for computation of the coefficient of restitution (refer to figure 3.5). (Can you derive this equation from equation 3.20 and equation 2.18?)

$$e = \sqrt{\frac{\text{bounce height}}{\text{drop height}}} \qquad (3.21)$$

The coefficient of restitution is a critical measure in most ball sports, since the "bounciness" of a ball and

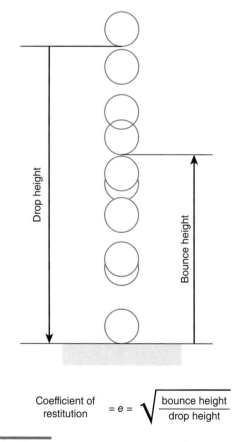

Coefficient of restitution $= e = \sqrt{\dfrac{\text{bounce height}}{\text{drop height}}}$

Figure 3.5 Determination of the coefficient of restitution from the drop and bounce heights.

the implement or surface it strikes will greatly affect the outcome of a competition. If bats had higher coefficients of restitution with baseballs, more home runs would be hit (and more pitchers would be injured by hit balls). If golf clubs had higher coefficients of restitution with golf balls, 300 yd drives might be common. Obviously, the rules makers must regulate the coefficient of restitution for the balls and implements involved in ball sports.

Most rules for ball sports directly or indirectly specify what the coefficient of restitution must be for the ball on the playing surface or implement. The U.S. Golf Association (USGA) rules forbid drivers having a coefficient

SAMPLE PROBLEM 3.2

A golf ball is struck by a golf club. The mass of the ball is 46 g, and the mass of the club head is 210 g. The club head's velocity immediately before impact is 50 m/s. If the coefficient of restitution between the club head and the ball is 0.80, how fast is the ball moving immediately after impact?

Solution:

Step 1: List the known quantities.

$$m_{ball} = 46 \text{ g}$$

$$m_{club} = 210 \text{ g}$$

$$u_{ball} = 0 \text{ m/s}$$

$$u_{club} = 50 \text{ m/s}$$

$$e = 0.80$$

Step 2: Identify the variable to solve for.

$$v_{ball} = ?$$

Step 3: Search for equations with the known and unknown variables in them.

$$m_1 u_1 + m_2 u_2 = m_1 v_1 + m_2 v_2$$

$$m_{ball} u_{ball} + m_{club} u_{club} = m_{ball} v_{ball} + m_{club} v_{club}$$

$$e = \left| \frac{v_1 - v_2}{u_1 - u_2} \right| = \frac{v_2 - v_1}{u_1 - u_2} = \frac{v_{club} - v_{ball}}{u_{ball} - u_{club}}$$

Step 4: We have two unknown variables, v_{club} and v_{ball}, which represent the postimpact velocities of the club and the ball. We also have two equations to use. If the number of independent equations is equal to the number of unknown variables, the unknown variables can be computed. We need to solve one of the equations for one of the unknown variables in terms of the other. Let's use the coefficient of restitution equation and solve for the postimpact velocity of the club. We want to manipulate the equation so that the postimpact velocity of the club, v_{club}, is on one side of the equation by itself.

$$e = \frac{v_{club} - v_{ball}}{u_{ball} - u_{club}}$$

$$e \, (u_{ball} - u_{club}) = v_{club} - v_{ball}$$

$$v_{club} = e \, (u_{ball} - u_{club}) + v_{ball}$$

Step 5: Now let's substitute this expression for the postimpact velocity of the club into the conservation of momentum equation.

$$m_{ball} u_{ball} + m_{club} u_{club} = m_{ball} v_{ball} + m_{club} v_{club}$$

$$m_{ball} u_{ball} + m_{club} u_{club} = m_{ball} v_{ball} + m_{club}$$
$$(e \, (u_{ball} - u_{club}) + v_{ball})$$

Step 6: Substitute known values and solve for the postimpact velocity of the ball.

$$(46 \text{ g})(0) + (210 \text{ g})(50 \text{ m/s}) = (46 \text{ g})v_{ball} + (210 \text{ g}) \times [0.80 \, (0 - 50 \text{ m/s}) + v_{ball}]$$

$$(210 \text{ g})(50 \text{ m/s}) = v_{ball} \, (46 \text{ g} + 210 \text{ g}) - (210 \text{ g})(0.8)(50 \text{ m/s})$$

$$(210 \text{ g})(50 \text{ m/s}) + (210 \text{ g})(0.8)(50 \text{ m/s}) = v_{ball} (256 \text{ g})$$

$$v_{ball} = \frac{(210 \text{ g})(90 \text{ m/s})}{256 \text{ g}}$$

$$v_{ball} = 74 \text{ m/s}$$

Step 7: Common sense check. This velocity is over 150 mi/h, but that seems about right when you think about how fast a golf ball rockets off the tee.

of restitution with the golf ball greater than 0.830. The National Collegiate Athletic Association men's basketball rules require the ball to bounce to a height between 49 and 54 in. or 1.24 and 1.37 m (measured to the top of the ball) when dropped from a height of 6 ft (1.83 m) (measured to the bottom of the ball). The rules of racquetball state that the ball must bounce to a height of 68 to 72 in. (1.73 to 1.83 m) if dropped from a height of 100 in. (2.54 m). What is the range of allowed values for the coefficient of restitution (equation 3.21) for a racquetball, according to this rule?

$$e = \sqrt{\frac{\text{bounce height}}{\text{drop height}}}$$

Low value:

$$e = \sqrt{\frac{68}{100}} = \sqrt{0.68} = 0.82$$

High value:

$$e = \sqrt{\frac{72}{100}} = \sqrt{0.72} = 0.85$$

According to the rules of racquetball, the coefficient of restitution of the ball must be between 0.82 and 0.85. The coefficient of restitution of a baseball with a wooden bat is about 0.55. The coefficient of restitution of a tennis ball on the court is about 0.73. How the ball bounces is determined by its coefficient of restitution.

Our exploration of Newton's first law of motion led us to the conservation of momentum principle and collisions. In the analysis of collisions, we considered the two colliding objects to be part of the same system and thus we ignored the force of impact—since it was an internal force. If we isolate only one of the objects involved in the collision, then this impact force becomes an external force, and Newton's first law is no longer applicable. What happens when the external forces acting on an object result in a net external force not equal to zero? Newton came up with an answer to this question in his second law of motion.

Newton's Second Law of Motion: Law of Acceleration

Mutationem motis proportionalem esse vi motrici impressae, et fieri secundum lineam rectam qua vis illa imprimitur. (Newton 1686/1934, p. 644)

This is Newton's second law of motion in Latin as originally presented in *Principia*. It is commonly referred to as the law of acceleration. Translated directly, this law states, "The change of motion of an object is proportional to the force impressed; and is made in the direction of the straight line in which the force is impressed" (Newton 1686/1934, p. 13). This law explains what happens if a net external force acts on an object. More simply stated, Newton's second law says that if a net external force is exerted on an object, the object will accelerate in the direction of the net external force, and its acceleration will be directly proportional to the net external force and inversely proportional to its mass. This can be stated mathematically as

$$\Sigma F = ma \qquad (3.22)$$

where

ΣF = net external force,

m = mass of the object, and

a = instantaneous acceleration of the object.

This is another vector equation, since force and acceleration are vectors. Newton's second law thus applies to the components of force and acceleration. Equation 3.22 can be represented by equations for the three dimensions (vertical, horizontal—forward and backward, and horizontal—side to side).

$$\Sigma F_x = ma_x \qquad (3.23)$$

$$\Sigma F_y = ma_y \qquad (3.24)$$

$$\Sigma F_z = ma_z \qquad (3.25)$$

Newton's second law expresses a cause-and-effect relationship. Forces cause acceleration. Acceleration is the effect of forces. If a net external force acts on an object, the object accelerates. If an object accelerates, a net external force must be acting to cause the acceleration. Newton's first law of motion is really just a special case of Newton's second law of motion—when the net force acting on an object is zero, its acceleration is also zero.

Any time an object starts, stops, speeds up, slows down, or changes direction, it is accelerating and a net external force is acting to cause this acceleration.

Newton's second law of motion explains how accelerations occur. Let's see if we can apply it. In chapter 2, we examined the motion of projectiles. The vertical acceleration of a projectile is governed by Newton's second law of motion. If the only force acting on a projectile is the downward force of gravity, then the acceleration of the projectile will also be downward and proportional to the force. Since the force of gravity is the object's weight *(W)*, using equation 3.24, we get the following result:

$$\Sigma F_y = ma_y$$

$$W = ma_y$$

$$W = mg$$

This is not new to us, since weight and the acceleration due to gravity was introduced in chapter 1 as equation 1.2. Let's consider other applications of Newton's second law that involve contact forces as well as gravity. Try self-experiment 3.4.

Self-Experiment 3.4

Try to find an elevator at your school or dorm, or look for one in a tall building in your town. Get on it and ride it up and down several times. What happens when you ride up an elevator? How does it feel when the elevator starts up? Do you feel heavier or lighter? When the elevator is between floors, do you feel any heavier or lighter? What about when the elevator comes to a stop at the upper floor? Do you feel heavier or lighter as the elevator slows down? You probably feel heavier as the elevator starts upward and lighter as the elevator slows to a stop. In between floors, you probably feel neither heavier nor lighter. Why is this so? Did you gain weight and then lose weight as the elevator sped up and slowed down?

Now, let's examine what happens to you during the elevator ride in self-experiment 3.4 using Newton's second law of motion. First let's draw a free-body diagram and determine what external forces act on you as you stand in the elevator. Figure 3.6 shows a free-body diagram of someone standing in an elevator. Gravity pulls downward on you with a force equal to your weight. Do any other forces act on you? What about the reaction force under your feet? The elevator floor exerts a reaction force upward on your feet. If you are not touching anything other than the floor of the elevator, then the only forces acting on you are gravity (your weight) and the reaction force from the floor. These are vertical forces, so if we want to know what your acceleration is or its direction, we can use equation 3.24:

$$\Sigma F_y = ma_y$$

$$\Sigma F_y = R + (-W) = ma_y$$

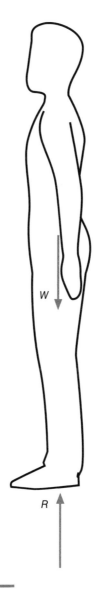

Figure 3.6 Free-body diagram of a person standing in an elevator.

where

ΣF_y = net external force in the vertical direction,

m = your mass,

a_y = your vertical acceleration,

W = your weight, and

R = the reaction force exerted on your feet by the elevator.

If the reaction force, *R,* is larger than your weight, you feel heavier and the net force acts upward, resulting in an upward acceleration. This is exactly what happens when the elevator speeds up in the upward direction; it accelerates you upward, and you feel heavier. If the reaction force, R, is equal to your weight, you feel neither heavier nor lighter, and the net force is zero, resulting in no acceleration. If you and the elevator were already moving upward, you would continue to move upward at constant velocity. If the reaction force, R, is less than your weight, you feel lighter and the net force acts downward, resulting in a downward acceleration. This is exactly what happens when the elevator slows down; you decelerate upward (slow down in the upward direction or accelerate downward), and you feel lighter. Let's confirm our analysis with some rough measurements in self-experiment 3.5.

Self-Experiment 3.5

You can crudely check the values of the reaction force, R, from the elevator floor that acts on your feet by taking a bathroom scale with you on your elevator ride. Stand on the scale while you are in the elevator. The scale indicates the force that you exert on it, which is equal to the force that it exerts on you—the reaction force, R. The scale reading goes up above your body weight as the elevator starts up and you accelerate upward. The scale reading then returns to your body weight as the elevator continues to move upward and you continue to move upward at constant velocity. The scale reading drops below your body weight as the elevator slows down and your acceleration is downward.

While you're on the elevator, let's see what happens when you ride the elevator down rather than up. The free-body diagram and equations used to analyze the upward elevator ride would be the same for the downward ride. When the elevator begins to descend, it speeds up downward. You feel lighter. The reaction force from the floor (the scale reading) is less than your body weight, so the net force is downward and you accelerate downward. As the elevator continues downward, it stops speeding up downward. You feel your normal weight. The reaction force from the floor (the scale reading) is equal to your body weight, so the net force is zero and you move at a constant velocity downward. As the elevator slows to a stop at the bottom floor, it slows down in the downward direction. You feel heavier. The reaction force from the floor (the scale reading) is greater than your body weight, so the net force is upward and you accelerate upward (your downward movement slows down).

The elevator example doesn't seem to have much to do with human movement in sport, but consider what force you must exert against a 10 lb (4.5 kg) dumbbell to lift it. The external forces acting on the dumbbell are the pull of gravity acting downward and the reaction force from your hand acting upward. The net force is thus the difference between these two forces, just as on the elevator. When does the lift feel most difficult? When does it feel easier? To start the lift, you must accelerate the dumbbell upward, so the net force acting on the dumbbell must be upward. The force you exert on the dumbbell must be larger than 10 lb. This is just like what occurs in the elevator example. Once you have accelerated the dumbbell upward, to continue moving it upward requires only that a net force of zero act on the dumbbell, and the dumbbell will move at constant velocity. The force you exert on the dumbbell must just equal 10 lb. As you complete the lift, you need to slow down the upward movement of the dumbbell, so the net force acting on the dumbbell is downward. The force you exert on the dumbbell must be less than 10 lb. When the dumbbell is held overhead, it is no longer moving, so the net force acting on the dumbbell is zero. The force you exert on the dumbbell must be equal to 10 lb.

Now let's consider how much force is required to accelerate something horizontally. Next time you are at the bowling alley, try this. Take a 16 lb (7.3 kg) bowling ball and set it on the floor. Try to get the ball rolling horizontally by pushing on it with only one finger. Were you able to? How much horizontal force was required to accelerate the ball horizontally? To accelerate the ball horizontally, you needed to exert only a very small force. This is easily done with one finger. Let's look at equation 3.23 to explain why this is so.

$$\Sigma F_x = ma_x$$

In the horizontal direction, the only other horizontal force acting on the ball is a small horizontal friction force exerted by the floor on the ball, so

$$\Sigma F_x = P_x + (-F_f) = ma_x$$
$$P_x - F_f = ma_x$$

where

P_x = pushing force from finger and

F_f = friction force from floor.

To accelerate the ball horizontally and start it moving, the pushing force from your finger just has to be slightly larger than the very small friction force exerted on the ball by the floor. In this case, the force you exert can be less than 1 lb (4.5 N) and the ball will accelerate.

Now try to lift the ball with the same finger (use one of the finger holes in the ball). Can you do it? If not, use your whole hand and all three finger holes. How much force is required to accelerate the ball upward? It is quite difficult to accelerate the ball upward using only the force from one finger. Why is so much more force required to

A 52 kg runner is running forward at 5 m/s when his foot strikes the ground. The vertical ground reaction force acting under his foot at this instant is 1800 N. The friction force acting under his foot is a 300 N braking force. These are the only external forces acting on the runner other than gravity. What is the runner's vertical acceleration as a result of these forces?

Solution:

Step 1: List the known quantities and the quantities that can be derived easily.

$m = 52$ kg

$R_x = 300$ N

$R_y = 1800$ N

$W = mg = (52$ kg$)(9.81$ m/s$) = 510$ N

Step 2: Identify the unknown variable the problem asks for.

$a_y = ?$

Step 3: Draw a free-body diagram of the runner.

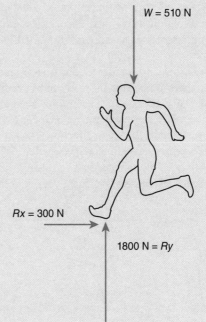

$W = 510$ N

$Rx = 300$ N

1800 N $= Ry$

Step 4: Search for equations with the known and unknown variables—Newton's second law works here.

$\Sigma F_y = ma_y$

$\Sigma F_y = (R_y - W) = ma_y$

Step 5: Substitute quantities and solve for the unknown variable.

1800 N $- 510$ N $= (52$ kg$)\, a_y$

$a_y = (1290$ N$)/(52$ kg$) = 24.8$ m/s^2 upward

Step 6: Common sense check.
The answer, 24.8 m/s^2, is about two and a half times the acceleration due to gravity (2.5 g's). This is about right for the impact phase of a running stride. The number is positive, so it indicates an upward acceleration. The runner's downward velocity is slowed down as the foot strikes the ground.

accelerate an object upward than to accelerate an object sideways? Look at equation 3.24 again.

$$\Sigma F_y = ma_y$$

In the vertical direction, the only other vertical force exerted on the ball is the force of gravity,

$$\Sigma F_y = P_y + (-W) = ma_y$$

$$P_y - W = ma_y$$

where

P_y = upward pulling force from finger (or hand) and

W = weight.

To accelerate the ball upward, you must exert an upward force larger than the weight of the ball, in this case, larger than 16 lb. In general, causing something to accelerate horizontally requires much less force (and thus less effort) than does causing something to accelerate upward. Figure 3.7 illustrates this.

We now know that a net force is needed to slow something down or speed it up. Slowing down and speeding up are examples of acceleration. Is a net force needed to change directions? Yes; because acceleration is caused by a net external force and a change in direction is an acceleration, a net force is needed to change direction. What happens when you run around a curve? You change your direction of motion. Because you change your direction

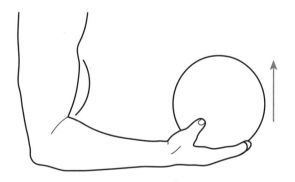

UGH!

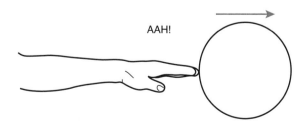

AAH!

Figure 3.7 It's much easier to accelerate something horizontally than vertically upward.

of motion, you are accelerating. A net force must act in the horizontal direction to cause this acceleration. Where does the external horizontal force come from? Think about what happens if you try to run around a curve in an ice rink. You can't, because there isn't enough friction. Friction is the external horizontal force that causes you to change direction as you run around a curve.

Impulse and Momentum

Mathematically, Newton's second law is expressed by equation 3.22:

$$\Sigma F = ma$$

This tells what happens only at an instant in time. The acceleration caused by the net force is an instantaneous acceleration. This is the acceleration experienced by the body or object at the instant the net force acts. This instantaneous acceleration will change if the net force changes. Except for gravity, most external forces that contribute to a net external force change with time. So the acceleration of an object subjected to these forces also changes with time.

In sports and human movement, we are often more concerned with the final outcome resulting from external forces acting on an athlete or object over some duration of time than with the instantaneous acceleration of the athlete or object at some instant during the force application. We want to know how fast the ball was going after the pitcher exerted forces on it during the pitching actions.

Newton's second law can be used to determine this. Looking at equation 3.22 slightly differently, we can consider what average acceleration is caused by an average net force:

$$\Sigma \bar{F} = m\bar{a} \tag{3.26}$$

where

$\Sigma \bar{F}$ = average net force and

$\bar{a}$ = average acceleration.

Because average acceleration is the change in velocity over time (equation 2.9),

$$\bar{a} = \frac{v_f - v_i}{\Delta t}$$

$$\bar{a} = \frac{\Delta v}{\Delta t}$$

equation 3.26 becomes

$$\Sigma \bar{F} = m\bar{a}$$

$$\Sigma \bar{F} = m\left(\frac{\Delta v}{\Delta t}\right) \tag{3.27}$$

Multiplying both sides by Δt, this becomes

$$\Sigma \bar{F} \, \Delta t = m \, \Delta v \qquad (3.28)$$

$$\Sigma \bar{F} \, \Delta t = m(v_f - v_i) \qquad (3.29)$$

This is the impulse–momentum relationship. **Impulse** is the product of force and the time during which the force acts. If the force is not constant, impulse is the average force times the duration of the average force. The impulse produced by a net force acting over some duration of time causes a change in momentum of the object upon which the net force acts. To change the momentum of an object, either its mass or its velocity must change. In sports and human movement, most objects we deal with have a constant mass. A change in momentum thus implies a change in velocity.

When Newton stated his second law of motion, he really meant momentum when he said motion. The change in momentum of an object is proportional to the force impressed.

The impulse–momentum relationship described mathematically by equation 3.29 is actually just another way of interpreting Newton's second law. This interpretation may be more useful to us in studying human movement. The average net force acting over some interval of time will cause a change in momentum of an object. We can interpret change in momentum to mean change in velocity, because most objects have constant mass. If we want to change the velocity of an object, we can produce a larger change in velocity by having a larger average net force act on the object or by increasing the time during which the net force acts.

> The average net force acting over some interval of time will cause a change in momentum of an object.

Using Impulse to Increase Momentum

The task in many sport skills is to cause a large change in the velocity of something. In throwing events, the ball (or shot, discus, javelin, or Frisbee) has no velocity at the beginning of the throw, and the task is to give it a fast velocity by the end of the throw. We want to increase its momentum. Similarly, in striking events, the racket (or bat, fist, club, or stick) has no velocity at the beginning of the swing (or stroke or punch), and the task is to give the implement a fast velocity just before its impact. Our bodies may be the objects whose momentum we want to increase in jumping events and other activities. In all of these activities, the techniques used may be explained in part by the impulse–momentum relationship. A large change in velocity is produced by a large average net

force acting over a long time interval. Because there are limits on the forces humans are capable of producing, many sport techniques involve increasing the duration of force application. Try self-experiment 3.6 to see how force duration affects performance.

Self-Experiment 3.6

Take a ball in your throwing hand and see how far (or how fast) you can throw it using only wrist motions. Move only your hand and keep your forearm, upper arm, and the rest of your body still. This is not a very effective technique, is it? Now try to throw the ball again, only this time use your elbow and wrist. Move only your hand and forearm, and keep your upper arm and the rest of your body still. This technique is better, but it still isn't very effective. Try it a third time using your wrist, elbow, and shoulder. Move only your hand, forearm, and upper arm, and keep the rest of your body still. This technique is an improvement over the previous one, but you could still do better. Try it a fourth time throwing as you normally would with no constraints. This throw was probably the fastest. In which throw were you able to exert a force against the ball for the longest amount of time? For the shortest amount of time?

During self-experiment 3.6, you exerted the largest impulse on the ball when you used your normal throwing technique. As a result, the ball's momentum changed very much, and the ball left your hand with the fastest velocity. The large impulse was the result of a relatively large average force being exerted on the ball for a relatively long time. You exerted the smallest impulse on the ball when you used only your wrist. The ball's momentum didn't change very much, and the ball left your hand with the slowest velocity. The small impulse was the result of a relatively small average force being exerted on the ball for a relatively short time. The normal throwing technique involved more limbs in the throwing action, and you were able to increase the time during which you could exert a force on the ball (and you were probably able to exert a larger average force). The end result was a faster throw. Due to a longer period of force application, the ball had more time to speed up, and thus its velocity at release was faster.

An important thing to remember about the impulse–momentum relationship (equation 3.29),

$$\Sigma \bar{F} \, \Delta t = m(v_f - v_i),$$

is that the average net force, $\Sigma \bar{F}$, in the impulse term is a vector, as are the velocities, v_f and v_i, in the momentum term. An impulse will cause a change in momentum, and thus a change in velocity, in the direction of the force. If you want to change the velocity of an object in a specific direction, the force you apply, or some component of that force, must be in that specific direction.

Which is the greater limitation on impulse—the force or the time? Try self-experiment 3.7 to help answer this question.

Self-Experiment 3.7

See how far (or fast) you can throw a very light object (such as a table tennis ball) compared to a very heavy object (such as a 16 lb [7.3 kg] shot). What limited your throwing performance with the lighter object? Was your strength the limiting factor (do you have to be exceptionally strong to throw a table tennis ball fast?) or was it technique (duration of force application)? What limited your throwing performance with the heavy object—strength (force) or technique (duration of force application)?

In self-experiment 3.7, the limiting factor for throwing the very light object was your technique, not your strength. More specifically, the duration of time during which you could exert a force on the ball was constrained. It was very short. The ball sped up so quickly that your hand had a difficult time keeping up with it and still exerting a force on it.

Conversely in this experiment, the limiting factor for throwing the very heavy object was your strength, not your technique. When you tried to throw (or put) the 16 lb shot, the limiting factor was not the duration of the force application but the size of the force itself. The force you exerted was definitely larger than the force exerted when you threw the table tennis ball, but the amount of force you exerted was constrained by your strength. If you were stronger, the force you could exert on the shot would have been larger, and the shot would have gone farther and faster.

When you threw the table tennis ball in self-experiment 3.7, the average net force, $\Sigma \bar{F}$, wasn't the problem; the time of its application, Δt, was. In the shot put, the time of application was long, but the amount of force applied was limited. In both instances, maximizing both quantities, $\Sigma \bar{F}$ and Δt, will result in the fastest throw or put. But in throwing a lighter object, technique (duration of force application) is more important for success, whereas in throwing heavy objects, the force applied is more important. Compare baseball pitchers and javelin throwers to shot-putters. Shot-putters are bigger and stronger. Their training and selection have been based on their ability to produce large forces ($\Sigma \bar{F}$ in impulse). Baseball pitchers and javelin throwers are not as strong. They're successful because their techniques maximize the duration of force application (Δt in impulse).

Now let's try an activity in which the force element of the impulse is constrained so that we are forced to emphasize duration of force application (Δt) in the impulse equation. Try self-experiment 3.8.

Self-Experiment 3.8

Fill several balloons with water so that each one is about the size of a softball. Take these balloons outside to an empty field or empty parking lot. Now, see how far you can throw one without having it break in your hand. If you exert too large a force against the balloon, it will break. To throw the balloon far, you must maximize the duration of force application during the throw while limiting the size of the force you exert against the balloon so that it doesn't break. Don't constrain your technique to what you perceive as normal throwing styles. Remember, the best technique will be the one in which you accelerate the balloon for the longest possible time while applying the largest (but non-balloon-breaking) force against the balloon.

Let's summarize what we've learned about impulse and momentum so far. The relationship is described mathematically by equation 3.29:

$$\Sigma \bar{F} \, \Delta t = m(v_f - v_i)$$

impulse = change in momentum

where

$\Sigma \bar{F}$ = average net force acting on an object,

Δt = interval of time during which this force acts,

m = mass of the object being accelerated,

v_f = final velocity of the object at the end of the time interval, and

v_i = initial velocity of the object at the beginning of the time interval.

In many sport situations, the goal is to impart a fast velocity to an object. The initial velocity of the object is zero, and the final velocity is fast, so we want to increase its momentum. We accomplish this by exerting a large force against the object for as long a time as possible (by exerting a large impulse). Techniques in sport activities such as throwing or jumping are largely based on increasing the time of force application to obtain a large impulse.

Techniques in sport activities such as throwing or jumping are largely based on increasing the time of force application to obtain a large impulse.

Using Impulse to Decrease Momentum

In certain other activities, an object may have a fast initial velocity and we want to decrease this velocity to a slow or zero final velocity. We want to decrease its momentum. Can you think of any situations like this? What about landing from a jump? Catching a ball? Being struck by a punch? Does the impulse–momentum relationship apply in an analysis of these situations? Yes. Let's try another activity. Try self-experiment 3.9 (it's best to try this on a warm day).

Self-Experiment 3.9

Fill a few more water balloons and take them outside. With a friend, play a game of catch with a water balloon. See how far apart you can get from one another and still catch the balloon intact. If you can't find a friend willing to take part in this activity, play catch by yourself. Throw a balloon up in the air and catch it without breaking it. Try to throw it higher and higher.

What does your catching technique look like when you try to catch a water balloon in self-experiment 3.9? To catch the balloon without breaking it, your arms have to "give" with the balloon. You start with your arms outstretched and then move them in the direction of the balloon's movement as you begin to catch the balloon.

This giving action increases the impact time, Δt, in the impulse–momentum equation, equation 3.29. This increase in the impact time decreases the average impact force, $\Sigma \overline{F}$, required to stop the balloon. This smaller average impact force is less likely to break the balloon.

Why does the balloon become more difficult to catch as you and your friend get farther apart or as you throw it higher and higher? As you move farther apart or throw the balloon higher, the impulse you exert on the balloon must be larger to create the greater change in momentum required to stop it. It becomes more difficult to create a large enough impulse to stop the balloon without allowing the impact force to exceed the point at which the balloon will break. The technique used in landing from a jump is similar to that used to catch a water balloon if you think of the jumper as the water balloon. Try self-experiment 3.10 to see.

Self-Experiment 3.10

Stand on a chair. Now, jump off the chair and land on your feet. How did you reduce the impact force? You flexed your knees, ankles, and hips. This increased the impact time—the time it took to slow you down. This increased Δt in the impulse–momentum equation (equation 3.29) and thus decreased the average impact force, $\Sigma \overline{F}$, since the change in momentum, $m(v_f - v_i)$, would be the same whether you flexed your legs or not.

What would happen if you landed stiff-legged? (**Don't try it!**) The impact time would be much smaller, and the average impact force would have to be much greater because the change in momentum would still be the same. This impact force might be great enough to injure you.

Let's examine the impulse–momentum equation (equation 3.29) again to see why this is so:

$$\Sigma \overline{F} \, \Delta t = m(v_f - v_i)$$

The right side of this equation is the same whether you jump from the chair and land stiff-legged or flex your legs. Your mass, m, does not change. Your final velocity, v_f, is the same for both conditions. This is your velocity at the end of the landing, which will be zero. Your initial velocity, v_i, is also the same for both conditions if you jump from the same height. This is your velocity when your feet first make contact with the ground. Your change in velocity, $(v_f - v_i)$, is the same for both conditions. So your change in momentum, $m(v_f - v_i)$, the right side of equation 3.29, is the same whether you land stiff-legged or flex your legs. But how can the average impact forces differ between conditions?

The change in momentum is the same for the two landing techniques, which means that the impulse, $\Sigma \overline{F} \Delta t$, the left side of equation 3.29, must also be the same for the two techniques. This doesn't mean that the average impact force, $\Sigma \overline{F}$, must be the same for both conditions, or that the impact time, Δt, must be the same. It means only that the product of the two, $\Sigma \overline{F}$ times Δt, must be the same for the two landing techniques. If the impact time, Δt, is short, the average impact force, $\Sigma \overline{F}$, must be large. If the impact time, Δt, is long, the average impact force, $\Sigma \overline{F}$, must be small. As long as the change in momentum, $m(v_f - v_i)$, is the same for both activities, changes in impact time, Δt, are accompanied by inversely proportional changes in average impact force, $\Sigma \overline{F}$.

In some sports and activities, athletes or participants may not be able to land on their feet and flex their legs to reduce the average impact force. How do high jumpers and pole-vaulters land? They land on their backs—not on the hard ground, but on soft landing pads. If they landed on their backs on the hard ground, the impact time would be short, and the large impact forces would definitely cause injury. How do landing pads prevent them from being injured? Landing pads are made of cushioned materials that slow the jumper down over a longer time. This longer impact time means that the average impact forces are lessened proportionally. Pole vault landing pads are thicker than high jump landing pads because vaulters fall from greater heights and have faster velocities at impact, so the impact time must be more greatly increased to reduce their average impact forces to safe levels.

Gymnasts don't usually land on their backs, but their landings from apparatus dismounts or floor exercise routines are worth more points if they allow less flexion at their knees, ankles, and hips. If these landings occurred on hard floors, there would be many more injuries in gymnastics. But the gymnasts don't land on hard floors; their apparatus dismounts are onto a padded surface. Likewise, their landings after floor exercise stunts are onto the elevated, padded, and sprung gymnastics floor. The gymnastics floor and the mats around gymnastics apparatus increase the time of impact during landings. This increased impact time decreases the average impact force. It should be noted that the increase in impact time is not of the same magnitude as that experienced by high jumpers and pole-vaulters, even though gymnasts may be jumping, and thus falling, from similar heights. Thus, gymnasts are more likely to experience injuries due to hard landings than are high jumpers.

Now think about other equipment that is used to cushion impacts by increasing the time of impact. You may have thought of the padding and equipment worn by football and hockey players, boxing gloves, baseball gloves, the midsole material used in running shoes, wrestling mats, and so on.

The impulse–momentum relationship provides the basis for techniques used in many sports and human

SAMPLE PROBLEM 3.4

A boxer is punching the heavy bag. The time of impact of the glove with the bag is 0.10 s. The mass of the glove and his hand is 3 kg. The velocity of the glove just before impact is 25 m/s. What is the average impact force exerted on the glove?

Solution:

Step 1: Identify the known quantities and quantities that can be assumed.

$\Delta t = 0.10$ s

$m = 3$ kg

$v_i = 25$ m/s

Assume that the glove comes to a stop at the end of the impact, that the punch is in the horizontal plane, and that no other significant horizontal forces act on the glove and hand.

$v_f = 0$ m/s

Step 2: Identify the unknown variable that the question asks for.

$\bar{F} = ?$

Step 3: Search for the appropriate equation that has the known and unknown variables in it (equation 3.29).

$\Sigma \bar{F} \Delta t = m(v_f - v_i)$

In this case, the average net force is the force exerted by the bag on the glove.

$\Sigma \bar{F} = \bar{F}$

Step 4: Substitute known quantities into the equation and solve for the unknown force.

$\bar{F} \Delta t = m(v_f - v_i)$

$\bar{F}\,(0.10 \text{ s}) = (3 \text{ kg})(0 - 25 \text{ m/s})$

$\bar{F} = -750$ N

The negative sign indicates that the force from the bag to the glove is in the opposite direction of the glove's initial velocity.

Step 5: Common sense check.
I don't know much about the impact forces in boxing, but 750 N seems reasonable. It's almost 170 lb.

movement skills. In throwing events, an athlete lengthens the duration of force application to increase the change of momentum of the object being thrown. This results in the thrown object having a faster release velocity. The same principle applies to jumping activities, but in this case the "thrown" object is the athlete. In catching, landing, and other impact situations, the goal may be to decrease the magnitude of the impact force. By increasing the duration of application of the impact force (increasing the impact time), the size of the average impact force is reduced. Now let's move on to Newton's third law of motion, which provides more insight into what a force is.

Newton's Third Law of Motion: Law of Action-Reaction

Actioni contrariam semper et aequalem esse reactionem: sive corporum duorum actiones in se mutuo semper esse aequales et in partes contrarias dirigi. (Newton 1686/1934, p. 644)

This is Newton's third law of motion in Latin as presented in *Principia*. It is commonly referred to as the law of action-reaction. Translated directly, this law states, "To every action there is always opposed an equal reaction: or the mutual actions of two bodies upon each other are always equal and directed to contrary parts" (Newton 1686/1934, p. 13). Newton used the words *action* and *reaction* to mean force. The term *reaction force* refers to the force that one object exerts on another. This law explains the origin of the external forces required to change motion . More simply stated, Newton's third law says that if an object (A) exerts a force on another object (B), the other object (B) exerts the same force on the first object (A) but in the opposite direction. So forces exist in mirrored pairs. The effects of these forces are not canceled by each other because they act on different objects. Another important point is that it is the forces that are equal but opposite, not the effects of the forces. Let's try self-experiment 3.11 for a better understanding of this law.

> If an object exerts a force on another object, the other object exerts the same force on the first object but in the opposite direction.

Self-Experiment 3.11

Go over to a wall and push hard against it. What happens when you push against the wall? The wall pushes back against you. The force the wall exerts against you is exactly equal to the force you exert against the wall, but it pushes in the opposite direction. Think about what you feel with your hand as you push against the wall. What direction is the force that you feel through your hand? The force that you feel is in fact the force of the wall pushing against you, not the force of you pushing against the wall. When you push or pull on something, what you feel is not the force that you are pushing or pulling with; it is the equal but opposite reaction force that is pushing or pulling on you.

When you pushed against the wall in self-experiment 3.11, why didn't you accelerate as a result of the force the wall exerted against you? At first, you might say that the force the wall exerted against you was canceled out by the force you exerted against the wall, so the net force was zero and no acceleration occurred. Would this be a correct explanation? No. The force you exerted against the wall does not act on you, so it can't counteract the effect of the force the wall exerts on you. What other forces act on you when you push against the wall? Gravity pulls down on you with a force equal to your weight. The floor pushes up against the soles of your shoes or your feet. And . . . a frictional force from the floor also acts against your feet. This frictional force opposes the pushing force from the wall and prevents you from accelerating as a result of the wall pushing against you.

What about forces that cause accelerations—do they come in pairs also? Let's try self-experiment 3.12.

Self-Experiment 3.12

Hold a ball in your hand. What happens when you push against it in a horizontal direction (as you do when you throw it)? The ball pushes back against you in the opposite direction with a force exactly equal to the force you exert against it. In this case, though, the ball accelerates as a result of the force. It still exerts a force against you while it accelerates. Again, think about what you feel with your hand as you throw the ball. The ball pushes against your hand. If you try to throw it harder, you feel a larger force against your hand. In which direction does this force act? It acts in the opposite direction of the ball's acceleration. It pushes back against your hand. The force that your hand exerts on the ball acts in the direction of the ball's acceleration.

Let's imagine this situation. You are lined up opposite an offensive lineman from a National Football League team. He weighs twice as much as you do. Your job is to push him out of the way. When you push against him, he pushes against you. Who pushes harder? According to Newton's third law of motion, the force that he pushes against you is exactly equal to the force that you push against him. What about the effects of these forces? Newton's second law of motion tells us that the effect of a force on a body is dependent on the mass of the body and the other forces acting on the body. The larger the mass, the smaller the effect. The smaller the mass, the larger the effect. Because his mass is so large, and the friction force beneath his feet is probably large as well, the effect of your pushing force on him will be small. Because your mass is small relative to his, and the friction force beneath your feet is probably less than the friction force beneath his feet, the effect of his pushing force on you will be greater.

Newton's third law of motion helps explain how forces act and what they act on. It gives us the basis for drawing free-body diagrams. It does not explain what the effects of forces will be, however. It tells us only that forces come in pairs, and that each force in a pair acts on a separate object.

Newton's Law of Universal Gravitation

Newton's law of universal gravitation gives us a better explanation of weight. This law was purportedly inspired by the fall of an apple on his family's farm in Lincolnshire while he was residing there during the plague years (see figure 3.8). He presented this law in two parts. First, he stated that all objects attract each other with a gravitational force that is inversely proportional to the square of the distance between the objects. Second, he stated

Figure 3.8 The alleged inspiration for Newton's law of universal gravitation.

that this force of gravity was proportional to the mass of each of the two bodies being attracted to each other. The universal law of gravitation can be represented mathematically as

$$F = G\left(\frac{m_1 m_2}{r^2}\right), \qquad (3.30)$$

where F is the force of gravity, G is the universal constant of gravitation, m_1 and m_2 are the masses of the two objects involved, and r is the distance between the centers of mass of the two objects.

Newton's universal law of gravitation was momentous because it provided a description of the forces that act between each object in the universe and every other object in the universe. This law, when used with his laws of motion, predicted the motion of planets and stars. The gravitational forces between most of the objects in sport are very small—so small that we can ignore them. However, one object that we must be concerned with in sport is large enough that it does produce a substantial gravitational force on other objects. That object is the earth. The earth's gravitational force acting on an object is the object's weight. For an object close to the earth's surface, several of the terms in equation 3.30 are constant. These terms are G, the universal constant of gravitation; m_2, the mass of the earth; and r, the distance from the center of the earth to its surface. If we introduce a new constant,

$$g = G\left(\frac{m_2}{r^2}\right), \qquad (3.31)$$

then equation 3.30 becomes

$$F = mg$$

or

$$W = mg$$

where W is the force of the earth's gravity acting on the object, or the weight of the object; m is the mass of the object; and g is the acceleration of the object caused by the earth's gravitational force. This is the same equation for weight we first saw in chapter 1. Now we know the basis for the equation.

Summary

The basics of linear kinetics, explaining the causes of linear motion, lie in Newton's laws of motion. Newton's first law explains that objects do not move or do not change their motion unless a net external force acts on them. An extension of the first law is the conservation of momentum principle. Newton's second law explains what happens if a net external force does act on an object. It will accelerate in the direction of the net external force, and its acceleration will be inversely related to its mass. The impulse–momentum relationship presents Newton's second law in a way that is more applicable to sports and human movement. The basis for the techniques used in many sport skills is in the impulse–momentum relationship. Increasing the duration of force application increases the change in momentum. Newton's third law explains that forces act in pairs. For every force acting on an object, there is another equal force acting in the opposite direction on another object. Finally, Newton's law of universal gravitation gives us the basis for the force of gravity.

KEY TERMS

coefficient of restitution (p. 96)
impulse (p. 103)
linear momentum (p. 90)
reaction force (p. 89)

REVIEW QUESTIONS

1. Can an object be in motion if no external forces act on it?

2. Can external forces act on an object and not cause acceleration? Explain.

3. When you turn a corner while walking, what force causes you to change directions and where does this force act on you?

4. Show how Newton's first law of motion can be derived from Newton's second law of motion.

5. Newton's third law states that for every action there is an equal but opposite reaction. When you are standing on the floor, the force of gravity pulls down on you. What is the equal but opposite reaction to the force of gravity?

6. Can an object change its direction of motion if no external forces act on it?

7. Mike's team and Pete's team are in a tug-of-war contest. Mike's team pulled harder initially, and now Pete's team, the rope, and Mike's team are all moving with constant velocity in the direction of the pulling force from Mike's team. Which team is pulling with greater force on the rope at this instant?

8. When you are lifting a weight upward, the weight often feels "heavier" near the beginning of the lift and "lighter" near the end of the lift, especially if the lift is done quickly. Why?

9. The weight of the average NFL football player increased by 10% percent in 20 years. Describe the mechanical advantages that a heavier player has over a lighter player.

10. Describe an example in sports or other human movement of Newton's first law of motion.

11. Describe an example in sports or other human movement of Newton's second law of motion.

12. Describe an example in sports or other human movement of conservation of momentum.

13. Describe an example in sports or other human movement of the impulse and momentum principle.

14. Describe an example in sports or other human movement of Newton's third law of motion.

PROBLEMS

1. Quentin is lifting a 10 kg dumbbell by pulling upward on it with a 108.1 N force. What is the acceleration of the dumbbell as a result of this force?

2. Grant accelerates a 1 kg ball 5.0 m/s² horizontally. The vertical acceleration of the ball is zero. The force of gravity is the only external force acting on the ball other than the forces Grant exerts on it.

   a. How large is the horizontal force Grant exerts on the ball at this instant to cause its 5.0 m/s² horizontal acceleration?

   b. How large is the vertical force Grant exerts on the ball at this instant?

3. During a fastball pitch, the maximum horizontal acceleration of the 146 g baseball is 1000 m/s². What horizontal force is the pitcher exerting on the baseball at this instant to cause this acceleration?

4. Tonya crashes into Nancy while they are practicing their figure skating routines. Tonya's mass is 60 kg and Nancy's mass is 50 kg. Just before the collision occurs, Tonya's velocity is 5 m/s and Nancy's velocity is 6 m/s in the opposite direction. During the collision, Tonya exerts an average force of 1000 N against Nancy.

    a. How big is the average force Nancy exerts against Tonya during the collision?

    b. If this is the only horizontal force acting on each skater during the collision, what average horizontal acceleration does each skater experience during the collision?

    c. If the collision is perfectly inelastic, how fast and in what direction will Tonya and Nancy be moving immediately after the collision?

5. Doris is trying to lift a stack of books off the table. The stack of books has a mass of 15 kg. What minimum upward force must Doris exert on the books to lift the books off the table?

6. Blanche, a 100 kg shot-putter, is putting the 4 kg shot in a track meet. Just before she releases the shot, Blanche is in the air (her feet are off the ground). At this instant, the only force she exerts against the shot is an 800 N horizontal force directed forward.

    a. What is Blanche's horizontal acceleration at this instant? $\frac{800}{100} = 8\,m/s^2$

    b. What is Blanche's vertical acceleration at this instant? $-9.81\,m/s^2$

    c. What is the shot's horizontal acceleration at this instant? $800/4 = 200\,m/s^2$

    d. What is the shot's vertical acceleration at this instant? $-9.81\,m/s^2$

7. Jay is gliding north on his cross-country skis across a flat section of snow at 7 m/s. Jay's mass is 100 kg. The coefficient of dynamic friction between the skis and the snow is 0.10. The coefficient of static friction between the skis and the snow is 0.12. The force of air resistance is 1.9 N acting backward (south) on Jay. If friction between the skis and snow and air resistance are the only horizontal forces acting on Jay, what is his horizontal acceleration?

8. Mary is trying to stop a 5 kg bowling ball that is rolling toward her. Its horizontal velocity is 8.0 m/s when she begins to stop it. It takes her 2.0 s to stop the ball. What average force did Mary exert on the ball during this 2.0 s to stop it?

9. A 0.43 kg soccer ball is stationary on the field when Darryl kicks it. Darryl's foot is in contact with the ball for 0.01 s during the kick. The horizontal velocity of the ball as it leaves his foot is 25 m/s. What is the average horizontal force exerted by Darryl's foot on the ball during the 0.01 s of contact (assume that friction between the ball and the field is zero)?

10. Peter, a 100 kg basketball player, lands on his feet after completing a slam dunk and then immediately jumps up again to celebrate his basket. When his feet first touch the floor after the dunk, his velocity is 5 m/s downward; when his feet leave the floor 0.50 s later, as he jumps back up, his velocity is 4 m/s upward.

    a. What is the impulse exerted on Peter during this 0.50 s?

    b. What is the average net force exerted on Peter during this 0.50 s?

    c. What is the average reaction force exerted upward by the floor on Peter during this 0.50 s?

11. Scott is rolling down a 30° slope on his skateboard. The total mass of Scott and the skateboard is 75 kg. The rolling friction between the skateboard wheels and the concrete is 9 N acting backward against the skateboard. The drag force due to air resistance is 11 N acting backward against Scott. What is Scott's acceleration?

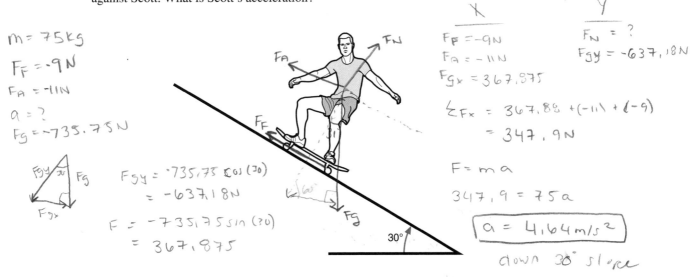

$$m = 75 kg$$
$$F_F = -9N$$
$$F_A = -11N$$
$$a = ?$$
$$F_g = -735.75N$$

$$F_{gy} = -735.75 \cos(30)$$
$$= -637.18N$$
$$F = -735.75 \sin(30)$$
$$= 367.875$$

30°

X
$$F_F = -9N$$
$$F_A = -11N$$
$$F_{gx} = 367.875$$
$$\Sigma F_x = 367.88 + (-11) + (-9)$$
$$= 347.9N$$
$$F = ma$$
$$347.9 = 75a$$
$$\boxed{a = 4.64 m/s^2}$$
down 30° slope

Y
$$F_N = ?$$
$$F_{gy} = -637.18N$$

12. To earn the *FIFA Approved* label, a soccer ball must bounce at least 135 cm high and no more than 155 cm high when dropped from a height of 200 cm onto a steel plate.

    a. What is the minimum coefficient of restitution of a *FIFA Approved* soccer ball?

    b. What is the maximum coefficient of restitution of a *FIFA Approved* soccer ball?

13. A warm hockey puck has a coefficient of restitution of 0.50, while a frozen hockey puck has a coefficient of restitution of only 0.35. In the NHL, the pucks to be used in games are kept frozen. During a game, the referee retrieves a puck from the cooler to restart play but is told by the equipment manager that several warm pucks were just put into the cooler. To check to make sure he has a game-ready puck, the referee drops the puck on its side from a height of 2 m. How high should the puck bounce if it is a frozen puck? 2 m

14. A 0.15 kg baseball collides with a 1.0 kg bat. The ball has a velocity of 40 m/s immediately before the collision. The center of mass of the bat also has a velocity of 40 m/s, but in the opposite direction, just before the collision. The coefficient of restitution between the bat and the ball is 0.50. Estimate how fast the baseball is moving as it leaves the bat following the collision.

15. Maddie is standing still when her dancing partner, Paul, begins to lift her up and throw her into the air. Maddie's mass is 40 kg. Paul exerts an average vertical force of 500 N for 1.0 s on Maddie during the lift and throwing motion.

    a. What is Maddie's vertical velocity when Paul releases her?

    b. If Maddie's center of gravity was 1.5 m above the floor when Paul released her, what peak height will she reach?

Motion Analysis Exercises Using MaxTRAQ

If you haven't done so already, review the instructions for downloading and using the educational version of the MaxTRAQ motion analysis software at the beginning of this book, then download and install the software. Once this is done, you are ready to try the following two-dimensional kinematic analyses using MaxTRAQ.

1. Open MaxTRAQ. Select Tools in the menu bar and then open Options under the Tools menu. In the Options submenu, select Video. To the upper right side of the Video window, under Video Aspect Ratio, make sure that Default-Used Preferred Aspect Ratio is selected. In the lower half of the right side of the Video window, under Deinterlace Options, select None. Click OK, then close MaxTRAQ and reopen it to have the deinterlace options take effect. Now, open the *Lacrosse Ball COR* video from within MaxTRAQ. This video was taken with a camera operated at 600 frames per second (fps), so one video frame is recorded every 1/600 s (0.00167 s). Events recorded in two consecutive video frames are separated in time by 1/600 s. MaxTRAQ doesn't recognize the frame rate of this video, so it assumes a frame rate of 30 fps. That means that the time shown is incorrect. To determine the time, you'll have to multiply the frame number by 1/600 s (or just divide by 600). When you open the video, you should see black background with a ruler on the top and a lacrosse ball at the left side of the video window. The camera that recorded this video was on its side, so horizontal is vertical in the video. Up is to the left, and down is to the right.

2. Make sure the the scaling/calibration tool is activated by clicking View on the menu bar, then selecting Tools from the drop-down menu, and making sure Show Scale is checked. Open the scaling tool by clicking on Tools on the menu bar, selecting Scale. In the Scaling Tool window that opens, set the gauge length to 64 mm—the diameter of a lacrosse ball—and then click OK. Place the cursor over the leftmost point on the ball and click the left mouse button once; then place the cursor over the rightmost point on the ball and click the left mouse button a second time. The scale should appear in the video window. Hide the scale by selecting View in the menu bar; then click Tools and uncheck Show Scale.

    a. What is the vertical velocity of the lacrosse ball just before impact? Use the digitizing function and digitize a point on the ball three frames before impact. Digitize the same point again one frame before impact. A good way to make sure you digitize the same point is to activate the grid in MaxTRAQ and then digitize the farthest left point on the ball that lies on one of the horizontal grid lines. To turn on the grid, click the button on the tool bar, or select Grid from the View menu. Determine the displacement of the ball between these two frames. Divide this displacement by the elapsed time (divide the displacement by the number of frames and multiply by 600).

    b. What is the vertical velocity of the lacrosse ball just after impact? Use the digitizing function to measure the position of the ball in the first frame when it is no longer in contact with the ground and again two frames after that. Determine the displacement of the ball between these two frames. Divide this displacement by the elapsed time.

    c. What is the coefficient of restitution (COR) of the lacrosse ball? Divide the velocity computed in question 1b by the velocity computed in question 1a. The absolute value of this number is the COR. A NOCSAE (National Operating Committee on Standards for Athletic Equipment)-approved lacrosse ball has a COR between 0.60 and 0.70.

3. Open the *Golf Ball COR* video from within MaxTRAQ. If you did not set the video aspect ratio and deinterlace options in exercise 1, do so now by following the instructions. This video was taken with a camera operated at 1200 frames per second (fps), so one video frame is recorded every 1/1200 s (0.00083 s). Events recorded in two consecutive video frames are separated in time by 1/1200 s. MaxTRAQ doesn't recognize the frame rate of this video, so it assumes a frame rate of 30 fps. That means that the time shown is incorrect. To determine the time, you'll have to multiply the frame number by 1/1200 s (or just divide by 1200). When you open the video, you should see a golf ball on a tee. Make sure the scaling/calibration tool is activated by clicking View on the menu bar, then selecting Tools from the drop-down menu, and making sure Show Scale is checked. Open the scaling tool by clicking on Tools on the menu bar and selecting Scale. In the Scaling Tool window that opens, set the gauge length to 316 mm, then click OK. Place the cursor over the upper left corner of the inner video window and click the left mouse button once; then place the cursor over the upper right corner of the inner video window and click the left mouse button a second time. The scale should appear in the video window. Hide the scale by selecting View in the menu bar; then click Tools and uncheck Show Scale.

   a. What is the horizontal velocity of the golf club head just before impact? Use the digitizing function to measure the position of the tail of the club head two frames before impact and one frame before impact. Determine the horizontal displacement of the club head between these two frames. Convert the displacement from mm to m by dividing by 1000 mm/m, then divide this displacement by the elapsed time 1/1200 s. Dividing by 1000 mm/m and then dividing by 1/1200 s is the same as multiplying by 1.2 m/mm/s. So, just multiply the displacement in mm by 1.2 m/mm/s to get the velocity in m/s.

   b. What is the horizontal velocity of the golf club head just after impact? Use the digitizing function to measure the position of the tail of the club head in one frame after impact and two frames after impact. Determine the horizontal displacement of the club head between these two frames and then compute velocity from this displacement.

   c. What is the horizontal velocity of the golf ball just after impact? Use the digitizing function to measure the position of the center of the ball one frame after impact and two frames after impact. Determine the horizontal displacement of the ball between these two frames and then compute velocity from this displacement. In the first frame after impact, the center of the ball is coincides with a point midway between the two black stripes on the ball. In the second frame after impact, the center of the ball coincides with a point at the right end of these two black stripes on the ball.

   d. Estimate the coefficient of restitution (COR) of the golf ball with the club head. Subtract the velocity computed in question 2b from the velocity computed in question 2c. Divide this difference by the velocity computed in question 2a. The absolute value of this number is the COR. A USGA-approved driver and golf ball cannot have a COR above 0.83. The driver used in this video is quite old and the ball is a range ball, so the COR for this driver and ball is much less than 0.83.

chapter 4

Work, Power, and Energy
Explaining the Causes of Motion Without Newton

objectives

When you finish this chapter, you should be able to do the following:

- Define mechanical work
- Distinguish the differences between positive and negative work
- Define energy
- Define kinetic energy
- Define gravitational potential energy
- Define strain energy
- Explain the relationship between mechanical work and energy
- Define power

A pole-vaulter runs down the runway, slowly lowering the pole as he approaches the takeoff. He is moving as fast as a sprinter when suddenly he drops the end of the pole into the vault box and jumps off the ground. The pole bends and bends and bends as the vaulter's forward movement slows and he swings upward. For a moment, it appears as if the pole will break. But then the pole begins to unbend. The unbending pole appears to fling the vaulter upward as he hangs on to it. The vaulter then swings into a handstand on the pole as the pole straightens to a vertical position. The vaulter pushes off the pole with one hand and soars up and over the crossbar to finish the vault. Wow! How is the vaulter able to convert the speed of his run-up into the height needed to get over the crossbar? The relationship between mechanical work and energy provides an answer to this question. This chapter introduces mechanical work and energy and covers material concerning their use in analysis of movement.

This chapter continues the

study of linear kinetics begun in the previous chapter. The explanations for the causes of motion given in this chapter do not rely on Newton's laws of motion, but rather on the relationships between work, energy, and power, which were discovered and developed by several different scientists in the two centuries following Isaac Newton's achievements. Theoretically, all we need to analyze and explain linear motion are Newton's laws of motion. But some analyses and explanations are easier if based on work and energy relationships rather than Newtonian mechanics. So this chapter provides more tools for analyzing and explaining sport skills.

Work

What is work? There are many definitions for work. *Webster's New World Dictionary* uses almost one full column (half a page) to list all the various definitions of work. In mechanics, however, **work** is the product of force and the amount of displacement in the direction of that force (it is the means by which energy is transferred from one object or system to another). Mathematically, this can be expressed as

$$U = F(d) \tag{4.1}$$

where

U = work done on an object (the letter W would be a better abbreviation for work, but we already used it to represent weight),

F = force applied to an object, and

d = displacement of an object along the line of action of the force.

Because work is the product of force and displacement, the units for work are units of force times units of length. These may be ft·lb or Nm. In the International System of Units, the **joule** (abbreviated with the letter J) is the unit of measurement for work; 1 J is equal to 1 Nm. The joule was named for James Prescott Joule (an English brewer who finally established the law of conservation of energy through practical experiments . . . more about this later).

⟳ Work is the product of force and displacement.

How can we best describe a force used to create displacement? Try self-experiment 4.1.

Self-Experiment 4.1

If you put this book on a table and push it so that it moves across the table, you have done work to the book. Try it. To quantify the amount of work you did, you would have to know what force you exerted against the book and how far you moved the book in the direction of the force (its displacement). Measuring the displacement is easy, but what about the force? When you pushed against the book, did you push with the same amount of force throughout the movement of the book or did the force change? The force probably varied somewhat, so it wasn't constant. If the force wasn't constant, what value should we use for the force, F, in equation 4.1? The force at the start of the movement? The force at the end of the movement? The peak force? How about the average force? This makes sense. The best value to describe a force that has many values during its application would be the average of these values—the average force, as in equation 4.2.

Equation 4.1 really describes only the work done by a constant force. The work done by a force whose magnitude varies is

⟳ $U = \bar{F}(d)$ **(4.2)**

where

    U = work done on an object,

    $\bar{F}$ = average force exerted on an object, and

    d = displacement of an object along the line of action of the average force.

To determine the amount of work done on an object, we need to know three things:

1. The average force exerted on the object

2. The direction of this force

3. The displacement of the object along the line of action of the force during the time the force acts on the object

Now let's look at an example. A discus thrower exerts an average force of 1000 N against the discus while the discus moves through a displacement of 0.6 m in the direction of this force (see figure 4.1). How much work did the discus thrower do on the discus?

    $U = \bar{F}(d)$

    $U = (1000 \text{ N})(0.6 \text{ m})$

    $U = 600 \text{ Nm} = 600 \text{ J}$

This was easy because the average force and displacement were given. Let's try something more difficult. A weightlifter bench presses a 1000 N barbell as shown in figure 4.2. He begins the lift with his arms extended and the barbell 75 cm above his chest. The lifter then lowers the barbell and stops it when it is 5 cm above his chest.

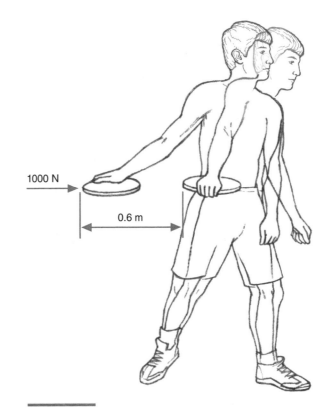

1000 N

0.6 m

Figure 4.1 A discus thrower does work on a discus by exerting an average force of 1000 N on the discus while moving it through a displacement of 0.6 m.

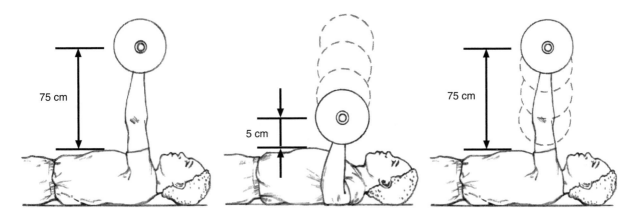

75 cm

5 cm

75 cm

Figure 4.2 The phases of a bench press.

He pauses there and then lifts the barbell upward away from his chest and back to the original starting position 75 cm above his chest. The average force exerted on the barbell by the lifter while lowering the weight is 1000 N upward. The average force exerted by the lifter while raising the weight is also 1000 N upward. (Could you determine this using Newton's laws?) So the average force exerted on the barbell by the lifter is 1000 N for the whole lift. How much work did the lifter do on the barbell from the start until the finish of the lift?

$$U = \bar{F}(d)$$

$$U = (1000 \text{ N})(d)$$

What was the displacement of the barbell? The starting and ending positions of the barbell were the same, so the displacement was zero.

$$U = (1000 \text{ N})(0) = 0$$

If the displacement was zero, the work done was also zero. Whoa! This doesn't seem correct! Certainly the lifter thinks he did work. He expended some calories in performing this lift. It is true that, physiologically, the lifter did some work, but mechanically, no work was done on the barbell because it was in the same position when the lift ended as when it started.

Was any work done during the raising of the barbell?

$$U = \bar{F}(d)$$

$$U = (1000 \text{ N})(d)$$

The displacement of the barbell during its raising was

$$d = \text{final position} - \text{initial position} = y_f - y_i \quad (4.3)$$

$$d = 75 \text{ cm} - 5 \text{ cm} = 70 \text{ cm upward.}$$

If we convert 70 cm into meters, we get

$$\frac{70 \text{ cm}}{100 \text{ cm/m}} = 0.70 \text{ m} = d.$$

So the work done was

$$U = (1000 \text{ N})(0.70 \text{ m}) = 700 \text{ Nm} = 700 \text{ J}.$$

The lifter did do work on the barbell when he raised it. He did 700 J of work. Then how can the total work he did on the barbell for the whole lift be zero? Let's determine the work done during the lowering of the barbell.

$$U = \bar{F}(d)$$

$$U = (1000 \text{ N})(d)$$

What was the displacement of the barbell during the lowering phase? Its initial position was 75 cm above

SAMPLE PROBLEM 4.1

A therapist is helping a patient with stretching exercises. She pushes on the patient's foot with an average force of 200 N. The patient resists the force and moves the foot 20 cm toward the therapist. How much work did the therapist do on the patient's foot during this stretching exercise?

Solution:

Step 1: Identify the known quantities.

$$\bar{F} = 200 \text{ N}$$

$$d = -20 \text{ cm}$$

Since displacement was in the opposite direction of the force, it is negative.

Step 2: Identify the variable to solve for.

Work done = U = ?

Step 3: Search for the appropriate equation that includes the known and unknown variables.

$$U = \bar{F}(d)$$

Step 4: Substitute the known quantities and solve the equation.

$$U = \bar{F}(d)$$

$$U = (200 \text{ N})(-20 \text{ cm}) = -4000 \text{ N cm} = -40 \text{ Nm} = -40 \text{ J}$$

the chest, and its final position after lowering was 5 cm above the chest, so:

$$d = \text{final position} - \text{initial position} = y_f - y_i$$

$$d = 5\text{ cm} - 75\text{ cm} = -70\text{ cm}$$

The displacement was −0.7 m, or 0.7 m downward. In which direction was the force exerted on the barbell? The force was upward, and the displacement was downward, so, using equation 4.2, the work is

$$U = \bar{F}(d)$$

$$U = (1000\text{ N})(-0.70\text{ m}) = -700\text{ Nm} = -700\text{ J}$$

The work done during the lowering of the barbell was *negative* 700 J of work. That sounds weird; how can work be negative? Mechanical work is negative if the force acting on an object is in the opposite direction of the motion (displacement) of the object.

Now it is more clear why zero work was done during the whole lift. If −700 J of work was done during the lowering of the barbell and +700 J of work was done during the raising of the barbell, the work done for the whole lift would be −700 J plus 700 J, or zero.

$$U_{whole\ lift} = U_{lowering} + U_{raising}$$

$$U_{whole\ lift} = -700\text{ J} + 700\text{ J}$$

$$U_{whole\ lift} = 0$$

Work can be positive or negative. Positive work is done by a force acting on an object if the object is displaced in the same direction as the force. A pitcher does positive work against a baseball when throwing it. A weightlifter does positive work against a weight when lifting or raising it. A gymnast does positive work when pulling up on the uneven bars. A high jumper does positive work when jumping off the ground.

Negative work is done by a force acting on an object when the object is displaced in the direction opposite the force acting on it. A first baseman does negative work against the ball when catching it. A weightlifter does negative work against a weight when lowering it. A gymnast does negative work when landing from a dismount. Friction does negative work on a skier sliding down a hill.

⤵ **Positive work is done by a force acting on an object if the object is displaced in the same direction as the force. Negative work is done by a force acting on an object when the object is displaced in the direction opposite the force acting on it.**

Muscles can also do mechanical work. When a muscle contracts, it pulls on its points of attachment. Positive work is done by a muscle when it contracts and its points of attachment move in the direction of the muscle force pulling on them. The force (muscle force) and the displacement (displacement at the point of muscle attachment) are in the same direction. The muscle shortens, and the muscle contraction is a concentric contraction.

Negative work is done by a muscle when it contracts and its points of attachment move in the opposite direction of the muscle force pulling on them. The force (muscle force) and the displacement (displacement at the point of muscle attachment) are in opposite directions. The muscle lengthens, and the muscle contraction is an eccentric contraction.

Not all muscle contractions produce mechanical work. A muscle can contract and do zero mechanical work. This occurs when a muscle contracts and its points of attachment do not move relative to each other. The displacement at the point of muscle attachment is zero. The muscle length remains unchanged, and the muscle contraction is an isometric contraction.

Energy

What is energy? Like *work, energy* is a term that has many meanings. In mechanics, **energy** is defined as the capacity to do work. There are many forms of energy: heat, light, sound, chemical, and so on. In mechanics, we are concerned primarily with mechanical energy, which comes in two forms: kinetic energy and potential energy. **Kinetic energy** is energy due to motion, whereas **potential energy** is energy due to position.

⤵ **Mechanical energy comes in two forms: kinetic energy, which is energy due to motion, and potential energy, which is energy due to position.**

Kinetic Energy

A moving object has the capacity to do work due to its motion. This capacity is the kinetic energy of the object. But how is kinetic energy quantified? What affects it? Let's try self-experiment 4.2 to see if we can get some insights into kinetic energy.

Self-Experiment 4.2

Close the book and lay it flat on a desk or table. Now take another book and give it a quick push so that it slides across the desk and strikes the first book. What happened? The book you slid across the desk did work on the first book. It exerted a force on the first book, and the first book was displaced by this force. The book sliding across the desk had the capacity to do work because it had kinetic energy—energy due to its motion.

Try this experiment again, only this time give the book a bigger push so it slides faster. Was more work done this time? Yes. Because the book was sliding faster, it had more kinetic energy and thus a greater capacity to do work. Try the experiment one more time, but this time slide a heavier book (but slide it just as fast). Was more work done this time? Somehow, the larger mass of the book meant that it had more kinetic energy and thus a greater capacity to do work. (And you also have now had the satisfaction of pushing this textbook off the desk several times!)

The kinetic energy of an object is affected by the mass and velocity of the object. If we made more precise measurements, we would discover that the kinetic energy is proportional to the square of the velocity. Mathematically, we define kinetic energy as follows:

$$KE = \frac{1}{2}mv^2 \qquad (4.4)$$

where

KE = kinetic energy,

m = mass, and

v = velocity.

The units for kinetic energy are units of mass times velocity squared, or $kg(m^2/s^2)$, but this is the same as $[kg(m/s^2)]m$, which is equivalent to Nm, which is a joule. The unit of measurement for kinetic energy is the same as the unit of measurement for work. To determine the kinetic energy of an object, we must know its mass and its velocity.

How much kinetic energy does a baseball thrown at 80 mi/h (35.8 m/s) have? A baseball's mass is 145 g (0.145 kg). To determine the ball's kinetic energy, use equation 4.4:

$$KE = \frac{1}{2}mv^2$$

$$KE = \frac{1}{2}(0.145 \text{ kg})(35.8 \text{ m/s})^2$$

$$KE = 92.9 \text{ J}$$

Determining the kinetic energy of an object is easier than determining the work done by a force, because we can measure mass and velocity more easily than we can measure force.

Potential Energy

Potential energy is the energy (capacity to do work) that an object has due to its position. There are two types of potential energy: gravitational potential energy, which is energy due to an object's position relative to the earth; and strain energy, which is due to the deformation of an object.

Gravitational Potential Energy

Gravitational potential energy is potential energy due to an object's position relative to the earth. The gravitational potential energy of an object is related to the object's weight and its elevation or height above the ground or some reference. Let's try self-experiment 4.3 to see a demonstration of gravitational potential energy.

Self-Experiment 4.3

Get a hammer and nail and a block of wood. Hold the nail on the block of wood. If you lift the hammer only a few inches above the nail and drop it, the hammer does not drive the nail into the wood very far. It does little work on the nail because it has little potential energy. If you lift the hammer much higher above the nail and let it swing down and strike the nail, it drives the nail much farther. The hammer does more work on the nail because it had greater potential energy (it was higher above the nail). Suppose you used a heavier hammer. Would it drive the nail farther if swung from the same height?

Mathematically, gravitational potential energy is defined as follows:

$$PE = Wh \qquad (4.5)$$

or

$$PE = mgh \qquad (4.6)$$

where

PE = gravitational potential energy,

W = weight,

m = mass,

g = acceleration due to gravity = 9.81 m/s^2, and

h = height.

The units for potential energy are units of force times units of length, or Nm, which is equivalent to joules, the same unit of measure as for kinetic energy and work. To determine an object's gravitational potential energy, we must know its weight and its height above the ground.

How much gravitational potential energy does a 700 N ski jumper have when taking off from the 90 m jump? We can use equation 4.5 to determine this, but what should we use for the height, h? The 90 m ski jump is 90 m above the base of the hill, but the takeoff point is on the hill itself and is only about 3 m above the ground at the side of the hill. Potential energy is a relative term—because height is measured relative to some point that should be referred to in describing potential energy. In this case, let's use the base of the hill as our reference point. The height is then 90 m, and equation 4.5 becomes

$$PE = Wh = (700\ N)(90\ m)$$

$$PE = (700\ N)(90\ m) = 63,000\ Nm$$

$$PE = 63,000\ J$$

Strain Energy

Another type of potential energy is also used in sport. **Strain energy** is energy due to the deformation of an object. When a fiberglass vaulting pole bends, strain energy is stored in the bent pole. Likewise, when an archer draws his bow or a diver deflects a diving board, strain energy is stored in the deformed bow or diving board. The greater the deformation of the object, the greater the strain energy stored in the object. Try self-experiment 4.4 to get a better feel for strain energy.

Self-Experiment 4.4

Take a rubber band and stretch it. By stretching it, you have given the rubber band strain energy. If you stretch it further, you increase the strain energy in the rubber band. The stiffer the object is, the greater its strain energy when it is deformed. Stretch a wider rubber band or two rubber bands parallel to each other. The strain energy with stretching the wider rubber band or two rubber bands is greater than the strain energy in the smaller or single stretched rubber band.

The strain energy of an object is related to its stiffness, its material properties, and its deformation. Mathematically, the strain energy of a material with a linear stress–strain relationship is defined as

$$SE = \frac{1}{2}k\,\Delta x^2 \tag{4.7}$$

where

> SE = strain energy,
>
> k = stiffness or spring constant of material, and
>
> Δx = change in length or deformation of the object from its undeformed position.

If the stiffness constant is expressed in N/m, then the strain energy is expressed in $(N/m)m^2$, or Nm, which is equivalent to joules—the same unit of measurement as for gravitational potential energy, kinetic energy, and work.

How much strain energy is stored in a tendon that is stretched 5 mm (0.005 m) if the stiffness of the tendon is 10,000 N/m?

$$SE = \frac{1}{2}k\,\Delta x^2$$

$$SE = \frac{1}{2}(10,000\ N/m)(0.005\ m)^2$$

$$SE = 0.125\ J$$

In human movement and sports, energy is possessed by athletes and objects due to their motion (kinetic energy), their position above the ground (gravitational potential energy), and their deformation (strain energy). We will be concerned primarily with the first two types of energy: kinetic energy and potential energy.

The Work-Energy Relationship

The definitions of work and energy indicate that a relationship exists between them. As a reminder, energy was defined earlier as the capacity to do work. The definition of work included this statement: "It is the means by which energy is transferred from one object or system to another." The unit of measure for work and energy is joules—the same for each quantity. This is another indication that work and energy are related.

Demonstrating the Work-Energy Relationship

How are work and energy related? Let's look at a previous example to reveal something about the relationship. Consider the example from earlier in this chapter, the weightlifter bench pressing a 1000 N barbell. During the lifting part of the exercise, the barbell was raised 70 cm and the work done by the lifter was 700 J. How much more energy did the barbell have after it was lifted? It didn't have any more kinetic energy because it was not moving before the lift started and it was not moving at the end of the lift. But its potential energy changed because it changed heights. What was the change in potential energy?

$$\Delta PE = PE_{final} - PE_{initial}$$

$$\Delta PE = Wh_{final} - Wh_{initial}$$

$$\Delta PE = W(h_{final} - h_{initial})$$

$$\Delta PE = 1000 \text{ N } (0.70 \text{ m})$$

$$\Delta PE = 700 \text{ J}$$

The change in the potential energy of the barbell was 700 J, the same as the work done to lift the barbell. Perhaps work done causes a change in potential energy. Or maybe work causes a change in total mechanical energy.

In the discus-throwing example we looked at earlier in this chapter (see figure 4.1), a discus thrower exerted an average force of 1000 N against the discus while the discus moved through a displacement of 0.60 m in the direction of the force. The work done by the thrower on the discus was 600 J. Let's assume that the force exerted by the thrower was constant and the displacement was horizontal, and that the discus was not moving at the start of the throw. If work does cause a change in total mechanical energy, the work done by the thrower on the discus would cause a change in the energy of the discus. Because the displacement of the discus was horizontal, there was no change in potential energy, so the work done by the thrower caused only a change in kinetic energy. The mass of a discus is 2 kg. Knowing that the work done was 600 J and that the initial velocity of the discus was zero, we can determine the velocity of the discus at the end of the period of work (v_f).

$$\text{Work done} = \Delta KE + \Delta PE + \Delta SE = \Delta KE + 0 + 0$$

$$600 \text{ J} = \Delta KE = KE_f - KE_i$$

$$600 \text{ J} = KE_f - 0$$

$$600 \text{ J} = \frac{1}{2}mv_f^2$$

$$(600 \text{ J}) = \frac{1}{2}(2 \text{ kg})v_f^2$$

$$\frac{2(600 \text{ J})}{2 \text{ kg}} = v_f^2$$

$$v_f = \sqrt{\frac{2(600 \text{ J})}{2 \text{ kg}}}$$

$$v_f = 24.5 \text{ m/s}$$

The velocity of the discus is 24.5 m/s according to the work–energy principle. Can we verify this using Newtonian mechanics? The constant 1000 N horizontal force acting on the discus is the net horizontal force acting on the discus. This force would cause the discus to accelerate at 500 m/s².

$$\Sigma F_x = ma_x$$

$$a_x = \frac{\Sigma F_x}{m} = \frac{1000 \text{ N}}{2 \text{ kg}} = 500 \text{ m/s}^2$$

The average velocity of the discus during the throwing action, $\bar{v}$, was the displacement of the discus during the throwing action, d, divided by the time of the throwing action, t. With a constant force acting on the discus, the velocity of the discus would increase linearly. The average velocity of the discus could also be computed by dividing the difference between the final (v_f) and initial (v_i) velocities of the discus by 2:

$$\bar{v} = \frac{d}{\Delta t}$$

$$\bar{v} = \frac{v_f - v_i}{2}$$

$$\bar{v} = \frac{0.6 \text{ m}}{\Delta t} = \frac{v_f - 0}{2}$$

We don't know the time or duration of the throwing action, but this would be the time it took to accelerate the discus from 0 m/s to its final velocity. We know the acceleration was 500 m/s². Average acceleration, $\bar{a}$, is change in velocity divided by time, so time can be computed by dividing change in velocity by acceleration:

$$\bar{a} = \frac{v_f - v_i}{\Delta t} = \frac{v_f - 0}{\Delta t}$$

$$500 \text{ m/s}^2 = \frac{v_f}{\Delta t}$$

$$\Delta t = \frac{v_f}{500 \text{ m/s}^2}$$

If we substitute this representation for time in the previous equation, we get

$$\bar{v} = \frac{0.6 \text{ m}}{\Delta t} = \frac{0.6 \text{ m}}{\left(\dfrac{v_f}{500 \text{ m/s}^2}\right)} = \frac{v_f}{2}$$

Solving for the final velocity gives us

$$\bar{v} = \frac{0.6 \text{ m}}{\left(\dfrac{v_f}{500 \text{ m/s}^2}\right)} = \frac{v_f}{2}$$

$$v_f = \frac{2(0.6 \text{ m})}{\left(\dfrac{v_f}{500 \text{ m/s}^2}\right)}$$

$$v_f^2 = 2(0.6 \text{ m})(500 \text{ m/s}^2)$$

$$v_f = 600 \text{ m}^2/\text{s}^2$$

$$v_f = \sqrt{600 \text{ m}^2/\text{s}^2}$$

$$v_f = 24.5 \text{ m/s}$$

Whew! This is the same result we got using the work–energy principle, but it took a lot more effort using Newton's laws. The computations involved in using the work–energy principle were much less complicated!

The examples of the bench press and the discus throw demonstrate the work–energy principle: The work done by the external forces (other than gravity) acting on an object causes a change in energy of the object. Mathematically, this relationship is shown in equation 4.8:

⟹ $$U = \Delta E \qquad (4.8)$$

$$U = \Delta KE + \Delta PE + \Delta SE$$

$$U = (KE_f - KE_i) + (PE_f - PE_i) + (SE_f - SE_i) \quad (4.9)$$

where

U = work done on an object by forces other than gravity,

ΔE = change in total mechanical energy,

KE_f = final kinetic energy,

KE_i = initial kinetic energy,

PE_f = final gravitational potential energy,

PE_i = initial gravitational potential energy,

SE_f = final strain energy, and

SE_i = initial strain energy.

⟹ **The work done by the external forces (other than gravity) acting on an object causes a change in energy of the object.**

Doing Work to Increase Energy

Why is the relationship between work and energy so important? In sports and human movement, we are often concerned with changing the velocity of an object. Changing velocity means changing kinetic energy, and the work–energy principle shows how kinetic energy can be changed by doing work. More work is done, and thus a greater change in energy occurs, if the average force exerted is large or the displacement in line with this force is long. This sounds very similar to what we discovered using the impulse–momentum relationship in chapter 3. Remember the impulse–momentum relationship?

$$\Sigma \bar{F} \Delta t = m(v_f - v_i)$$

Impulse = change in momentum

Using the impulse–momentum relationship as the basis for technique analysis, the creation of a large change in velocity requires that a large force be applied over a long time. The work–energy principle indicates that production of a large change in kinetic energy (and thus a large change in velocity) requires application of a large force over a long distance.

Think back to the example of throwing a ball. If you throw only with your wrist, you are able to exert a force on the ball through a small displacement. The work done is small, and as a result, the ball's change in kinetic energy is small. The ball's velocity when it leaves your hand is slow.

⟹ **A large change in kinetic energy requires that a large force be applied over a long distance.**

If the wrist and elbow are involved in the throwing motion, you are able to exert a force on the ball through a larger displacement. The work done on the ball is larger, and as a result, the ball's change in kinetic energy is larger. So the ball's velocity when it leaves your hand is faster.

When you involve your whole arm, your trunk, and your legs in the throwing motion, you are able to apply a force to the ball through a much larger displacement. The work done on the ball is much larger, and as a result, the ball's change in kinetic energy is much larger as well. The ball's velocity when it leaves the hand is much faster (more than 100 mi/h [44.7 m/s] for some major league pitchers).

Something similar to this actually occurred in the evolution of shot-putting technique. The rules for shot putting indicate that the put must be made from a 7 ft (2.13 m) diameter circle. The shot-putter must begin the shot put from a standstill, with no body part touching anything outside of this circle. The putter must complete the put without touching anything outside of the circle until the judge has ruled it a fair put. Only then is the putter allowed to step outside of the "throwing ring," but only through the rear half. The size of the ring thus limits how much work the athlete can do to the shot by constraining the distance over which the putter can exert a force on the shot.

Early in the 20th century, shot-putters began their put from the rear of the ring. The initial stance was similar to that shown in figure 4.3.

The athlete's shoulders were aligned approximately 45° to the throwing direction, meaning that the right-handed athlete faced slightly to the right. The putter would hop across the ring on his right leg and put the shot.

Gradually, the technique evolved, and the putter's shoulders were turned more and more toward the rear of the circle in the initial stance. The greater shoulder rotation allowed the putter to start from a position that allowed for greater displacement of the shot before release. Finally, in the 1950s, Parry O'Brien began putting the shot from an initial position in the rear of the ring facing the opposite direction of the put. (Figure 4.4 shows an athlete using this technique in his initial stance at the back of the ring.) This stance put him in a position where he could maximize the displacement of the shot in the direction of his force application. He could also involve stronger muscle groups and have a larger force act on the shot during the putting action. The work done on the shot was thus increased. This increased the change in energy (potential and kinetic), which resulted

SAMPLE PROBLEM 4.2

A pitcher exerts an average horizontal force of 100 N on a 0.15 kg baseball during the delivery of the pitch. His hand and the ball move through a horizontal displacement of 1.5 m during this period of force application. If the ball's horizontal velocity was zero at the start of the delivery phase, how fast would the ball be going at the end of the delivery phase when the pitcher released it?

Solution:

Step 1: Identify the known quantities.

$m = 0.15$ kg

$\bar{F} = 100$ N

$d = 1.50$ m

$v_i = 0$

Step 2: Identify the unknown variable to solve for.

$v_f = ?$

Step 3: Search for appropriate equations with the known and unknown variables in them.

$U = \Delta E$

$\bar{F}(d) = \Delta KE = KE_f - KE_i = \frac{1}{2}m(v_f^2 - v_i^2)$

Step 4: Substitute the known quantities and solve for the unknown variable.

$(100 \text{ N})(1.5 \text{ m}) = (0.15 \text{ kg})(v_f^2 - 0)/2$

$v_f^2 = 2 (100 \text{ N})(1.5 \text{ m})/(0.15 \text{ kg}) = 2000$ Nm/kg

$v_f = 44.7$ m/s

Step 5: Common sense check.
This is very fast—the speed of a major league fastball.

in a greater height and velocity of the shot at release. The result was a longer put.

Doing Work to Decrease (or Absorb) Energy

The work–energy principle can also be used to explain the techniques used in transferring (or absorbing) energy from an object. When you catch a ball, its kinetic energy is reduced (or absorbed) by the negative work you do on it. Similarly, your muscles do negative work on your limbs and absorb their energy when you land from a jump or fall. The average force you must exert to absorb energy in catching a ball or landing from a jump or fall depends on how much energy must be absorbed and over how long a distance you can apply the force. If this force is too great, it may injure you. You attempt to decrease it by "giving" with a ball when you catch it or by flexing at your knees, ankles, and hips when you land from a jump or fall. Remember how you caught water balloons in the previous chapter. These actions increase the distance

Figure 4.4 An athlete using modern shot-putting technique begins in a position turned farther away from the putting direction.

Photo courtesy of University of Michigan, Bentley Historical Library

Figure 4.3 Starting the shot put from the rear of the ring early in the 20th century.

over which the force acts, thus decreasing the average value of the force.

The safety and protective equipment used in many sports uses the work–energy principle to reduce potentially damaging impact forces. The landing pads used in gymnastics, high jumping, and pole vaulting all increase the displacement of the athlete during the impact period as the kinetic energy of the athlete is decreased (absorbed). The impact force is thus decreased because the displacement during the impact is increased. The sand in a long-jumping pit does the same thing when you jump into it, as does the water in a pool when you dive into it, the midsole material in your running shoes when you run on them, the padding in a boxing glove when you punch with it, the air bag in a car when you crash into it, and so on. All of these materials may be referred to as "shock absorbing," but they are actually energy-absorbing materials.

Conservation of Mechanical Energy

The work–energy relationship is also useful when we examine situations in which no external forces act other than gravity. In these situations, no work can be done because no external forces act. If no work can be done, the total mechanical energy of the object in question is conserved; it cannot change, and equation 4.8 becomes

A rider in an equestrian event falls off her horse and strikes her head on the ground. Luckily, she is wearing a helmet. At the first instant of impact the vertical velocity of the rider's head is 5.8 m/s. The mass of the rider's head is 5 kg. The helmet is lined with energy-absorbing material that is 3.0 cm thick. The thickness of the hard outer shell of the helmet, as well as the innermost padding used for fit and comfort, allows the head to displace 1.5 cm during the impact while crushing and compressing the energy-absorbing material and padding as well as the dirt that her head hits. What is the average force exerted by the helmet on the rider's head during the impact from the fall?

Solution:

Step 1: Identify the known quantities.

$$m = 5.0 \text{ kg}$$

$$d = 1.5 \text{ cm}$$

$$v_i = 5.8 \text{ m/s}$$

$$v_f = 0$$

Step 2: Identify the unknown variable to solve for.

$$\bar{F} = ?$$

Step 3: Search for appropriate equations with the known and unknown variables in them.

$$U = \Delta E$$

This equation includes the change in all energies (potential, kinetic, and strain), but the changes in potential energy and strain energy of the head are tiny compared to the change in kinetic energy, so we'll just compute the change in kinetic energy.

$$\bar{F}(d) = \Delta KE = KE_f - KE_i = \frac{1}{2}m(v_f^2 - v_i^2)$$

Step 4: Substitute the known quantities and solve for the unknown variable.

$$\bar{F} (0.015 \text{ m}) = (5.0 \text{ kg})((5.8 \text{ m/s})^2 - 0))/2$$

$$\bar{F} (0.015 \text{ m}) = 84.1 \text{ J}$$

$$\bar{F} = \frac{84.1 \text{ J}}{5.0 \text{ kg}}$$

$$\bar{F} = 5607 \text{ N}$$

Step 5: Common sense check.
This is a large impact force. Since this is an average impact force, the peak impact force is larger. A good estimate for a peak impact force is twice the average impact force. This puts the peak impact force at

$$2 \times 5607 \text{ N} = 11{,}214 \text{ N}.$$

This is a huge force. The criteria for a catastrophic head injury are often expressed in terms of the peak acceleration of the head. This acceleration is expressed in terms of the acceleration due to gravity, or g. A peak acceleration above 300 g will cause a catastrophic head injury or death. In this example, we can determine the acceleration from the 11,214 N force acting on a 5 kg head.

$$F = ma$$

$$11{,}214 \text{ N} = (5 \text{ kg}) a$$

$$a = \frac{11{,}214 \text{ N}}{5 \text{ kg}}$$

$$a = 2243 \text{ m/s}^2$$

Dividing this value by g = 9.81 m/s² will give us acceleration in terms of g:

$$a = \frac{2243 \text{ m/s}^2}{9.81 \text{ m/s}^2}$$

$$a = 229 \text{ g}$$

The acceleration is below 300 g but still high enough to cause a head injury, though probably not a catastrophic one. The acceleration threshold for a concussion is estimated to be 80 g. So, although the helmet may have protected our equestrian athlete from a catastrophic head injury, it certainly did not protect her from a less severe head injury such as a concussion.

$$U = \Delta E$$

$$U = 0 = \Delta KE + \Delta PE + \Delta SE$$

$$0 = (KE_f - KE_i) + (PE_f - PE_i) + (SE_f - SE_i)$$

$$(KE_i + PE_i + SE_i) = (KE_f + PE_f + SE_f)$$

$$E_i = E_f \qquad (4.10)$$

The total mechanical energy of the object is constant if no external forces other than gravity act on the object. This principle may be useful for examining projectile motion. Gravity is the only external force that acts on a projectile. If this is true, the total mechanical energy of a projectile does not change during its flight. Let's consider dropping a ball as an example. Just before you let go of it, the ball has potential energy but no kinetic energy. During the ball's fall, its potential energy decreases because its height decreases. At the same time, though, its kinetic energy increases as it is accelerated downward by gravity. This increase in kinetic energy is exactly matched by the decrease in potential energy, so the total mechanical energy of the ball remains the same.

⟳ **The total mechanical energy of an object is constant if no external forces other than gravity act on the object.**

Let's try some numbers in this example. Suppose you drop a 1 kg ball from a height of 4.91 m (that's about 16 ft). When you first let go of the ball, its potential energy (measured relative to the ground) would be

$$PE_i = Wh = mgh$$

$$PE_i = (1 \text{ kg})(9.81 \text{ m/s}^2)(4.91 \text{ m})$$

When you first let go of the ball, its velocity is zero, so its kinetic energy would be zero. The instant before the ball strikes the ground, it has no height above the ground, so its potential energy would be zero. The kinetic energy of the ball at this time would be

$$KE_f = \frac{1}{2}mv_f^2$$

$$KE_f = \frac{1}{2}(1 \text{ kg})v_f^2$$

If we use equation 4.10, we can determine how fast the ball is going just before it hits the ground:

$$E_i = E_f$$

$$(KE_i + PE_i + SE_i) = (KE_f + PE_f + SE_f)$$

$$PE_i = KE_f$$

$$mgh = \frac{1}{2}mv_f^2$$

$$v_f^2 = \frac{2mgh}{m}$$

$$v_f^2 = 2gh \qquad (4.11)$$

$$v_f^2 = 2(9.81 \text{ m/s}^2)(4.91 \text{m})$$

$$v_f = \sqrt{2(9.81 \text{ m/s}^2)(4.91 \text{ m})}$$

$$v_f = 9.81 \text{ m/s}$$

Just before the ball hits the ground, its velocity is 9.81 m/s downward. Hmm. The falling ball was a projectile. We could have computed its final velocity using the projectile equations from chapter 2. Look at equation 2.18 from chapter 2. It describes the final velocity of a falling object.

$$v_f^2 = 2g\Delta y$$

Now compare this equation to equation 4.11, which we just derived from the conservation of energy principle.

$$v_f^2 = 2gh$$

They are really the same equation and just an extension of equation 2.15, which describes the vertical velocity of a projectile if its initial vertical velocity is not zero.

$$v_f^2 = v_i^2 + 2g\Delta y$$

The v^2 term in equation 2.15 comes from the initial kinetic energy a projectile may have due to its initial vertical velocity. This equation can be derived from the conservation of energy principle if we start with an initial kinetic energy due to the vertical velocity of the projectile as well as an initial height.

The conservation of mechanical energy principle gives us another tool for analyzing and understanding projectile motion. It can also allow us to analyze other situations in which no work is done. For instance, in the pole vault, if the vaulter does no work during the vault itself, her total mechanical energy at the instant of takeoff should equal the total mechanical energy at bar clearance. In this case, the vaulter's kinetic energy at takeoff is transformed into strain energy as the pole bends, and this strain energy is in turn transformed into potential energy as the vaulter is lifted by the unbending pole. How high a pole-vaulter can vault thus has a lot to do with how fast the vaulter can run.

Power

The ability of an athlete to increase the displacement of an object (or body part) while exerting a force affects performance in many skills. Success in these skills thus requires the athlete to exert a large amount of work on an object (or body part). In some sports, excelling requires not just the ability to do a large amount of work, but also the ability to do that work in a short time. *Power* is the mechanical term that describes this ability. Like *work* and *energy, power* is another word that you have some familiarity with and that has numerous meanings. In mechanics, **power** is the rate of doing work, or how much work is done in a specific amount of time. Mathematically, power is defined as

$$P = \frac{U}{\Delta t}$$ (4.12)

where

P = power,

U = work done, and

Δt = time taken to do the work.

Power can be thought of as how quickly or slowly work is done. The SI units for power are **watts** (abbreviated with the letter W), named after the Scottish inventor James Watt; 1 W equals 1 J/s. You may be familiar with watts because light bulbs, amplifiers, and other electrical devices are rated in watts. Another unit of measurement for power is horsepower, but the watt is the unit of measure for power in the International System of Units.

⟳ **Power can be thought of as how quickly or slowly work is done.**

If we examine equation 4.12 more closely, another way of defining power can be derived:

$$P = \frac{U}{\Delta t}$$

$$P = \frac{\bar{F}(d)}{\Delta t} = \bar{F}\left(\frac{d}{\Delta t}\right)$$

$$P = \bar{F}\bar{v}$$ (4.13)

Power can be defined as average force times average velocity along the line of action of that force.

The concept of power is useful in biomechanics for several reasons. The best way to explain one use of power is by example. Suppose you have to move a stack of books from one table to another, and you want to finish this task as quickly as possible. This means that you want to maximize your power output. Numerous strategies are available to you, from moving the books one at a time to moving them all at once. The amount of work done to

the books will be the same, but the time it takes (and thus the power output) may differ. Carrying the entire stack at once would require a large force, and the movement would be slow. Carrying a few books at a time in several trips wouldn't require as much force, and each trip would be quicker. In the first case, you are exerting large forces but moving with a slow velocity. In the second case, you are exerting smaller forces but moving with a faster velocity. The combination of force and velocity determines the power output. Does the larger force in the first case make up for its decreased velocity, or does the larger velocity in the second case make up for its decreased force? What is the tradeoff between force and velocity?

You are faced with similar questions in certain sports and activities. How do you choose which gear to use while pedaling your bicycle? Do you use a high gear, which requires larger pedal forces and a slower pedaling rate, or do you use a lower gear, which requires smaller pedal forces but a faster pedaling rate? When you're running, how do you choose your stride length and stride rate? Do you use a long stride, which requires larger forces and a slower stride rate, or do you use a short stride, which requires smaller forces and a faster stride rate? These questions are difficult to answer because of the number of variables involved, so you have to experiment. One clue to answering these questions may come from studying muscles.

Because the power we produce in our movements ultimately originates in our muscles, the power production characteristics of muscles may provide some insight into the questions just raised. As a muscle's velocity of contraction increases, its maximum force of contraction decreases. So a muscle contracting slowly can produce greater force than the same muscle contracting at a faster rate. If the muscle's velocity of contraction is multiplied by its maximum force of contraction for that velocity, the muscle's power output for each velocity can be determined. The maximum power output occurs at a velocity approximately one-half the muscle's maximum contraction velocity. This would mean that the best gear to use in cycling may not be the highest or the lowest, but one in between. The best stride length may not be the longest or the shortest, but one in between. The best way to move that stack of books may not be to carry them all at once or one at a time, but several at once. The best choice of bicycling gear, stride length, and so on may be the one in which your muscles contract at a velocity corresponding to their velocity of maximal power output. The mechanics of muscular contraction are discussed further in chapter 11.

Another reason that power is an important topic in the study of human movement is that it is actually a constraint on human movement. What does that mean? Consider an Olympic weightlifter performing a clean and jerk. The forces he exerts on the barbell and the fast movement of the barbell indicate that the power output of the lifter is quite large—but only for a brief interval of time. If the time interval were longer, would the lifter be able to produce as much power? The duration of the activity influences the sustainable power output of an individual. A sprinter can maintain a high power output for only a short time (0-60 s). A middle-distance runner's power output is smaller but sustained for a longer time (1-7 min). A marathon runner's power output is smaller still, but it is sustained for a much longer time (2-4 h). Figure 4.5 shows the theoretical relationship between maximum power output and the duration of that power output for humans. This relationship shows the mechanical constraint placed on humans by their power-generating system (their metabolic system).

Summary

In this chapter, we learned the mechanical definitions of work, energy, and power. Work done by a force is the force times the displacement of the object along the line of action of the force acting on it. Energy was defined as the capacity to do work. Mechanically, energy takes two forms: potential energy, which is energy due to position or deformation, and kinetic energy, which is energy due to motion. Potential energy could be due to the position of the object in a gravitational field, called gravitational potential energy, or to its bending, stretch, or deformation, called strain energy. The work done by a force (other than

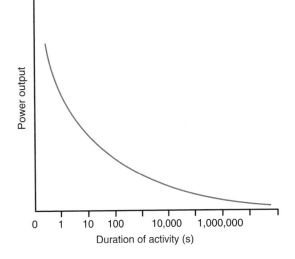

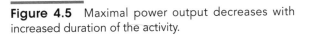

Figure 4.5 Maximal power output decreases with increased duration of the activity.

gravity) on an object causes a change in energy of the object. If a force does work, a change in energy results. Likewise, if a change in energy is observed, a force did work to cause this change in energy.

Power is defined as the rate of doing work. An alternative definition of power is average force times average velocity. In human movement, the maximum power output a human is capable of producing is related to the duration of the activity involved. This has to do with the metabolic capabilities of the human body.

KEY TERMS

energy (p. 119)
gravitational potential
 energy (p. 120)
joule (p. 116)

kinetic energy (p. 119)
potential energy (p. 119)
power (p. 128)

strain energy (p. 121)
watts (p. 128)
work (p. 116)

REVIEW QUESTIONS

1. The sand in a long-jump pit serves two purposes. First, it enables measurement of the distance of the jump by marking the jumper's landing. Second, it cushions the landing of the jumper. Give a mechanical explanation for how the sand "cushions" the landing of the jumper.

2. Why is a windup or backswing important in throwing and striking activities?

3. Why is a follow-through important in throwing and striking activities?

4. Most major league pitchers tend to be taller and longer limbed than other players. Is this an advantage in throwing and striking activities? Explain.

5. Why do safety helmets for bicycling, lacrosse, football, ice hockey, and so on have to be so thick?

6. As speed increases, is there a proportional increase in stopping distance—that is, if you double your speed, does your minimum stopping distance also double? Explain.

7. If you drop a basketball and a tennis ball together so that the basketball bounces on the floor and the tennis ball bounces on top of the basketball, what will happen? Explain.

8. How does an air bag protect you in an accident?

9. Why is running on loose sand more fatiguing than running on a rubberized track?

10. Why is it more appropriate for young children to play with softer and lighter balls and implements?

11. What advantages might a sprinter with well-designed prosthetic legs (below the knee) have over a sprinter with legs? Assume the prosthetic devices have been optimized for sprinting.

12. Imagine that you are in a room on the second floor of a burning building. The only way out is through a window, and the window's sill is 5 m above the ground. Describe what actions you would take to minimize your chance of injury if you exited the building through the second-story window. Explain the mechanical basis for each action.

PROBLEMS

1. How much kinetic energy does a 2 kg discus have if its velocity is 20 m/s?

2. Which of the balls or implements shown in table 2.3 (p. 65) has

   a. the greatest kinetic energy?

   b. the least kinetic energy?

3. At the top of a giant swing on the gymnastics high bar, Candy's velocity is 1 m/s, and she is 3.5 m high. If Candy's mass is 50 kg, what is her total mechanical energy at this instant?

4. You are slowly forcing your opponent's wrist to the table in an arm-wrestling match.

   a. Are you doing any mechanical work against your opponent? If so, is the work positive or negative?

   b. What type of contractions do your arm and shoulder muscles produce?

   c. Is your opponent doing any work against you? If so, is the work positive or negative?

   d. What type of contractions do your opponent's arm and shoulder muscles produce?

5. An archer draws his compound bow and shoots an arrow. The 23 g arrow leaves the bow with a velocity of 88 m/s. The power stroke of the bow is 57 cm; that is, the bowstring exerts force on the arrow through a displacement of 57 cm. The peak draw weight of the bow is 312 N. (This is the maximum force that the archer has to exert on the bowstring.)

   a. How much kinetic energy does the arrow have after release?

   b. How much work does the bowstring do on the arrow?

   c. What average force does the bowstring exert on the arrow?

6. During the pitching motion, a baseball pitcher exerted an average horizontal force of 60 N against the 0.15 kg baseball while moving it through a horizontal displacement of 2.0 m before he released it.

   a. How much work did the pitcher do to the baseball as a result of this force?

   b. Was the work done positive or negative?

   c. If the baseball's velocity at the start of the pitching action was zero, how fast was the ball moving at the instant of release?

7. A baseball strikes the catcher's glove with a horizontal velocity of 40 m/s. The mass of the baseball is 0.15 kg. The displacement of the baseball due to the deformation of the catcher's glove and the movement of the catcher's hand is 8 cm from the instant it first makes contact with the glove until it stops.

   a. How much kinetic energy does the baseball possess just before it strikes the glove?

   b. How much work does the catcher do on the baseball during the catch?

   c. Is the work done positive or negative?

   d. What is the average impact force exerted by the glove on the baseball?

8. Which bowling ball has more energy, a 5 kg ball rolling at 4 m/s or a 6 kg ball rolling at 3 m/s?

9. To test the impact performance of a 10 cm thick gymnastics mat, a 20 kg cylindrical mass is dropped onto the mat from a height of 1 m above the mat. During the impact, the cylinder's vertical velocity reaches zero at the instant the mat has compressed to only 4 cm thick.

   a. How much kinetic energy does the cylinder have at the instant just before it contacts the mat?

   b. How much work does the mat do to stop the fall of the cylinder?

   c. What average vertical force does the mat exert on the cylinder during the impact, from contact until the cylinder's vertical velocity is zero?

10. A gymnast falls from the high bar and lands on a 10 cm thick gymnastics mat. The gymnast strikes the back of his head against the mat during his landing. His head is moving at 7 m/s when it first strikes the mat. The mass of his head is 5 kg. The impact ends when the gymnast's head comes to a stop after deflecting the mat 6.5 cm.

    a. How much kinetic energy does the gymnast's head have at the instant just before it contacts the mat?

    b. How much work does the mat do to stop the motion of the gymnast's head?

    c. What average impact force does the mat exert on the gymnast's head during the impact?

    d. Estimate the peak impact force exerted by the mat on the gymnast's head.

    e. Estimate the peak acceleration of the gymnast's head.

    f. Express this peak acceleration in g's.

11. Jon snatched 100 kg. In a snatch, the barbell is moved from a stationary position on the floor to a stationary position over the athlete's head. Only 0.50 s elapsed from the first movement of the barbell until it was overhead, and the barbell moved through a vertical displacement of 2.0 m. What was the Jon's average power output during the lift?

12. In a vertical jump-and-reach test, 60 kg Nellie jumps 60 cm, while 90 kg Ginger jumps 45 cm. Assuming both jumps took the same amount of time, which jumper was more powerful?

13. Zoe is pole-vaulting. At the end of her approach run, she has a horizontal velocity of 8 m/s and her center of gravity is 1.0 m high. If Zoe's mass is 50 kg, estimate how high she should be able to vault if her kinetic and potential energies are all converted to potential energy.

14. Mike is a 70 kg pole-vaulter. He falls from a peak height of 5.90 m after pole-vaulting over a crossbar set at that height. He lands on a thick mat. When Mike first makes contact with the mat, his center of gravity is only 1.0 m high. During Mike's impact with the mat, the mat compresses. At the point of maximum compression, Mike's vertical velocity reaches zero, and his center of gravity is only 0.5 m high. What average force did the mat exert on Mike during this impact?

If you haven't done so already, review the instructions for downloading and using the educational version of the MaxTRAQ motion analysis software at the beginning of this book, then download and install the software. Once this is done, you are ready to try the following two-dimensional kinematic analyses using MaxTRAQ.

1. Open MaxTRAQ. Select Tools in the menu bar and then open Options under the Tools menu. In the Options submenu, select Video. To the upper right side of the Video window, under Video Aspect Ratio, make sure that Default-Used Preferred Aspect Ratio is selected. In the lower half of the right side of the Video window, under Deinterlace Options, select BOB, use Odd lines first, and Stretch Image Vertically. Click OK. Close MaxTRAQ and reopen it to have the deinterlace options take effect. Next, open the *High Jump* video from within MaxTRAQ. Make sure the scaling/calibration tool is activated by clicking View on the menu bar, then selecting Tools from the drop down menu, and making sure Show Scale is checked. Open the scaling tool by clicking on Tools on the menu bar and select Scale. In the Scaling Tool window that opens, set the gauge length to 480 cm and click OK. Place the cursor over the lower left corner of the video window and click the left mouse button once, then place the cursor over the lower right corner of the video window and click the left mouse button a second time. The scale should appear in the video window. Hide the scale by selecting View in the menu bar; then click Tools and uncheck Show Scale.

   a. What is the vertical velocity of the high jumper at the instant when her takeoff foot touches the ground? Advance the video to two frames before touchdown of the left foot. Activate the digitizing function and digitize the anterior edge of the stripe on the jumper's shorts at the waistband. We'll use this point as a very rough estimate for the location of the jumper's center of gravity. Advance the video two frames to the instant when the jumper's foot first contacts the ground, and digitize this point again. Determine the vertical displacement of the jumper between these frames. Divide this displacement by the elapsed time (divide the displacement by the number of frames, 2, and multiply by 60) to compute vertical velocity.

   b. What is the vertical displacement of the high jumper from the instant her left foot touches the ground to the instant her left foot leaves the ground? Advance the video to the first frame when the left foot contacts the ground. Activate the digitizing function, and digitize the anterior edge of the stripe on the jumper's shorts at the waistband. Advance the video to the first frame when the jumper's left foot is no longer in contact with the ground, and digitize this point again. Determine the vertical displacement of the jumper between these two frames.

   c. What is the vertical velocity of the high jumper at the instant when her takeoff foot leaves the ground? Advance the video to the last frame when the jumper's left foot is in contact with the ground. Activate the digitizing function, and digitize the anterior edge of the stripe on the jumper's shorts at the waistband. Advance the video two frames and digitize this point again. Determine the vertical displacement of the jumper between these frames. Divide this displacement by the elapsed time (divide the displacement by the number of frames, 2, and multiply by 60) to compute vertical velocity.

   d. If the jumper's mass is 70 kg, what change in potential energy occurs from touchdown to takeoff?

   e. If the jumper's mass is 70 kg, what change in kinetic energy occurs as a result of her change in vertical velocity from touchdown to takeoff?

   f. How much work was done by the jumper in the vertical direction during the takeoff phase, from touchdown until takeoff of the left foot?

   g. Estimate the size of the average ground reaction force that acts on the jumper's left foot during the takeoff phase, from touchdown until takeoff of the left foot.

chapter 5

Torques and Moments of Force

Maintaining Equilibrium or Changing Angular Motion

objectives

When you finish this chapter, you should be able to do the following:

- Define torque (moment of force)

- Define static equilibrium

- List the equations of static equilibrium

- Determine the resultant of two or more torques

- Determine if an object is in static equilibrium, when the forces and torques acting on the object are known

- Determine an unknown force (or torque) acting on an object, if all the other forces and torques acting on the object are known and the object is in static equilibrium

- Define center of gravity

- Estimate the location of the center of gravity of an object or body

© Alexander Yakovlev/fotolia.com

You're sitting in the gym contemplating your next lat pull on the weightlifting machine when you notice that the stack of weight you're about to lift doesn't look heavy enough to be the 90 lb indicated by the label on the weights. But, when you sit down on the bench and pull down on the bar, it feels like 90 lb. How do the pulleys, levers, and cables of the machine make a relatively small stack of metal plates feel like 90 lb? The answer has to do with the torques created by the weight stack and those pulleys, levers, and cables. This chapter introduces the concepts of torque and center of gravity and also adds to the concept of static equilibrium.

This chapter is about torques. Torques cause changes in angular motion. The movements of our limbs at joints are controlled by the torques produced by muscles. Muscles create torques about joints, and these torques control and cause the movements of the limbs and the entire body. Torques are important even if you aren't moving. Equilibrium and balance are affected not only by forces but by torques as well. In this chapter, you will learn about torques, equilibrium of forces and torques, center of gravity, and stability. All of these concepts are related to torque in some way.

What Are Torques?

The turning effect produced by a force is called a **torque**. The torque produced by a force may also be called a **moment of force**. Occasionally, this term is simplified further and shortened to *moment*. One way to think of torque is to think of it as an angular or rotary force. To get a greater understanding of how forces produce torques, let's try self-experiment 5.1.

Self-Experiment 5.1

a. Place a book flat on a table or desk. Using two fingers (or a pencil), strike the book on its side to create a force directed through the center of gravity of the book (see figure 5.1a). (If you could balance the book on a pencil point, the book's center of gravity would lie vertically above this point of balance. A more complete discussion of center of gravity appears later in this chapter. For now, consider the center of gravity of the book as the center of the book.) What happens as a result of the force you exerted on the book? The book moves linearly. The net force acting on the book (your pushing force minus friction) was directed through the book's center

of gravity, and this net force caused a linear acceleration of the book. Little if any turning or rotation of the book occurred. Evidently, this type of force does not produce a turning effect or torque on the book.

b. Now try the experiment again, only this time hit the book with your fingers or a pencil so that the force is not directed through the book's center of gravity (see figure 5.1b). What happened? The linear motion of the book changed (the book's center of gravity translated), but the book rotated also. In this case, the net force acting on the book was not directed through the book's center of gravity, and this net force caused a linear acceleration of the book as well as a rotation of the book. This type of force did create a torque on the book.

c. Repeat the experiment a third time, but use both hands (or two pencils) this time. Strike the book on the top left side with the fingers of your left hand while simultaneously striking the book on the bottom of the right side with the fingers of your right hand (see figure 5.1c). Try to create the same size force against the book with each hand, but have these forces act in opposite directions. What happened? In this case, the book rotated, but its center of gravity barely moved, if at all. The forces acting on the book caused the book to rotate, but not translate. The combination of these two forces created a torque on the book.

⤵ **The turning effect produced by a force is called a torque.**

An examination of the results of these self-experiments leads to some generalizations. In self-experiment 5.1a, the resultant force acting on the book was directed through the center of gravity of the book. An external force directed through the center of gravity of an object is called a **centric force**. The effect of a centric force is

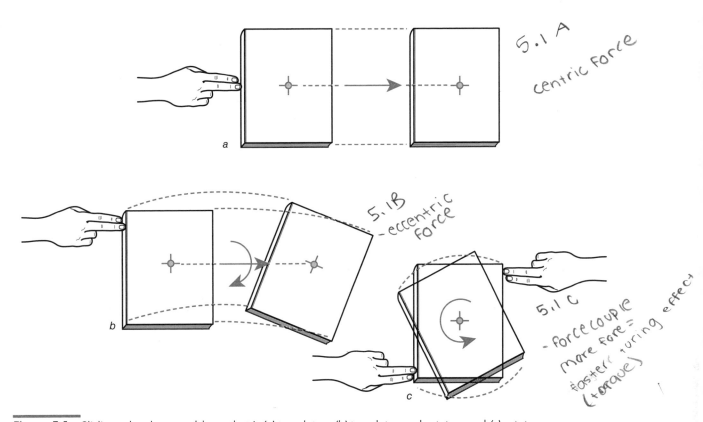

[handwritten: 5.1 A centric force]

[handwritten: 5.1 B - eccentric force]

[handwritten: 5.1 C - force couple more fore = faster (turning effect) (+ torque)]

Figure 5.1 Sliding a book on a table so that it *(a)* translates, *(b)* translates and rotates, and *(c)* rotates.

to cause a change in the linear motion of the object, as predicted by Newton's second law and as shown in this first experiment.

In self-experiment 5.1b, the resultant force acting on the book was not directed through the center of gravity of the book. An external force not directed through the center of gravity of an object is called an **eccentric force**. (Eccentric in this case refers to a type of force, not a type of muscular contraction.) The effect of an eccentric force is to cause a change in the linear and angular motions of an object. Both motions were observed in this second experiment. The change in linear motion is explained by Newton's second law. The torque produced by the eccentric force caused the book to rotate.

In self-experiment 5.1c, a pair of forces acted on the book. These forces were equal in size but opposite in direction and noncolinear. A pair of such forces is called a **force couple**. The effect of a force couple is to cause a change only in the angular motion of an object. The resultant of the two forces in a force couple is a force of zero, so according to Newton's first and second laws, no change in linear motion occurs, and none was observed in this third experiment. (Any translation of the book that occurred was small and due to the fact that the forces you exerted on the book were not exactly equal in size and opposite in direction.) The torque produced by the force couple caused the book to rotate.

In general, then, a centric force will cause or tend to cause a change in the linear motion of an object; an eccentric force will cause or tend to cause changes in the linear and angular motions of an object; and a force couple will cause or tend to cause only a change in the angular motion of an object. In this chapter we further examine the turning effect or torque produced by the eccentric force in the second case and by the force couple in the third case.

Mathematical Definition of Torque

What affects the size of a torque and how is it quantified? Intuitively, you would think that the size of the force that produces the turning effect would influence the size of the torque. To verify this, try self-experiment 5.1b again several times and increase the size of the eccentric force you exert on the book each time. As you increased the size of the force, the turning effect increased, as evidenced by the increase in the amount of rotation caused. Do the same with self-experiment 5.1c. Try it again several times and increase the size of the forces in the force couple each time. Again, as the size of the forces in the force couple increased, the amount of rotation increased, so the turning effect increased. Intuition was correct in this case; torque is directly related to the size of the force that creates it.

The torque produced by a force does not depend on the force alone, however. If this were the case, a turning effect would have been produced by the centric force you exerted on the book in self-experiment 5.1a. The only difference between self-experiments 5.1a and 5.1b was the line of action of the force you applied to the book. What else besides force affects torque? Self-experiment 5.2 may help answer this question.

Self-Experiment 5.2

Now do self-experiment 5.1a again. Strike the book so that the force your fingers create is directed through the center of gravity of the book. Do it again, only this time direct the force so its line of action is just off center, causing the force to be an eccentric force. Repeat the experiment a number of times; each time, strike the book so the line of action of the force is farther and farther away from the center of gravity of the book. Try to keep the size of the force the same in each trial. What happens? As you strike the book farther and farther from its center of gravity, the torque created by the force becomes larger and larger, causing the book to rotate more and more.

To reinforce this concept, try self-experiment 5.1c again. Apply a pair of forces (a force couple) to the book at either end of the book and in opposite directions. Do it again, only this time move your hands so the distance between them and the forces they apply to the book is not as great. Repeat this a number of times, so that each time the lines of application of the forces move closer and closer together until finally the forces are colinear. What happens? As the distance between the lines of action of each of the forces gets shorter and shorter, the torque produced by the force couple gets smaller and smaller and disappears completely when the forces become colinear.

Torque is influenced by the position and orientation of the line of action of the force as well as by its size. The torque produced by a force is directly proportional to the size of the force as well as the distance between the line of action of the force and the point about which the object tends to rotate (the axis of rotation). In the case of the force couple, the turning effect is again directly proportional to the size of the forces and the distance between the lines of action of these forces. Now we can expand our definition of torque. A torque is the turning effect produced by a force and is equal to the product of the magnitude of the force and the distance between the line of action of the force and the axis of rotation of the object (or the axis about which the object will tend to rotate). This distance between the line of action of the force and the axis of rotation is the perpendicular distance between the line of action of the force and a line parallel to it that passes through the axis of rotation. This distance

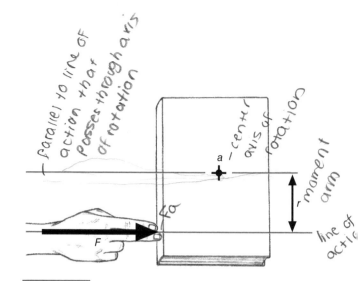

Figure 5.2 The moment arm (r) of a force (F) is the perpendicular distance between the line of action of the force and a parallel line passing through the axis of rotation (a).

is sometimes referred to as the perpendicular distance and is more often termed the **moment arm** of the force. Figure 5.2 shows the force and the moment arm of the force for self-experiment 5.1b.

Mathematically, then, torque is defined as follows:

$$T = F \times r \tag{5.1}$$

where

T = torque (or moment of force),

F = force, and

r = moment arm (or perpendicular distance).

The units of measurement for torque are thus units of force (newtons in the SI system) times units of length (meters in the SI system). So torque is measured in newton meters, which are abbreviated Nm.

To fully describe a torque, you must describe the size of the torque, the axis about which the turning effect is created, and the sense of the turning effect about that axis (clockwise or counterclockwise). So torque is a vector quantity, because the turning effect is around a specific axis that is directed in a specific direction. As with forces, once a direction is specified (or axis, in the case of a torque), a positive (+) or negative (−) sign is used to indicate the sense of the force or torque along (or around) that line (or axis). The conventional approach is to indicate counterclockwise torques as positive (+) and clockwise torques as negative (−). As with forces, torques that act about the same axis may be added or subtracted algebraically. Before getting further into the mathematics, let's look at examples of how we use torques.

Examples of How Torques Are Used

First, let's consider a door. How do you open or close a door? You probably grab its handle (or doorknob) and

Axis of
rotation =
corner hinges

push or pull. This force that you exert on the door creates a torque that acts about the axis of rotation of the door, an axis passing vertically through the door's hinges. This torque causes the door to start swinging open or closed, as shown in figure 5.3. To experience the effect of the length of the moment arm in the door example, try self-experiment 5.3.

Self-Experiment 5.3

Why do you suppose doorknobs or door handles are located on the opposite side of the door from the hinges? This location makes the moment arm, r in the torque equation, large so that the force required to create a large enough torque to swing open the door is small. Try opening or closing the door by pushing on it with a force directed closer to the hinges. You have to push with a greater force to create the same torque. As you move your hand closer and closer to the hinges, the force required to move the door gets bigger and bigger, because the moment arm of the force is getting smaller and smaller.

A given size torque can be created with a large force and a small moment arm or with a small force and a large moment arm. Because the amount of force humans can exert is generally limited, we use large moment arms when we want to create large torques. How do the tools shown in figure 5.4 increase the torque we are capable of producing?

All of these tools have handles that increase the length of the moment arm of the force, thus increasing the torque applied to the screw, nut, and so on. In the case of the wrench or pliers, holding these tools farther out on the handle increases the torque in the same way, by increasing the moment arm of the force. Other everyday things we turn and thus apply torques to include steering wheels (why do heavy trucks have larger-diameter steering wheels than cars?), bicycle handlebars, jar tops, knobs and dials on appliances, bicycle cranks, light switches, clothespins, and staplers.

How is torque used in sport? In rowing, canoeing, and kayaking, torque is applied by the athlete to the oar or paddle to cause it to rotate. In golf, baseball, and tennis, torque is applied to the club, bat, or racket to swing these implements. In any sport in which we turn, spin, or swing something (including our bodies), torque must be created to initiate these turns, spins, and swings. The holds used in wrestling provide great examples of torques that are used to turn an opponent. Consider the half nelson, as shown in figure 5.5.

In this hold, your opponent is prone, and you try to turn him over onto his back by putting your hand under his shoulder and onto the back of his head. If you then push down on his head with your hand and use your arm to lift up under his shoulder, you produce a force couple that creates a torque about a longitudinal axis through your opponent. This torque creates a turning effect that tends to turn your opponent over. To counter this, your opponent could create a torque with his other arm by abducting it so that it is perpendicular to his body. Pushing down onto the mat with this arm and hand creates a torque about the same axis but in the opposite direction.

Muscular Torque

The examples given so far are examples of external torques that act on the body or other objects. What about within the body? What creates the turning effects that cause our limbs to rotate about our joints? Muscles create the torques that turn our limbs. A muscle creates a force that pulls on its points of attachment to the skeletal system when it contracts. The line of action (or line of pull) of a muscle force is along a line joining its attachments and is usually indicated by the direction of its tendons. The bones that a muscle attaches to are within the limbs on either side of a joint, or two or more joints in some cases. When a muscle contracts, it creates a pulling force

External torques

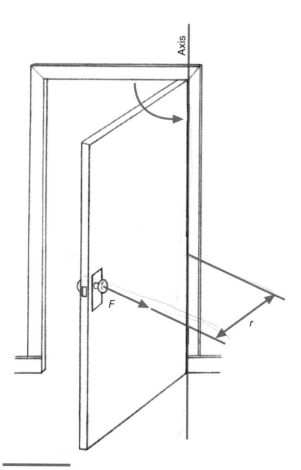

Figure 5.3 The torque created by the pulling force on the doorknob causes the door to swing open.

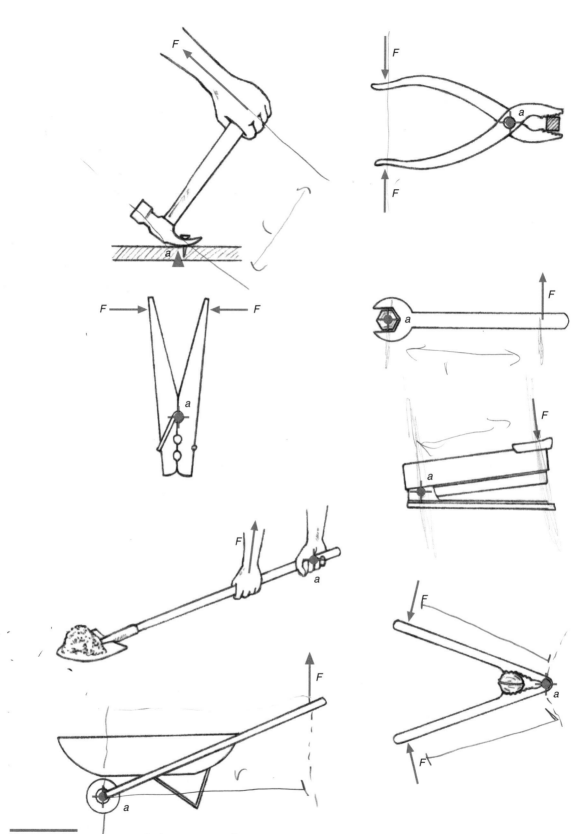

Figure 5.4 Common tools that you exert forces (*F*) on, thus creating torques around the axes (*a*).

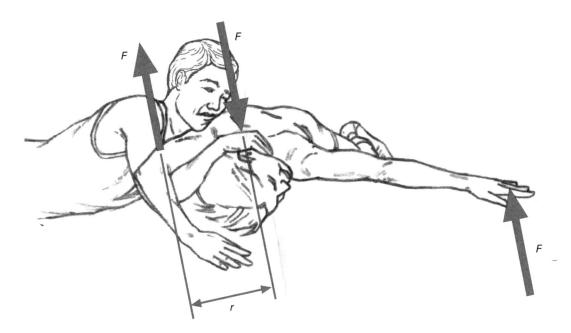

Figure 5.5 The forces exerted by a wrestler using a half nelson on his opponent create a torque that tends to turn the opponent over. The opponent can counter this torque by pushing down on the mat with his outstretched arm.

on these limbs. Because the line of action of the muscle force is some distance from the joint axis, a moment arm exists, and torques about the joint axis are produced by the muscle force on the limbs on either side of the joint where the muscle attaches. The torque produced by the muscle on the distal limb will tend to rotate that limb in one direction about an axis through the joint, and the torque produced by the muscle on the proximal limb will tend to rotate that limb in the opposite direction about the same axis. Figure 5.6 shows how the force produced by the biceps brachii muscle creates a torque that tends to rotate the forearm around the elbow joint.

What happens to the torque on the forearm produced by the biceps brachii muscle as the forearm is moved from full extension to 90° of flexion at the elbow joint? Can the muscle create the same torque throughout this range of motion? If the muscle produces the same force and the moment arm of the muscle stays the same throughout the range of motion, the torque produced will not change. Let's try self-experiment 5.4 to see if the moment arm of the muscle stays the same length.

Self-Experiment 5.4

Hold a heavy book in your hand and extend your right forearm. Now use the thumb and index finger of your left hand to pinch your right elbow. Your index finger should be able to feel the tendon of your right biceps brachii, and your thumb should feel the back of your elbow, the olecranon process of your ulna. Now flex your forearm at the elbow. Does the distance between your thumb

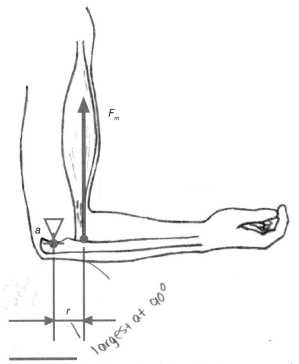

Figure 5.6 The biceps brachii exerts a torque around the axis of the elbow joint by producing a force (F_m) with a moment arm (*r*) around the joint.

and forefinger change? It gets bigger as the degree of elbow flexion approaches 90°. This distance is a crude measurement of the moment arm of the biceps brachii muscle about the elbow joint. The moment arm of the biceps brachii is largest when the elbow is at 90° and gets smaller as the elbow is flexed or extended away from this position. Figure 5.7 shows how the moment arm of the biceps brachii muscle changes with elbow joint position.

The results of self-experiment 5.4 show that the biceps brachii's ability to create a torque about the elbow joint is dependent on the position of the elbow joint because the moment arm of the muscle changes as the elbow is flexed and extended. A similar situation exists for most of our muscles and the joints they cross. Changing the angle at the joint changes the moment arm of the muscles that cross that joint. This partially explains why our muscles are apparently stronger in some joint positions than others.

Strength Training Devices and Torque

Strength training exercises with free weights or weight-lifting machines provide many examples of torques that are exerted on our limbs by nonmuscular forces. Consider an arm curl exercise. In this exercise, you hold a dumbbell in your hand and lift it by flexing at the elbow joint. We already saw how contraction of the biceps brachii muscle produces a torque around the elbow joint. As you lift the

dumbbell, its weight produces a torque around the elbow joint as well, but this torque tends to rotate the forearm in the opposite direction. It creates a torque that would tend to extend your forearm at the elbow joint.

We already saw how the biceps brachii torque could change as the arm moved through its range of motion at the elbow joint, because the moment arm of the biceps brachii changed with changing joint position. What happens to the torque produced around the elbow joint by the dumbbell when an arm curl exercise is performed? The dumbbell doesn't get heavier, but the torque gets larger as the elbow is flexed. This occurs because the line of action of the dumbbell's weight moves farther from the elbow joint as the exercise is performed, thus increasing the moment arm. When the forearm is horizontal and the elbow is at 90°, the moment arm of the dumbbell is at maximum and the torque is greatest. Flexion beyond this position results in a smaller moment arm and thus a smaller torque, as does extension beyond this position. For most exercises involving free weights, the torques produced by the weights vary as the moment arms of these weights change during the movement.

Weightlifting machines may or may not have this characteristic. To analyze the torques a weightlifting machine exerts on you, you must first identify the resistance force. Often cables or chains are used to redirect the line of action of the force of gravity acting on the weight stack. The direction of this cable (where it attaches to the arm or cam of the part of the machine you actually move) indicates the line of action of the resistance force. You would then determine the moment arm by measuring the perpendicular distance between this line of action and the axis of rotation of the arm or cam of the device.

Let's examine a leg extension machine, which is common in many weight training facilities (see figure 5.8). The resistance force is provided by a stack of weights. This force is then transmitted to the arm of the machine via a cable attached to the middle of the arm. As the exercise is performed, the cable pulls on the machine arm. A torque is created about the axis of the machine arm because the cable pulls some distance away from this axis. As the exercise is performed, the line of action of the cable force changes, and the moment arm gets smaller as the leg reaches full extension at the knee. The resistance torque produced by this machine is thus largest at the starting position with the knee at 90° and gets smaller as the leg extends. Could you think of a way to redesign this machine so that the torque remains constant throughout the exercise? The design goal of some exercise machines on the market is to provide a constant resistance torque throughout the range of motion of the exercise. Other machines are designed to provide a resistance torque that varies in proportion to the changes in the moment arm of the muscle being exercised as the exercise is performed.

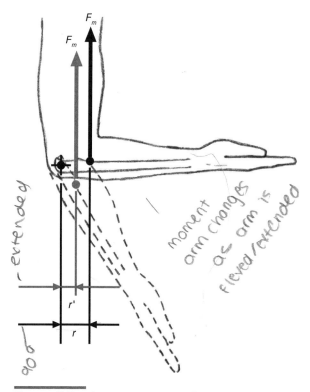

Figure 5.7 The moment arm of the biceps brachii muscle decreases from *r* to *r'* as the elbow extends from 90°.

In these machines, the design goal is to have the muscle produce a constant force throughout the range of motion of the exercise. This is the premise behind the design of the Nautilus-type weight machines.

By now you should have an understanding of what a torque is and how it is quantified. The fact that torques can be added (and subtracted) was briefly noted, but we have not discussed the effects that multiple torques have when acting on the same object. Can torques be added together to form an equivalent net torque? What if the torques act in opposite directions? Must there be equilibrium of torques (as is the case for forces) for an object to be in equilibrium?

Forces and Torques in Equilibrium

In chapter 1, the concept of static equilibrium was introduced during the discussion of forces. If an object is at rest, it is described as being in a state of static equilibrium. For an object to be in static equilibrium, the external forces acting on it must sum to zero (i.e., the net external force must equal zero). Our experiment with a force couple acting on a book earlier in this chapter indicated that a net force of zero is not the only condition of static equilibrium. A net force of zero ensures that no change will occur in the linear motion of an object, but it does

not constrain the object's angular motion. The net torque acting on an object must be zero to ensure that no changes occur in the angular motion of the object. For an object to be in static equilibrium, the external forces must sum to zero and the external torques (about any axis) must sum to zero as well. Mathematically, these conditions are expressed as

$$\sum F = 0$$

$$\sum T = 0 \qquad (5.2)$$

where

$\sum F$ = net external force and

$\sum T$ = net torque.

> For an object to be in static equilibrium, the external forces must sum to zero and the external torques (about any axis) must sum to zero as well.

Net Torque

Earlier in this chapter, when we came up with a mathematical definition of torque, we also said that torques that acted around the same axis could be added or subtracted algebraically, just like forces that act in the same

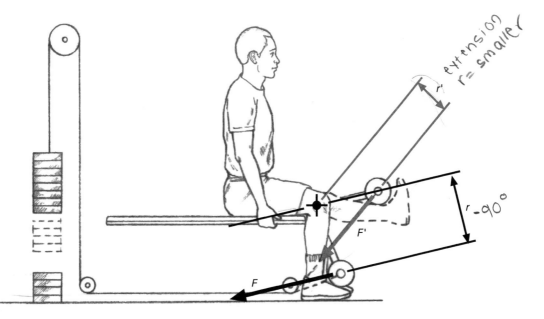

Figure 5.8 A leg extension machine. The torque varies with position due to the change in the size of its moment arm (*r*).

Jeff is pushing on a door with a horizontal force of 200 N. The moment arm of this force around the hinges of the door is 60 cm. Ted is pushing in the opposite direction on the other side of the door. The moment arm of his pushing force is 40 cm. How large is the force that Ted pushes with if the door is in static equilibrium?

Solution:

Step 1: List the known quantities.

$F_j = 200$ N

$r_j = 60$ cm

$r_t = 40$ cm

Step 2: Identify the unknown variable to solve for.

$F_t = ?$

Step 3: Draw a free-body diagram.

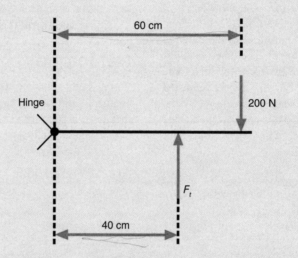

Step 4: Search for equations with the known and unknown variables.

$\sum T = 0$

Measure torques about the hinge joint.

$\sum T = 0 = (F_j)(r_j) + (F_t)(r_t) = 0$

$\sum T = 0 = 200(60) + F(40)$

$\sum T = 0 = 12000 + F(40)$

Step 5: Substitute known quantities and solve the equation for the unknown quantity.

$\sum T = 0 = (200 \text{ N})(60 \text{ cm}) + (F_t)(40 \text{ cm}) = 0$

$-F = \dfrac{12000}{40}$

$F_t = -(200 \text{ N})(60 \text{ cm})/(40 \text{ cm}) = -300$ N

$= -300$ N

The negative sign indicates that Ted's force is in the opposite direction of Jeff's force.

opposite direction of other force

Step 6: Common sense check.
The force Ted exerts should be larger than the force Jeff exerts due to Ted's smaller moment arm.

direction. In a planar situation, then, we compute a net torque by summing the torques that act on an object. To investigate net torques, try self-experiment 5.5.

Self-Experiment 5.5

For this experiment and the examples that follow, let's use a simple system. Find a ruler, a rubber eraser, and 10 same-denomination coins—we'll use pennies as our example. Balance the ruler on the edge of the eraser. If you don't have an eraser, find something with a flat surface about 1/4 in. (0.6 cm) wide that you can balance the ruler on. If the ruler is 12 in. (30 cm) long it probably balances with the eraser at 6 in. (15 cm). Now place a penny on the ruler 5 in. (13 cm) to the left of the eraser. Does the ruler stay balanced (is it in a state of static equilibrium)? No. Why not?

The penny in self-experiment 5.5 created a counterclockwise torque about the eraser that caused the ruler to rotate counterclockwise. The moment arm of the penny was 5 in. The force created by the penny was the weight of the penny. The torque created about the eraser was one pennyweight ("p") times 5 in., or 5 penny inches of torque. In this case, the net torque acting on the ruler about an axis through the eraser was caused only by the weight of the penny.

$$\sum T = \sum(F \times r)$$

$$\sum T = (-1 \text{ penny})(-5 \text{ in.})$$

$$\sum T = +5 \text{ penny inches (p·in.)}$$

This torque is positive because it tended to cause a rotation of the ruler in the counterclockwise direction.

Now place a penny on the right side of the ruler 3 in. (about 8 cm) from the eraser. What is the net torque acting on the ruler about an axis through the eraser in this case?

The net torque is

$$\sum T = \sum(F \times r)$$

$$\sum T = (-1 \text{ p})(-5 \text{ in.}) + (-1 \text{ p})(+3 \text{ in.})$$

$$\sum T = +5 \text{ p·in.} + (-3 \text{ p·in.})$$

$$\sum T = +5 \text{ p·in.} - 3 \text{ p·in.}$$

$$\sum T = +2 \text{ p·in.}$$

The net torque is still counterclockwise (+), but it has been reduced to 2 p·in. by the clockwise torque created by the penny on the right side of the ruler.

Now place another penny on the right side of the ruler 2 in. (5 cm) to the right of the eraser. What happens? The ruler balances on the eraser. It is in static equilibrium.

This seems odd, because there are twice as many pennies on the right side of the ruler. Maybe an analysis of the free-body diagram shown in figure 5.9 will help explain the situation.

What is the net torque acting on the ruler about an axis through the eraser? The net torque is

$$\sum T = \sum(F \times r)$$

$$\sum T = (-1 \text{ p})(-5 \text{ in.}) + (-1 \text{ p})(+3 \text{ in.}) + (-1 \text{ p})(+2 \text{ in.})$$

$$\sum T = +5 \text{ p·in.} + (-3 \text{ p·in.}) + (-2 \text{ p·in.})$$

$$\sum T = +5 \text{ p·in.} - 3 \text{ p·in.} - 2 \text{ p·in.}$$

$$\sum T = 0$$

The ruler balances (it is in static equilibrium) because the net torque acting on it is zero.

Let's try another example. Clear the pennies off the ruler. Now, stack four pennies on the ruler 3 in. (about 8 cm) to the right of the eraser. What net torque do these pennies create about the eraser? Four pennies times 3 in. is 12 p·in. of torque in the clockwise direction about an axis through the eraser. If you had only two pennies left to use to balance the ruler, where would you stack them?

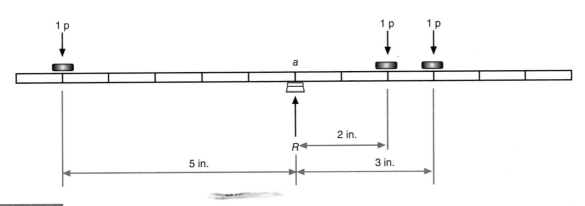

Figure 5.9 Free-body diagram of a ruler balanced on an eraser with pennies on the ruler. *R* is the reaction force exerted by the support of the ruler, and *a* is the axis. (The weight of the ruler is omitted in this diagram.)

Because we now know that the net torque has to be zero if the ruler is balanced (and in static equilibrium), we can solve this problem mathematically as follows:

$$\sum T = \sum (F \times r) = 0$$

$$0 = (-2 \text{ p})(r) + (-4 \text{ p})(+3 \text{ in.})$$

$$(-2 \text{ p})(r) = (4 \text{ p})(+3 \text{ in.})$$

$$r = \frac{(4 \text{ p})(3 \text{ in.})}{-2 \text{ p}} = -6 \text{ in.}$$

Two pennies will create a counterclockwise torque of 12 p·in. (to counter the 12 p·in. of clockwise torque created by the four pennies) if you stack them 6 in. to the left of the eraser (the negative sign before the 6 in. indicates that the moment arm is to the left of the axis). Two pennies times 6 in. equals 12 p·in. of torque in the counterclockwise direction.

Muscle Force Estimates Using Equilibrium Equations

Now let's examine how the conditions for static equilibrium allow us to estimate the forces our muscles produce to lift or hold up objects. Suppose you are holding a 20 lb (9 kg) dumbbell in your hand and your elbow is flexed 90° so that your forearm is parallel to the floor. If the moment arm of this dumbbell is 12 in. (30 cm) about the elbow joint axis, what torque is created by this dumbbell about your elbow joint axis? Using equation 5.1, the torque is

$$T = F \times r$$

$$T = (-20 \text{ lb})(-12 \text{ in.})$$

$$T = +240 \text{ lb·in.} \quad \text{~about elbow joint counter clockwise}$$

The dumbbell creates a torque of 240 lb·in. about the elbow joint axis in the counterclockwise direction.

To hold up this dumbbell, your elbow flexor muscles must create a clockwise torque equal to the counterclockwise torque created by the dumbbell (if the weight of the forearm and hand are ignored). If the moment arm of these muscles is 1 in. (2.5 cm), what force must they pull with to hold the dumbbell in the position described? Because the forearm must be in static equilibrium, the problem is solved as follows (let F_m represent the muscle force):

$$\sum T = \sum (F \times r) = 0 \quad \text{muscle force}$$

$$\sum (F \times r) = (-20 \text{ lb})(-12 \text{ in.}) + F_m (-1 \text{ in.}) = 0$$

$$F_m (-1 \text{ in.}) = -(-20 \text{ lb})(-12 \text{ in.})$$

$$F_m = \frac{-(-20 \text{ lb})(-12 \text{ in.})}{-1 \text{ in.}} = +240 \text{ lb}$$

Wow! Your elbow flexor muscles must create a force of 240 lb (about 1068 N) just to hold up a 20 lb dumbbell! This seems to be too large a force, but our muscles are arranged so that their moment arms about the joints are short. This means that they must create relatively large forces to produce practically effective torques about the joints.

More Examples of Net Torque

Now let's look at some examples of net torques in sport. Consider a pole-vaulter. What external forces act on a pole-vaulter? Gravity pulls downward on the vaulter with a force equal to his weight, and the pole exerts forces on each of the vaulter's hands where he grips it. Figure 5.10 shows a free-body diagram of the pole-vaulter with values of the external forces estimated as well.

What net torque acts on this vaulter around an axis through his center of gravity? Is the vaulter in equilibrium? The 500 N force acting at the vaulter's left hand has a moment arm of 0.50 m about his center of gravity. This force creates a clockwise torque about the vaulter's center of gravity. The 1500 N force acting at the vaulter's right hand has a moment arm of 1.00 m about his center of gravity. The torque created by this force about the vaulter's center of gravity is also clockwise. The vaulter's weight of 700 N acts through his center of gravity, and thus the moment arm of this weight is zero, so it does not create any torque about the vaulter's center of gravity.

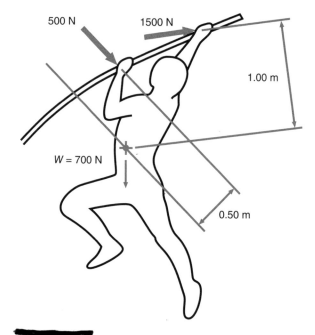

500 N

1500 N

1.00 m

W = 700 N

0.50 m

Figure 5.10 Free-body diagram of a pole-vaulter just after takeoff. The reaction forces from the pole create clockwise torques on the vaulter.

Mathematically, the net torque about a transverse axis through the vaulter's center of gravity is

$$\sum T = \sum(F \times r) = (-500 \text{ N})(0.50 \text{ m})$$
$$- (1500 \text{ N})(1.00 \text{ m})$$

$$\sum T = -250 \text{ Nm} - 1500 \text{ Nm}$$

$$\sum T = -1750 \text{ Nm}$$

The negative sign on this net torque indicates that it acts in a clockwise direction. The 1750 Nm net torque produces a turning effect that tends to rotate the vaulter clockwise onto his back, as if he were doing a backward somersault.

Now let's consider what happens to the vaulter later in the vault, when he is in the position shown in the free-body diagram in figure 5.11.

What is the net torque acting on the vaulter in this situation? The 300 N force acting on the vaulter's left hand has a moment arm of 0.50 m and still creates a clockwise (−) torque about the vaulter's center of gravity. The 500 N force acting on the vaulter's right hand also has a moment arm of 0.50 m, but it now creates a counterclockwise (+) torque about the vaulter's center of gravity. The net torque is

$$\sum T = \sum(F \times r) = -(300 \text{ N})(0.50 \text{ m})$$
$$+ (500 \text{ N})(0.50 \text{ m})$$

$$\sum T = -150 \text{ Nm} + 250 \text{ Nm}$$

$$\sum T = +100 \text{ Nm}$$

The positive sign on this net torque indicates that it acts in the counterclockwise direction. The turning effect produced by the forces acting on the vaulter thus tends to rotate the vaulter counterclockwise, as if he were doing a forward somersault or to slow down the vaulter's clockwise rotation. The latter is likely the case, since the pole vaulter was rotating clockwise earlier in the vault. This counterclockwise torque will stop the clockwise rotation as the vaulter lines up his or her body with the pole in preparation for going over the crossbar. It will eventually cause the vaulter to rotate counterclockwise. This counterclockwise rotation is helpful as the vaulter's body rotates over the crossbar during the clearance.

What Is Center of Gravity?

By now you should understand what a torque is and how it is created by external forces as well as muscular forces, how to determine a net torque, and what the conditions of static equilibrium are. The expression "center of gravity" has been used several times in this chapter. You probably have heard this expression before and already have some ideas about what it means. In this section, the concept of center of gravity is defined and explained.

Center of gravity is the point in a body or system around which its mass or weight is evenly distributed or balanced and through which the force of gravity acts. Center of mass is the point in a body or system of bodies at which, for certain purposes, the entire mass may be assumed to be concentrated. For bodies near the earth, this coincides with the center of gravity. Because all of the human movement activities we are concerned with occur on or near the earth, center of gravity and center of mass are equivalent terms and may be used interchangeably. The center of gravity is an imaginary point in space. It is not a physical entity; its location is not marked on an object. The center of gravity is a useful concept for analysis of human movement because it is the point at which the entire mass or weight of the body may be considered to be concentrated. So the force of gravity acts downward through this point. If a net external force acts on a body, the acceleration caused by this net force is the acceleration of the center of gravity. If no external forces act on an object, the center of gravity does not accelerate. When we are interpreting and applying Newton's laws of motion, it is the center of gravity of a body whose motions are ruled by these laws. It is thus important to know how to locate or estimate the location of the center of gravity of a body or object.

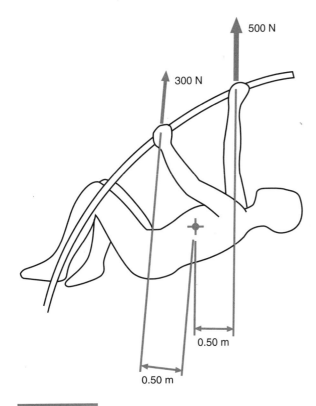

500 N

300 N

0.50 m

0.50 m

Figure 5.11 Free-body diagram of a pole-vaulter midway through the vault. The reaction forces from the pole create torques in opposite directions on the vaulter.

⟳ The center of gravity is the point at which the entire mass or weight of the body may be considered to be concentrated.

Locating the Center of Gravity of an Object

The definition of center of gravity gives an indication of how to find its location: The center of gravity is the point in the body around which its weight is balanced. The center of gravity is the point of balance. What does this mean? Every object can be considered to be composed of many smaller elements. These smaller elements may represent the parts that make up the object. In the human body, these elements might be represented by the limbs, trunk, and head (i.e., two hands, two forearms, two upper arms, two feet, two shanks, two thighs, one torso, one neck, and one head make up the human body). At the most basic level, these elemental parts could represent molecules or atoms. The force of gravity pulls downward on each of these smaller elements. The sum or resultant of these forces is the total weight of the object. This weight acts through a point about which the moments of the weights of each of the elemental parts (the torques created by their weights) sum to zero, no matter what position the object is in. Self-experiment 5.6 demonstrates variations in center of gravity using the ruler, eraser, and pennies you gathered earlier in this chapter.

Self-Experiment 5.6

Take out the ruler, eraser, and pennies again. First, balance the ruler on the edge of the eraser. By our definition of center of gravity as "the point of balance," the center of gravity of the ruler must lie above the points of support provided by the edge of the eraser. The counterclockwise torque created by the weight of the ruler to the left of the eraser balances the clockwise torque created by the weight of the ruler to the right of the eraser. Now, make two stacks of pennies with four pennies in each stack. Place one stack 1 in. (2.5 cm) to the left of the eraser and the other stack 1 in. to the right of the eraser so that the ruler remains balanced. The center of gravity of the pennies and the ruler is still above the eraser.

Now slide the right stack of pennies 1 in. to the right and move the eraser so that the ruler remains in equilibrium. Which way did you have to move the eraser? Which way did the center of gravity move? To keep the ruler balanced, you had to move the eraser to the right a little. The center of gravity of the ruler and pennies also moved to the right.

Now, move the right stack of pennies all the way to the right end of the ruler and move the eraser so that the ruler remains balanced. You again had to move the eraser to the right, so the center of gravity of the system (ruler and pennies) moved to the right. You didn't have to move the eraser as far as you had to move the pennies, though. If some of the elemental parts of an object move or change position, the center of gravity of the object moves as well, although not as far.

Now, take three pennies off of the left stack and move the eraser so that the system remains in balance. You had to move the eraser to the right again, so the center of gravity moved to the right as well. If an elemental part of an object is removed from the object, the center of gravity of the object moves away from the point of removal.

Now, add three pennies to the right stack and move the eraser so that the ruler remains balanced. The eraser is now farther to the right, and the center of gravity of the system has also moved farther to the right. If mass is added to an object, the center of gravity of the object moves toward the location of the added mass. Observe how many pennies are on either side of the center of gravity of the system now. The weights on either side of the center of gravity do not have to be equal—one penny does not equal seven pennies—but the torques created by these weights about the center of gravity must equal each other.

Mathematical Determination of the Center-of-Gravity Location

If the weights and locations of the elemental parts that make up an object are known, the center-of-gravity location can be computed mathematically. The definition of center of mass indicated that it is the point at which the entire mass (or weight) may be considered to be concentrated. By this definition, a ruler with six pennies distributed on it at 2 in. (5 cm) intervals is equivalent to a ruler with six pennies stacked on it at one location if that location is the center of gravity of the first ruler. Let's look more closely at this. Suppose the first ruler had pennies placed on it at the 1, 3, 5, 7, 9, and 11 in. marks. This ruler and its equivalent are shown in figure 5.12.

If you closed your eyes and picked up one of these rulers by the end, and then the other ruler by its end, they would feel identical. You wouldn't be able to tell which ruler had the pennies distributed on it at 2 in. intervals and which ruler had the pennies all stacked in one place. The rulers would weigh the same, and you would have to create the same torque on the end of each ruler to hold it up with one hand. This is the key to determining the center-of-gravity location mathematically—the two rulers create the same torque about their ends. The sum of the moments of force created by each of the elemental weights, the pennies in this case, equals the moment of force created by the total weight stacked at

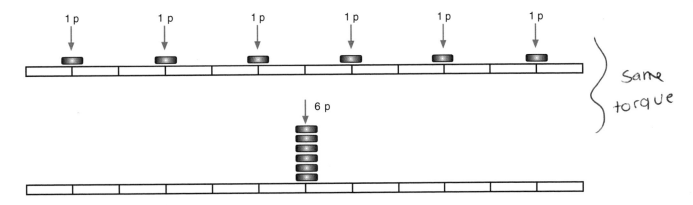

Figure 5.12 A ruler with six pennies on it, one placed every 2 in., feels the same and is equivalent to a ruler with six pennies stacked at the center of the ruler.

[handwritten: same torque]

the center-of-gravity location. Mathematically, this can be expressed as

[handwritten arrow]
$$\sum T = \sum(W \times r) = (\sum W) \times r_{cg} \qquad (5.3)$$

where

    W = the weight of one element,

    r = the moment arm of an individual element,

    $\sum W$ = the total weight of the object, and

    r_{cg} = the moment arm of the entire weight of the object (the location of the center of gravity of the object relative to the axis about which the moments of force are being measured).

Equation 5.3 can be used to solve for the location of the center of gravity, r_{cg}. Let's use the end of the first ruler shown in figure 5.12 as the axis about which the moments of force will be measured (any axis may be used in this calculation, so long as the same one is used on either side of the equation). *[handwritten: To find location of COG]*

$$\sum(W \times r) = (\sum W) \times r_{cg}$$

$$\sum(W \times r) = (1\ p)(1\ in.) + (1\ p)(3\ in.) + (1\ p)(5\ in.)$$
$$+ (1\ p)(7\ in.) + (1\ p)(9\ in.) + (1\ p)(11\ in.)$$ *[handwritten: distance × #]*

$$\sum(W \times r) = 36\ p{\cdot}in.$$

$$(\sum W) \times r_{cg} = (1\ p + 1\ p + 1\ p + 1\ p + 1\ p + 1\ p) \times r_{cg}$$
$$= (6\ p) \times r_{cg}$$

$$\sum(W \times r) = 36\ p{\cdot}in. = (6\ p) \times r_{cg} = (\sum W) \times r_{cg}$$

$$\frac{36\ p{\cdot}in.}{6\ p} = r_{cg}$$

$$r_{cg} = 6\ in.$$

[handwritten: $36 p{\cdot}in = 6p \times r_{cg}$; $\frac{6p}{6p}$; $r_{cg} = 6 in$]

So the center of gravity of the ruler is 6 in. from the end of the ruler. If all six pennies were stacked at this point rather than distributed across the ruler at 2 in. intervals, this ruler would feel the same as the first ruler.

To determine the location of the center of gravity mathematically, we use the relationship between the sum of the moments of force created by the elemental weights and the moment of force created by the sum of the elemental weights (i.e., they are equal). Stated more simply, the sum of the moments equals the moment of the sum. Try self-experiment 5.7.

Self-Experiment 5.7

Let's try another example. Take out the ruler, eraser, and pennies again. Place three pennies on the ruler at the 1 in. (2.5 cm) mark and seven pennies on the ruler at the 8 in. (20 cm) mark. Without picking up the ruler, can you determine where its center of gravity is?

Mathematically, we can solve for the location of the center of gravity as we did in the previous example. Let's measure the moments about the end of the ruler for convenience (remember, we can measure these moments about any axis, so we choose an axis that is convenient). *[handwritten: add up pennies (objects)]*

$$\sum(W \times r) = (\sum W) \times r_{cg}$$

$$\sum(W \times r) = (3\ p)(1\ in.) + (7\ p)(8\ in.)$$

$$= (3\ p + 7\ p) \times r_{cg} = (\sum W) \times r_{cg}$$

$$3\ p{\cdot}in. + 56\ p{\cdot}in. = (10\ p) \times r_{cg}$$

$$59\ p{\cdot}in. = (10\ p) \times r_c$$

$$\frac{59\ p{\cdot}in.}{10\ p} = r_{cg}$$

$$r_{cg} = 5.9\ in.$$

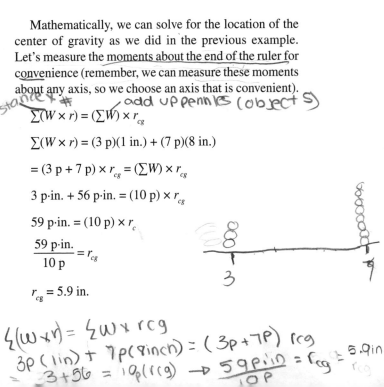

[handwritten: $\sum(W \times r) = \sum W \times r_{cg}$; $3p(1in) + 7p(8inch) = (3p + 7p)\ r_{cg}$; $3 + 56 = 10p(r_{cg}) \rightarrow \frac{59p{\cdot}in}{10p} = r_{cg} = 5.9in$]

A weightlifter has mistakenly placed a 20 kg plate on one end of the barbell and a 15 kg plate on the other end of the barbell. The barbell is 2.2 m long and has a mass of 20 kg without the plates on it. The 20 kg plate is located 40 cm from the right end of the barbell, and the 15 kg plate is located 40 cm from the left end of the barbell. Where is the center of gravity of the barbell with the weight plates on it?

Solution:

Step 1: Draw a free-body diagram of the barbell and plates.

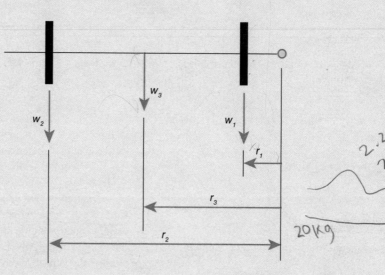

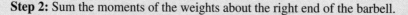

Step 2: Sum the moments of the weights about the right end of the barbell.

$$\sum T = W_1 r_1 + W_2 r_2 + W_3 r_3 = (m_1 g) r_1 + (m_2 g) r_2 + (m_2 g) r_3$$

$$\sum T = g\, (m_1 r_1 + m_2 r_2 + m_2 r_3) = g\, [(20\ kg)(0.4\ m) + (15\ kg)(1.8\ m) + (20\ kg)(1.1\ m)]$$

$$\sum T = g\, (57\ kg{\cdot}m)$$

Step 3: Equate this to the moment of the total weight about the right end of the barbell and solve for r_{cg}.

$$\sum T = W_{total}\, r_{cg} = (m_{total} g) r_{cg} = g\, (57\ kg{\cdot}m)$$

$$g\, (55\ kg)\, r_{cg} = g\, (57\ kg{\cdot}m)$$

$$r_{cg} = (57\ kg{\cdot}m)/55\ kg$$

$$r_{cg} = 1.04\ m$$

The center of gravity is 1.04 m from the right end of the barbell.

Step 4: Common sense check.
The center of gravity is to the right of the center of the barbell (1.04 m vs. 1.10 m). This makes sense because the larger plate is on the right side.

148

The center of gravity is 5.9 in. (15 cm) from the end of the ruler. Pick up the ruler 5.9 in. from its end and see if it will balance on your finger at that location. If it does, you've verified that 5.9 in. is the center-of-gravity location.

With the pennies and ruler, we found the center of gravity along one dimension only. For more complex objects, the center-of-gravity location is defined by three dimensions, because most objects occupy space in three dimensions. To determine the center-of-gravity location for an object in three dimensions, the procedure we used in the previous two examples is repeated for each dimension, with gravity assumed to be acting in a direction perpendicular to that dimension.

Center of Gravity of the Human Body

Now let's consider the human body and the location of its center of gravity. The human body is not a rigid object, so the location of its center of gravity depends on the position of its limbs, just as the location of the center of gravity of the ruler and pennies depended on the positions of the pennies on the ruler. Let's assume you are standing up with your arms at your sides, as shown in figure 5.13.

Because your left and right sides are symmetrical, your center of gravity lies within the plane that divides your body into left and right halves (the midsagittal plane). If

you lift your left arm away from your side, your center of gravity shifts to the left. Although your body is not symmetrical from front to back, the center of gravity lies within a plane that divides your body into front and back halves (the frontal plane). This plane passes approximately through your shoulder and hip joints and slightly in front of your ankle joints. If you raise your arms in front of you, your center of gravity will be moved forward slightly.

The center-of-gravity location in these two dimensions, front to back and side to side, is easy to approximate if you are in anatomical position. The location of the center of gravity in the vertical dimension is more difficult to estimate. The center of gravity from top to bottom lies within a plane that passes horizontally through your body 1 to 2 in. (2.5 to 5 cm) below your navel, or about 6 in. (15 cm) above your crotch. This plane is slightly higher than half of your standing height, about 55% to 57% of your height. If you reach overhead with both arms, your center of gravity will move upward slightly (about 2 or 3 in. [5 to 8 cm]). Someone with long legs and muscular arms and chest has a higher center of gravity than someone with shorter, stockier legs. A woman's center of gravity is slightly lower than a man's because women have larger pelvic girdles and narrower shoulders relative to men. A woman's center of gravity is approximately 55% of her height from the ground, whereas a man's center of gravity is approximately 57% of his height from the ground. Infants and young children have higher centers of gravity relative to their height because of their relatively larger heads and shorter legs.

The center of gravity of the human body will move from the position just described if the body parts change their positions. If the limbs, trunk, head, and neck are considered the elemental units of the body, then their positions determine the center-of-gravity location of the body. You can estimate the center-of-gravity location of the body for any body position using the following procedure. First, imagine the body in the standing position shown in figure 5.13 and locate the center of gravity for this position using the information given in the previous paragraphs. Then consider the movement that each limb had to make to get into the position being described. For each limb movement, the center of gravity of the entire body shifts slightly in the direction of that movement. How much it moves depends on the weight of the limb that moved and the distance it moved. Raising one leg out in front of you will move your center of gravity forward and upward farther than will raising one arm out in front of you. Practice estimating the location of the body's center of gravity for each of the positions shown in figure 5.14.

For the high jumper in the arched position over the crossbar shown in figure 5.14*a*, her center of gravity is

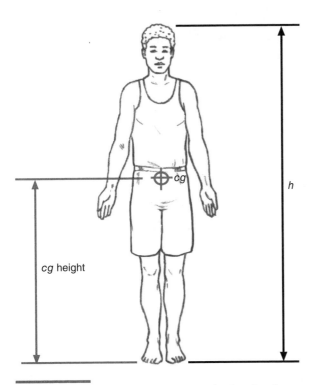

h

cg

cg height

Figure 5.13 In anatomical position, the height of your center of gravity is 55% to 57% of your standing height.

a

b

© Rogan Thomson/Action Plus/Icon SMI

c

Figure 5.14 Where is the athlete's center of gravity located in each of these body positions?

actually outside of her body, as it may be for the diver in the piked position shown in figure 5.14*b* and the pole-vaulter shown in figure 5.14*c*. The flexibility and complexity of the human body allows you to assume positions in which your center of gravity may actually lie outside of your body.

Center of Gravity and Performance

Now let's consider how your limb movements may actually affect your performance of a skill by affecting your center-of-gravity position. In a vertical jump-and-reach test, the objective is to jump up and reach as high as possible with one hand. Try the various jump-and-reach test techniques described in self-experiment 5.8.

Self-Experiment 5.8

Find a room with a very high ceiling and a clear space next to a wall. You may have to go outside to perform this experiment. First, from a standing position, jump up and reach up to touch the wall as high as you can with one hand. Keep your other hand down at your side. See how high you can touch on the wall. This might be difficult to do while you are jumping, so have a friend watch to see how high you reach. Try this three times to get a good idea about how high you can jump and reach using this technique.

Next, jump from a standing position again and reach as high as you can with one hand, but reach overhead with your other hand as well. Try this technique three times also. Did you reach as high?

Now, jump from a standing position again and reach as high as you can with both hands as you did in the previous case, but this time, when you are in the air, lift your knees and legs so your heels touch your buttocks. Try this technique three times also. Did you reach as high?

Using the first technique, you probably reached the highest. Using the second technique, your reach was probably 1 or 2 in. (2.5 to 5 cm) lower. And your reach with the third technique was probably 4 to 6 in. (10 to 15 cm) lower. Why?

With all three techniques, your center of gravity probably reached the same height if you jumped off the ground with the same effort each time. But wait—wouldn't your center of gravity be higher when you raised both arms over your head, and higher still when you lifted your legs up as well? Well, we did say that your center of gravity moved upward when any part of your body moved upward, but we should have said that center-of-gravity movement was a relative movement. In other words, raising your arms over your head causes your center of gravity to move higher in relation to your other body parts, for instance, your head. When you raise both arms over your head, your center of gravity moves closer to your head. In absolute terms (i.e., movement measured in relation to a fixed point on the earth), this could mean that (1) your center of gravity actually did move upward and higher off the ground; (2) your center of gravity stayed the same height off the ground and your head and other parts moved downward or closer to the ground to compensate for other parts moving upward; or (3) some combination of (1) and (2).

The explanation for what happened in the jumping tests comes from our early discussion of the center-of-gravity concept. The application and interpretation of Newton's laws to the motion of a complex object are simplified by the center-of-gravity concept. It is the center of gravity whose motion these laws govern. When you jumped up in the air during the jump-and-reach tests, the only external force acting on you was the force of gravity (your weight). This force causes your center of gravity to accelerate downward at a constant rate. You were a projectile once your feet left the ground. The motion of your limbs when you were in the air could not alter the motion of your center of gravity because they were not pushing or pulling against anything external to your body. The net force acting on you was still only the downward pull of gravity. The path that your center of gravity followed was not changed by your limb actions, but these actions caused the motion of your other limbs and trunk to change. Lifting your arms caused something else to move downward in order for your center of gravity to continue moving along its parabolic path. When you raised your arms and legs, your head and trunk got lower to compensate for this movement.

Here's another way to interpret or explain why you were able to reach the highest with only one hand above your head versus with both arms, or with both arms and both legs raised. First, let's assume that the center of gravity reached the same peak height. If this is the case, then to maximize the reach of one hand (its height above the ground), you want to maximize the vertical distance between the center of gravity and the outstretched arm. Raising both arms, or both arms and both legs, moves the center of gravity closer to the head and thus closer to the reaching hand. Keeping all of the limbs and body parts (with the exception of the reaching arm and hand) as low as possible relative to the center of gravity maximizes the distance from the reaching hand to the center of gravity. Figure 5.15 demonstrates this graphically.

A basketball player jumping up to block a shot can reach higher if only one hand reaches up and the other arm and the legs are not lifted in relation to the trunk. This is also true for a volleyball player trying to block a spike, although the volleyball player may be more effective reaching up with both hands. The player's hands won't reach as high, but they will block a greater area. In basketball, the direction of the shot is known to the defender—it's toward the goal, so only one hand is needed to block a shot. In volleyball, the direction of the spike is not known, so two hands are used to cover a greater area.

Let's consider one final jumping activity. How do basketball players, dancers, figure skaters, and gymnasts appear to "hang in the air"? During some of their jumps,

Figure 5.15 Three different vertical jump techniques result in three different reach heights, but the height of the center of gravity (*cg*) above the ground may be similar.

Labels on figure: Hand to *cg* distance; *cg* height; Reach height #3; Reach height #2; Reach height #1

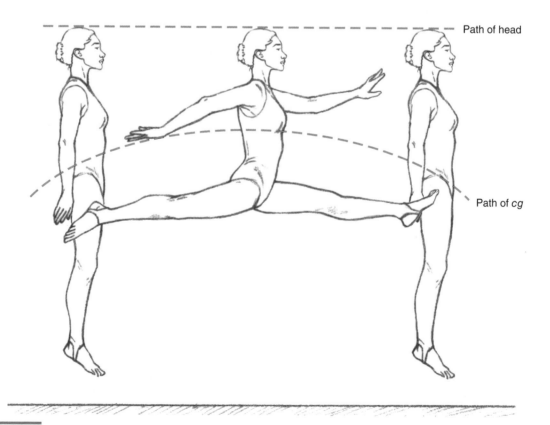

Path of head

Path of *cg*

Figure 5.16 During a leap, the jumper's head stays at the same level but the center of gravity follows a parabolic path.

it appears that their bodies are suspended in the air for a short time, rather than rising and falling in a parabolic path as a projectile should. Can you explain how these athletes "defy gravity"? Perhaps figure 5.16 will help.

If we follow the path of the jumper's center of gravity, it does indeed rise and fall through a parabolic curve. But the jumper's head and trunk appear to be suspended at the same height during the middle stage of the leap. During this time, the jumper's legs and arms rise and then fall. These movements account for the rise and fall of the center of gravity, so the head and trunk do not rise appreciably.

Center of Gravity and Stability

Our final topic concerning the center of gravity and performance in sports and human movement is stability. What is stability? When you say that something or someone is stable, what characteristic are you speaking of? There are a variety of definitions for the word *stable*. It can mean a structure for housing domestic animals, typically horse – oops, that's a definition for the noun stable. We're looking for the definition for the adjective stable. Stable can mean not easily moved or thrown off balance; firm; steady. This is probably similar to your own definition of the word *stable*. Stable can also mean capable of returning to equilibrium or original position after having been displaced. This is a mechanical definition of *stable*. **Stability** is thus the capacity of an object to return to equilibrium or to its original position after it has been displaced.

> Stability is the capacity of an object to return to equilibrium or to its original position after it has been displaced.

In many sports and human movement activities, the athletes or performers do not want to be moved from a particular stance or position. They want to be in a very stable position. Wrestlers, football linemen, and even basketball players are more successful at certain skills if they adopt stable positions. In other sports, success may be determined by how quickly an athlete is able to move out of a position. The receiver of a serve in tennis or racquetball, a sprinter, a swimmer, a downhill skier, and a goalie in soccer are more successful if their stance is less stable.

Factors Affecting Stability

What factors affect stability? How can you make yourself more or less stable? For some insights, try balancing this book on a table. In what position is it most easily balanced (i.e., which is its most stable position)? The book is most stable when it is lying flat on the table. This is the orientation that puts its center of gravity at the lowest position. This is also the position that offers a base of support with the largest area. If the book weighed more, would it be more or less stable? A heavier book would be more stable. Apparently, the stability of an object is affected by the height of the center of gravity, the size of the base of support, and the weight of the object. The **base of support** is the area within the lines connecting the outer perimeter of each of the points of support. Figure 5.17 shows examples of various stances and their bases of support. Which of these stances are most stable? Which of these stances are least stable?

> The stability of an object is affected by the height of the center of gravity, the size of the base of support, and the weight of the object.

What is the mechanical explanation for the fact that center-of-gravity height, base of support size, and weight affect stability? Let's examine the forces and moments of force that act on a book when a toppling force is exerted on it.

Stand a book on its edge and exert a horizontal force against it to tip it over. If the book remains in static equilibrium, the net force and torque acting on the book must be zero. The external forces acting on the book include the book's weight, *W*, acting through its center of gravity, *cg*; the toppling force, *P*; and the reaction force, *R*. If we examine the situation in which the toppling force is just large enough that the book is almost starting to move, the free-body diagram shown in figure 5.18 is appropriate.

If moments of force are measured about an axis, a, through the lower left corner of the book, the sum of moments about this point is zero:

$$\sum T_a = 0$$

$$0 = (P \times h) - (W \times b)$$

$$P \times h = W \times b \tag{5.4}$$

where

P = toppling force,

h = moment arm of the toppling force,

W = weight of the book (object),

b = moment arm of the book (object),

F_f = friction force (in figure 5.18), and

R = normal contact force (in figure 5.18).

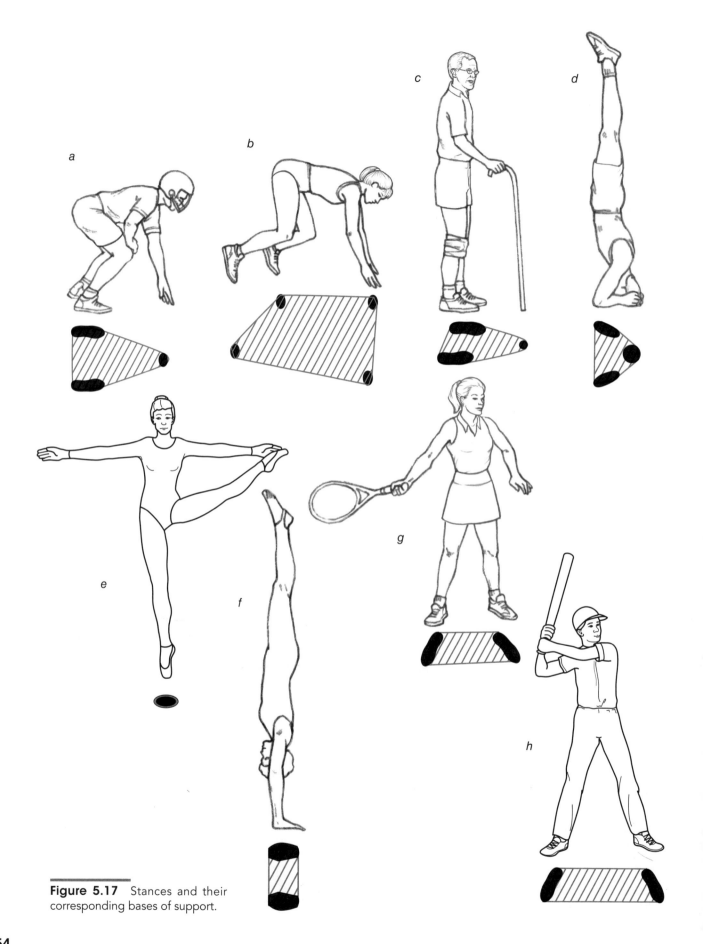

Figure 5.17 Stances and their corresponding bases of support.

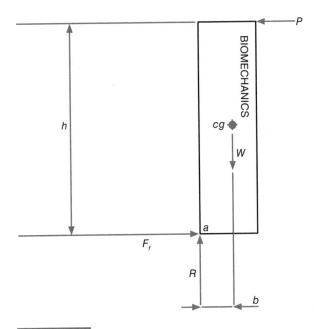

Figure 5.18 Free-body diagram of a book standing on its edge with a toppling force, *P*, exerted on it.

The terms on the left side of equation 5.4 are ones that are minimized to increase stability, and the terms on the right are ones that are maximized to increase stability. On the right side of equation 5.4, two variables appear: *W*, the weight of the object, and *b*, the moment arm of this weight. Increasing the weight will increase the stability because the moment of force keeping the object upright will be larger. Likewise, increasing the moment arm of the object's weight will increase stability. This dimension, *b*, is related to the size of the base of support, but it may be smaller or larger depending on which direction the toppling force comes from. Figure 5.19 demonstrates this for objects of various shapes.

The triangular block is less stable and more likely to be tipped over if the toppling force is directed to the left (figure 5.19a) rather than to the right (figure 5.19b). Stability is directional. An object can be more stable in one direction than in another. It is not the size of the base of support that affects stability, but the horizontal distance between the line of gravity and the edge of the base of support in the direction in which the toppling force is pushing or pulling.

On the left side of equation 5.4, the toppling force, *P*, is a factor that is not related to any characteristic of the object. The moment arm of this toppling force, *h*, is a characteristic of the object. This distance is related to the height of the object, which is also related to the height of the center of gravity. So a lower center of gravity, which implies a lower height and a shorter moment arm for the toppling force, increases stability.

Stability and Potential Energy

This explanation of why height or center of gravity affects stability is rather weak. It is actually an explanation of why the height of the object might affect stability. A better way of explaining why center-of-gravity height affects stability uses the concepts of work and potential energy. Consider the block shown in figure 5.20.

As long as the center of gravity of the block is to the left of the lower right corner, the weight creates a righting moment of force in opposition to the toppling moment of force created by the force *P*. But when the block is tipped past the configuration shown in figure 5.20b, where the center of gravity lies directly over the supporting corner, the moment of force created by the weight changes direction and becomes a toppling moment that causes the block to topple, as shown in figure 5.20c.

To move the block from its stable position (figure 5.20a) to the brink of instability (figure 5.20b), the center of gravity had to be raised a distance, Δ*h*. Work was required to do this, and the potential energy of the block increased.

Now let's examine what happens if the center of gravity is higher or lower. Figure 5.21 shows three blocks of the same shape and weight but with differing center-of-gravity heights.

The figure shows the vertical displacement, Δ*h*, that each block undergoes before it topples. The higher the center of gravity, the smaller this vertical displacement, thus the smaller the change in potential energy and the smaller the amount of work done. So a block with a lower center of gravity is more stable because more work is required to topple it.

If the distance from the line of gravity to the edge of the base of support about which toppling will occur (the moment arm of the weight) is increased, the vertical displacement that the center of gravity goes through before the object topples also increases, so the object is more stable. Figure 5.22 shows two blocks with identical center-of-gravity heights but different horizontal distances from the line of gravity to the edge of the base of support.

The most stable stance or position an object or person can be in is the one that minimizes potential energy. Moving to any other position increases potential energy and requires work to be done on the object or person. Positions that place the center of gravity below the points of support are more stable than positions that place the center of gravity above the base of support.

> The most stable stance or position that an object or person can be in is the one that minimizes potential energy.

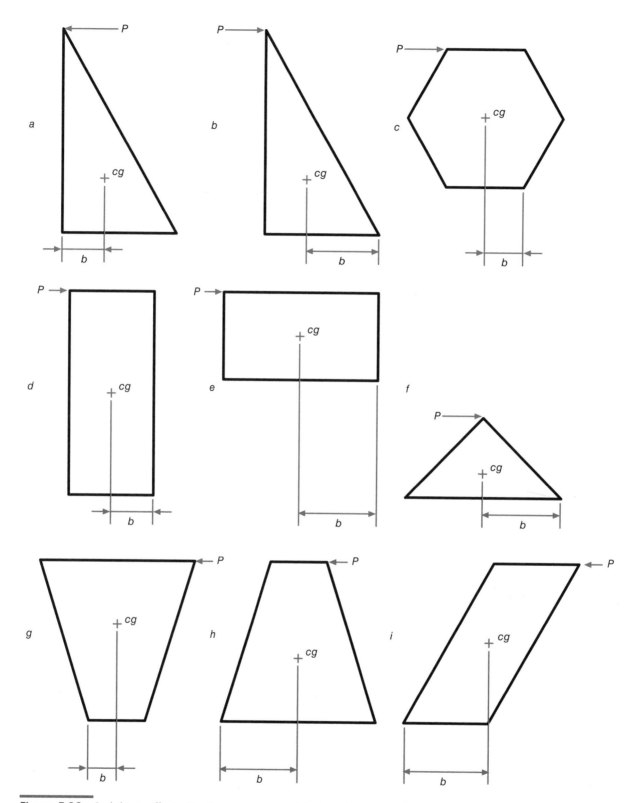

Figure 5.19 Stability is affected by the moment arm (b) of an object's weight around the edge of the base of support toward which the toppling force (P) pushes or pulls.

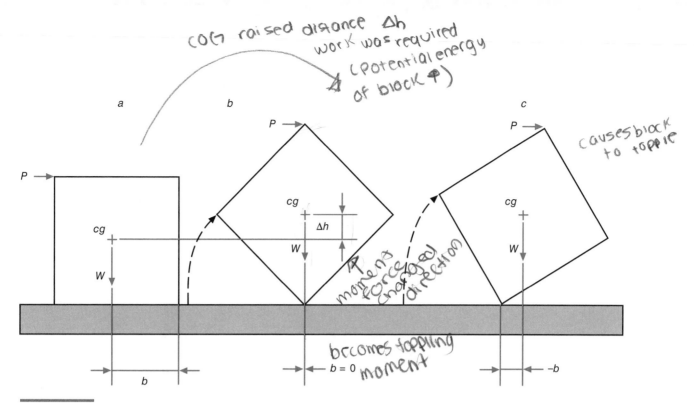

Handwritten annotations:

COG raised distance Δh work was required (potential energy of block↑)

causes block to topple

moment a force changed direction

becomes toppling moment

b = 0

Figure 5.20 As an object is toppled, the center of gravity is raised and the righting moment of the weight decreases. In (a), the weight creates a righting moment (W × b); in (b), the weight creates no moment (W × 0); and in (c), the weight creates a toppling moment [W × (−b)].

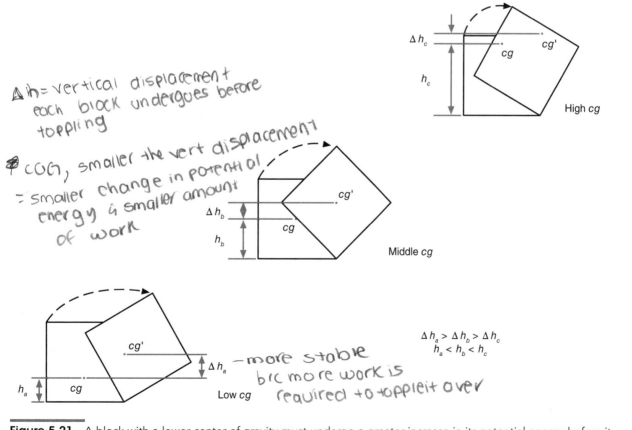

Handwritten annotations:

Δh = vertical displacement each block undergoes before toppling

↓ COG, smaller the vert displacement = smaller change in potential energy & smaller amount of work

High cg

Middle cg

Low cg — more stable b/c more work is required to topple it over

$\Delta h_a > \Delta h_b > \Delta h_c$
$h_a < h_b < h_c$

Figure 5.21 A block with a lower center of gravity must undergo a greater increase in its potential energy before it topples.

if COG = ↑ or ↓

each block = same shape & weight by different COG heights

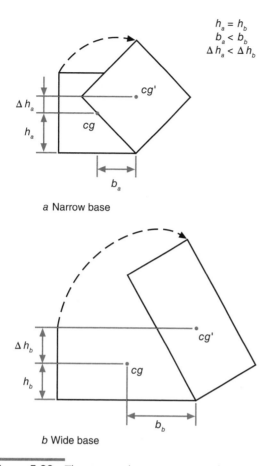

$$h_a = h_b$$
$$b_a < b_b$$
$$\Delta h_a < \Delta h_b$$

a Narrow base

b Wide base

Figure 5.22 The greater the moment arm of the weight, the greater the vertical displacement of the center of gravity, Δh, before toppling occurs.

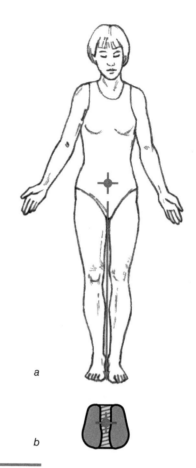

a

b

Figure 5.23 Line of gravity (a) and base of support (b) for normal stance.

If the center of gravity is below the base of support, after any displacement of the object, the object will return to its original equilibrium position. This is an example of stable equilibrium. A gymnast hanging from the horizontal bar is an example of stable equilibrium. When the center of gravity lies above the base of support, stability is maintained only as long as the line of gravity falls within the base of support (see figure 5.23).

Center of Gravity, Stability, and Human Movement

Our discussion of stability has focused primarily on rigid objects with fixed centers of gravity and bases of support. The human body is not rigid, and its center-of-gravity position and base of support can change with limb movements. Humans can thus control their stability by changing their stance and body position. Before examining how athletes manipulate their centers of gravity and bases of support to affect stability, let's look at a few activities.

How do you initiate a walking step? You don't just lift your foot and place it in front of you. You lean forward until your line of gravity falls in front of your feet and you lose your stability. You begin to fall forward, and you step with one foot to catch your fall and reestablish your stability. So walking could be described as a series of falls and catches!

In athletic activities, athletes may want to maximize their stability in general or in a specific direction, or they may want to minimize stability (increase their mobility). During the first period of a wrestling match, the two wrestlers are standing, and each is trying to take the other down. The direction of the toppling force is unknown; the other wrestler may be pulling or pushing forward, backward, or to the left or right. To maximize stability (while still maintaining the ability to move), the wrestler crouches to lower his center of gravity and widens his base of support by placing his feet slightly wider than shoulder-width apart in a square stance (figure 5.24a) or by placing one foot in front of the other, again slightly farther than shoulder-width apart, in a staggered stance (figure 5.24*b*).

When the wrestler is in a defensive position on his belly and trying not to be turned over onto his back, he maximizes his stability by sprawling his limbs to the sides to maximize the size of his base of support and to lower his center of gravity as much as possible (figure 5.25).

When force is expected from a specific direction, the base of support is widened in that direction. If a heavy medicine ball were thrown to you, the most stable stance for you to be in to catch the ball would be a staggered stance, with one foot in front of the other in line with the direction of the throw and your body leaning toward the front foot (figure 5.26).

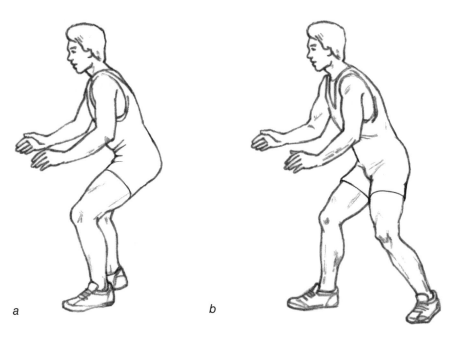

a   b

Figure 5.24 Two wrestling stances, square (a) and staggered (b), that represent compromises between stability and mobility.

Figure 5.25 The defensive sprawl of a wrestler maximizes his stability by minimizing the height of his center of gravity and maximizing his base of support.

Figure 5.26 A staggered stance provides directional stability for catching.

Staggered stance (handwritten marginal note)

This type of stance might be adopted by competitors at the start of a tug-of-war, except that they would shift their weight over the rear foot. Boxers also use staggered stances, as do tennis players, baseball batters, and so on. This is a popular stance in many sports, not only because it allows momentum to be reduced or increased by force application over a long time, but also because it is a more stable position.

Unless some other part of our body or an implement we use touches the ground, the size of our base of support is limited by our shoe size and the stances being adopted. In sport, skis increase stability forward and backward. In physical rehabilitation and medicine, crutches, canes, walkers, and so on are used to increase the base of support and stability of people who are injured, sick, or infirm.

In some activities, stability is minimized to enhance quick movement. For instance, in a track sprint start, in the set position, the sprinter raises her center of gravity and moves it forward to the edge of the base of support over her hands. At the starter's signal, lifting her hands off the track puts her line of gravity well in front of her base of support, and the sprinter falls forward. A similar strategy is used in swimming starts.

Summary

The turning effect created by a force is a moment of force, also called a torque. A torque is equal to the magnitude of the force times the perpendicular distance between the line of action of the force and the axis of rotation about which the torque is being measured. This perpendicular distance is called the moment arm of the force. Torque is a vector quantity. Its direction is defined by the orientation of the axis of rotation and the sense of the torque (clockwise or counterclockwise) about that axis of rotation. The units for torque are units of force times units of distance, or Nm in the SI system.

For an object to be in balance (in a state of equilibrium), the external forces acting on it must sum to zero and the moments of those forces must also sum to zero.

The center of gravity of an object is its balance point or the point about which the moments of force created by the weights of all of the parts of the object sum to zero. To locate an object's center of gravity, you can balance it, suspend it, or spin it. In the human body, the center of gravity lies on the midline left to right and front to back and between 55% and 57% of a person's height above the ground when the person is standing in anatomical position.

Stability is affected by the position of an object's center of gravity above the ground and relative to the edges of the support base. Stability increases as the center of gravity is lowered and moved farther from any edge of the base of support. Increasing weight also increases stability.

KEY TERMS

base of support (p. 153)

center of gravity (p. 145)

centric force (p. 134)

eccentric force (p. 135)

force couple (p. 135)

moment arm (p. 136)

moment of force (p. 134)

stability (p. 153)

torque (p. 134)

REVIEW QUESTIONS

1. Why is it recommended that you flex your knees and hips (squat down) when lifting a heavy load off the floor, rather than just flexing your hips and back (leaning over)?

2. Which exercise is easier to do, a straight-arm pullover with a 45 lb barbell or a bent-arm pullover with a 45 lb barbell? Why? In a pullover exercise, the lifter lies supine on a bench with his arms holding a barbell and extended past the end of the bench and his head. He then rotates his arms at the shoulder joints and moves the barbell to a position directly over his chest. In a straight-arm pullover, the arms stay extended at the elbow, while in a bent-arm pullover, the elbow is flexed.

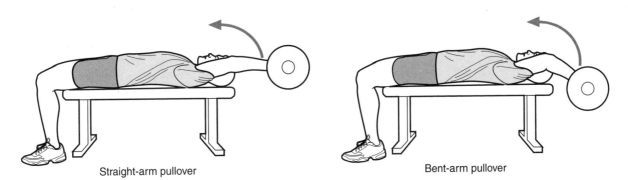

Straight-arm pullover   Bent-arm pullover

3. You are doing abdominal crunches on a bench whose incline can be adjusted. As you increase the incline of the bench, your feet get higher than your hips. What effect does the increase in the incline of the bench have on the abdominal crunch exercise?

4. Stand with your back and heels against a wall. Now try to lean forward and pick something up off the floor while still keeping your heels against the wall. Can you do this without falling over? Why is this so difficult or impossible to do?

5. Why is it more difficult to get up out of a chair whose seat is lower to the ground than if the seat is higher?

6. John jumps up in the air to block a shot in basketball. After he is off the ground, John raises both hands over his head. Explain how this action affects the flight path of his center of gravity.

7. Describe the best position to be in when you compete in a tug-of-war.

8. In basketball, which is easier to do—a one-handed dunk or a two-handed dunk? Explain.

9. Can a high jumper clear a cross bar even if her center of gravity doesn't go above the height of the cross bar? Explain your answer.

10. Explain how larger handles on doors, valves, and so on are helpful to a person with limited dexterity and hand strength.

PROBLEMS

1. The Achilles tendon inserts on the calcaneus (the heel bone) at a distance of 8 cm from the axis of the ankle joint. If the force generated by the muscles attached to the Achilles tendon is 3000 N, and the moment arm of this force about the ankle joint axis is 5 cm, what torque is created by these muscles about the ankle joint?

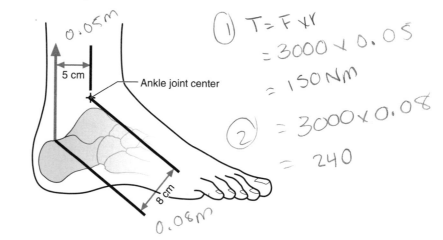

0.05m

5 cm

Ankle joint center

8 cm

0.08m

① T= F ✗ r
  = 3000 ✗ 0.05
  = 150 Nm

② = 3000 ✗ 0.08
  = 240

2. Leigh is doing a knee extension exercise using a 100 N weight strapped to her ankle 40 cm from her knee joint. She holds her leg so that the horizontal distance from her knee joint to the weight is 30 cm.

   a. For this position, what torque is created by the dumbbell about her knee joint axis?

   b. If the moment arm of the knee extensor muscles is 4 cm about the knee joint axis, what amount of force must these muscles produce to hold the leg in the position described? Ignore the weight of the leg. 750N

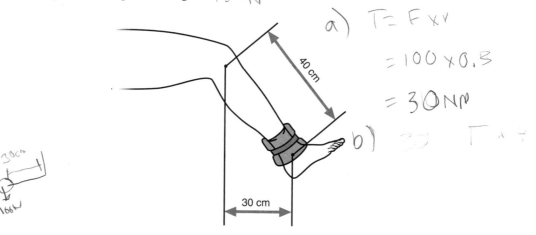

40 cm

30 cm

30cm

166w

a) T= F ✗ r
   = 100 ✗ 0.3
   = 30 Nm

b)

3. Katherine is attempting a biceps curl with a 100 N dumbbell. She is standing upright, and her forearm is horizontal. The moment arm of the dumbbell about Katherine's elbow joint is 30 cm.

   a. What torque is created by the dumbbell about the elbow joint? T= 100 ✗ 0.03   30 Nm

   b. What torque is created by Katherine's elbow flexor muscles to hold the dumbbell in this position? Ignore the weight of her arm and hand. 7= 100 ✗ 0.03

   c. If the elbow flexor muscles contract with a force of 1000 N and hold the barbell in static equilibrium, what is the moment arm of the elbow flexor muscles about the elbow joint?

30 = 1000 r
r = 0.03 m

T= F ✗ r
30 = 100 0 ✗
r

1000 r / 1000

0.03 m = r
3 cm

4. Cole is trying a biceps curl and attempting to lift a 150 N dumbbell. The moment arm of this weight about his elbow joint is 25 cm. The force created by the elbow flexor muscles is 2000 N.

0.25m

$T = 150(0.25)$
$= 37.5 \, Nm$

$37.5 = F(0.02)$
$F = 1875N$ yes he can.

The moment arm of the elbow flexor muscles is 2 cm. Is Cole able to lift the weight with this amount of force in his flexor muscles?

5. Matt has a portable basketball goal in his driveway. He has the basket set 8 ft (2.4 m) high so he can practice dunking the ball. He slams the ball through the hoop and then hangs onto the rim. This exerts a downward force of 600 N on the front of the rim. The front of the rim is 1.1 m in front of the front edge of the portable basketball goal's base. The mass of the whole portable basketball goal is 70 kg. The center of gravity of the portable basketball goal is 1.0 m behind the front edge of its base.

   a. How much torque is produced around the front of the goal base by the 600 N force Matt exerts on the front of the rim? $600(1.1) = 660 \, Nm$

   b. How much torque would be needed to tip the goal? $T = (70(9.81) \times 1.0) = 686.7 \, Nm$

   c. What is the largest vertical force that can be exerted on the front edge of the rim before the portable basketball goal begins to tip?

   $686.7 = F(1.1)$
   $F = 624 \, N$

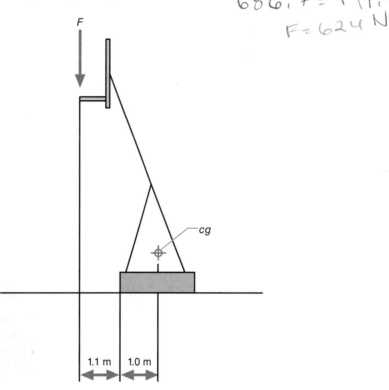

6. A baseball batter sets up in the batting box and holds his bat still over the plate for a moment. The batter holds the 0.91 kg bat in a horizontal position with only one hand. The hand is 10 cm from the knob end of the bat. The bat is 89 cm long, and its center of gravity is 60 cm from the knob end of the bat.

   a. How large is the vertical force exerted by the batter's hand on the bat? $(0.91 \times 9.81) = 8.9$

   b. How large is the torque exerted by the batter's hand on the bat? $T = (8.9)(0.6)$

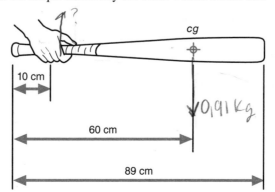

7. A pole-vaulter is holding a vaulting pole parallel to the ground. The pole is 5 m long. The vaulter grips the pole with his right hand 10 cm from the top end of the pole and with his left hand 1 m from the top end of the pole. Although the pole is quite light (its mass is only 2.5 kg), the forces that the vaulter must exert on the pole to maintain it in this position are quite large. How large are they? (Assume that the vaulter exerts only vertical—up or down—forces on the horizontal pole and that the center of gravity of the pole is located at the center of its length.)

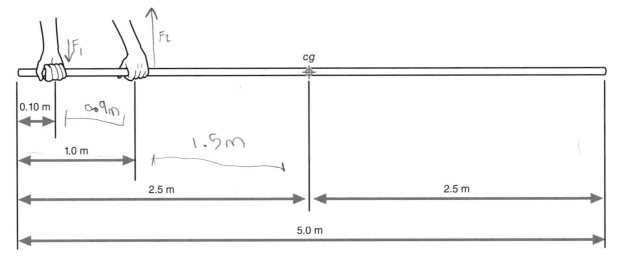

8. The patellar tendon attaches to the tibia at the tibial tuberosity. The tibial tuberosity is 7 cm from the center of the knee joint (a). The patellar tendon pulls at an angle of 35° to a line passing through the tibial tuberosity and the center of the knee joint. If the patellar tendon produces an extension torque around the knee joint equal to 400 Nm, how large is the force in the patellar tendon?

$2.5 kg (9.81)$
$= 24.525$

7. $\sum T = 0 = F_1(0) + F_2 (0.9) - W(2.4)$

$0 = F_2 (0.9) - 24.525(2.4m)$

$= 0.9F - 58.86 Nm$

$\dfrac{58.86}{0.9} = \dfrac{0.9F}{0.9}$

$F = \boxed{65.4 N}$

$0 = F_1 + F_2 + W$

$0 = 65.4 - 24.5 + F$

$\boxed{-40.9 N} = F$

$\dfrac{400 Nm}{0.07m} = 5714.29 N$

$Sin 35 = \dfrac{5714.29 N}{H}$

$H = 9962.55$

$\dfrac{400 Nm}{0.07m} = 5714.29 N$

9. A 60 kg gymnast holds an iron cross position on the rings. In this position, the gymnast's arms are abducted 90° and his trunk and legs are vertical. The horizontal distance from each ring to the gymnast's closest shoulder is 0.60 m. The gymnast is in static equilibrium.

$60 kg (9.8)$
$= 588.6 N$

a. What vertical reaction force does each ring exert on each hand?

b. What torque is exerted by the right ring about the right shoulder joint?

c. How much torque must the right shoulder adductor muscles produce to maintain the iron cross position?

d. If the moment arm of the right shoulder adductor muscles about the shoulder joint is 5 cm, how much force must these muscles produce to maintain the iron cross?

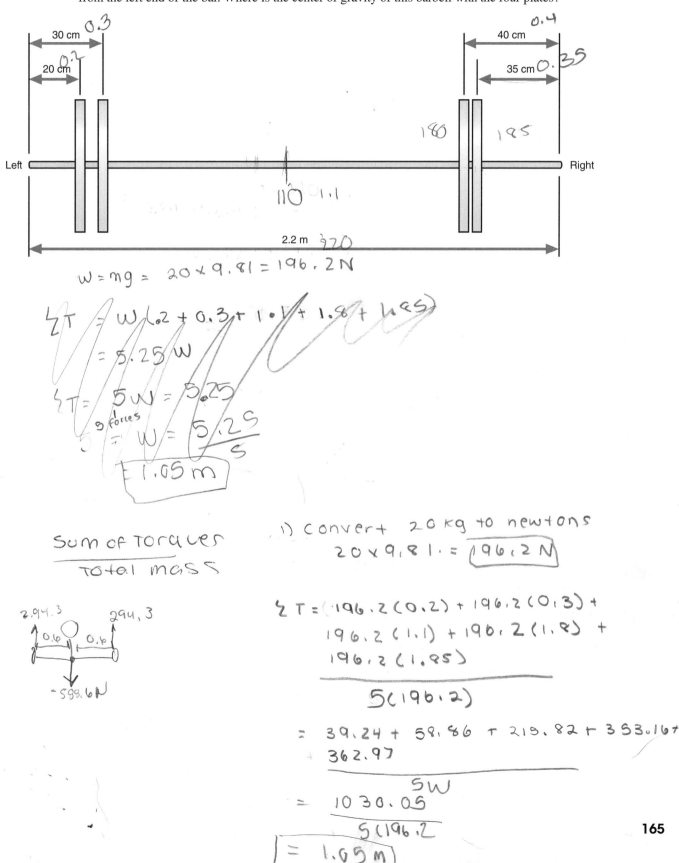

10. A barbell is loaded with two 20 kg plates on its right side and two 20 kg plates on its left side. The barbell is 2.2 m long, and its unloaded mass is 20 kg. The two 20 kg plates on the right side are locked in place at 35 cm and 40 cm from the right end of the bar. The two 20 kg plates on the left side have slipped. One is 30 cm from the left end of the bar, and the other is 20 cm from the left end of the bar. Where is the center of gravity of this barbell with the four plates?

30 cm 0.3

20 cm 0.2

40 cm 0.4

35 cm 0.35

180   185

Left

Right

110   1.1

2.2 m   220

$W = mg = 20 \times 9.81 = 196.2 \text{ N}$

$\Sigma T = W(.2 + 0.3 + 1.1 + 1.8 + 1.85)$

$= 5.25 W$

$\Sigma T = 5W = 5.25$
9 forces

$W = \dfrac{5.25}{5}$

$= 1.05 \text{ m}$

$\dfrac{\text{Sum of Torques}}{\text{Total mass}}$

2.94.3   294.3

0.6   0.6

-588.6 N

1) convert 20 kg to newtons
$20 \times 9.81 = 196.2 \text{ N}$

$\Sigma T = \dfrac{196.2(0.2) + 196.2(0.3) + 196.2(1.1) + 196.2(1.8) + 196.2(1.85)}{5(196.2)}$

$= \dfrac{39.24 + 58.86 + 215.82 + 353.16 + 362.97}{5W}$

$= \dfrac{1030.05}{5(196.2}$

$= 1.05 \text{ m}$

chapter 6

Angular Kinematics
Describing Objects in Angular Motion

© Chung Jin Mac | Dreamstime.com

objectives

When you finish this chapter, you should be able to do the following:

- Define relative and absolute angular position, and distinguish between the two

- Define angular displacement

- Define average angular velocity

- Define instantaneous angular velocity

- Define average angular acceleration

- Define instantaneous angular acceleration

- Name the units of measurement for angular position, displacement, velocity, and acceleration

- Explain the relationship between average linear speed and average angular velocity

- Explain the relationship between instantaneous linear velocity and instantaneous angular velocity

- Define tangential acceleration and explain its relationship to angular acceleration

- Define centripetal acceleration and explain its relationship to angular velocity and tangential velocity

- Describe the anatomical position

- Define the three principal anatomical planes of motion and their corresponding axes

- Describe the joint actions that can occur at each of the major joints of the appendages

A hammer thrower steps into the throwing circle. After a few windups, the thrower begins rotating as he swings the hammer around and moves across the circle. The turning rate increases as he approaches the front of the circle. Suddenly he lets go of the hammer, and it becomes a fast-moving projectile. The 16 lb (7.3 kg) steel ball seems to fly forever before it buries itself into the ground with a thud 250 ft (76 m) away. Wow! How did the thrower's rotary motion cause the hammer to move so fast and go so far linearly when he let go of it? To answer this question, you need to know something about angular kinematics and its relationship to linear kinematics. Both of these topics are discussed in this chapter.

This chapter is concerned with angular kinematics, the description of angular motion. Recall that in chapter 2, we described kinematics as part of dynamics, which is a branch of mechanics. In chapter 2, we also learned about linear kinematics. Angular kinematics is the other branch of kinematics. Angular motion occurs when all points on an object move in circular paths about the same fixed axis. Angular motion is important because most human movements are the result of angular motions of limbs about joints. An understanding of how angular motion is measured and described is important.

Angular Position and Displacement

Before further discussion, a definition of *angle* is needed. What is an angle? An angle is formed by the intersection of two lines, two planes, or a line and a plane. The term *angle* refers to the orientation of these lines or planes to each other (figure 6.1).

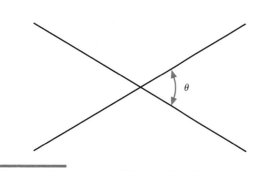

Figure 6.1 An angle (*θ*) formed by the intersection of two lines.

⟳ **An angle is formed by the intersection of two lines, two planes, or a line and a plane.**

In figure 6.1, the Greek letter *θ* (theta) is used to represent the angle formed by the intersections of the lines and planes. Greek letters are used to represent many of the terms used in angular kinematics.

Angular Position

Angular position refers to the orientation of a line with another line or plane. If the other line or plane is fixed and immovable relative to the earth, the angular position is an **absolute angular position**. The angle your forearm makes with a horizontal plane describes the absolute angular position of your forearm, because a horizontal plane is a fixed reference. These absolute angular positions are our primary concern in the first part of this chapter.

If the other line or plane is capable of moving, the angular position is a **relative angular position**. The angle your forearm makes with your upper arm describes a relative angular position of your forearm, or your elbow joint angle. Angles formed by limbs at joints describe relative angular positions of limbs. Anatomists have developed special terms for describing the relative positions and movements of limbs at joints. These anatomical terms are presented in the last part of this chapter.

What units of measurement are used for angles? You probably are most familiar with measuring angles in degrees, but there are units of measurement for angles besides degrees, just as there are units of measurement for linear distance besides feet. If we wanted to measure the absolute angle of line AB, as shown in figure 6.2*a*, with a horizontal plane, we might imagine a horizontal

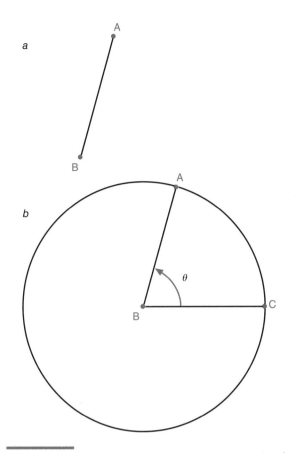

Figure 6.2 A circle is used in describing the angle of a line (a) if the center of the circle coincides with the intersection of the lines defending the angle (b).

line BC with the same length as AB. We could then draw a circle using the length of AB as the radius and the point B as the center (figure 6.2b). The angle of ABC could then be described as the fraction of the circle created by the pie ABC. An angle of 1° represents 1/360 of a circle because there are 360° in a circle.

Another way to describe an angle is to measure how many radii are in the arc length AC if one radius is equal to the length of line segment BC or AB. In other words, if the length BC represents 1 radius, how long is the arc from A to C measured using this radius as the unit of measure? Mathematically, an angle measured in radius units (we'll call these **radians**) is

$$\theta = \frac{arc\ length}{r} = \frac{\ell}{r} \qquad (6.1)$$

where

θ = angular measurement in radians,

ℓ = arc length, and

r = radius.

The radian (abbreviated rad) unit of measure for an angle is really a ratio of arc length divided by the radius.

Figure 6.3 graphically shows the definition of an angle of 1 rad, π (pi) rad, and 2π rad. If you recall from geometry, the circumference of a circle is $2\pi r$, so there are 2π rad in a circle or 2π rad in 360°. These conversions are as follows:

$$\frac{360°}{circle} = \frac{2\pi\ rad}{circle} = \frac{6.28\ rad}{circle}$$

$$\frac{360°}{2\pi\ rad} = \frac{57.3°}{rad}$$

Angular Displacement

Angular displacement is the angular analog of linear displacement. It is the change in absolute angular position experienced by a rotating line. Angular displacement is thus the angle formed between the final position and the initial position of a rotating line. (We often speak of the angular displacement of an object when the object is not a line. To measure such a displacement, choose any two points on the object. Imagine a line connecting these two points. If the object is rigid, the angular displacement of this line segment is identical to the angular displacement of the object.)

⊃ **Angular displacement is the change in absolute angular position experienced by a rotating line.**

As with linear displacement, angular displacement has direction associated with it. How is angular direction described? *Clockwise* and *counterclockwise* are common

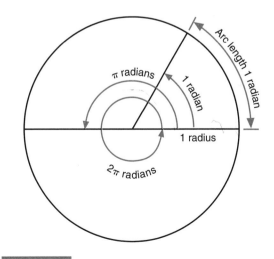

Figure 6.3 The relationships among the radius of a circle, the arc length along a circle, and an angle measured in radians.

terms used to describe the direction of rotation. The hands of a clock rotate in a clockwise direction when you view a clock face from the front. If you could view a clock face from the back, would you still see the hands rotating in a clockwise direction? Because your viewpoint changed, the clock hands would now appear to be rotating counterclockwise! They didn't change the direction of rotation; you just changed your viewing position. If you describe an angular displacement as clockwise, a person must know your viewing position to be certain of the direction of the angular displacement.

One way of overcoming this possible source of confusion is to first identify the axis of rotation and the plane in which the part rotates. The axis of rotation is always perpendicular to the plane in which the motion occurs. This axis of rotation is like the axle of a bicycle wheel, and the spokes of the wheel lie in the plane of motion. Along the axis of rotation, establish a positive direction. If you situate the thumb of your right hand so that it points in the positive direction along the axis of rotation, the direction in which your fingers curl is the positive direction of rotation. This is called the right-hand thumb rule.

Now consider the direction of rotation of the hands of a clock. The plane of their motion is the clock face, and the axis of rotation is a line through the plane of the clock face. If the positive direction along this axis is established as pointing out of the clock face toward you, the positive direction of rotation is counterclockwise as you face the clock. Check this by pointing your right thumb out away from the clock in the positive direction of the axis of rotation. Then observe the direction in which your fingers curl—counterclockwise (see figure 6.4).

Most screws, nuts, and bolts have right-handed threads. They follow the right-hand thumb rule. If you point your right thumb in the direction you want the screw or nut to move, your right fingers curl in the direction in which the screw or nut must be turned. This sign convention also applies to torques and measures of angular position.

Now let's consider how angular displacement is measured. A pitcher is being evaluated for shoulder joint range of motion. The measurement begins with the pitcher's arm at his side, as shown in figure 6.5. The pitcher then raises his arm away from his side as far as he can (he abducts his shoulder). What is the angular displacement of his arm?

The axis of rotation is the anteroposterior axis, a line through the shoulder joint with the positive direction pointing out of the page toward us. The plane of motion is the frontal plane—the plane formed by his arms, legs, and trunk. If the initial position of the arm is 5° from the vertical and its final position is 170° from the vertical, the angular displacement is

$$\Delta\theta = \theta_f - \theta_i \qquad (6.2)$$

$$\Delta\theta = 170° - 5°$$

$$\Delta\theta = +165°$$

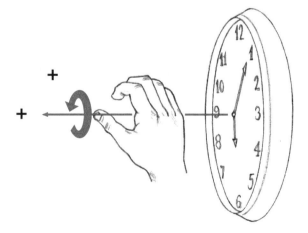

Figure 6.4 The right-hand thumb rule. The direction in which the fingers curl indicates the positive angular direction if the right thumb points in the positive linear direction along the axis of rotation.

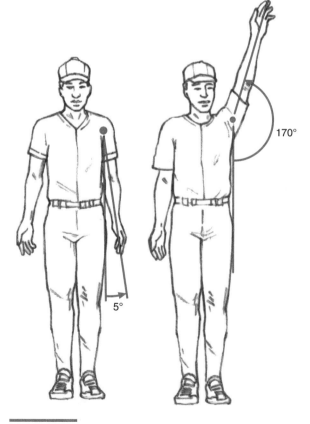

Figure 6.5 Angular displacement of a pitcher's arm at the shoulder joint around the anteroposterior axis.

where

$\Delta\theta$ = angular displacement,

θ_f = final angular position, and

θ_i = initial angular position.

The displacement is positive because the rotation is in the direction in which the fingers of your right hand curl if your thumb is pointed in the positive direction of the axis of rotation (anteriorly and away from the shoulder).

As a coach or teacher, you will rarely measure angular displacements precisely, but in some sports, angular displacement is an important part of the skill. The number of twists or somersaults done in diving, gymnastics, or figure skating is a measure of angular displacement and plays an important role in how many points a judge awards. The angular displacement of a swing (range of motion) in golf or tennis affects the manner in which the ball is hit in these sports.

Angular and Linear Displacement

In chapter 5, we discovered that our muscles must produce very large forces to lift modest loads. The reason for this is that most muscles attach to bones close to the joint; thus they have small moment arms about the joint. Because of the small moment arm, large forces must be produced by the muscles to produce modest torques about the joints. The muscles are at a mechanical disadvantage for producing torque. Is there any advantage to this arrangement? Try self-experiment 6.1 for some insight into the advantages.

Self-Experiment 6.1

Place your forearm on the desk or table in front of you. Now flex at the elbow and bring your hand off the table. Move it toward your shoulder as far as it can go while keeping your elbow on the desktop as shown in figure 6.6.

All the parts of the arm underwent the same angular displacement, but which moved farther, your hand (point A in figure 6.6) or the point of attachment of your biceps muscle (point B in figure 6.6)? Obviously your hand moved farther. The linear distance it traveled (arc length AA, ℓ_a, in figure 6.6) and its linear displacement (chord AA, d_a, in figure 6.6) are larger than the linear distance traveled (arc length BB, ℓ_b) and linear displacement (chord BB, d_b) of the biceps insertion.

How far any point on the arm moves when you flex your arm, as in self-experiment 6.1, is dependent on

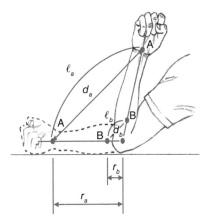

Figure 6.6 The distance that the hand or wrist (A) moves (ℓ_a or d_a) when your elbow (B) flexes is greater than the distance that the insertion point of the biceps moves (ℓ_b or d_b). The ratio of these distances to each other is the same as the ratio of the radius r_a to the radius r_b.

how far that point is from the elbow. This relationship is actually apparent in the definition of an angle measured in radians, as shown in equation 6.1:

$$\Delta\theta = \frac{arc\ length}{r} = \frac{\ell}{r}$$

For our example, then, let's use an angular displacement measured in radians, and

$$\Delta\theta = \frac{\ell}{r} = \frac{\ell_a}{r_a} = \frac{\ell_b}{r_b}$$

$$\frac{\ell_a}{\ell_b} = \frac{r_a}{r_b} \tag{6.3}$$

If the angular displacement of the arm was 1 rad, r_a was 10 in. (25 cm), and r_b was 1 in. (2.5 cm), then

$$\Delta\theta = \frac{\ell}{r}$$

$$\ell_a = \Delta\theta r_a = (1\ rad)(10\ in.) = 10\ in.$$

$$\frac{\ell_a}{\ell_b} = \frac{r_a}{r_b}$$

$$\ell_b = \ell_a\left(\frac{r_b}{r_a}\right) = 10\ in.\left(\frac{1\ in.}{10\ in.}\right) = 1\ in.$$

In this example, the hand moves 10 times as far as the insertion of the biceps tendon. The linear distance (arc length) traveled by a point on a rotating object is directly proportional to the angular displacement of the object and the radius, the distance that point is from the axis of rotation of the object. If the angular displacement is measured in radians, the linear distance traveled (the arc length) is equal to the product of the angular displacement and the radius. This is true only if the angular displacement is measured in radians. This relationship is expressed mathematically in equation 6.4:

$$\ell = \Delta\theta r \quad\quad (6.4)$$

where

ℓ = arc length,

$\Delta\theta$ = angle measured in radians, and

r = radius.

The relationship between angular displacement and arc length as expressed in equation 6.4 provides the basis for why staggered starting positions are used in track races around a curve. Without a staggered start, runners in outside lanes would run farther since the radius of the curve they must run is larger. The larger radius causes the length of their lane—the arc length—to be longer. Staggering the starting positions accounts for the larger arc lengths in the outer lanes.

Not only is arc length proportional to the radius, but the linear displacement of a point on a rotating object is also directly proportional to the distance that point is from the axis of rotation (the radius). This linear displacement is also related to the angular displacement, but it is not

directly proportional to the angular displacement. The relationship is more complex.

The relationship between linear displacement and radius is easily seen in figure 6.7. The linear displacement and radii form sides of similar triangles, as shown in figure 6.7. From this, the following relationship can be established:

$$\frac{r_a}{r_b} = \frac{d_a}{d_b} \quad\quad (6.5)$$

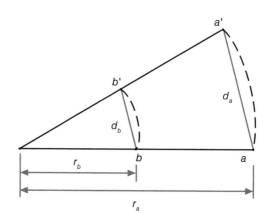

Figure 6.7 The ratio of the linear displacements of two points on a rotating object (d_a/d_b) is equal to the ratio of the radii of these two points from the axis of rotation (r_a/r_b).

SAMPLE PROBLEM 6.1

The golfer's hands move through an arc length of 10 cm during a putt. What arc length does the head of the putter move through if the hands are 50 cm from the axis of rotation and the putter head is 150 cm from the axis of rotation?

Solution:

Step 1: Identify the known quantities and inferred relationships.

ℓ_{hands} = 10 cm

r_{hands} = 50 cm

r_{putter} = 150 cm

The hands, the arms, and the putter move together as one pendulum, so

$\Delta\theta_{hands} = \Delta\theta_{putter}$

Step 2: Identify the unknown variable to solve for.

ℓ_{putter} = ?

Step 3: Search for equations with the unknown and known variables.

$\ell = \Delta\theta r$

$\Delta\theta = \ell/r = \Delta\theta_{hands} = \Delta\theta_{putter}$

$(\ell_{hands})/(r_{hands}) = (\ell_{putter})/(r_{putter})$

Step 4: Substitute known quantities and solve for the unknown quantity.

$(10\ cm)/(50\ cm) = (\ell_{putter})/(150\ cm)$

$\ell_{putter} = (150\ cm)(10\ cm)/(50\ cm) = 30\ cm$

Step 5: Common sense check.
The putter head moves farther than the hands, and 30 cm is about a foot. This is a reasonable movement for the putter head.

where

r = radius and

d = linear displacement (or chord length).

One advantage of muscle insertions close to joints should now be clear. The muscle has to contract and shorten only a short distance to produce a large movement (linear displacement) at the end of the limb. Because the distance a muscle can shorten is limited to approximately 50% of its resting length, the range of motion at a joint would also be further limited if muscles inserted farther away from the joints.

This concept is useful when we use sport implements. Small movements of the hands produce large linear displacements at the end of a putter, a badminton racket, a fishing pole, a hockey stick, and so on.

Angular Velocity

Angular velocity is defined as the rate of change of angular displacement. Its units of measurement are radians per second (rad/s), degrees per second (°/s), revolutions per minute (rpm), and so on. Angular velocity is abbreviated with the Greek letter omega (ω). Angular velocity is a vector quantity, just like linear velocity, so it has direction associated with it. The direction of an angular velocity is determined using the right-hand thumb rule, as with angular displacement. Because angular velocity is a vector, a change in the size of the angular velocity or in the direction of its axis of rotation results in a change in angular velocity.

> ⟳ Angular velocity is defined as the rate of change of angular displacement.

Average angular velocity is computed as the change in angular position (angular displacement) divided by time. Mathematically,

$$\bar{\omega} = \frac{\Delta\theta}{\Delta t} = \frac{\theta_f - \theta_i}{\Delta t} \qquad (6.6)$$

where

$\bar{\omega}$ = average angular velocity,

$\Delta\theta$ = angular displacement,

Δt = time,

θ_f = final angular position, and

θ_i = initial angular position.

If we are concerned with how long it takes for something to rotate through a certain angular displacement, **average angular velocity** is the important measure. If we are concerned with how fast something is rotating at a specific instant in time, **instantaneous angular velocity** is the important measure. Instantaneous angular velocity is an indicator of how fast something is spinning at a specific instant in time. The tachometer on a car's instrument panel gives a measure of the engine's instantaneous angular velocity in revolutions per minute.

The average angular velocity of a batter's swing may determine whether or not she contacts the ball, but it is the bat's instantaneous velocity at ball contact that determines how fast and how far the ball will go. Similar situations exist in all racket sports and striking activities. To gymnasts, divers, and figure skaters, average angular velocity is the more important measure. It determines whether or not they will complete a certain number of twists or somersaults before landing or entering the water.

Angular and Linear Velocity

In several sports, especially ball sports, implements are used as extensions to the athletes' limbs. Golf, tennis, squash, lacrosse, racquetball, badminton, field hockey, and ice hockey are examples of these sports. One advantage of using these implements was explained earlier—they amplify the motion (displacement) of our limbs. Now compare the velocities of the ball (or shuttlecock or puck) in each of these sports if it were thrown by hand versus if it were struck (or thrown) with the respective stick (or racket or club). Which would be faster? The implements enable us to impart faster linear velocities to the ball (or shuttlecock or puck) in each of these sports. Go back to chapter 2 and take a look at table 2.3 again. Notice that six of the top seven fastest-moving objects are projectiles struck with an implement. This faster linear velocity is another advantage of using a stick, racket, or club. How is this accomplished?

Deriving the Relationship Between Linear and Angular Velocity

The relationship between angular displacement and linear distance traveled provides the answer. Consider a swinging golf club. All points on the club undergo the same angular displacement, and thus the same average angular velocity, because they all take the same time to undergo that displacement. But a point on the club closer to the club head (and farther from the axis of rotation) moves through a longer arc length than a point farther from the club head (and closer to the axis of rotation). The two points travel their respective arc lengths in the same time. The point farther from the axis of rotation must have a faster linear speed because it moves a longer distance but in the same time. Mathematically, this relationship can be derived from the relationship between angular displacement and linear distance traveled (arc length), shown in equation 6.4.

$$\ell = \Delta\theta r$$

Dividing both sides by the time it takes to rotate through the displacement gives us

$$\frac{\ell}{\Delta t} = \frac{\Delta\theta}{\Delta t}$$

$$\bar{s} = \bar{\omega}r \qquad (6.7)$$

where

ℓ = arc length,

r = radius,

$\Delta\theta$ = angular displacement (measured in radians),

t = time,

$\bar{s}$ = average linear speed, and

$\bar{\omega}$ = average angular velocity (measured in radians per second).

The average linear speed of a point on a rotating object is equal to the average angular velocity of the object times the radius (the distance from the point on the object to the axis of rotation of the object).

At an instant in time, this relationship becomes

$$v_T = \omega r \qquad (6.8)$$

where

v_T = instantaneous linear velocity tangent to the circular path of the point,

ω = instantaneous angular velocity (measured in radians per second), and

r = radius.

> **The average linear speed of a point on a rotating object is equal to the average angular velocity of the object times the radius.**

The instantaneous linear velocity of a point on a rotating object is equal to the instantaneous angular velocity of the rotating object times the radius. The direction of this instantaneous linear velocity is perpendicular to the radius and tangent to the circular path of the point. The instantaneous linear velocities for the two points on the golf club are shown in figure 6.8. For a demonstration of the relationship between linear and angular velocities, try self-experiment 6.2.

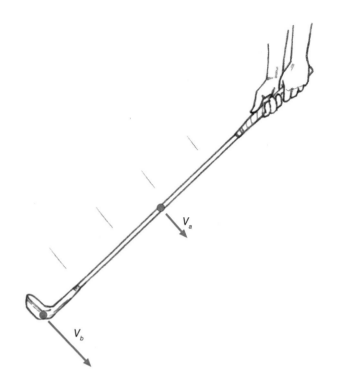

Figure 6.8 The linear velocity of the club head (v_b) is faster than the linear velocity of a point on the shaft (v_a) because the club head is farther from the axis of rotation.

Self-Experiment 6.2

Take out a ruler and place it on a table. Line up five same-denomination coins in a row on the table alongside the edge of the ruler. Fix one end of the ruler by pushing down on it with your finger. This point will be the pivot or axis of rotation of the ruler. Now draw back the other end of the ruler and then strike it so that it swings forward and hits the coins. Which coin went farther and faster? The coin farthest from the pivot end of the ruler went farther and faster. The average angular velocity of all points on the ruler was the same, but the radius, r, for the part of the ruler that hit this coin was largest, so this coin had the largest linear velocity.

Let's try another experiment. Place the ruler on top of a pencil so that the pencil is under the ruler at the 2 in. (5 cm) mark. Put a coin on the ruler at the 11 in. (28 cm) mark and another coin of the same denomination at the 4 in. (10 cm) mark. Now bring your hand downward and strike the ruler at the 0 in. end, flipping the coins into the air. Which coin flew higher and moved faster? The coin farthest from the axis of rotation (the pencil) went higher and faster. Now do some experimenting on your own to see how different locations of the axis of rotation affect the flight of one coin.

Applying the Relationship Between Linear and Angular Velocity

The sticks, clubs, and rackets used in the sports mentioned at the beginning of this section all use this relationship between linear and angular velocity and radius. The linear velocity of a point farther from the axis of rotation is faster if the angular velocity is the same. The woods, the longest clubs in a golf bag, are thus used for imparting a faster velocity to the ball to make it travel farther, whereas the shorter clubs are used for closer shots. If a batter can swing a longer bat with the same angular velocity as a shorter one, he'll be able to hit the ball farther with the longer bat. Gripping a tennis (or racquetball or squash) racket farther down on the grip (farther from the racket face) increases the radius of the swing and the linear velocity of the racket's striking surface (if the player is still able to swing the racket with the same angular velocity).

The implements used in these sports and other striking activities effectively lengthen the limbs of the performer, but the techniques used by the performer may increase the effective radius of the striking implements even further.

Consider a golf swing. Where is the axis of rotation for the club during the swing? Actually, the club and arms act together as one long lever, and the axis of rotation at ball contact lies along the vertebral column in the trunk, as shown in figure 6.9. The effective radius of the club head is thus much longer than just the length of the club itself.

Consider other sports and human movement activities. Where is the axis of rotation and how long is the effective radius during serving in tennis? While someone is chopping wood with a long-handled ax? While swinging a baseball bat? While bowling? While throwing a ball? While throwing a punch in boxing? While kicking a ball? While spiking a volleyball? While executing a badminton smash? While hitting a kill shot in racquetball? In many of these examples, the axis of rotation lies within the body, often between the shoulders.

The relationship among linear and angular velocities and radius of rotation also explains another advantage of the insertion of our muscles close to the joints. Consider the insertion of the biceps on the forearm. If it inserts only 1 in. from the elbow joint, and the hand is 10 in. from the joint, when the arm rotates, the linear velocity of the hand will be 10 times as fast as the linear velocity of the biceps insertion. Because the velocity of muscular

SAMPLE PROBLEM 6.2

The sweet spot on a baseball bat is 120 cm from the axis of rotation during the swing of the bat. (This seems long at first, since the bat is shorter than 120 cm. But the bat's axis of rotation is outside of the bat and passes through the body of the batter.) If the sweet spot on the bat moves 40 m/s, how fast is the angular velocity of the bat?

Solution:

Step 1: Identify the known quantities.

$r_{bat} = 120$ cm $= 1.2$ m

$v_{bat} = 40$ m/s

Step 2: Identify the unknown variable to solve for.

$\omega = ?$

Step 3: Search for the equation with the known and unknown variables.

$v = \omega r$

Step 4: Substitute the known quantities and solve for the unknown variable.

40 m/s $= \omega$ (1.2 m)

$\omega = (40$ m/s$) / (1.2$ m$) = 33.3$ rad/s

Step 5: Common sense check.
33.3 rad/s is about 5 rev/s—this seems fast, but a 40 m/s velocity at the sweet spot on the bat is fast also.

Axis

Effective radius

Figure 6.9 The axis of rotation and effective radius of the club head during a golf swing.

contraction is limited, the insertion of muscles close to the joint allows a relatively slow contraction velocity of the muscle to be amplified at the end of a limb. Our hands and feet can move at linear velocities much faster than the maximum velocities of the muscles that control the movements of the arm and leg.

Angular Acceleration

Angular acceleration is the rate of change of angular velocity. Its units of measurement are radians per second per second (rad/s/s or rad/s^2), degrees per second per second (°/s/s or °/s^2), or some unit of angular velocity per unit of time. Angular acceleration is abbreviated with the Greek letter α (alpha). Just like linear acceleration, angular acceleration is a vector quantity; it has direction and size. The right-hand thumb rule is used to describe directions of angular acceleration vectors.

> Angular acceleration is defined as the rate of change of angular velocity.

Average angular acceleration is computed as the change in angular velocity divided by time. Mathematically,

$$\bar{\alpha} = \frac{\Delta\omega}{\Delta t} = \frac{\omega_f - \omega_i}{\Delta t} \qquad (6.9)$$

where

$\bar{\alpha}$ = average angular acceleration,

$\Delta\omega$ = change in angular velocity,

Δt = time,

ω_f = final angular velocity, and

ω_i = initial angular velocity.

Angular acceleration occurs when something spins faster and faster or slower and slower, or when the spinning object's axis of spin changes direction.

> Angular acceleration occurs when something spins faster and faster or slower and slower, or when the spinning object's axis of spin changes direction.

Angular and Linear Acceleration

When the angular velocity of a spinning object increases, the linear velocity of a point on the object increases as well. The angular and linear accelerations of a point on a spinning object are related.

Tangential Acceleration

The component of linear acceleration tangent to the circular path of a point on a rotating object is called the **tangential acceleration** of that point. Remember that a line is tangent to a circle if the line intersects the circle at just one point. A line from this point to the center of the circle—a radial line—is perpendicular to the tangent line. Tangential acceleration is related to the angular acceleration of the object in the following way:

> $$a_T = \alpha r \qquad (6.10)$$

where

a_T = instantaneous tangential acceleration,

α = instantaneous angular acceleration (measured in rad/s^2), and

r = radius.

A point on a rotating object undergoes a linear acceleration tangent to its rotational path and equal to the angular acceleration of the object times the radius. Figure 6.10 shows the direction of the tangential acceleration of an object moving along a curved path.

Centripetal Acceleration

Equation 6.10 indicates that a linear acceleration tangent to the path of rotation of a point occurs if the rotating object is being accelerated angularly. What if no angular acceleration occurs? Does a point on a rotating object experience any linear acceleration if the object spins at a constant angular velocity with no angular acceleration? Yes. Remember that linear acceleration occurs if something speeds up, slows down, or changes direction. A point on an object spinning at constant angular velocity doesn't speed up or slow down, but because the point follows a circular path, it is constantly changing direction and is thus experiencing a constant linear acceleration.

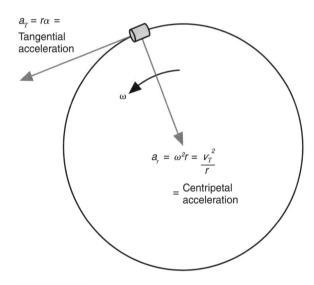

$a_T = r\alpha =$
Tangential acceleration

ω

$a_r = \omega^2 r = \dfrac{v_T^2}{r}$

$=$ Centripetal acceleration

Figure 6.10 The directions of tangential and centripetal accelerations for an object moving in a circular path.

In which direction does this linear acceleration occur? Tie a weight tightly onto a sturdy string and spin it around so that it circles with a constant angular velocity, as shown in figure 6.11.

Which way must you pull on the string to keep the weight traveling in a circle? You pull on the string with a force directed toward the center of the circle, toward the axis of rotation. This must be the force that causes the linear acceleration of the weight, so the linear acceleration must be directed toward the center of the circle as well. This acceleration is called **centripetal acceleration** (or radial acceleration), and the force causing it is called **centripetal force**.

Mathematically, centripetal acceleration can be defined using two different equations:

$$a_r = \frac{v_T^2}{r} \tag{6.11}$$

$$a_r = \omega^2 r \tag{6.12}$$

where

a_r = centripetal acceleration,

v_T = tangential linear velocity,

r = radius, and

ω = angular velocity.

Centripetal acceleration is the linear acceleration directed toward the axis of rotation (see figure 6.10). It is directly proportional to the square of the tangential linear velocity and the square of the angular velocity. If angular velocity is held constant, centripetal acceleration is directly proportional to the radius of rotation. If the tangential linear velocity is held constant, centripetal

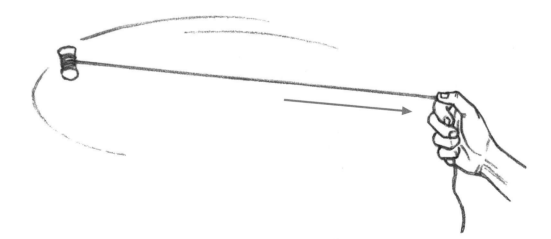

Figure 6.11 Tie a weight on a string and swing it around in a circle. Which way do you pull on the string?

acceleration is inversely proportional to the radius of rotation.

> **Centripetal acceleration is the linear acceleration directed toward the axis of rotation.**

When you run around a curve on the inside lane of a running track, you experience more centripetal acceleration and thus require more friction on your shoes to exert centripetal force on yourself than you would if you ran at the same linear speed in the outside lane (see figure 6.12). In these two cases, tangential linear velocity would be the same, but the radius of rotation would be larger in the outside lane. Equation 6.11 ($a_r = v^2_T/r$) would be the appropriate equation to use to evaluate this situation because the linear velocities are the same.

On the other hand, a hammer thrower swinging a hammer with a 1.0 m chain (see figure 6.13a) would have to pull on the chain with greater force than if he swung a hammer with a 0.75 m chain at the same angular velocity

(see figure 6.13b). The centripetal acceleration is greater (and thus the centripetal force provided by the thrower must be greater) for the hammer with the 1.0 m chain than for the one with the shorter chain because the radius of rotation is larger for the hammer with the 1.0 m chain. Equation 6.12 ($a_r = \omega^2 r$) would be the appropriate equation to use to evaluate this situation because the angular velocities are the same.

You can try this for yourself by turning in a circle while holding this book at arm's length and then pulling it closer to you while still turning at the same angular velocity. You have to exert a greater force on the book when it is farther from the axis of rotation.

Anatomical System for Describing Limb Movements

The first part of this chapter dealt with strictly mechanical terminology for describing angular motion. Anatomists use their own terminology for describing the relative

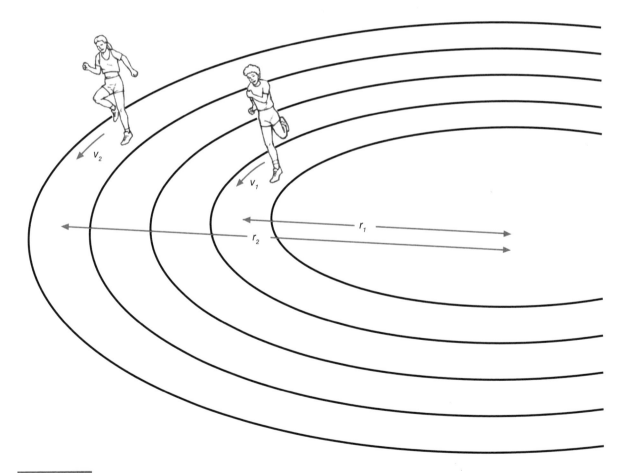

Figure 6.12 A runner in the inside lane must exert more centripetal force via friction than a runner in the outside lane if both runners have the same linear velocity ($a_{r1} > a_{r2}$ if $v_1 = v_2$). The centripetal acceleration of the runner in the inside lane is greater due to the smaller radius ($r_1 < r_2$).

Figure 6.13 A hammer thrower using a hammer with a 1.0 m chain (a) must exert a greater centripetal force than a thrower using a 0.75 m chain (b) if the two hammers rotate with the same angular velocity ($a_{r1} > a_{r2}$ since $\omega_1 = \omega_2$ and $r_1 > r_2$). The centripetal acceleration of the hammer with the 1.0 m chain is greater due to the larger radius.

angular positions and movements of the limbs of the body. You probably already have some knowledge of anatomy and anatomical terminology. This part of the chapter presents the system used by anatomists and other human movement professionals to describe relative positions and movements of the body and its parts.

Anatomical Position

Try to describe where a specific freckle or mole or hair is on your body. It's a difficult task. You probably identified the location of the freckle or mole or hair by describing where it was in relation to some body part. A similar situation occurs if you try to describe the movement of a limb. To describe the location or movement of a body part, the other parts of the body are used as a reference. But the human body can adopt many different positions, and the orientation of the limbs may change as well, so a common reference position of the body must be used. The most commonly used reference position of the human body is called **anatomical position**. Early anatomists suspended cadavers in this position to study them more easily. The body is in anatomical position when it is standing erect, facing forward, with the feet aligned parallel to each other, toes forward, arms and hands hanging straight below the shoulders at the sides, fingers extended, and palms facing forward. Anatomical position is the standard reference position for the body when we describe locations, positions, or movements of limbs or other anatomical structures. The body is shown in anatomical position in figure 6.14.

> Anatomical position is the standard reference position for the body when we describe locations, positions, or movements of limbs or other anatomical structures.

Planes and Axes of Motion

Anatomists have developed names to identify specific planes that pass through the body. Each plane has a corresponding axis that passes perpendicularly through the plane. These planes are useful to anatomists in describing planes of dissections or imaginary dissections. The planes are also useful in describing relative movements of body parts, with the axes used to describe the lines around which these motions occur.

Anatomical Planes

A plane is a flat two-dimensional surface. A **sagittal plane**, also called an anteroposterior plane, is an imaginary plane running anterior (front) to posterior (back) and superior (top) to inferior (bottom), dividing the body into right and left parts. A **frontal plane**, also called a coronal or lateral plane, runs side to side and superior to inferior, dividing the body into anterior and posterior parts. A **transverse plane**, or horizontal plane, runs from side to side and anterior to posterior, dividing the body into superior and inferior parts. All sagittal planes are perpendicular to all frontal planes, which are perpendicular to all transverse planes.

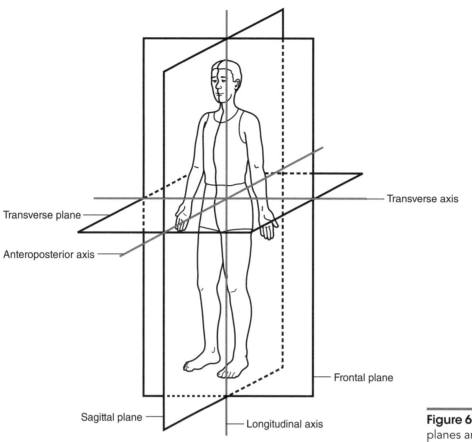

Figure 6.14 Cardinal anatomical planes and axes of the body.

Many sagittal planes can be imagined as passing through the body, but they are all parallel to each other. Likewise, many frontal or transverse planes can be imagined as passing through the body. A **cardinal plane** is a plane that passes through the midpoint or center of gravity of the body. The center of gravity is the point on which the body would balance if it were supported at only one point. The cardinal sagittal plane (midsagittal or median plane) is the plane that divides the body into equal right and left halves. The cardinal sagittal, frontal, and transverse planes of the body are shown in figure 6.14.

Biomechanically, the anatomical planes may be useful for locating anatomical structures, but their greatest worth is for describing limb movements. How can a plane be useful for describing movement? Movements of most limbs occur as rotations of the limbs. Rotations occur around specific axes of rotation and within specific planes of movement. Descriptions of the limb movements relative to each other are thus facilitated by identification of the axis of rotation around which the limb moves and the plane within which the limb moves.

⤵ **Biomechanically, the anatomical planes may be useful for locating anatomical structures, but their greatest worth is for describing limb movements.**

Anatomical Axes

We've defined several specific anatomical planes. What are the specific anatomical axes? The anatomical axes correspond to lines that are perpendicular to the previously defined anatomical planes. An **anteroposterior axis** (sagittal, sagittal–transverse, or cartwheel axis) is an imaginary line running from anterior to posterior and perpendicular to the frontal planes. The anteroposterior axis is often abbreviated as AP axis. An AP axis is defined by the intersection of a transverse plane with a sagittal plane, so it may also be called a sagittal–transverse axis. A **transverse axis** (lateral, frontal, mediolateral, frontal–transverse, or somersault axis) is an imaginary line running from left to right and perpendicular to the sagittal planes. A transverse axis is defined by the intersection of a transverse plane with a frontal plane, so it may also be called a frontal–transverse axis. A **longitudinal axis** (vertical, frontal–sagittal, or twist axis) is an imaginary line running from top to bottom and perpendicular to the transverse planes. A longitudinal axis is defined by the intersection of a frontal plane with a sagittal plane, so it may also be called a frontal–sagittal axis. All AP axes are perpendicular to all transverse axes, which are perpendicular to all longitudinal axes. An infinite number of these axes pass through the body. Examples of AP, transverse, and longitudinal axes are shown in figure 6.14.

Identifying Planes and Axes of Motion

Now let's see how axes and planes are used to describe human motion. Imagine a bicycle wheel. The wheel turns about an axle. The line along and through the axle of the wheel defines the axis of rotation of the wheel. This is the axis around which the wheel rotates. The spokes of the wheel are perpendicular to the axle or axis of rotation. Thus the spokes must lie in the plane of motion of the wheel. Now let's look at an example from the human body. Stand in anatomical position and flex (bend) your right arm at the elbow without moving your upper arm, as shown in figure 6.15. Think of your forearm as a spoke of the bicycle wheel and your elbow as the axle. What plane does your forearm lie in throughout the movement? It is moving within a sagittal plane. What axis is perpendicular to a sagittal plane? A transverse axis is perpendicular to a sagittal plane. The movement of your forearm and hand is in a sagittal plane and around a transverse axis.

Now let's describe the method we use to identify the plane or axis of motion for any limb movement. First, one principle should be noted. If you can identify either the plane or the axis of motion, the other is easily identified. If the plane within which the motion occurs is known, there is only one axis around which the motion can occur, and that is the axis perpendicular to the plane of motion. Likewise, if the axis around which the motion occurs is known, there is only one plane within which the motion can occur, and that is the plane perpendicular to the axis of motion. Table 6.1 lists each of the three planes of motion along with the corresponding axis that is perpendicular to that plane of motion.

Table 6.1 Anatomical Planes of Motion and Their Corresponding Axes of Motion

Plane	Axis
Sagittal	Transverse
Frontal	Anteroposterior (AP)
Transverse	Longitudinal

⟳ If you can identify either the plane or the axis of motion, the other is easily identified.

Before describing techniques for determining the plane of motion, a more precise and accurate definition of a plane may be helpful. Just as two points geometrically define a line in space, three noncolinear points or two intersecting lines define a plane in space. If a table or chair has only three legs, each of those legs will always touch a plane, such as the floor, because it takes only three points to define a plane. Even if the legs are different lengths and the table tips, three legs will always touch the floor. A table with four legs will always have three legs on one plane, but the fourth leg will touch that plane only if its end lies on the plane defined by the other three legs. If three legs are the same length and the fourth leg is a different length, only three legs will touch the floor.

Most of our limbs are longer in one dimension than in the others, so they can be thought of as long cylinders or even line segments. These line segments move by swinging or rotating around joints. If you can imagine the line segment defined by the limb at the start of motion and the line segment defined by the limb at any other instant during its motion, these line segments will intersect at the joint (if it's a single-joint motion). The plane of motion is defined by these two line segments. For example, stand in anatomical position. Imagine the line segment defined by a line drawn from your right shoulder to your right wrist. Now abduct at the shoulder (lift your right arm up and to the side, laterally away from your body, until it is shoulder height) as shown in figure 6.16. Imagine a line segment drawn through your shoulder and wrist now. What planes do both line segments fall within? When you were in anatomical position, your right arm was in a sagittal plane and a frontal plane. When you completed the movement, your arm was in a transverse plane and a frontal plane. Your arm was in the frontal plane at the start and end of the movement (and throughout the move-

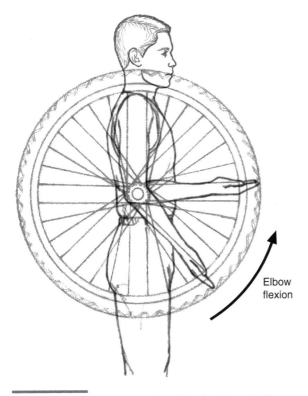

Elbow flexion

Figure 6.15 Imagining a bicycle wheel can help you identify the plane and axis of motion.

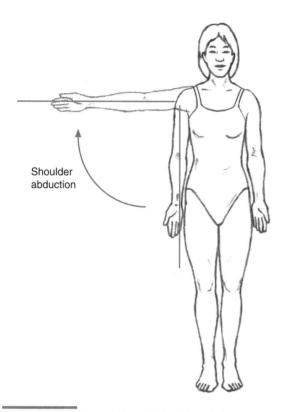

Shoulder abduction

Figure 6.16 Frontal view of shoulder abduction.

ment), so the movement occurred within a frontal plane. Because frontal planes are perpendicular to AP axes, the movement occurred around an AP axis.

Here's another procedure for determining the plane of motion. If you could view the movement from any vantage point, what vantage point would give you the best view so that you could always see the entire length of the moving limb? What vantage point would you view from so that the moving limb was not moving toward or away from you, but across your field of view? If the best viewpoint is from in front or behind, the view of the body is a frontal view and the movement is in a frontal plane. If the best viewpoint is from the left or right side, the view of the body is a sagittal view and the movement is in a sagittal plane. If the best viewpoint is from above or below (admittedly difficult vantage points to view from, but imagine the view if you could get into a position above or below a person), the view of the body is a transverse view and the movement is in a transverse plane.

For example, if you viewed the movement described in the previous example (abduction of the right arm at the shoulder) from a position in front of (anterior to, as shown in figure 6.16) or behind (posterior to) the person moving, you would see the full length of the arm throughout the movement, and the arm wouldn't move away from or toward you. If you viewed the motion from

the side (sagittally, as shown in figure 6.17a), the full length of the arm would be in view at the start; but as the arm moved upward, it would appear to shorten until you would only see the hand when the arm reached shoulder height. The arm would move toward you. If you viewed the motion from above (transversely, as shown in figure 6.17b), only the shoulder would be in view at the start; but the full length of the arm would be in view when it reached shoulder height. The arm would move toward you. The best view was from the front, so the view was a frontal view and the movement was in a frontal plane. The axis perpendicular to a frontal plane is an AP axis.

In some limb movements, the length of the limb doesn't swing around an axis, but the limb rotates or twists about its length. In this case, it is easier to first determine the axis of rotation of the limb and then determine its plane of motion. The axis of rotation of a limb twisting about its length is defined by the direction of the line running the length of the limb from the proximal to the distal end. If this line is parallel to a longitudinal axis, it is a longitudinal axis. If it is parallel to an AP axis, it is an AP axis. If it is parallel to a transverse axis, it is a transverse axis.

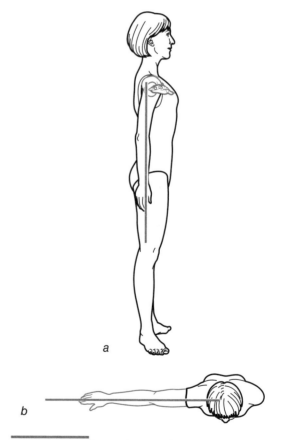

a

b

Figure 6.17 Sagittal (a) and transverse (b) views of shoulder abduction.

For example, stand in anatomical position and turn the palm of your right hand toward your side and then to the rear as shown in figure 6.18. Your arm twisted around an axis through its length. A line drawn from the proximal to the distal end of your arm is a vertical line and parallel to a longitudinal axis. The axis of rotation for this movement was a longitudinal axis. A longitudinal axis is perpendicular to a transverse plane, so the twisting motion of your right arm occurred in the transverse plane.

So far, the examples we've used have all been movements that occurred within one of the three planes of motion we defined. Sagittal, transverse, and frontal planes are primary planes of motion. Other planes exist that are not primary planes of motion, but movements can occur within these planes. For example, what plane do your arms move in when you swing a golf club as shown in figure 6.19? The best viewpoint wouldn't be directly in front or directly above but in front and above. The plane of motion is a diagonal plane between the transverse plane and the frontal plane, and the axis of motion is a diagonal axis between a longitudinal and an AP axis. An infinite number of diagonal planes and axes exist within or around which our limbs can move. The principal planes and axes give us standard planes and axes from which other, diagonal planes and axes can be described.

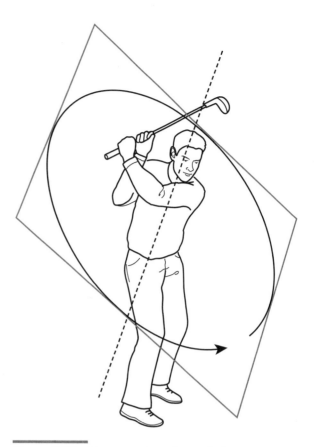

Figure 6.19 Movement in a diagonal plane.

Joint Actions

We can identify the plane and action of limb motions, but what are the terms that describe the limb motions? Human movement description uses terminology that describes the relative movements of two limbs on either side of a joint (relative angular motion) rather than terminology that describes individual limb movements (absolute angular motion). The terms thus describe joint actions, the relative angular movements of the limbs on the distal and proximal sides of a joint. From anatomical position, the joint actions that occur when limbs move around transverse joint axes and within sagittal planes are flexion, extension, hyperextension, plantar flexion, and dorsiflexion. The joint actions that occur when limbs move around AP joint axes and within frontal planes are abduction, adduction, radial deviation (radial flexion), ulnar deviation (ulnar flexion), inversion, eversion, elevation, depression, lateral flexion to the right, and lateral flexion to the left. The joint actions that occur when limbs move around longitudinal axes and within transverse planes are internal (inward or medial) rotation, external (outward or lateral) rotation, pronation, supination, horizontal abduction (horizontal extension),

Figure 6.18 Movement around a longitudinal axis.

horizontal adduction (horizontal flexion), rotation left, and rotation right.

> Human movement description uses terminology that describes the relative movements of two limbs on either side of a joint (relative angular motion) rather than terminology that describes individual limb movements (absolute angular motion).

Movements Around Transverse Axes

Flexion, extension, and hyperextension are joint actions occurring at the wrist, elbow, shoulder, hip, knee, and intervertebral joints. Starting from anatomical position, **flexion** is the joint action that occurs around the transverse axes through these joints and causes limb movements in sagittal planes through the largest range of motion. **Extension** is the joint action that occurs around the transverse axes through these joints and causes the opposite limb movements in sagittal planes that return the limbs to anatomical position. **Hyperextension** is the joint action that occurs around the transverse axes and is a continuation of extension past anatomical position. Elbow flexion thus occurs when the forearm is moved forward and upward and the angle between the forearm and the upper arm at the anterior side of the elbow joint gets smaller. Elbow extension occurs when the forearm is returned to anatomical position. Elbow hyperextension

would occur if the forearm could continue extending past anatomical position. Figure 6.20, *a* through *g,* shows the flexion, extension, and hyperextension joint actions that occur at the wrist, elbow, shoulder, hip, knee, trunk, and neck joints, respectively.

Dorsiflexion and plantar flexion are joint actions that occur at the ankle. Starting from anatomical position, **dorsiflexion** is the joint action that occurs around the transverse axis through the ankle joint and causes the foot to move in a sagittal plane such that it moves forward and upward toward the leg. When you lift your toes off the ground and put your weight on your heels, you are dorsiflexing at your ankles. **Plantar flexion** is the joint action that occurs around the transverse axis through the ankle joint and causes the opposite movement of the foot in a sagittal plane so that the foot moves downward away from the leg. When you stand on your toes, you are plantar flexing at your ankles. Dorsiflexion and plantar flexion are also shown in figure 6.20*g.*

Movements Around Anteroposterior Axes

Abduction and adduction are joint actions occurring at the shoulder and hip joints. Starting from anatomical position, **abduction** is the joint action that occurs around the AP axes through these joints and causes limb movements

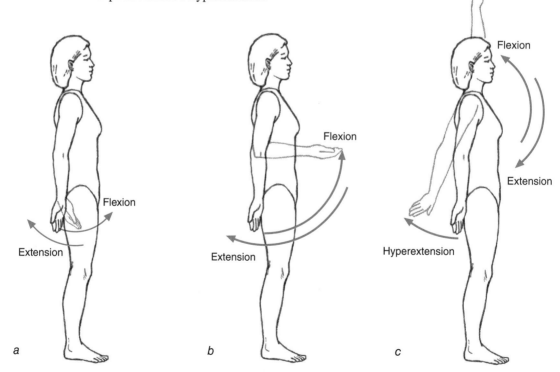

Figure 6.20 Sagittal plane joint actions at the wrist, elbow, shoulder, hip, knee, trunk, and neck and ankle.

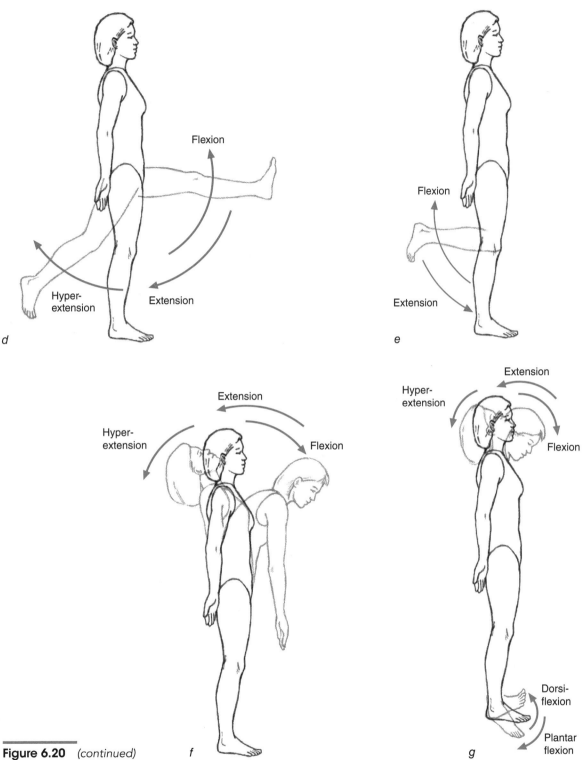

Figure 6.20 (continued)

in frontal planes through the largest range of motion. **Adduction** is the joint action that occurs around the AP axes through these joints and causes limb movement in a frontal plane back toward anatomical position. Shoulder abduction thus occurs when the arm is moved upward and laterally away from the body. Shoulder adduction occurs

when the arm is returned to anatomical position. Figure 6.21, *a* and *b,* shows the abduction and adduction joint actions that occur at the shoulder and hip joints.

Ulnar deviation (adduction or ulnar flexion) and radial deviation (abduction or radial flexion) are joint actions occurring at the wrist joint (see figure 6.21*c*). Starting

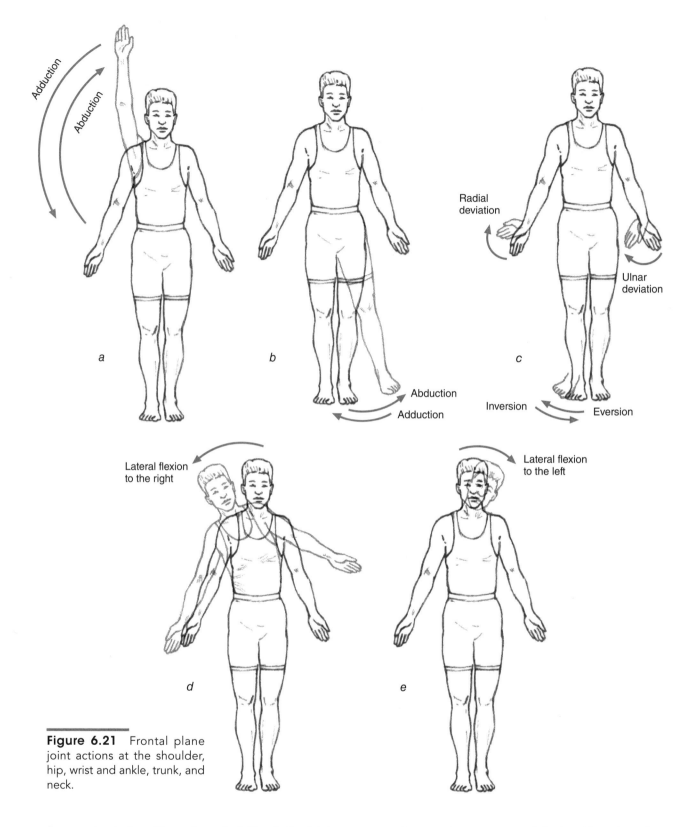

Figure 6.21 Frontal plane joint actions at the shoulder, hip, wrist and ankle, trunk, and neck.

from anatomical position, **ulnar deviation** is the joint action that occurs around the AP axis through the wrist and causes hand movement in a frontal plane toward the little finger. **Radial deviation** is the joint action that occurs around the AP axis through the wrist joint and causes the opposite movement of the hand in a frontal plane, moving it laterally toward the thumb.

Inversion and eversion are frontal plane movements that occur at the ankle joint (see figure 6.21c). These joint actions occur around an AP axis through the foot. Starting from anatomical position, **inversion** occurs when the medial side of the sole of the foot is lifted. The return to anatomical position and the movement of the foot beyond anatomical position where the lateral side of the sole of the foot is lifted is **eversion**.

The movements of the scapula (the shoulder blade) and shoulder girdle occur primarily in the frontal plane as well. These movements include abduction (movement of the scapula away from the midline) and adduction (movement of the scapula toward the midline), **elevation** (superior movement of the scapula) and **depression** (inferior movement of the scapula), and **upward rotation** (such that the medial border moves inferiorly and the shoulder joint moves superiorly) and **downward rotation** (such that the medial border moves superiorly and the shoulder joint moves inferiorly).

Lateral flexion to the left or right also occurs in the frontal plane. Lateral flexion of the trunk to the left occurs when you lean to the left, and lateral flexion to the right occurs when you lean to the right. Likewise, lateral flexion of the neck to the left occurs when you lean your head toward your left shoulder, and lateral flexion of the neck to the right occurs when you lean your head toward your right shoulder. Lateral flexion is shown in figure 6.21, *d* and *e*.

Movements Around Longitudinal Axes

Internal rotation (inward or medial rotation) and **external rotation** (outward or lateral rotation) are joint actions occurring at the shoulder and hip joints. Starting from anatomical position, internal rotation is the joint action that occurs around the longitudinal axes through these joints and causes limb movements in the transverse plane such that the knees turn inward toward each other or the palms of the hands turn toward the body. External rotation is the joint action that occurs around the longitudinal axes through these joints and causes the opposite limb movements in transverse planes and returns the limbs to anatomical position or moves them beyond anatomical position. Figure 6.22, *a* and *b*, shows the internal and external rotation joint actions that occur at the hip and shoulder joints.

Pronation and supination are joint actions occurring at the radioulnar joint in the forearm. Starting from anatomical position, **pronation** is the joint action that occurs around the longitudinal axis of the forearm and through the radioulnar joint and causes the palm to turn toward the body. This motion is similar to internal rotation at the shoulder joint except that it occurs at the radioulnar joint. **Supination** is the joint action that occurs around the longitudinal axis through the radioulnar joint and causes the opposite limb movement in a transverse plane and returns the forearm and hand to anatomical position or moves them beyond anatomical position. Figure 6.22*d*

also shows the supination and pronation actions that occur at the radioulnar joint.

Horizontal abduction (horizontal extension or transverse abduction) and **horizontal adduction** (horizontal flexion or transverse adduction) are joint actions occurring at the hip and shoulder joints. These joint actions do not commence from anatomical position. First, hip or shoulder flexion must occur and continue until the arm or thigh is in the transverse plane. Horizontal abduction is then the movement of the arm or leg in the transverse plane around a longitudinal axis such that the arm or leg moves away from the midline of the body. The return movement is horizontal adduction. These joint actions are also shown in figure 6.22*c*.

Rotations of the head, neck, and trunk also occur around a longitudinal axis. Turning your trunk so that you face left is rotation to the left, and turning your trunk so that you face right is rotation to the right. These actions are shown in figure 6.22, *e* and *f*.

Circumduction is a multiple-axis joint action that occurs around the transverse and AP axes. Circumduction is (a) flexion combined with abduction and then adduction or (b) extension or hyperextension combined with abduction and then adduction. The trajectory of a limb being circumducted forms a cone-shaped surface. If you abduct your arm at the shoulder joint and then move your arm and forearm so that your hand traces the shape of a circle, the joint action occurring at the shoulder joint is circumduction.

Just as there are diagonal planes and axes, joint actions may also occur in diagonal planes around diagonal axes. Such diagonal joint actions may be combinations of joint actions if they occur at multiple-axis joints, or they may be one of the joint actions described earlier if the joint or limb has been moved into a diagonal plane by the action at a more proximal joint. Each of the terms describing joint actions really specifies the direction of relative angular motion at a joint. The joint actions and the corresponding planes and axes of motion for these actions are summarized in table 6.2.

⟲ **Each of the terms describing joint actions really specifies the direction of relative angular motion at a joint.**

Summary

Angular kinematics is concerned with the description of angular motion. Angles describe the orientation of two lines. Absolute angular position refers to the orientation of an object relative to a fixed reference line or plane, such as horizontal or vertical. Relative angular position refers to the orientation of an object relative to a nonfixed reference line or plane. Joint angles are relative, whereas limb positions may be relative or absolute. The angular

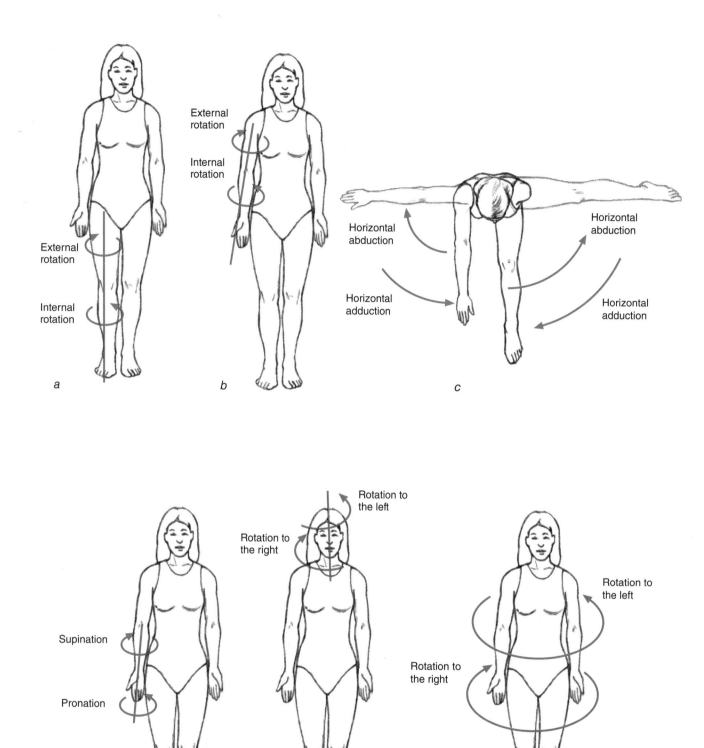

Figure 6.22 Transverse plane joint actions at the hip, shoulder, shoulder and hip, radioulnar joint, neck, and trunk.

Table 6.2 Joint Actions and Their Corresponding Planes and Axes of Motion

Plane of motion	Axis	Joint actions
Sagittal	Transverse	Flexion Extension Hyperextension Plantar flexion Dorsiflexion
Frontal	Anteroposterior	Abduction Adduction Ulnar deviation Radial deviation Inversion Eversion Elevation Depression Upward rotation Downward rotation Lateral flexion to the left Lateral flexion to the right
Transverse	Longitudinal	Internal rotation External rotation Pronation Supination Horizontal abduction Horizontal adduction Rotation to the right Rotation to the left

movements of limbs around joints are described with terminology developed by anatomists using the anatomical position of the body as a reference. The three principal anatomical planes (sagittal, frontal, and transverse) along with their corresponding axes (transverse, anteroposterior, and longitudinal) are also useful for describing movements of the limbs.

When an object rotates, it undergoes an angular displacement. To define the angular displacement, the axis and plane of rotation must be known. The direction of the angular displacement (and all other angular motion and torque vectors) is then established using the right-hand thumb rule. The definitions of angular displacement, angular velocity, and angular acceleration are similar to those for their linear counterparts.

The linear displacement and distance traveled by a point on a rotating object are directly proportional to the radius of rotation. The linear distance traveled equals the product of the angular displacement measured in radians times the radius of rotation.

Tangential linear velocity and acceleration of a point on a rotating object are directly proportional to the radius as well. The tangential linear velocity is equal to the product of the angular velocity times the radius of rotation. Lengthening the radius while maintaining the angular velocity is an important principle in a variety of striking skills. Tangential linear acceleration is equal to the product of angular acceleration times the radius of rotation.

Centripetal acceleration (also called radial acceleration) of an object rotating in a circular path is the component of linear acceleration directed toward the axis of rotation. It is directly proportional to the square of the tangential linear velocity or the square of the angular velocity. Centripetal force is the force exerted on the rotating object to cause the centripetal acceleration.

KEY TERMS

abduction (p. 184)
absolute angular position (p. 168)
adduction (p. 185)
anatomical position (p. 179)
angular acceleration (p. 176)
angular displacement (p. 169)
angular position (p. 168)
angular velocity (p. 173)
anteroposterior axis (p. 180)
average angular velocity (p. 173)
cardinal plane (p. 180)
centripetal acceleration (p. 177)
centripetal force (p. 177)
circumduction (p. 187)
depression (p. 187)

dorsiflexion (p. 184)
downward rotation (p. 187)
elevation (p. 187)
eversion (p. 187)
extension (p. 184)
external rotation (p. 187)
flexion (p. 184)
frontal plane (p. 179)
horizontal abduction (p. 187)
horizontal adduction (p. 187)
hyperextension (p. 184)
instantaneous angular
  velocity (p. 173)
internal rotation (p. 187)
inversion (p. 187)

lateral flexion (p. 187)
longitudinal axis (p. 180)
plantar flexion (p. 184)
pronation (p. 187)
radial deviation (p. 186)
radians (p. 169)
relative angular position (p. 168)
sagittal plane (p. 179)
supination (p. 187)
tangential acceleration (p. 176)
transverse axis (p. 180)
transverse plane (p. 179)
ulnar deviation (p. 186)
upward rotation (p. 187)

REVIEW QUESTIONS

1. Most skeletal muscles in our limbs attach close to the joints.
   a. What are the advantages of this arrangement?
   b. What is the disadvantage of this arrangement?
2. In golf, the longest club is the driver, and the shortest club is the pitching wedge. Why is it easier to hit the ball farther with a driver than with a pitching wedge?
3. What advantages does a longer-limbed individual have in throwing and striking activities? ↑ velocity & ↓ F for T
4. Explain how step length (a linear kinematic variable) might increase if flexibility exercises increase the range of motion at the hip joint, thus increasing the angular displacement (an angular kinematic variable) of the hip joint during a step.
5. Why is sprinting around a curve in lane 1 (the inside lane) more difficult to do than sprinting around the curve in lane 8 (the outside lane)?
6. What is the plane of motion for most of the joint actions that occur during sprint running? What is the corresponding axis of motion for these joint actions?
7. When you swing a baseball bat, what is the plane of motion for the action occurring at your leading shoulder? What is the axis of motion? What joint action occurs at the leading shoulder during the swing?
8. During the delivery phase of a baseball pitch, what joint actions occur at the shoulder and elbow joints of the throwing arm?
9. What joint action occurs at the shoulder joint and what are the plane and axis of motion during the pulling-up phase of a wide-grip pull-up?

Bar

Wide-grip
pull-up

10. How do the actions at the shoulder joint differ between a wide-grip pull-up and a narrow-grip pull-up?

Bar

Narrow-grip pull-up

11. When you spike a volleyball, what joint action occurs at the elbow joint of your hitting arm?

PROBLEMS

1. Fred examines the range of motion of Oscar's knee joint during his rehabilitation from a knee injury. At full extension, the angle between the lower leg and thigh is 178°. At full flexion, the angle between the lower leg and thigh is 82°. During the test, Oscar's thigh was held in a fixed position and only the lower leg moved. What was the angular displacement of Oscar's leg from full extension to full flexion? Express your answer in (a) degrees and (b) radians.

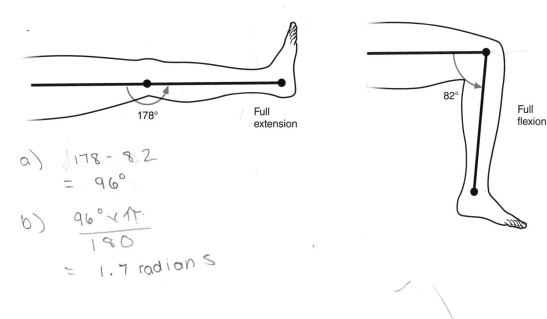

178°   Full extension

82°   Full flexion

a) $178 - 82$
   $= 96°$

b) $\dfrac{96° \times \pi}{180}$
   $= 1.7$ radians

$\frac{3(2\pi)}{0.8} = \omega$

2. Adelaide performs a triple twisting jump while figure skating. She rotates around her longitudinal axis three times while she is in the air. The time it takes to complete the jump from takeoff to landing is 0.8 s. What was Adelaide's average angular speed in twisting for this jump?

3. Becky is sparring in a Tae Kwon Do class. Her opponent executes a roundhouse kick. The average angular velocity of his kicking leg and foot is 20 rad/s. The angular displacement of his foot to Becky's head is 5 radians. How much time does Becky have to move if she wants to avoid being kicked in the head? $\frac{5}{20} = 0.25 \text{ sec}$

4. When Josh begins his discus throwing motion, he spins with an angular velocity of 5 rad/s. Just before he releases the discus, Josh's angular velocity is 25 rad/s. If the time from the beginning of the throw to just before release is 1 s, what is Josh's average angular acceleration? $25 - 5 = 20 \text{ rad/s}$ $\frac{20}{1 s} = 20$

5. The tendon from Lissa's knee extensor muscles attaches to the tibia bone 1.5 in. (4 cm) below the center of her knee joint, and her foot is 15 in. (38 cm) away from her knee joint. What arc length does Lissa's foot move through when her knee extensor muscles contract and their point of insertion on the tibia moves through an arc length of 2 in. (5 cm)?

6. During Charlie's golf drive, the angular velocity of his club is zero at the top of the backswing and 20 rad/s at the bottom of the downswing just before ball impact. The downswing lasts 0.20 s, and the distance from the club head to the axis of rotation is 2.0 m at the bottom of the downswing.

   a. What is the average angular acceleration of the club during the downswing? $\frac{\omega}{t} = \frac{20}{0.2} = 100 \text{ rad/}$

   b. What is the linear velocity of the club head just before impact with the ball? $\lambda = \Delta \omega r = 20 \times 2 = 40$

7. A hammer thrower spins with an angular velocity of 1200°/s. The distance from his axis of rotation to the hammer head is 1.2 m. $\frac{1200 \times \pi}{60} = 21 \text{ rad}$

   a. What is the linear velocity of the hammer head? $\lambda = \Delta \omega r = 21 \times 1.2 = 25.1 \text{ m/s}$

   b. What is the centripetal acceleration of the hammer head? $a_c = \omega^2 r = 21^2 (1.2) 529.2 \text{ m/s}^2$

8. During the delivery phase of fastball pitch, the arm internally rotates at the shoulder. The angular velocity of this internal rotation peaks at 120 rad/s. At this instant, the elbow angle is 90°, so the angular velocity of the forearm is also 120 rad/s. The baseball in the pitcher's hand is 35 cm from this axis of rotation through the shoulder joint. At this instant the linear velocity of the baseball is 45 m/s.

   a. How much of the baseball's total linear velocity is due to the 120 rad/s angular velocity of the forearm?

   b. What is the centripetal acceleration of the baseball at this instant? $a_c = \omega^2 r$

   c. How large is the force exerted by the pitcher on the baseball to cause this acceleration? The baseball's mass is 145 g.

9. Julie runs around the curve of a track in lane 1 while Monica runs around the curve in lane 8. The radius of lane 8 is twice as big as the radius of lane 1. If Julie has to run 50 m to get fully around the curve in lane 1, how far does Monica have to run to get fully around the curve in lane 8?

10. A hook in boxing primarily involves horizontal flexion of the shoulder while maintaining a constant angle at the elbow. During this punch, the horizontal flexor muscles of the shoulder contract and shorten at an average speed of 75 cm/s. They move through an arc length of 5 cm during the hook, while the fist moves through an arc length of 100 cm. What is the average speed of the fist during the hook?

11. A baseball pitcher pitches a fastball with a horizontal velocity of 40 m/s. The horizontal distance from the point of release to home plate is 17.50 m. The batter decides to swing the bat 0.30 s after the ball has been released by the pitcher. The average angular velocity of the bat is 12 rad/s. The angular displacement of the bat from the batter's shoulder to hitting positions above the plate is between 1.5 and 1.8 rad.

   a. Will the bat be in a hitting position above the plate when the ball is above the plate? Assume the pitch is in the strike zone.

   b. Assume that the batter does hit the ball. If the bat's instantaneous angular velocity is 30 rad/s at the instant of contact, and the distance from the sweet spot on the bat to the axis of rotation is 1.25 m, what is the instantaneous linear velocity of the sweet spot at the instant of ball contact?

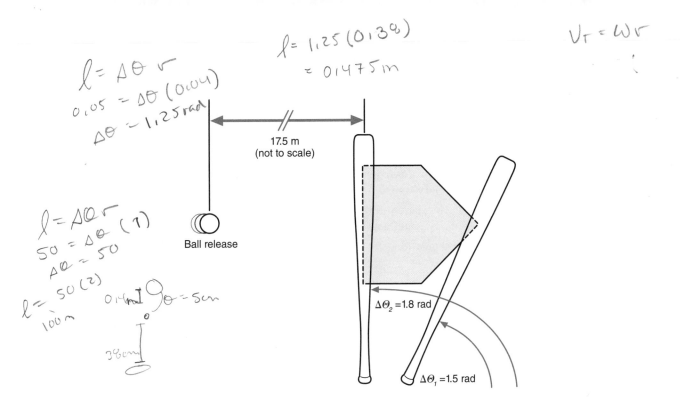

Handwritten annotations:

$\ell = \Delta\theta \, r$

$0.05 = \Delta\theta (0.04)$

$\Delta\theta = 1.25 \text{ rad}$

$\ell = 1.25 (0.138)$

$= 0.1475 \text{ m}$

$V_T = \omega r$

$\ell = \Delta\theta \, r$ (?)

$50 = \Delta\theta \cdot 50$

$\Delta\theta = 50$

$\ell = 50 (2)$

$\ell = 100 \text{ m}$

0.14 m $\quad 9\theta = 5 \text{ cm}$

38 cm

17.5 m
(not to scale)

Ball release

$\Delta\theta_2 = 1.8 \text{ rad}$

$\Delta\theta_1 = 1.5 \text{ rad}$

Motion Analysis Exercises Using MaxTRAQ

If you haven't done so already, review the instructions for downloading and using the educational version of the MaxTRAQ motion analysis software at the beginning of this book, then download and install the software. Once this is done, you are ready to try the following two-dimensional kinematic analyses using MaxTRAQ.

1. Open MaxTRAQ. Select Tools in the menu bar and then open Options under the Tools menu. In the Options submenu, select Video. To the upper right side of the Video window, under Video Aspect Ratio, make sure that Default-Used Preferred Aspect Ratio is selected. In the lower half of the right side of the Video window, under Deinterlace Options, select BOB, use Odd lines first, and Stretch Image Vertically. Click OK. Close MaxTRAQ and reopen it to have the deinterlace options take effect. Next, open the *Giant Swing* video from within MaxTRAQ. Make sure the scaling/calibration tool is activated by clicking View on the menu bar, then selecting Tools from the drop down menu, and making sure Show Scale is checked. Open the scaling tool by clicking on Tools on the menu bar and selecting Scale. In the scaling tool window that opens, set the gauge length to 305 cm. Now place the cursor over the lower left corner of the blue stripe in the center of the back wall and click the left mouse button once, then place the cursor over the lower right corner of the blue stripe in the center of the back wall and click the left mouse button a second time. The scale should appear in the video window. Hide the scale by selecting View in the menu bar; then click Tools and uncheck Show Scale.

   a. What is the absolute angular position (measured in degrees from a right horizontal) for the gymnast at the start of the first swing, in frame 1? Digitize the gymnast's hip joint center (the middle of her shorts) and the end of the bar. The end of the bar is at the same level as the bottom of the blue stripe on the wall in the background. Because you are digitizing more than one point in the same frame, you will need to increase the number of points listed in the digitizing controls panel on the right to two. You can do this using the arrow buttons at the bottom of the list of points. Before selecting each point, make sure that that point is highlighted in the list to the right. Compute the difference in x-coordinates for the two points Δx, and the difference in y-coordinates for the two points Δy. When computing the difference, subtract the bar coordinate from the hip coordinate. Divide Δy by Δx and take the arctangent of the result. The answer is the angle that the line from the bar to the hip forms with horizontal. Be sure your calculator angle mode is set to degrees.

b. What is the absolute angular position (measured with a right horizontal) for the gymnast at the bottom of the first swing, in frame 61?

c. What is the absolute angular position (measured with a right horizontal) for the gymnast at the start of the second swing, in frame 122?

d. What is the absolute angular position (measured with a right horizontal) for the gymnast at the bottom of the second swing, in frame 184?

e. What is the absolute angular position (measured with a right horizontal) for the gymnast at the start of the third swing, in frame 237?

f. What is the average angular velocity of the gymnast during first half of her first giant swing from frame 1 to 61?

g. What is the average angular velocity of the gymnast during the second half of her first giant swing from frame 61 to 122?

h. What is the average angular velocity of the gymnast during first half of her second giant swing from frame 122 to 184?

i. What is the average angular velocity of the gymnast during the second half of her second giant swing from frame 184 to 237?

j. What is the average angular velocity of the gymnast during her first giant swing from frame 1 to 122?

k. What is the average angular velocity of the gymnast during her second giant swing from frame 122 to 237?

l. What actions does the gymnast do to increase her angular velocity from the first giant swing to the second giant swing?

m. Estimate the instantaneous angular velocity of the gymnast at the bottom of the second giant swing. Digitize the gymnast's hip joint center and the bar end in frame 183 and again in frame 185. Compute the angular position of the gymnast in these two positions. Then compute the angular displacement. Estimate the instantaneous angular velocity by dividing this angular displacement by the time between these two frames (multiply the angular displacement by 60 and divide by 2).

n. Estimate the instantaneous angular velocity of the gymnast at the top of the giant swing—at the end of the second giant swing and the beginning of the third giant swing. Digitize the gymnast's hip joint center and the bar end in frame 236 and again in frame 238. Compute the angular position of the gymnast in these two positions. Then compute the angular displacement. Estimate the instantaneous angular velocity by dividing this angular displacement by the time between these two frames (multiply the angular displacement by 60 and divide by 2).

o. Why is the gymnast's angular velocity so much faster at the bottom of the giant swing compared to the top of the giant swing?

chapter 7

Angular Kinetics
Explaining the Causes of Angular Motion

objectives

When you finish this chapter, you should be able to do the following:

- Define moment of inertia

- Explain how the human body's moment of inertia may be manipulated

- Explain Newton's first law of motion as it applies to angular motion

- Explain Newton's second law of motion as it applies to angular motion

- Explain Newton's third law of motion as it applies to angular motion

- Define angular impulse

- Define angular momentum

- Explain the relationship between angular impulse and angular momentum

© Olga Besnard/Dreamstime.com

A diver jumps off the 3 m platform. She is in a stretched-out position (a layout) and barely rotating at first. Then she flexes at the hips and folds herself in half (she pikes), and her rotation speeds up as if by magic. She spins around three and a half times in this position as she falls toward the water. When she is just above the water, her spinning seems to stop as she opens up into a layout position and enters the water with barely a splash. How did the diver control her rate of spin (her angular velocity) during the dive? It seemed as if she could make herself spin faster or slower at will. The information in this chapter should help us answer this question.

This chapter is about the subbranch of

mechanics called kinetics. Specifically, it is about angular kinetics, or the causes of angular motion. Many of the concepts developed in linear kinetics have counterparts in angular kinetics. The concepts you learned in chapter 3 are thus important for a good understanding of the concepts introduced in this chapter. An understanding of angular kinetics will help explain why a discus thrower spins a discus; why a long jumper "runs" in the air; how figure skaters, divers, and gymnasts can speed up or slow down their rate of spin when they're not in contact with the ground; and also some aspects of the techniques used in a variety of other skills in sports and human movement.

Angular Inertia

Earlier in the book, inertia was defined as the property of an object that resists changes in motion. Linear inertia was quantified as an object's mass. It is more difficult to speed up, slow down, or change the direction of a more massive object because it has more linear inertia. The property of an object to resist changes in its angular motion is **angular inertia** or **rotary inertia**. It is more difficult to speed up or slow down the rotation or change the axis of rotation of an object with more angular inertia. What factors affect the angular inertia of an object? How is angular inertia quantified?

⟳ **The property of an object to resist changes in its angular motion is angular inertia or rotary inertia.**

To learn some things about angular inertia, try self-experiment 7.1.

Self-Experiment 7.1

Hold this book so that its spine is parallel to the floor and throw it up in the air so that it turns one revolution about an axis perpendicular to its cover (see figure 7.1). Now find a heavier book (or a lighter one) and do the same trick. Which book was easier to flip? Which book has less angular inertia? The lighter book was easier to flip. Its angular inertia is smaller.

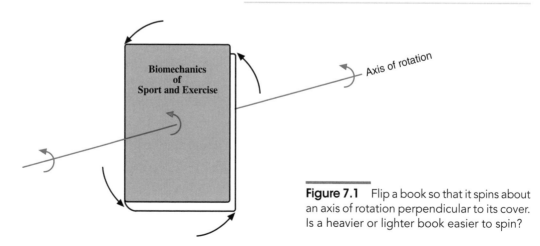

Figure 7.1 Flip a book so that it spins about an axis of rotation perpendicular to its cover. Is a heavier or lighter book easier to spin?

The results of self-experiment 7.1 indicate that angular inertia is affected by the mass of an object. In fact, it is directly proportional to the mass of the object. If you consider the various sport implements and tools that you swing, you'll note that the heavier implements are more difficult to start swinging or stop swinging. A heavier bat is harder to swing than a lighter bat. Is mass the only property of an object that affects its angular inertia? Try self-experiment 7.2 to continue our investigation of angular inertia.

Self-Experiment 7.2

Open this book to page 10 (or some page close to the beginning of the book), and place it on a table with the open pages facing down. Apply a force couple to the book with your fingers and spin it on the table. Try this several times to see how large a torque you must apply to rotate the book through an angular displacement of 180° or one-half a revolution. Now open up the book to the middle pages and repeat the experiment. Try spinning it several times. In which configuration did the book turn more easily—when it was opened to the front or when it was opened to the middle? When the book was opened to page 10, it was easier to rotate. It took more torque to cause the same rotation of the book when it was opened to the middle. In this configuration, the book had more angular inertia.

In self-experiment 7.2, what characteristic of the book changed to cause this difference in angular inertia between the two spinnings? The mass of the book stayed the same, but where the mass was located relative to the axis of rotation changed. When the book was opened to page 10, most of the mass of the book was on one side and close to the center of gravity and the axis of rotation. When the book was opened to the middle, about half of the book's mass was on each side of the spine. The center of gravity and axis of rotation were then through the spine of the book, and more mass was located farther from the axis of rotation. Angular inertia is affected by mass and how the mass is distributed relative to the axis of rotation. Again, consider the various sport implements and tools that you swing; the longer implements are more difficult to start or stop swinging. A longer bat is harder to swing than a shorter one.

Mathematical Definition of Angular Inertia

The quantity that describes angular inertia is called **moment of inertia.** Moment of inertia is abbreviated with the letter I. Theoretically, an object may be considered to be composed of many particles of mass. The moment of inertia of such an object about an axis through its center of gravity can be defined mathematically as follows:

$$I_a = \Sigma m_i r_i^2 \qquad (7.1)$$

where

I_a = moment of inertia about axis a through the center of gravity,

Σ = summation symbol,

m_i = mass of particle i, and

r_i = radius (distance) from particle i to axis of rotation through the center of gravity.

Each particle provides some resistance to change in angular motion. This resistance is equal to the mass of the particle times the square of the radius from the particle to the axis of rotation. The sum of all the particles' resistances to rotation is the total moment of inertia of the object. The units of measurement for moment of inertia are units of mass times units of length squared, or kg·m² in SI units.

Moment of inertia may also be represented mathematically as

$$I_a = m k_a^2 \qquad (7.2)$$

where

I_a = moment of inertia about axis a through the center of gravity,

m = mass of the object, and

k_a = radius of gyration about axis a through the center of gravity.

The **radius of gyration** is a length measurement that represents how far from the axis of rotation all of the object's mass must be concentrated to create the same resistance to change in angular motion as the object had in its original shape.

Whereas linear inertia is dependent on only one variable (mass), angular inertia is dependent on two variables: mass and the distribution of the mass. These two variables do not have equal effects on the moment of inertia. The influence of mass on angular inertia is much less than the influence of the distribution of the mass because in equation 7.2 the radius of gyration is squared and mass is not. Doubling the mass of an object would double its moment of inertia, but doubling its radius of gyration would quadruple its moment of inertia. A batter who switches to a longer bat will have a more difficult time swinging it than if he switched to a heavier one.

Moments of Inertia About Eccentric Axes

Equations 7.1 and 7.2 define the moment of inertia of an object about an axis through its center of gravity. If an object is unconstrained and free to rotate about any axis, it will rotate about an axis through its center of gravity. But when we swing implements (rackets, bats, sticks, and so on), we force the rotation of the implement around another axis—an eccentric axis. An eccentric axis is one that does not pass through the implement's center of gravity. How does the moment of inertia change? The mass of the implement is definitely farther from the axis of rotation, so the moment of inertia should increase. The moment of inertia of an object around an axis not through its center of gravity is defined by the following equation:

$$I_b = I_{cg} + mr^2 \qquad (7.3)$$

where

I_b = moment of inertia about axis b,

I_{cg} = moment of inertia about axis through the center of gravity and parallel to axis b,

m = mass of object, and

r = radius = distance from axis b to parallel axis through the center of gravity.

An object's moment of inertia about an axis not through the center of gravity of the object is larger than its moment of inertia about a parallel axis through its center of gravity. The increase in size of the moment of inertia is equal to the mass of the object times the square of the distance from the axis of rotation to a parallel axis through its center of gravity.

No matter what axis an object's moment of inertia is measured around, the distance from the mass to the axis of rotation is the dominant influence on the size of the moment of inertia. So, when you qualitatively assess the moment of inertia of an object, this distance—where the mass is relative to the object's axis of rotation—will give you the most information about the object's resistance to change in rotation.

SAMPLE PROBLEM 7.1

A 0.5 kg lacrosse stick has a moment of inertia about a transverse axis through its center of gravity of 0.10 kg·m². When a player swings the stick at an opponent, the axis of rotation is through the end of the stick, 0.8 m from the center of gravity of the stick. What is the moment of inertia relative to this swing axis?

Solution:

Step 1: Identify the known quantities.

$m = 0.5$ kg

$I_{cg} = 0.10$ kg·m²

$r = 0.80$ m

Step 2: Identify the unknown variable to solve for.

$I_{swing} = ?$

Step 3: Search for equations with the known and unknown variables (equation 7.3).

$I_b = I_{cg} + mr^2$

$I_{swing} = I_{cg} + mr^2$

Step 4: Substitute the known quantities and solve for the unknown variable.

$I_{swing} = 0.10$ kg·m² $+ (0.5$ kg$)(0.8$ m$)^2 = 0.42$ kg·m²

Step 5: Common sense check.
Wow, the moment of inertia increased by a factor larger than four. But this makes sense because the mass is much farther away from the axis of rotation.

Moments of Inertia About Different Axes

While an object can have only one linear inertia (mass), it might have more than one angular inertia (moment of inertia), because it can rotate about many different axes of rotation.

> An object may have more than one moment of inertia because an object may rotate about more than one axis of rotation.

Try self-experiment 7.3.

Self-Experiment 7.3

Take this book and flip it so that it rotates about an axis perpendicular to the front and back covers of the book as you did earlier (see figure 7.1). Now take the book, hold it so that the covers are parallel to the floor, and flip it about an axis parallel to the spine of the book as shown in figure 7.2. Did it take the same effort to flip the book in both cases? No. The book was harder to flip in the first case. Its moment of inertia is larger about an axis perpendicular to its covers than about an axis parallel to the spine.

Any object, including this book, has an infinite number of possible axes of rotation, so it also has an infinite number of moments of inertia. If an object is not symmetrical about all planes through its center of gravity, then there will be one axis of rotation about which the moment of inertia is largest, and there will be one axis of rotation about which the moment of inertia is smallest. These two axes will always be perpendicular to each other and are the **principal axes** of the object. The third principal axis of an object is the axis perpendicular to these first two axes.

For this book, the largest moment of inertia is about an axis perpendicular to the book cover, and the smallest

moment of inertia is about an axis parallel to the spine of the book. These are two of the principal axes of the book. The third principal axis is perpendicular to these two and is parallel to the pages of the book and perpendicular to the spine. The three principal axes are shown in figure 7.3.

The principal axes of the human body depend on the position of the limbs. For anatomical position (standing with the arms at the sides), the principal axes are shown in figure 7.4. These axes correspond to the anteroposterior, transverse, and longitudinal axes. We cartwheel around the anteroposterior axis, somersault around the transverse axis, and twist around the longitudinal axis.

Manipulating the Moments of Inertia of the Human Body

A rigid object has many different moments of inertia because it may have many axes of rotation. But for any one axis of rotation, only one moment of inertia is associated with that axis. The human body is not a rigid object, though, because humans can move their limbs relative to each other. These movements may change the distribution of mass about an axis of rotation, thus changing the moment of inertia about that axis. A human's moment of inertia about any axis is variable. There is more than one value for the moment of inertia about an axis. This means that humans can manipulate their moments of inertia. A figure skater spinning about a longitudinal axis with her arms at her sides can more than double her moment of inertia about this longitudinal axis by abducting her arms to shoulder level.

A diver in a layout position and somersaulting about a transverse axis can reduce her moment of inertia to less than half that of the layout position by tucking as shown in figure 7.5.

A sprinter flexes his leg at the knee and hip when he angularly accelerates his leg during the swing-through. This reduces the moment of inertia of the leg about the hip joint, where the axis of rotation is located (see figure 7.6).

In pole vaulting, the vaulter rotates about the top end of the pole where he grips it, and the pole itself (with the vaulter on it) rotates about the bottom end of the pole. At

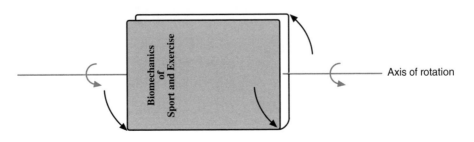

Axis of rotation

Biomechanics of Sport and Exercise

Figure 7.2 Spin a book about an axis of rotation parallel to its spine. Is it easier or harder to spin about this axis than about an axis perpendicular to its cover?

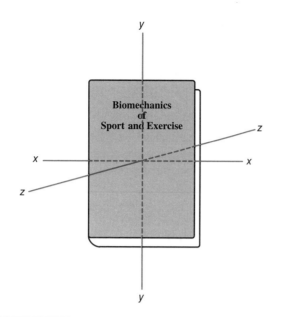

Figure 7.3 The three principal axes of rotation for a book.

© Franck Faugere/DPPI/Icon SMI

Figure 7.5 Tucking reduces the diver's moment of inertia about her transverse axis.

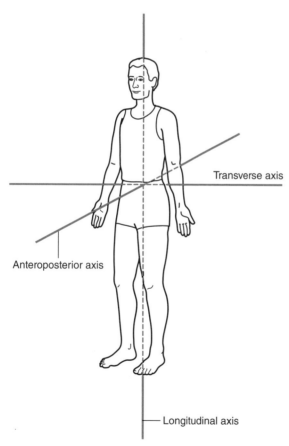

Transverse axis

Anteroposterior axis

Longitudinal axis

Figure 7.4 The three principal axes of rotation for the human body in anatomical position.

Figure 7.6 Flexing the knee and hip of the recovery leg reduces the moment of inertia of the leg about the sprinter's hip joint.

200

the start of the vault, the vaulter is concerned with having the pole rotate about its bottom end so that he can land safely on the landing mat. To make this rotation easier, the moment of inertia about the pole's axis of rotation (its bottom end) must be kept small. The vaulter stays in an extended position just after he leaves the ground, as shown in figure 7.7a. This keeps his center of gravity closer to the bottom end of the pole. The pole bends, which also shortens the distance between the vaulter's center of gravity and the bottom end of the pole. Both of these actions decrease the moment of inertia of the pole about its bottom end and facilitate the rotation of the pole toward the landing mat. Once the vaulter is sure the pole is going to undergo an angular displacement that will be large enough to place him over the bar and the landing mat, he wants to increase his rotation on the pole and decrease the rotation of the pole. He accomplishes this by tucking and moving his center of gravity toward his top handgrip, which is the axis of rotation for his body (see figure 7.7b). This decreases the moment of inertia of his body about his top handgrip and facilitates his rotation about that axis. On the other hand, this action increases the moment of inertia of the pole about its axis of rotation, because the vaulter's center of gravity has moved farther from the bottom end of the pole. This slows down the rotation of the pole toward the landing mat.

Dancers, divers, gymnasts, skaters, and other athletes regularly alter the moments of inertia of their entire bodies or the body parts they are swinging to perform stunts or skills more effectively.

Sport equipment designers are also aware of the effects an implement's moment of inertia may have on a skill. Downhill skiers use longer skis than slalom skiers use. The longer skis give the downhill skier greater stability—something desirable for a skier going 60 mi/hr down a mountain. Slalom skiers need more maneuverable skis—skis with a smaller moment of inertia about the turning axis. Slalom skis are thus shorter. In archery, compound bows have rods that protrude from the bow. These stabilizers increase the moment of inertia of the bow, thus helping the archer steady the bow when the arrow is released (see figure 7.8).

Figure 7.7 Vaulter in extended position (a) and tucked position (b). The moments of inertia of the vaulter about his top handgrip and the bottom of the pole differ in these positions.

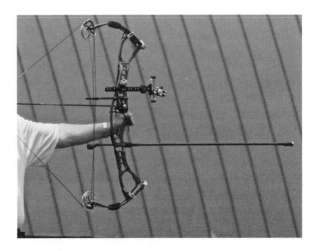

Figure 7.8 The stabilizers on a compound bow help the archer steady the bow during arrow release.

Moments of Inertia and Linear Velocity

In chapter 6, the advantage of using a longer striking implement was discussed and described mathematically by equation 6.8. Recall that

$$v_T = \omega r$$

where

v_T = instantaneous linear velocity tangent to the circular path of the point,

ω = instantaneous angular velocity (measured in radians per second), and

r = radius.

The longer radius of rotation of the striking surface of an implement causes a larger linear velocity at the striking end of the implement if the angular velocity is the same as the angular velocity of a shorter implement. The *if* at the end of the previous statement is a big qualifier. Unfortunately, as the length of an implement increases, its moment of inertia increases also. Due to the increase in angular inertia, it becomes more difficult to accelerate the implement angularly to achieve the same angular velocity. Increasing the length of a swinging implement to achieve a faster linear velocity at the end of the implement may not always yield the desired outcome unless the increase in moment of inertia caused by the increase in length is accounted for in some way.

Golf club designers have accommodated the larger moment of inertia about the swing axis caused by the longer lengths of woods and drivers by making them lighter and less massive than the shorter irons. The smaller mass of a longer club causes a reduction in the club's moment of inertia that almost balances the increase in moment of inertia caused by the longer length (see figure 7.9).

Most golf club designers have also manipulated the moment of inertia about the twist axis of the clubs. During a long backswing and downswing, a golfer may inadvertently twist the club shaft so that the club face is not square to the ball at impact. The ball may not go where the golfer wants it to if this happens. An increase in the moment of inertia of the club about its longitudinal axis throughout the shaft would increase the club's resistance to twisting. Designers have increased the moment of inertia about the twist axis by distributing more of the mass of the club head toward the heel and toe of the club, as shown in figure 7.10, or by increasing the dimensions of the club head. This has resulted in cavity-backed irons and oversized woods. This quest to increase

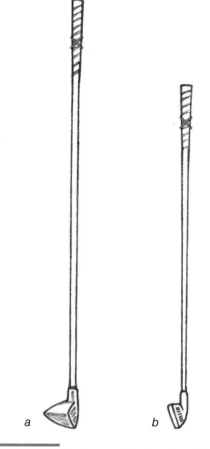

Figure 7.9 The longer club, the driver *(a)*, has a greater moment of inertia than the nine-iron *(b)*, despite its smaller mass.

the moment of inertia of the club head led to larger and larger club heads in drivers. Some of the club heads of these oversized drivers exceeded 500 cc in volume. The U.S. Golf Association became so concerned about the effect that these huge club heads was having on the game that in 2004, it implemented a new rule that limited the maximum volume of a club head to 460 cc.

By now, you should have a good understanding of angular inertia and moment of inertia. They are important elements in the angular versions of Newton's laws of motion, but before we discuss those versions, it's important to understand another important element they include.

Angular Momentum

In chapter 3, the concept of momentum was introduced. Linear momentum is the product of mass and velocity.

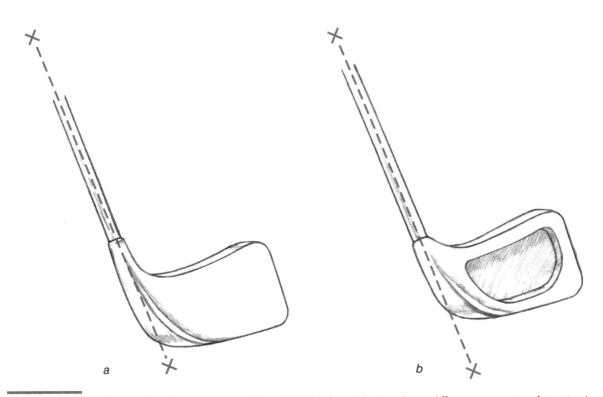

Figure 7.10 Traditional (a) and perimeter-weighted (b) club head designs have different moments of inertia about the twist axis (x–x) of the club.

Mathematically, linear momentum is described by equation 3.6 as

$$L = mv$$

where

L = linear momentum,

m = mass, and

v = linear velocity.

Linear momentum quantifies the linear motion of an object. **Angular momentum** quantifies the angular motion of an object.

Angular Momentum of a Rigid Body

Angular momentum is the angular analog of linear momentum, so it is the product of the angular analog of mass (moment of inertia) times the angular analog of linear velocity (angular velocity). Mathematically, then, the angular momentum of a rigid body is

$$H_a = I_a \omega_a \qquad (7.4)$$

where

H_a = angular momentum about axis a,

I_a = moment of inertia about axis a, and

ω_a = angular velocity about axis a.

Angular momentum is abbreviated with the letter H. The units for angular momentum are kilogram meters squared per second (kg·m²/s). Angular momentum is a vector quantity, just like linear momentum, so it has size and direction. The direction of angular momentum is the same as the direction of the angular velocity that defines it. The right-hand thumb rule is used to determine direction.

Linear momentum depends on two variables: mass and velocity. But the mass of most objects does not change; it is not variable. Changes in linear momentum thus depend on changes in only one variable: velocity. Angular momentum also depends on two variables: moment of inertia and angular velocity. For rigid objects, changes in angular momentum also depend on changes in only one variable—angular velocity—because the moment of inertia of a rigid object does not change. For nonrigid objects, however, changes in angular momentum may result from changes in angular velocity or changes in moment of inertia, or both, because angular velocity and moment of inertia are both variable.

Angular Momentum of the Human Body

Equation 7.6 mathematically defines angular momentum and seems simple, but does it adequately explain how the angular momentum of the human body is determined? If the body acts as a rigid object, so that all the body segments rotate with the same angular velocity, this angular velocity is used as ω and the moment of inertia of the whole body is used as I_a in equation 7.4.

What if some limbs rotate at different angular velocities than other limbs? How is the angular momentum of the body determined then? Mathematically, the angular momentum about an axis through the center of gravity of a multisegment object such as the human body is defined by equation 7.5 (and illustrated in figure 7.11):

$$H_a = \Sigma(I_i\omega_i + m_i r^2_{i/cg}\omega_{i/cg}) \qquad (7.5)$$

where

H_a = angular momentum about axis a through the center of gravity,

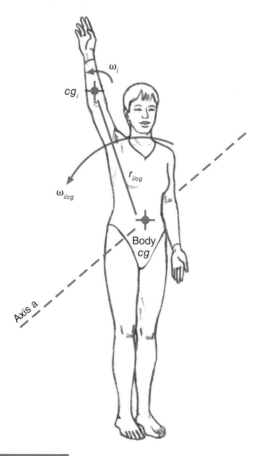

Figure 7.11 Terms used to compute the angular momentum of the human body (for only one segment, the right forearm).

Σ = summation symbol,

I_i = moment of inertia of segment i about its own center of gravity,

ω_i = angular velocity of segment i,

m_i = mass of segment i,

$r_{i/cg}$ = distance from center of gravity of segment i to center of gravity of entire body, and

$\omega_{i/cg}$ = angular velocity of $r_{i/cg}$ about the center of gravity of entire body.

Wow! Equation 7.5 is too complicated to understand easily or use in qualitative analyses in coaching or teaching. Instead, the following gross approximation is a more useful representation of the angular momentum of the entire body:

$$H_a \cong \Sigma(H_i) \cong \Sigma(I_{i/cg}\omega_i) \qquad (7.6)$$

$$= (I\omega)_{r.arm} + (I\omega)_{l.arm} + (I\omega)_{r.leg} + (I\omega)_{l.leg} + (I\omega)_{trunk}$$

where

H_a = angular momentum about axis a through the center of gravity,

Σ = summation symbol,

H_i = angular momentum of segment i about the entire body's center of gravity,

$I_{i/cg} = I$ = moment of inertia of segment i about the entire body's center of gravity, and

ω_i = angular velocity of segment i.

In other words, the sum of the angular momenta of all the body segments gives an approximation of the angular momentum of the entire body.

Let's examine the body's actions during running. The left arm swings backward as the right arm swings forward. The left leg swings forward as the right leg swings backward. What is the angular momentum of the runner about a transverse axis (left to right) through the runner's center of gravity? Using the right-hand thumb rule, with the positive direction of the axis to the left, the angular momentum of the left arm is positive, the angular momentum of the right arm is negative, the angular momentum of the right leg is positive, and the angular momentum of the left leg is negative. The trunk isn't rotating, so its angular momentum is zero. If we use equation 7.6 to approximate the total angular momentum of the body, it would appear that the angular momenta of the arms sum to zero (they cancel each other out) and the angular momenta of the legs sum to zero also. The total angular momentum of the body is zero.

Angular Interpretation of Newton's First Law of Motion

For linear motion, Newton's first law states that every body continues in its state of rest, or of uniform motion in a straight line, unless it is compelled to change that state by forces impressed upon it. This law is sometimes referred to as the law of inertia. We learned in chapter 3 that this law may also be interpreted as the conservation of momentum principle. The momentum of a body is constant unless a net external force acts on it.

The angular equivalent of Newton's first law may be stated as follows: The angular momentum of an object remains constant unless a net external torque is exerted on it. For a rigid object whose moments of inertia are constant, this law implies that the angular velocity remains constant. Its rate of rotation and its axis of rotation do not change unless an external torque acts to change it. Try self-experiment 7.4 to demonstrate this law.

> ⟲ **The angular momentum of an object remains constant unless a net external torque is exerted on it.**

Self-Experiment 7.4

Throw your pen up in the air and give it a flip as you release it so it rotates end over end. As the pen falls, does its angular velocity change? Does its rotation speed up or slow down after you've released it? Does the axis of rotation change direction? The answer to all of these questions is no. The angular momentum of the pen was constant because no external torque acted on it once it was in the air. What about the force of gravity? Doesn't it create an external torque on the pen? The force of gravity (the weight of the pen) acts through the center of gravity of the pen. Once it leaves your hand, the pen becomes a projectile, and the axis of rotation of a projectile is through its center of gravity. The force of gravity can't create a torque on the pen since the moment arm of the force of gravity is zero (see figure 7.12).

For a single rigid body, the angular version of Newton's first law seems rather simple. But what about a system of linked rigid bodies? How does the angular version of Newton's first law apply to the human body? The angular momentum of the human body is constant unless external torques act on it. Mathematically, this can be represented as

$$H_i = I_i\omega_i = I_f\omega_f = H_f \qquad (7.7)$$

where

H_i = initial angular momentum,

H_f = final angular momentum,

I_i = initial moment of inertia,

I_f = final moment of inertia,

ω_i = initial angular velocity, and

ω_f = final angular velocity.

Because the body's moment of inertia is variable and can be changed by altering limb positions, the body's angular velocity also changes to accommodate the changes in the moment of inertia. In this case, Newton's first law does not require that the angular velocity be constant, but rather that the product of the moment of inertia times the angular velocity be constant if no external torques act. For this to occur, any increases in moment of inertia created by

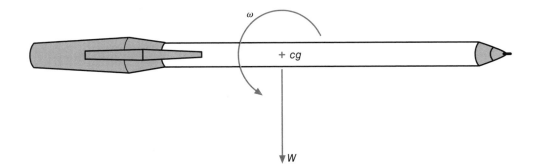

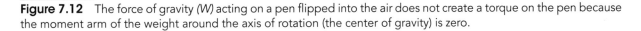

Figure 7.12 The force of gravity (W) acting on a pen flipped into the air does not create a torque on the pen because the moment arm of the weight around the axis of rotation (the center of gravity) is zero.

the person moving limbs farther from the axis of rotation cause decreases in angular velocity to keep angular momentum constant. Likewise, any decreases in moment of inertia created by the person bringing limbs closer to the axis of rotation require increases in angular velocity to keep angular momentum constant. Even though angular momentum remains constant in situations in which no external torques act, athletes may change their angular velocities by changing their moments of inertia.

⊃ **Newton's first law does not require that the angular velocity be constant.**

Conservation of angular momentum is well demonstrated by a figure skater doing a spin. The torque created by friction between the ice and the skates is minimal and may be ignored. As a skater begins a spin, one leg and both arms may be held up and away from the body. The skater thus has a large moment of inertia about the longitudinal axis. As the spin progresses, though, the skater adducts the arms and legs, bringing them closer to the body, thus reducing the moment of inertia. Angular momentum remains the same, so the reduction in the moment of inertia must be accompanied by an increase in angular velocity, which is exactly what happens to the skater.

Gymnasts, figure skaters, dancers, divers, and other athletes use this principle of conservation of angular momentum to control their angular velocities when somersaulting and twisting. A gymnast tucks her body to speed rotation. A figure skater abducts his arms to slow down his spin. A dancer adducts her arms to speed

SAMPLE PROBLEM 7.2

A 60 kg diver has an angular velocity of 6 rad/s about her transverse axis when she leaves the board in a layout position. When she tucks, her angular velocity increases to 24 rad/s. If her moment of inertia is 15 kg·m² in the layout position, what is her radius of gyration in the tuck position?

Solution:

Step 1: Identify the known quantities and quantities that can be computed.

$m = 60$ kg

$\omega_{layout} = 6$ rad/s

$\omega_{tuck} = 24$ rad/s

$I_{layout} = 15$ kg·m²

Step 2: Identify the unknown variable to solve for.

$k_{tuck} = ?$

Step 3: Search for the equation with the known and unknown variables (equation 7.7).

$H_i = I_i \omega_i = I_f \omega_f = H_f$

$(I\omega)_{layout} = (I\omega)_{tuck}$

$(I\omega)_{layout} = (mk^2\omega)_{tuck}$

Step 4: Substitute the known quantities and solve for the unknown quantity.

$(15$ kg·m²$)(6$ rad/s$) = (60$ kg$)(24$ rad/s$) k^2_{tuck}$

$k^2_{tuck} = (15$ kg·m²$)(6$ rad/s$) / (60$ kg$)(24$ rad/s$) = 0.0625$ m²

$k_{tuck} = 0.25$ m

Step 5: Common sense check.
This makes sense. The radius of gyration is small in the tuck position.

up her spin. A diver extends from a pike to slow down somersaulting.

Another way athletes make use of this principle is not to control the angular velocity of their entire body, but to control the individual angular velocities of their limbs or trunk. This situation is best described by combining equation 7.6, which describes an approximation of the angular momentum of the entire body when its segments act independently, with equation 7.7, which explains the angular version of Newton's first law. Combining these gives us

$$H_i \cong [(I\omega)_{r.arm} + (I\omega)_{l.arm} + (I\omega)_{r.leg} + (I\omega)_{l.leg} + (I\omega)_{trunk}]_{initial}$$

$$\cong [(I\omega)_{r.arm} + (I\omega)_{l.arm} + (I\omega)_{r.leg} + (I\omega)_{l.leg} + (I\omega)_{trunk}]_{final} \cong H_f \quad (7.8)$$

where

H_i = initial angular momentum of entire body,

H_f = final angular momentum of entire body,

I = moment of inertia of segment about the entire body's center of gravity, and

ω = angular velocity of segment.

Consider the hurdler shown in figure 7.13. When the hurdler leaves the ground during the hurdling stride, his angular momentum about a longitudinal axis through his center of gravity is zero. Because the hurdler becomes a projectile once his foot leaves the ground, no external torques act on him, and his angular momentum remains constant, in this case at zero. During the flight over the hurdle, the hurdler must swing his trailing leg (his left leg in this case) over the hurdle very quickly. When he does this, the trailing leg has a large angular momentum because it has a large angular velocity and a large moment of inertia about the longitudinal axis through the hurdler's center of gravity. But the hurdler's total angular momentum is still zero because no external torque has acted on him. This means that the angular momentum of some other body part or parts must be in the opposite direction to keep the angular momentum of the entire body zero. The hurdler's right arm swings in the opposite direction, but its angular momentum is not large enough to counter the angular momentum of the trailing leg. The trunk also turns slightly in the opposite direction of the trailing leg. Because the trunk is bent over, it has a large moment of inertia relative to the longitudinal axis. The angular momentum of the trunk along with the angular momentum of the right arm are enough to counter the angular momentum of the trailing leg.

Figure 7.13 The angular momentum of the hurdler's trail leg must be countered by the angular momentum of the arms and trunk.

If the hurdler swings the trailing leg too far to the left (i.e., if the trailing leg is too extended when the hurdler swings it over the hurdle), it will have a very large angular momentum due to its large moment of inertia. In such a case, the arms and trunk must swing too far in the opposite direction to counter this angular momentum, and the hurdler is likely to fall. The technique that better hurdlers use is to flex the trailing leg and keep it as close to the body as possible when clearing the hurdle. Another advantage is that a more tightly tucked trailing leg has a smaller moment of inertia about the hip and can thus be moved more quickly. Similar strategies are used by long jumpers, runners, skiers, gymnasts, bronco riders, and other athletes to prevent loss of balance when one body part has too much angular momentum relative to the other parts.

Angular Interpretation of Newton's Second Law of Motion

Newton's second law of motion describes what happens when external forces do act on an object. It states that the change in motion of an object is proportional to the force impressed and is made in the direction of the straight line in which the force is impressed. Stated more simply, this law says that if a net external force is exerted on an object, the object will accelerate in the direction of the net external force, and its acceleration will be proportional to the net external force and inversely proportional to its mass. Recall from chapter 3 that this law is stated mathematically by equation 3.18 as

$$\Sigma F = ma$$

where

ΣF = net external force,

m = mass of the object, and

a = linear acceleration of the object.

The angular analog of Newton's second law of motion may be stated as follows: The change in angular momentum of an object is proportional to the net external torque exerted on it, and this change is in the direction of the net external torque. The net external torque exerted on an object is proportional to the rate of change in angular momentum. For a rigid object with constant moments of inertia, we can state this law more simply by substituting torque for force, angular acceleration for acceleration, and moment of inertia for mass in the linear version of this law: If a net external torque is exerted on an object, the object will accelerate angularly in the direction of the net external torque, and its angular acceleration will be directly proportional to the net external torque and inversely proportional to its moment of inertia. Mathematically, for a rigid object with constant moments of inertia, this law is stated as

$$\Sigma T_a = I_a \alpha_a \tag{7.9}$$

where

ΣT_a = net external torque about axis a,

I_a = moment of inertia of the object about axis a, and

α_a = angular acceleration of the object about axis a.

If the external torques acting on an object do not sum to zero, the object will experience an angular acceleration in the direction of the net torque. Its angular velocity will speed up or slow down, or its axis of rotation will change direction. If an object's angular velocity or axis of rotation changes, a net external torque must be acting on the object to cause the angular acceleration.

> The change in angular momentum of an object is proportional to the net external torque exerted on it, and this change is in the direction of the net external torque. The net external torque exerted on an object is proportional to the rate of change in angular momentum.

For a nonrigid object with a variable moment of inertia, equation 7.9 does not apply. In this case, the net external torque equals the rate of change of momentum. Mathematically, this can be expressed for average net torque as follows:

$$\Sigma \overline{T}_a = \frac{\Delta H_a}{\Delta t} = \frac{\left(H_f - H_i\right)}{\Delta t} \tag{7.10}$$

where

$\Sigma \overline{T}_a$ = average net external torque about axis a,

ΔH_a = change in angular momentum about axis a,

H_f = final angular momentum about axis a,

H_i = initial angular momentum about axis a, and

Δt = change in time.

A net external torque acting on a nonrigid object with variable moments of inertia will cause a large and quick change in angular momentum if the net torque is large, and a small and slow change in angular momentum if the net torque is small, provided the torques act for equal time intervals. The change in angular momentum may be seen as (1) a speeding up or slowing down of the object's angular velocity, (2) a change in the direction of the axis of rotation, or (3) a change in the moment of inertia. The angular acceleration of the object or a change in its moment of inertia does not necessarily indicate the presence of a net external torque, because the total angular momentum of a nonrigid object may remain constant even if it angularly accelerates or if its moment of inertia changes.

As with the linear version of Newton's second law, the angular version indicates only what happens at an instant in time when a net torque acts. In most sports and human movements, we are more concerned with the final outcome of external torques acting on an athlete or implement over some duration of time. The impulse–momentum relationship provides us with this information.

Angular Impulse and Angular Momentum

Linear impulse is the product of force times the duration of that force application. The impulse produced by a net force causes a change in momentum of the object that the net force acts on, as indicated by equation 3.25:

$$\Sigma \bar{F} \Delta t = m(v_f - v_i)$$

Impulse = change in momentum

where

$\Sigma \bar{F}$ = average net force acting on an object,

Δt = interval of time during which this force acts,

m = mass of object being accelerated,

v_f = final velocity of the object at the end of the time interval, and

v_i = initial velocity of the object at the beginning of the time interval.

The angular analog of the impulse–momentum relationship may be derived from equations 7.10 and 3.25:

$$\Sigma \bar{T}_a = \frac{\Delta H_a}{\Delta t} = \frac{\left(H_f - H_i\right)}{\Delta t}$$

$$\Sigma \bar{T}_a \Delta t = (H_f - H_i)_a \qquad (7.11)$$

Angular impulse = Change in angular momentum

where

$\Sigma \bar{T}_a$ = average net external torque about axis a,

Δt = time interval during which this force acts,

ΔH_a = change in angular momentum about axis a,

H_f = final angular momentum about axis a, and

H_i = initial angular momentum about axis a.

In many sport skills, the athlete must cause a change in the angular momentum of the entire body or an implement or individual body part. The angular impulse–momentum relationship shown in equation 7.11 indicates how this is accomplished. A larger external torque acting over a longer duration will create a larger change in angular momentum. Larger torques can be created by using longer moment arms, as discussed in chapter 5. Increasing the time of application of a torque seems straightforward at first, but it may be more difficult than it seems.

Consider how a dancer or gymnast initiates a spin or twist on the floor. The axis of rotation for the twisting action is a longitudinal axis that passes through one foot on the floor. The other foot pushes on the floor to create a frictional force that creates a torque about the longitudinal axis. This foot must be some distance away from the twisting foot to maximize the moment arm and the torque created. If the moment of inertia of the dancer is small as the torque is created, the dancer begins to turn faster and faster, and the foot must stop pushing on the floor because it begins to rotate with the rest of the body. How can the dancer prolong the pushing action to create a larger impulse and thus a larger change in angular momentum? If the dancer has a larger moment of inertia as the torque is created, she begins to turn as a result of the torque; but she doesn't turn as fast, so she has more time to push with her foot before her body has turned far enough that the foot must come off the floor. This prolonging of the torque application by increasing the moment of inertia to slow down the spin results in a larger impulse and thus a larger change in momentum. After her foot comes off the floor, the dancer can reduce her moment of inertia and increase her angular velocity.

Discus throwers use a similar technique. At the start of their throwing action, they have a large moment of inertia, but at the release of the discus, the moment of inertia is smaller. In activities in which the goal is to spin very fast, the athlete should start with a large moment of inertia during the torque production period so that the duration of torque application is maximized. This increases the impulse and thus the change in angular momentum that occurs. Once the torque production phase is complete, the performer reduces the moment of inertia, and the angular velocity increases accordingly.

Angular Interpretation of Newton's Third Law of Motion

Newton's third law of motion states that for every action there is an equal but opposite reaction. A clearer statement of this law is that for every force exerted by one body on another, the other body exerts an equal force back on the first body but in the opposite direction. The angular version of this law states that for every torque exerted by one object on another, the other object exerts an equal torque back on the first object but in the opposite direction.

A point often in need of clarification regarding the linear version of this law is that it is the *forces* that are equal but opposite in direction, not the *effects* of the forces. The effects of forces are often accelerations, and the accelerations of the objects that these forces act on depend on the masses of the objects as well as any other forces that act on the objects.

In interpreting the angular version of Newton's third law, we must remember that the torques acting on the two objects have the same axis of rotation. Also, the effects of these torques are different because they act on different objects. The effects of the torques depend on the moments of inertia of the objects and whether or not any other torques act on the objects.

One angular example of Newton's third law involves muscles. As discussed in chapter 5, muscles produce torques about joints by creating forces on the limbs on either side of the joint. The vastus group of the quadriceps femoris are knee extensor muscles. When these muscles contract, they create a torque on the lower leg that causes it to rotate (or tend to rotate) in one direction, and they create an equal but opposite torque on the thigh that causes it to rotate (or tend to rotate) in the opposite direction. These two opposite rotations produce extension at the knee joint.

Let's look at another example of how Newton's third law is used. Tightrope walkers use long poles to help them stay on the tightrope, as shown in figure 7.14. How does a long pole help an acrobat stay balanced on the tightrope? One way is that it can lower the center of gravity of the acrobat and pole system as well as increase its moment of inertia. These effects both increase the stability of the acrobat. The most important use of the pole can be explained using Newton's third law. Suppose the tightrope walker begins to fall clockwise to his left. What should he do with his pole to help him regain his balance?

If the acrobat exerted a clockwise torque on the pole and moved it clockwise in the same direction of his fall, the pole would exert an equal but oppositely directed torque on him. This torque would act counterclockwise on the acrobat, opposite his direction of falling, and possibly move him into a position where he could regain his balance. If the acrobat moved the pole counterclockwise in the opposite direction of the fall, the reaction torque exerted on the acrobat would be in the clockwise direction and would move him faster in the direction he was falling! You can try a safer example of this for yourself by doing self-experiment 7.5.

Self-Experiment 7.5

Stand with your toes just behind a line on the floor. Imagine that the line on the floor is the edge of a cliff. Imagine that you are standing right on the edge when you lean forward too far and start to fall, so that your body rotates counterclockwise (or in a forward somersaulting direction). Go ahead, try it. Which way did you swing your arms to try to catch yourself as you started to fall? You probably swung your arms back, upward, and then forward so that they circled in a counterclockwise direc-

© Zuma Press/Icon SMI

Figure 7.14 Long poles are used by tightrope walkers to maintain their balance. By creating a torque on the pole in one direction, the acrobat causes an equal but opposite torque to be created on him.

tion (unless you thought too hard before trying this). Your shoulder muscles exerted a counterclockwise torque on your arms to cause this rotation. The reaction to this was a clockwise torque exerted by these same muscles on your trunk. This clockwise torque would tend to rotate your trunk and legs clockwise or slow down their counterclockwise rotation. This action might have saved you from falling off the cliff.

The effects of equal but opposite torques produced by muscles on different limbs can be observed in various sports when athletes attempt to maintain balance or position. When a gymnast loses his balance on the balance beam, you often see frantic arm swinging, which is his attempt to regain his balance. Similar arm flailing can be observed in ski jumpers if their forward rotation is too much. Recreational skiers can often be seen doing this when a bump gives them some unexpected angular momentum.

Summary

The basics of angular kinetics, the causes of angular motion, are explained by angular interpretations of Newton's laws of motion. We must understand angular analogs of inertia and momentum to make these interpretations. Angular inertia, called moment of inertia, is an object's resistance to change in its angular motion. Mathematically, it is defined as the product of mass times radius of gyration squared. Radius of gyration is a length dimension representing, on average, how far an object's mass is located from an axis of rotation. Objects have many different moments of inertia, one for each of their possible axes of rotation. Angular momentum, like linear momentum, is a measure of an object's motion. Angular momentum is the product of moment of inertia and angular velocity. It is a vector quantity that is specific for an axis of rotation. Angular momentum may be constant even if angular velocity varies, so long as the variation in angular velocity is accompanied by an inverse variation in moment of inertia. The angular interpretation of Newton's first law says that objects do not change their

angular momentum unless a net external torque acts on them. The angular interpretation of Newton's second law explains what happens if a net external torque does act on an object. A rigid object will accelerate angularly in the direction of the net external torque, and its angular acceleration will be inversely related to its moment of inertia. For objects with variable moments of inertia, the impulse–momentum relationship is a more applicable angular interpretation of Newton's second law. A net external torque acting over some duration of time causes a change in angular momentum in the same direction as the net external torque. An angular interpretation of Newton's third law explains that torques act in pairs. For every torque, there is an equal torque acting on another object but in the opposite direction.

Table 7.1 gives a mathematical comparison of the angular kinetic parameters and principles, along with their linear counterparts. This table is an excellent summary of the elements of this chapter.

Table 7.1 Comparison of Linear and Angular Kinetic Quantities

Quantity	Symbol and equation for definition	SI units
Linear		
Inertia (mass)	m	kg
Force	F	N
Linear momentum	$L = mv$	kg·m/s
Impulse	$\Sigma \bar{F} \Delta t$	N·s
Angular		
Moment of inertia	$I = \Sigma mr^2 = mk^2$	kg·m²
Torque of moment of force	$T = F \times r$	Nm
Angular momentum	$H = I\omega$	kg·m²/s
Angular impulse	$\Sigma \bar{T} \Delta t$	Nm·s

KEY TERMS

angular impulse (p. 209)
angular inertia (p. 196)
angular momentum (p. 203)
moment of inertia (p. 197)
principal axes (p. 199)
radius of gyration (p. 197)
rotary inertia (p. 196)

REVIEW QUESTIONS

1. Refer to the four dancers described below. The dancers have identical heights, weights, and limb lengths. Assume that each dancer is in anatomical position unless stated otherwise.

   Anna is standing on her left foot. Her right leg is abducted approximately 30° at the hip. Her left arm is abducted 90° at the shoulder.

   Bryn is standing on her right foot. Her left leg is flexed 90° at the knee. Each arm is slightly abducted at the shoulder and fully flexed at the elbow.

   Catherine is standing on her right foot. Her left leg is abducted 60° at the hip. Each arm is abducted 90° at the shoulder.

   Donna is standing on both her feet. Her left arm is flexed 180° at the shoulder so it is outstretched over her head. Her right arm is slightly adducted past anatomical position so it crosses in front of her body.

   a. Which dancer has the largest moment of inertia about her anteroposterior (AP) axis?

   b. Rank the dancers' positions from largest to smallest moment of inertia about the longitudinal axis.

   c. If each dancer is rotating with the same angular velocity about the longitudinal axis, which dancer has the greatest angular momentum about the longitudinal axis?

   d. If each dancer is rotating with the same angular momentum about the longitudinal axis, which dancer is spinning the fastest about her longitudinal axis?

| Anna | Bryn | Catherine | Donna |

2. Which sprinting technique is more effective—flexing the knee of the recovery leg (the leg that is swinging forward) more during the swing-through, or flexing the knee of the recovery leg less during the swing-through? Why?

More knee flexion   Less knee flexion

3. Ann performs three different dives and rotates through the same number of revolutions in the same amount of time in all three dives. In all the dives, Ann's positions at takeoff and just before water entry are identical layout positions. In dive 1, she is in a layout position in midflight. In dive 2, she is in a tuck position in midflight. In dive 3, she is in a pike position in midflight.

    a. In which dive does Ann have the largest angular momentum about the transverse axis?

    b. In which dive does Ann have the smallest angular momentum about the transverse axis?

    c. In which dive will Ann be rotating the fastest about the transverse axis just before water entry?

    d. In which dive will Ann be rotating the slowest about the transverse axis just before water entry?

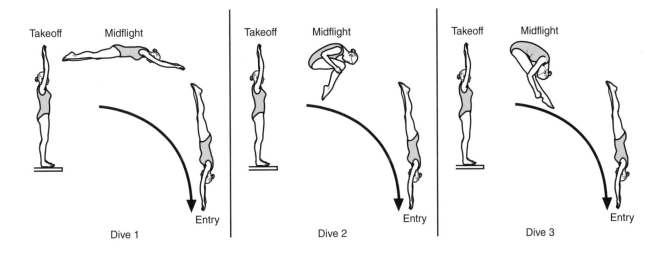

Takeoff   Midflight   Takeoff   Midflight   Takeoff   Midflight

Entry   Entry   Entry

Dive 1   Dive 2   Dive 3

4. Why are most gymnasts small in stature?

5. Heidi jumps up in the air to spike the volleyball. She leaves the ground without any angular momentum. As Heidi swings her arm forward to spike the ball, what happens to her legs if her trunk and nonhitting arm do not rotate?

6. Harvey has skied over a mogul and become airborne. He has some angular momentum that is rotating him forward (clockwise). If he continues to rotate, he will probably fall on his face in the snow when he lands. In which direction should Harvey rotate his arms to try to prevent a fall?

7. Why do runners lean when running around a curve?

8. What is the advantage of a banked running track over a flat running track?

9. A gymnast is performing her routine on the balance beam. She is standing on her left leg and facing forward along the long axis of the beam. Her right leg is flexed 90° at the hip and with the knee extended. Her left arm is fully abducted 180° at the shoulder, and her right arm is abducted 120° at the right shoulder. She begins to fall to the right. What should she do with her arms to help regain her balance?

10. During the descent phase of a giant swing on the horizontal bar, a gymnast stretches himself out and tries to maximize the distance between his grip on the bar and his center of gravity. During the ascent phase of a giant swing, the gymnast tries to shorten this distance slightly. How do these actions contribute to the successful execution of a giant swing?

11. A freestyle aerial skier completes a helicopter 1080 in which she rotates three times around her longitudinal axis while in the air. While performing the trick, she adducts her arms and keeps her poles close to her sides. Just before landing, she abducts her arms. How do these actions—adducting her arms during the trick and abducting her arms just before landing—contribute to the successful execution of the trick?

PROBLEMS

1. A tennis racket has a mass of 0.350 kg. The radius of gyration of the racket about its longitudinal (twist) axis is 7.2 cm, and its radius of gyration about its swing axis is 20 cm.

    a. What is the moment of inertia of this racket about its longitudinal axis?

    b. What is the moment of inertia of this racket about its swing axis?

2. The moment of inertia of the club head is a design consideration for a driver in golf. A larger moment of inertia about the vertical axis parallel to the club face provides more resistance to twisting of the club face for off-center hits. The mass of one club head is 200 g and its moment of inertia is 5000 g cm². What is the radius of gyration of this club head?

3. A 1 kg baseball bat has a moment of inertia around a transverse axis through its center of gravity of 650 kg·cm². What is the moment of inertia of the bat about an axis through the handle of the bat if this axis is 50 cm from the center of gravity of the bat?

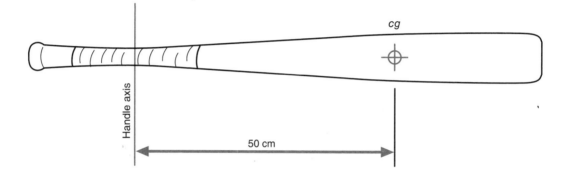

4. The moment of inertia of the lower leg and foot about an axis through the knee joint is 0.20 kg·m². What is the moment of inertia of the leg and foot about the knee joint if a 0.50 kg shoe is worn on the foot? Assume that the shoe's mass is all concentrated in one point 45 cm from the knee joint.

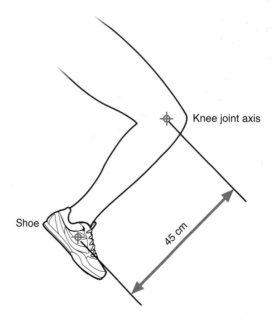

5. Lily is a 50 kg diver. At the instant of takeoff, her angular momentum about her transverse axis is 20 kg·m²/s. Her radius of gyration about the transverse axis is 0.4 m at this instant. During the dive, Lily tucks and reduces her radius of gyration about the transverse axis to 0.2 m.

   a.   At takeoff, what is Lily's moment of inertia about her transverse axis?

   b.   At takeoff, what is Lily's angular velocity about the transverse axis?

   c.   After Lily tucks, what is her moment of inertia about her transverse axis?

   d.   After Lily tucks, what is her angular velocity about the transverse axis?

6. Kristen is spinning on the ice at 20 rad/s about her longitudinal axis when she abducts her arms and doubles her radius of gyration about her longitudinal axis from 30 cm to 60 cm. If her angular momentum is conserved, what is her angular velocity about her longitudinal axis after she increases her radius of gyration?

7. The average net torque Justin exerts on a discus about its axis of spin is 100 Nm during a throw. The mass of the discus is 2 kg, and its radius of gyration about the spin axis is 12 cm. If the discus is not spinning at the start of Justin's throwing action, and the throwing action lasts for 0.20 s, how fast is the discus spinning when Justin releases it?

8. Tom's leg angularly accelerates 3000°/s² around the hip joint during a roundhouse kick in the transverse plane. The moment of inertia of the leg around the axis of rotation for this kick is 0.75 kg·m². How large is the torque that produces this acceleration?

9. Doug is driving a golf ball off the tee. His downswing takes 0.50 s from the top of the swing until ball impact. At the top of the swing, the club's angular velocity is zero; at the instant of ball impact, the club's angular velocity is 30 rad/s. The swing moment of inertia of the club about the grip is 0.220 kg·m². What average torque does Doug exert on the golf club during the downswing?

10. Sarah's twist angular momentum increases from 0 to 50 kg·m²/s in 0.25 s as she initiates a twisting jump on the ice. During this 0.25 s, her moment of inertia about her twist axis is 2.2 kg·m².

   a.   How large is the average torque that produces this change in angular momentum?

   b.   How fast is Sarah's twist angular velocity at the end of the 0.25 s?

chapter 8

Fluid Mechanics

The Effects of Water and Air

objectives

When you finish this chapter, you should be able to do the following:

- Define relative and absolute motion
- Define buoyancy and buoyant force
- Define pressure
- Define fluid
- Explain how a fluid exerts forces on an object moving through it
- Identify the components of fluid forces
- Define drag force
- Distinguish between surface drag and form drag
- Define lift force
- Explain Bernoulli's principle
- Explain the Magnus effect
- Identify the various factors that determine the effect fluid forces have on an object

© Nuralya/Dreamstime.com

You are standing in the batter's box, waiting for the next pitch. The pitcher winds up and throws. The ball is coming at you, and it looks as though the pitcher has thrown the ball right to the middle of the strike zone. You start your swing, anticipating the crack of the bat hitting the ball. Whiff! Whomp! "Strike three, you're out!" Rather than feeling and hearing the crack of the bat hitting the ball, all you feel is the bat slicing through air and all you hear is the thud of the ball hitting the catcher's mitt. Just as you were visualizing the ball sailing over the left-field fence, it seemed to swerve down and to the left. What happened? How could a ball that seemed to be on a perfect trajectory toward the strike zone (and your bat) swerve out of your bat's path? The answer to this question comes from fluid mechanics.

This chapter is about the branch of

mechanics called fluid mechanics. Specifically, it is about the forces that fluids exert on objects in them or moving through them. Unlike solids, liquids and gases can flow and change shape quickly and easily without separating, so they are classified as fluids. The fluids we are most concerned about in sport biomechanics are air and water. Air is the medium we move through in all land-based sports and human activities, and water is the medium we move through in all aquatic sports and activities.

In swimming and other aquatic activities, fluid forces are large, and their importance to success in these activities is obvious. In many land-based activities, fluid forces (air resistance) may be so small that they can be ignored. But in other land-based activities, fluid forces may be large enough to affect the movements of bodies or implements or so large that they determine the outcome of a movement skill. Consider the importance of air resistance in the following activities: sprint running, baseball pitching, cycling, sailboarding, discus throwing, sailing, speed skating, downhill ski racing, hang gliding, and skydiving. In the last two of these activities, one's life depends on air resistance!

Because fluid forces are fundamental for success in aquatic sports and activities, as well as certain land-based activities, a basic understanding of fluid forces is desirable. Information presented in this chapter will assist you in gaining a basic understanding of fluid forces.

Buoyant Force: Force Due to Immersion

Two types of forces are exerted on an object by a fluid environment: a **buoyant force** due to the object's immersion in the fluid and a dynamic force due to its relative motion in the fluid. The dynamic force is usually resolved into two components: drag and lift forces. The buoyant force, on the other hand, always acts vertically. A buoyant force acts upward on an object immersed in a fluid. You are probably most familiar with this principle in connection with objects in water. For a demonstration, try self-experiment 8.1.

Two types of forces are exerted on an object by a fluid environment: a buoyant force due to the object's immersion in the fluid and a dynamic force due to its relative motion in the fluid.

Self-Experiment 8.1

Fill a bathtub or a deep sink with water. Take a basketball (or any other large inflatable ball) and try to push it into the water. Now try it again with a tennis ball (or racquetball). Which is easier to push under water? It's easier to push the tennis ball under water. You may also notice that the farther you push the ball into the water, the larger the force you have to push with. The force that pushed upward on the tennis ball or basketball was buoyant force. Buoyant force seems to be related to the size of the object immersed in water and to how much of the object is immersed.

Pressure

Let's see if we can explain the cause of this buoyant force. Suppose you are in a pool of still water. As you dive deeper and deeper into the pool, the water exerts greater and greater pressure on you. The **pressure** the water exerts is due to the weight (force) of the water above you. But the water pressure doesn't just act downward on you; the water below you pushes upward on you, and the water to either side pushes laterally on you. Water pressure acts in all directions with the same magnitude, as long as you stay at the same level. The deeper you go, the greater the pressure. Pressure is defined as force per unit area. One cubic meter (m^3) of water weighs about 9800 N, so at a depth of 1 m the water pressure is 9800 N/m^2. At a depth of 2 m, the 2 m^3 of water above a 1 m square weighs 19,600 N, so the pressure is 19,600 N/m^2. Figure 8.1a illustrates how water pressure increases linearly with depth.

Imagine removing a cube of this water from 1 m below the surface, as shown in figure 8.1b. Draw a free-body diagram of the cube of water as shown in figure 8.1c. What vertical forces act on the cube of water? The water above the cube exerts 9800 N/m^2 pressure on top of the cube. This pressure acting on the square-meter top face of the cube creates a 9800 N resultant force acting downward on the cube. Let's represent this force with R_u, for upper force, and with an arrow acting downward on the cube. The water below the cube exerts 19,600 N/m^2

pressure on the bottom of the cube. This pressure acting on the square-meter bottom face of the cube creates a 19,600 N resultant force acting upward on the cube. Let's represent this force with R_l for lower force, and with an arrow acting upward on the cube. Some forces also act on the sides of the cube, as shown, but because we are concerned only with the vertical forces, we won't try to quantify these lateral forces. Do any other vertical forces act on the cube of water? The cube of water weighs something, about 9800 N, because gravity exerts a force on it. Let's represent this force with W_w, for weight, and with an arrow acting downward on the cube. If the cube is in equilibrium and not accelerating, then according to Newton's second law,

$$\sum F = R_l + (-R_u) + (-W_w) = 0 \qquad (8.1)$$

$$\sum F = 19,600 \text{ N} - 9800 \text{ N} - 9800 \text{ N} = 0$$

Sure enough, the net vertical force acting on the cube of water is zero, so the water is in equilibrium. What happens if we take out a cube at a depth of 4 m? At 4 m, the pressure is 39,200 N/m^2 (4 m × 9800 N/m^2), so the force R_u is 39,200 N. At 5 m, the pressure is 49,000 N/m^2

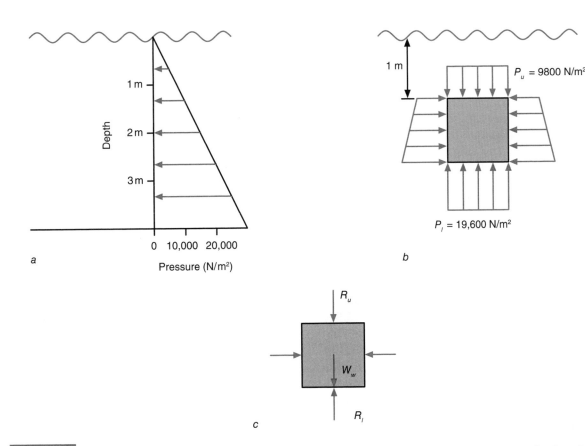

Figure 8.1 Graphical representations of the relationship between water pressure and depth (a); a 1.0 m^3 cube of water in 1.0 m below the water (b); and a free-body diagram of the forces acting on the cube of water (c).

($5 \text{ m} \times 9800 \text{ N/m}^2$), so the force R_l is 49,000 N. The cube still weighs the same, so equation 8.1 becomes

$$\sum F = R_l + (-R_u) + (-W_w) = 0$$

$$\sum F = 49{,}000 \text{ N} - 39{,}200 \text{ N} - 9800 \text{ N} = 0$$

No matter what depth we choose, the difference in pressures at the top and bottom of the cube accounts for the weight of the cube of water, and equation 8.1 is valid.

Buoyant force is the difference between the force acting upward on the cube, R_l, and the force acting downward on the cube, R_u. If we let F_b, for buoyant force, represent this difference, equation 8.1 becomes

$$\sum F = R_l + (-R_u) + (-W_w) = 0$$

$$\sum F = F_b + (-W_w) = 0 \qquad (8.2)$$

$$F_b = W_w$$

The size of the buoyant force is equal to the weight of the volume of fluid displaced by the object. This principle was discovered by the Greek mathematician Archimedes more than 2000 years ago. He lived from 287 to 212 B.C. It is called Archimedes' principle. For simplicity, you may think of buoyant force as similar to a reaction force from the water, but the reaction force depends on how much water is pushed out of the way by the object.

⟳ **The size of the buoyant force is equal to the weight of the volume of fluid displaced by the object.**

What happens if we fill the hole made by the cube of water we took out with a cube made of some material? The pressure on top of the cube will still be the same and will result in the same force, and the pressure on the bottom of the cube will still be the same and will result in the same force. The only difference will be in the weight of the cube, represented by W_c. Applying Newton's second law, equation 8.2 becomes

$$\sum F = F_b + (-W_c) = ma$$

If the cube of material weighs more than the cube of water, the buoyant force acting upward is less than the weight acting downward, and the cube will accelerate downward. If the cube of material weighs less than the cube of water, the buoyant force acting upward is more than the weight acting downward, and the cube will accelerate upward. If you let go of the basketball after submerging it, this would be demonstrated clearly. If the cube of material weighs the same as the cube of water, the buoyant force acting upward equals the weight acting downward, and the cube is in equilibrium. For something

to float, then, the buoyant force must equal the weight of the object.

Specific Gravity and Density

Whether or not something floats is determined by the volume of the object immersed and the weight of the object compared to the weight of the same volume of water. **Specific gravity** is the ratio of the weight of an object to the weight of an equal volume of water. Something with a specific gravity of 1.0 or less will float. Another measure that can be used to determine if a material will float is density. **Density** is the ratio of mass to volume. The density of water is about 1000 kg/m³. The density of air is only about 1.2 kg/m³.

$$\rho = \frac{m}{V} \qquad (8.3)$$

where

ρ = density,

m = mass, and

V = volume.

Buoyancy of the Human Body

Muscle and bone have densities greater than 1000 kg/m³ (specific gravities greater than 1.0), whereas fat has a density less than 1000 kg/m³ (specific gravity less than 1.0). These differences in density are the basis for the underwater weighing techniques used to determine body composition.

Someone who has low body fat can still float, because the lungs and other body cavities may be filled with air or other gases that are much less dense than water. A forced exhalation of the air out of the lungs may cause a lean person to sink, however. The volume of your chest increases when you inhale and decreases when you exhale, so you actually have some control over the total density of your body. To increase your buoyancy, increase your volume by inhaling forcibly. To decrease your buoyancy, decrease your volume by exhaling forcibly.

Most people are able to float if they take a large breath and hold it. But why is it so difficult to float with your legs horizontal? If you try to float on your back with your legs horizontal, you will notice that your legs will drop or your entire body will rotate toward a more vertical orientation, as shown in figure 8.2.

The reason the body rotates to a more vertical position or the legs drop is that the buoyant force and the force of gravity are not colinear. The line of action of the buoyant force is through the center of volume of the body rather than through the center of gravity. For many objects (the basketball, for instance), these two points coincide, but

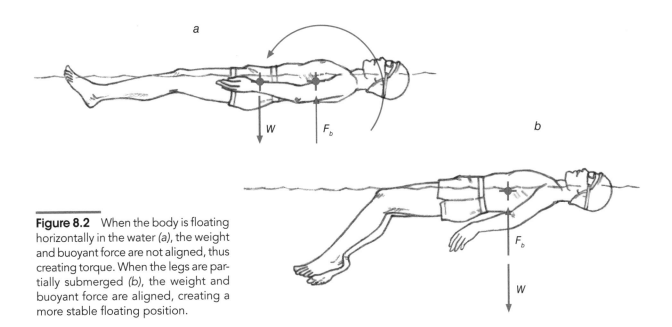

Figure 8.2 When the body is floating horizontally in the water (a), the weight and buoyant force are not aligned, thus creating torque. When the legs are partially submerged (b), the weight and buoyant force are aligned, creating a more stable floating position.

for objects with parts that have different densities, this may not be the case. Because your legs (mostly muscle and bone) are more dense than your abdomen and chest (lots of cavities filled with air or other gases), the center of volume of your body is farther toward your head than your center of gravity is. The conditions for equilibrium are not met when you try to hold your legs horizontal, because the buoyant force and the gravitational force create a torque. The net torque acting on the body is not zero, so rotation occurs until these forces align and the torque is zero.

Buoyant force is a significant force for activities in the water, but what about land-based activities? Does the air exert a buoyant force? To answer this question, consider a helium balloon, a hot-air balloon, or a blimp. What force keeps these things from falling? Air does exert a buoyant force, but the size of this force is very small unless the volume of air displaced is very large. The density of air (at sea level) is approximately 1.2 kg/m³ (a weight density of 11.8 N/m³). A normal-sized 80 kg human being has a volume of approximately 1/12 of a cubic meter, so the buoyant force exerted by air on an 80 kg human body is approximately 1 N = (1/12 m³) (11.8 N/m³). This is about 0.2 lb, an amount so small compared to our weight that we can ignore it.

Dynamic Fluid Force: Force Due to Relative Motion

Buoyant force is the vertical force exerted on an object immersed in a fluid. It is present whether the object is at rest or is moving relative to the fluid. When an object moves within a fluid (or when a fluid moves past an object immersed in it), dynamic fluid forces are exerted on the object by the fluid. The dynamic fluid force is proportional to the density of the fluid, the surface area of the object immersed in the fluid, and the square of the relative velocity of the object to the fluid. Equation 8.4 summarizes this relationship:

$$F \propto \rho A v^2 \tag{8.4}$$

where

$\propto$ = is proportional to,

F = dynamic fluid force,

ρ = fluid density,

A = surface area of the object, and

v = relative velocity of the object with respect to the fluid.

In equation 8.4, fluid density and object surface area are linear terms; an increase in either will cause a proportional increase in the dynamic fluid force. If the area doubles, the dynamic fluid force doubles. If the fluid becomes three times as dense, the dynamic fluid force becomes three times larger. The relative velocity term in equation 8.4 is not linear, it is squared; so if the relative velocity doubles, the dynamic fluid force quadruples (2^2). If the relative velocity becomes three times as fast, the dynamic fluid force becomes nine (3^2) times larger. Figure 8.3 demonstrates these relationships. Relative velocity is obviously the most important factor in determining dynamic fluid forces, so before we go further, an explanation of relative velocity is in order.

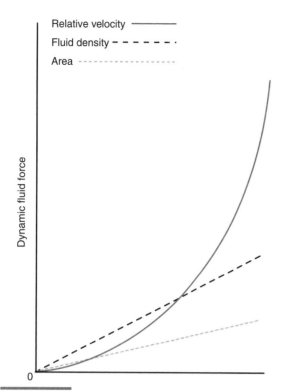

Relative velocity ———
Fluid density – – – – – –
Area · · · · · · · · · · · · · · ·

Dynamic fluid force

0

Figure 8.3 The relationship between dynamic fluid force and object area, fluid density, and relative velocity.

🔁 The dynamic fluid force is proportional to the density of the fluid, the surface area of the object immersed in the fluid, and the square of the relative velocity of the object to the fluid.

Relative Velocity

When we consider the dynamic fluid forces acting on an object, we must take into account the velocity of the object as well as the velocity of the fluid itself. Relative velocity is used to represent the effects of these two absolute velocities. Relative velocity is the difference between the object's velocity and the fluid's velocity. Suppose you were standing still on a running track and the wind was blowing in your face at 5 m/s (see figure 8.4a). The relative velocity is the difference between your velocity and the wind's velocity: 0 m/s – (–5 m/s) = 5 m/s. (Let's assume the positive direction is the direction that you were facing. Because the wind was blowing opposite to this direction, we indicate its velocity with a negative sign.) Now suppose there was no wind and you were running down the track at 5 m/s. The relative velocity is the difference between your velocity and the wind's velocity: 5 m/s – 0 m/s = 5 m/s (see figure 8.4b). The wind that you feel on your body would be the same

in both cases. In one case, the air was moving at 5 m/s, and in the other case, you were moving at 5 m/s; but the fluid forces produced by the air were identical.

Suppose the wind was blowing in your face at 5 m/s and you were running at 5 m/s (see figure 8.4c). What is the relative velocity? The relative velocity is again the difference between your velocity and the wind's velocity: 5 m/s – (–5 m/s) = 10 m/s. What if the wind was blowing at your back at 5 m/s and you were running at 5 m/s (see figure 8.4d)? The relative velocity would be 5 m/s – 5 m/s = 0 m/s.

Quantitatively, to determine relative velocity, we take the difference between the object's absolute velocity and the fluid's absolute velocity. Qualitatively, you can consider relative velocity as follows. If you are sitting on the object that is moving through the fluid, how fast does the fluid move past you? The velocity of the fluid moving past you is the relative velocity of the fluid with respect to the object.

Drag Force

For practical reasons, the dynamic fluid force that results from motion within a fluid is commonly resolved into two components: the drag force and the lift force. Figure 8.5 illustrates the resolution of the dynamic fluid force into lift and drag components.

🔁 The dynamic fluid force that results from motion within a fluid is commonly resolved into two components: the drag force and the lift force.

Drag force, or drag, is the component of the resultant dynamic fluid force that acts in opposition to the relative motion of the object with respect to the fluid. A drag force will tend to slow down the relative velocity of an object through a fluid if it is the only force acting on the object. Drag force is defined by equation 8.5:

$$F_D = \frac{1}{2}C_D\rho A v^2 \tag{8.5}$$

where

F_D = drag force,

C_D = coefficient of drag,

ρ = fluid density,

A = reference area (usually the cross-sectional area of the object perpendicular to the relative velocity), and

v = relative velocity of the object with respect to the fluid.

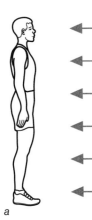  

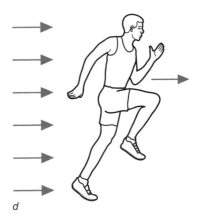

Figure 8.4 Relative velocity between a runner and the wind.

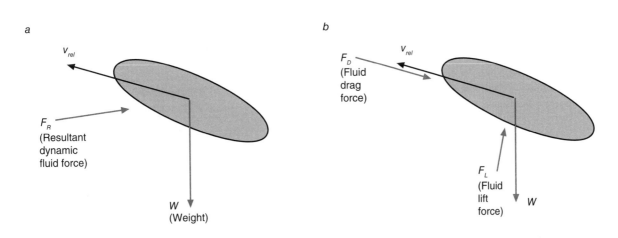

Figure 8.5 The resultant dynamic fluid force (F_R) acting on an object (a); the drag (F_D) and lift (F_L) components of this force (b).

Qualitatively, drag force can be considered in the following way. Drag is the backward force that the molecules of the fluid exert on the object moving relative to the fluid. According to Newton's third law, an equal but opposite force is exerted by the object on the molecules of the fluid. The size of the drag force is thus proportional to the acceleration (slowing down) of the fluid molecules as they pass the object, as well as to the mass of the molecules that are slowed down. The greater the decrease in velocity of the molecules and the faster the rate of this decrease, the greater the total drag.

Drag forces are produced by two different means: surface drag and form drag. **Surface drag** may be thought of as equivalent to the sum of the friction forces acting between the fluid molecules and the surface of the object (or between the fluid molecules themselves). **Form drag** may be thought of as equivalent to the sum of the impact forces resulting from the collisions between the fluid molecules and the object.

⟳ Drag forces are produced by two different means: surface drag and form drag.

Surface Drag

Surface drag is also called skin friction or viscous drag. As a fluid molecule slides past the surface of an object, the friction between the surface and the molecule slows down the molecule. On the opposite side of this molecule are fluid molecules that are now moving faster than this molecule, so these molecules are also slowed down as they slide past the molecules closest to the object. These molecules, in turn, slow down the molecules next to them. So the surface drag is proportional to the total mass of the molecules slowed down by the friction force and the average rate of change of velocity of these molecules.

The size of the surface drag is affected by the factors included in equation 8.5: the coefficient of drag, the density of the fluid, the cross-sectional area of the object, and the square of the relative velocity. The coefficient of drag is influenced by several other factors associated with surface drag. The roughness of the surface is one of these factors. Rougher surfaces create larger friction forces between the fluid molecules and the object. Thus the equipment used and the clothing worn by athletes in many activities are smooth so as to reduce surface drag (figure 8.6). Examples of items that reduce surface

Figure 8.6 Athletes in certain sports wear skintight apparel to reduce surface drag.

drag in sport include the sleek, skintight suits worn by speed skaters, cyclists, downhill skiers, lugers, and even sprint runners. Swimmers shave their bodies to reduce surface drag.

The coefficient of drag is also influenced by the viscosity of the fluid associated with surface drag. **Viscosity** is a measure of the internal friction between the layers of molecules of a fluid or the resistance of a fluid to shear forces. Slower-flowing fluids are more viscous than faster-flowing fluids. Motor oil is more viscous than water, which is more viscous than air. Larger friction forces between the fluid and the surface of an object moving through it are created by fluids with greater viscosity, so surface drag increases with increased viscosity. Because most athletes have no control over the properties of the fluid they move through, examples of ways to make changes in fluid viscosity in sport are rare.

Form Drag

Besides surface drag, form drag is the other means by which drag is produced. Form drag is also called shape drag, profile drag, or pressure drag. As a fluid molecule first strikes the surface of an object moving through it, it bounces off; but then the molecule strikes another fluid molecule and is pushed back toward the surface of the object. Thus the molecule will tend to follow the curvature of the object's surface as the object moves past it. Because a change in direction is an acceleration, the object must exert a force on the molecule to change its direction. The molecule, in turn, exerts an equal but opposite force on the object as it changes direction. The larger the changes in direction, the larger the forces exerted.

On the leading surfaces of the object, the forces exerted by the fluid molecules have components directed toward the rear of the object. These forces contribute to the form drag. On the trailing surfaces of the object, the forces exerted by the fluid molecules have components directed toward the front of the object. These forces reduce the form drag; however, these forces are present only if the fluid molecules stay close to the surface of the object and press against it. This occurs during what is known as **laminar flow** as shown in figure 8.7a. The lines and arrows in this figure represent the paths followed by molecules of the fluid as they pass by the object. But if the change in the surface curvature is too large or the relative velocity is too fast, the impact forces between fluid molecules are not large enough to deflect the molecules back toward the surface, and the molecules separate from the surface. **Turbulent flow** results, and the molecules no longer press against the surface of the object. If the fluid molecules are not pressing against the trailing surface of the object, then nothing is, and a vacuum may be created behind the object. Less force (or no force at all)

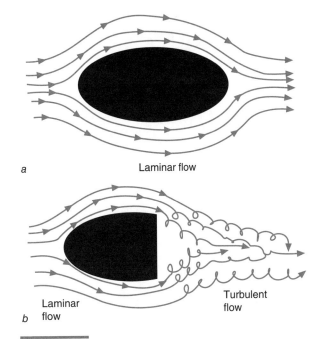

Figure 8.7 Examples of objects experiencing laminar flow (a) and laminar and turbulent flow (b).

is thus exerted against the trailing surfaces of the object. The difference between these forward- and backward-directed forces is form drag. Form drag thus increases as the amount of turbulent flow increases. Figure 8.7b shows an example of an object experiencing laminar and turbulent flow.

As with surface drag, the size of the form drag is affected by the factors included in equation 8.5: drag coefficient, density of the fluid, cross-sectional area of the object (frontal area), and the square of the relative velocity. The drag coefficient is influenced by several other factors associated with form drag as well. The shape of the object has the greatest influence on the form drag contribution to the drag coefficient. To minimize turbulent flow, the fluid molecules have to follow the shape of the object. If the molecules must change direction quickly, large forces are required to cause this acceleration, so the reaction forces that the molecules exert against the object are large, or turbulence occurs. To reduce form drag, then, the surface of an object should curve gently or be "streamlined," without any protrusions or rough spots. An aerodynamic or streamlined shape is one that is long in the direction of flow so that the surface curvature of the object is gentle and not sudden. The tail of the streamlined object essentially fills up the vacuum or empty space created by the turbulent flow. An illustration of nonstreamlined and streamlined shapes is presented in figure 8.8, along with examples of how athletes adopt streamlined positions or use equipment with streamlined shapes.

Figure 8.8 Examples of nonstreamlined and streamlined shapes that affect the magnitude of form drag.

Like surface drag, form drag is influenced by surface texture. A rough surface causes turbulent flow at lower velocities than a smooth surface does. Because turbulence increases form drag, in most cases an object with a rougher surface will experience greater form drag than will an object with a smoother surface.

Sometimes, however, a rougher surface actually decreases form drag. The reason for this paradox is the turbulent flow induced by the rough surface. When turbulent and laminar flow are present, form drag is large due to the difference in pressure (and thus forces) pushing against the leading and trailing surfaces of the object.

When turbulent flow occurs at the leading surface of the object and all the flow around the object is turbulent, form drag actually decreases. How does this happen? When the fluid flow around the object is completely turbulent, a layer of turbulent fluid surrounds the object and is carried along with it. The flow of the rest of the fluid is laminar. This laminar-flowing fluid doesn't exert much drag force on the object and its layer of turbulent fluid. Figure 8.9 illustrates this concept. The dimples on a golf ball reduce the form drag that the ball experiences by creating a layer of turbulent air around it. The fuzz on a tennis ball and the stitches on a baseball may produce the same effect.

Strategies for Reducing Drag Force

In many sports, athletes want to minimize drag forces to maximize their performance. Equipped with our knowledge of the sources of drag forces, let's summarize the ways athletes can reduce drag force by examining the variables in equation 8.5, which is repeated here:

$$F_D = \frac{1}{2} C_D \rho A v^2$$

where

F_D = drag force,

C_D = coefficient of drag,

ρ = fluid density,

A = reference area (usually the cross-sectional area of the object perpendicular to the relative velocity), and

v = relative velocity of the object with respect to the fluid.

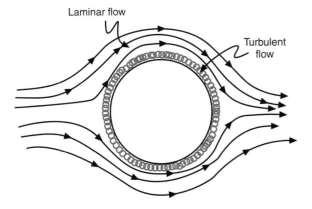

Laminar flow

Turbulent flow

Figure 8.9 A rough surface might decrease form drag by creating a layer of turbulent flow completely surrounding the object on the leading and trailing edges. Outside of this turbulent layer is laminar flow.

Four variables appear in equation 8.5: drag coefficient, fluid density, area of the object, and relative velocity. Fluid density is usually not considered to be a variable under the athlete's control. However, record attempts in cycling and in track and field events have been specifically planned at high-altitude locations where air density is lower. An athlete may also attempt to control this variable by choosing warm-weather competitions, because warmer air is less dense than cooler air, and humid air is less dense than dry air.

Athletes do have several means of reducing the drag coefficient of their bodies and equipment. One way is to make the body surfaces and clothing (or equipment) smoother. Another way is to streamline the shape of the body or equipment. For the athlete, positions that put the body or parts of the body in an elongated orientation relative to the fluid flow will reduce form drag. Cyclists, swimmers, speed skaters, downhill skiers, and lugers all manipulate their positions to reduce form drag. A third way is to reduce the total surface area exposed to the flow, which reduces surface drag. Swimmers (and other aquatic athletes) may also have one other means of reducing the drag coefficient: choosing pools with warmer water. Warmer water is less viscous (and thus creates less surface drag) than cooler water.

The contributions of form drag and surface drag to the total drag force vary with velocity. Form drag accounts for most of the drag force at faster velocities, whereas surface drag accounts for most of the drag force at slower velocities. One must consider this difference when deciding on strategies to use in reducing drag. Exposing a smaller surface area to the fluid flow will reduce surface drag, yet streamlining a shape to reduce form drag usually involves an increase in surface area. If you are moving (or your equipment is moving) through the fluid at fast speeds (greater than 10 m/s or 20 mi/h [32 km/h] in the air), go for the streamlined shape. Otherwise, try to reduce the surface area.

> Form drag accounts for most of the drag force at faster velocities, whereas surface drag accounts for most of the drag force at slower velocities.

The cross-sectional (or frontal) area of the athlete or equipment may also be controlled to some extent. Basically, the athlete wants to reduce the frontal area exposed to the flow. Most of the streamlined positions taken to reduce form drag in the preceding examples also reduce frontal area. Besides making the surface smoother, skin-tight clothing also reduces frontal area. Compare a baggy T-shirt and shorts to a Lycra bodysuit. Athletes orient the implements they use in various sports to minimize frontal area. A football is thrown so that the airflow is along its long axis, and similar orientations are used with a javelin, discus, or Frisbee. These orientations minimize drag.

Relative velocity is the last variable in equation 8.5. Because this term is squared, it has the greatest effect on drag force, so it is the most important variable that the athlete can control. But how can athletes control this variable to minimize drag—when it's their velocity that they want to maximize? One way is to time a performance. Running velocity is an important determinant of success for long jumpers, triple jumpers, and pole-vaulters. The rules for track and field put limits on the time an athlete has to initiate a jump. It is usually 1 min, and most athletes use this time to wait for a tailwind to pick up (or for a headwind to slow down). For the same reason, track and field meet promoters often choose to run the straightaway races in the direction of the prevailing winds. However, records in the long jump, the triple jump, and the short sprints and hurdles do not count if the wind velocity in the direction of the jump or sprint is greater than 2.0 m/s.

Runners, cyclists, and swimmers use another way to reduce the relative velocity. Runners and cyclists rely on a teammate or opponent to lead the race, and they draft the leader. Swimmers may ride the wake of another swimmer who is leading them. In these cases, the fluid has been disturbed by the person leading. The relative velocity of the fluid passing the trailing athlete is thus much lower, and the drag forces exerted on this athlete are smaller. Athletes must run or ride very close to the athlete in front of them to maximize this effect, however. An athlete who leads a cycling or running race is doubly disadvantaged by this tactic. First, the leading athlete must work harder because the drag force is larger due to the faster relative velocity of the fluid. And second, the athletes behind the leader (if they can stay close enough) don't have to work as hard because the drag force is lower due to the slower relative velocity of the fluid. These athletes are thus more likely to sprint past the leader at the end of the race!

Strategies for Increasing Drag Force

In some instances, athletes want to increase drag force. They manipulate their bodies or use implements that increase form or surface drag or both. Paddles and oars are designed to increase drag force in order to propel the boat or shell. Parachutes are designed to increase drag force to an amount large enough to reduce the downward velocity of a parachutist to a safe speed for landing. Kite surfers use kites similar to parachutes to propel them over waves and across the water. Injured athletes may "run" in a pool as part of their rehabilitation routine or as cross-training even if they are healthy. The shoes shown in figure 8.10 are designed for underwater running exercise. Notice the "gills" on the sides of the shoes. The "gills" increase the resistance that the athlete must overcome as the foot is

Figure 8.10 The "gills" or scoops on the AQx water exercise shoes create a drag force as the foot is pulled through the water.

pulled through the water. This resistance is produced by the drag force resulting from the shoe's "gills." Swimmers also use drag force to propel them through the water, but the drag force might also be combined with another dynamic fluid force, lift force, to produce a resultant propulsive force on the swimmer.

Lift Force

Lift force is the dynamic fluid force component that acts perpendicular to the relative motion of the object with respect to the fluid. Rather than opposing the relative motion of the object through the fluid, the effect of lift force is to change the direction of the relative motion of the object through the fluid. The word *lift* implies that the lift force is directed upward, but this is not necessarily the case. A lift force can be directed upward, downward, or in any direction. The possible directions of the lift force are determined by the direction of flow of the fluid. The lift force must be perpendicular to this flow. Lift force is defined by equation 8.6:

$$F_L = \frac{1}{2} C_L \rho A v^2 \qquad (8.6)$$

where

F_L = lift force,

C_L = coefficient of lift,

ρ = fluid density,

A = reference area (usually the cross-sectional area of the object perpendicular to the relative motion), and

v = relative velocity of the object with respect to the fluid.

Qualitatively, lift force can be considered in the following manner. Lift is caused by the lateral deflection of fluid molecules as they pass the object. The object exerts a force on the molecules that causes this lateral deflection (an acceleration, because the molecules change direction). According to Newton's third law, an equal but opposite lateral force is exerted by the molecules on the object. This is the lift force. Lift force is thus proportional to the lateral acceleration of the fluid molecules and the mass of the molecules that are deflected. Figure 8.11 graphically illustrates lift force , and you can demonstrate lift force to yourself the next time you are in a moving car by trying self-experiment 8.2.

> Lift force is the dynamic fluid force component that acts perpendicular to the relative motion of the object with respect to the fluid.

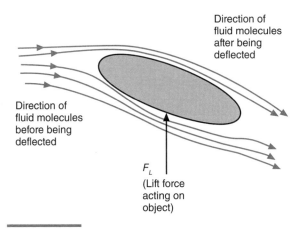

Figure 8.11 The fluid molecules passing by an object are deflected laterally. The change in direction is a lateral acceleration caused by the force exerted by the object. The reaction to this force is the lift force acting on the object.

Self-Experiment 8.2

Reach your hand out the window of a moving car. With your fingers extended and palm facing the ground, slowly supinate your forearm until your palm faces upward. As you do this, you'll notice forces that try to push your hand backward (drag force) or push your hand upward (lift force). These forces will change as the orientation of your hand to the airflow changes or as the velocity of the airflow changes. Now clench your fingers together into a fist and try the same thing. Are the lift forces the same?

The example of your hand in the airflow beside the car demonstrates several factors that influence lift force. Again, relative velocity is the most important. Increasing the car's velocity (and thus the relative velocity of the air) greatly increases the lift force, but the shape and orientation of your hand greatly affect lift force as well. These two factors are partly accounted for in equation 8.6 by the frontal area of the object. The other variable that takes these factors into account is the coefficient of lift.

Some common examples of the use of lift forces, shown in figure 8.12, may further clarify this force. Wings (airfoils) on planes use lift force to keep the aircraft from falling. Likewise, submerged wings (hydrofoils) on some boats (hydroplanes) use lift force to lift the hull of the boat out of the water. A rudder on a boat or plane uses lift force to change the direction of the craft. Propellers on boats and airplanes use lift forces to actually propel the craft. The spoilers on the rear of some race cars use lift force to push the rear-drive wheels against the ground for better traction. Ski jumpers use long, wide skis and position themselves so that the lift forces generated keep them in the air longer and they jump farther. When you tread water, you scull your hands in a horizontal plane, and the lift forces generated by your hands help keep your head above water. The propulsion techniques used in various swimming strokes are a resultant of the lift and drag forces acting on the hands. Sails on sailboats and sailboards use lift forces to propel them forward when they are tacking.

In most of these examples, the object that generates the lift force is longer in the dimension parallel to the flow and shorter in the dimension perpendicular to the flow. Also, lift is generated in most of these examples when the longer dimension of the object is not aligned parallel to the fluid flow. Some objects generate lift even though they are not longer in one dimension than another (e.g., a spinning ball), or even if their longest dimension appears to align with the fluid flow (e.g., an airfoil). How do these objects deflect the fluid laterally to generate lift forces?

Bernoulli's Principle

In 1738, Daniel Bernoulli (1700-1782), a Swiss mathematician, discovered that faster-moving fluids exert less pressure laterally than do slower-moving fluids. This principle, called **Bernoulli's principle**, may explain why some objects are able to generate lift forces even if they are not longer in one dimension than another or if their longest dimension is aligned with the flow.

> Faster-moving fluids exert less pressure laterally than do slower-moving fluids.

Figure 8.12 Examples of objects that create lift forces.

Let's examine an airfoil to see how Bernoulli's principle works. An airfoil is an example of an object whose longest dimension appears to align with the fluid flow, even while it creates lift. Figure 8.13*a* shows an airfoil in cross section. One surface (the upper surface of an airplane wing) of the airfoil is more curved than the other. The curvature is very gentle, and the airfoil is streamlined. When the airfoil is oriented such that its long dimension aligns with the fluid flow, its streamlined shape produces laminar flow (and minimal drag). Imagine two rows of four air molecules each as they approach the airfoil (figure 8.13*b*). When they strike the leading edge of the airfoil, one row travels along the upper, curved surface of the airfoil, and the other row travels along the lower, flat surface. If the flow is laminar, each molecule on top reaches the trailing edge of the airfoil at the same time as the corresponding molecule on the bottom. The molecules on the upper surface travel farther to get to the trailing edge of the airfoil, but they get there at the same time as the molecules on the lower surface. The molecules on top are moving faster; so according to Bernoulli, the lateral pressure exerted by the faster-moving molecules is less than that exerted by the slower-moving molecules, and an upward lift force is generated. The faster-moving upper molecules have

a downward component to their velocity as they slide off the trailing edge of the airfoil, so some downward deflection of the air occurs.

Now let's try to determine the basis for Bernoulli's principle. Back to the two rows of molecules on the airfoil. When the first molecules reach the trailing edge of the airfoil, the molecules will be distributed along the surfaces of the airfoil as shown in figure 8.13*c*. All of these molecules will exert some force against the surface of the airfoil due to the pressure from adjacent molecules. The forces exerted by the molecules on the lower surface of the airfoil will be directed normal to this surface, or upward. The forces exerted by the molecules on the upper surface of the airfoil will be directed perpendicular to this surface. But the upper surface is curved, so these forces will be directed downward as well as backward or forward, as shown in figure 8.13*d*. If the resultant force that each molecule exerts is the same from the upper to the lower surfaces, the net upward force exerted by the molecules on the lower surface is greater than the net downward force exerted by the molecules on the upper surface, because the forces from the molecules on the upper surface have forward or backward components as well. The difference in these net forces is the lift force that is generated.

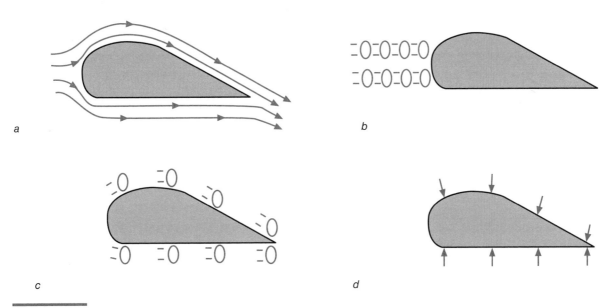

a   b   c   d

Figure 8.13 A possible explanation for Bernoulli's principle using an airfoil as an example.

Spin and the Magnus Effect

In 1852, the German scientist Gustav Magnus noticed that lift forces are also generated by spinning balls. This effect is called the **Magnus effect**, and a lift force caused by a spin is called a Magnus force. But how can an object that doesn't have any broad, relatively flat surface deflect the air laterally to cause a lift force? Let's examine what happens to the air molecules as they approach and pass a ball with topspin. Figure 8.14 is an illustration of a ball (moving left to right across the page) with topspin.

> A lift force caused by a spin is called a Magnus force.

Notice that on a ball with topspin, the upper surface of the ball has a forward velocity (to the right in figure 8.14) relative to the center of the ball, and the lower surface has a backward velocity (to the left) relative to the center of the ball. The air molecules, on the other hand, all have a backward velocity (to the left) relative to the center of the ball. When molecules strike the lower surface of the ball, they don't slow down as much because this surface is moving in the same direction as the molecules (backward or to the left) relative to the center of the ball. When the molecules strike the top surface of the ball, they are slowed down more because this surface is moving in the opposite direction (forward or to the right) relative to the center of the ball. The velocity of the air over the top surface of the ball is lower than the velocity of the air over its bottom surface. According to Bernoulli's principle, then, less pressure will be exerted by the faster-moving molecules on the bottom surface of the ball. This difference in pressure results in a lift force acting downward on the ball, as shown in figure 8.14.

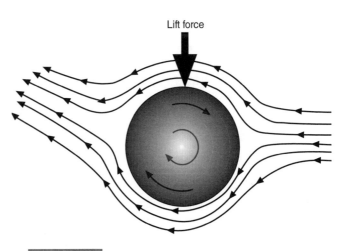

Lift force

Figure 8.14 A ball with topspin has a lift force that acts downward on the ball.

We can also explain the lift force by imagining how the slower-moving molecules get bunched up on the ball's upper surface, whereas the faster-moving molecules spread out on the ball's lower surface. The bunched-up molecules exert more pressure (and thus more force) on the ball than do the spread-out molecules. Imagine a crowd of people exiting a theater or arena through only one or two doors. As they approach the doors, they bunch together more and slow down, and more bumping occurs. This is analogous to what happens to the molecules on top of the ball. More bumping translates to more pressure.

You may also notice that the faster molecules on the bottom of the ball have an easier time staying on the surface of the ball. They don't break away from the ball until they are past the most rearward point on the ball. Thus the molecules are deflected upward (and the ball is deflected downward).

Lift force due to the Magnus effect is responsible for the curved flights of balls observed in a variety of sports. Backspin will keep a ball aloft longer, whereas topspin will cause a ball to drop to the ground sooner. Sidespin will cause a ball to swerve right or left. Golf clubs are designed to impart backspin on the ball so that it stays in the air longer and thus carries farther. Many golfers unintentionally impart sidespin to the ball that causes it to hook or slice to the left or right. Professional golfers may use this sidespin intentionally. Baseball pitchers use sidespin, topspin, or backspin in their repertoire of pitches to make the path of the pitched ball less predictable to the batter. Soccer players use sidespin to cause their corner kicks or penalty kicks to swerve around opposing players. Tennis players use topspin to force their shots to drop into the court sooner. Perhaps the most striking example of the Magnus effect in a ball sport is in table tennis, where the spin of the ball causes wide swerves in its flight.

Center of Pressure

The resultant of the lift and drag forces that act on an object is the **dynamic fluid force**. This force is actually the result of pressures exerted on the surfaces of the object. The theoretical point of application of this force to the object is called the **center of pressure**. If the resultant force acting at the center of pressure is not on a line passing through the center of gravity of the object, a torque is produced that will cause the object to rotate. A Frisbee or discus or football thrown with no spin is more likely to wobble because of this torque. An airplane whose cargo or passengers are loaded too far to the rear will have a dangerous torque act on it that will tend to tip the nose of the plane upward. The International Association of Athletics Federations (IAAF) specifications adopted in 1986 for men's javelins and in 1999 for women's javelins essentially forced the center of pressure to be behind the grip (and the center of gravity) of the javelin. The torque produced by the dynamic forces acting on the javelin during flight causes it to rotate so that its tip strikes the ground. Figure 8.15 illustrates this effect.

> The resultant of the lift and drag forces that act on an object is the dynamic fluid force. The theoretical point of application of this force to the object is called the center of pressure.

Effects of Dynamic Fluid Forces

The dynamic fluid forces that act on an object moving through a fluid have been explained in the previous pages. But athletes are more concerned with the effects of these forces than with the forces themselves. Newton's second law is expressed by equation 3.18 as

$$\sum F = ma$$

where

$$\sum F = \text{net force,}$$

$$m = \text{mass of the object, and}$$

$$a = \text{acceleration of the object.}$$

The effect of a force is the acceleration that would be caused if this were the net force acting, or

$$\sum F = ma$$

$$\frac{\sum F}{m} = a$$

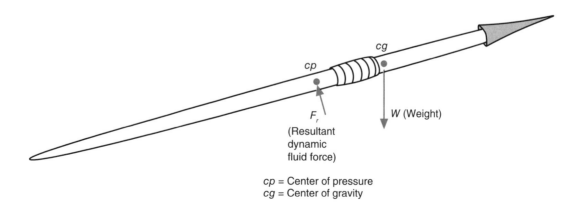

cp = Center of pressure
cg = Center of gravity

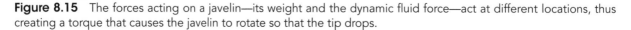

Figure 8.15 The forces acting on a javelin—its weight and the dynamic fluid force—act at different locations, thus creating a torque that causes the javelin to rotate so that the tip drops.

The effects of dynamic fluid forces can be determined from equation 8.4:

$$F \propto \rho A v^2$$

$$a = \frac{\Sigma F}{m} \propto \frac{\rho A v^2}{m}$$

$$a \propto \frac{\rho A v^2}{m}$$

Accelerations are the effects of dynamic fluid forces, and these accelerations are related to all of the factors that the dynamic fluid forces are related to; however, the acceleration is inversely proportional to the mass of the object. This means that two objects similar in size and shape will experience the same dynamic fluid forces, but the more massive object will experience less acceleration. The wind blowing in the face of a 50 kg distance runner will have a greater effect than the same wind blowing in the face of a 70 kg distance runner. It's easier to throw a curveball with a Nerf ball than with a real baseball. The Magnus force on a table tennis ball has a much greater effect than the same Magnus force on a racquetball. The lift force acting on a Frisbee is similar to the lift force acting on a discus, but the effects are much different. The relative effects of lift and drag forces on projectiles can be seen in figure 8.16. This figure shows the trajectories of a variety of balls of similar volume. The initial velocities, angles, and heights of projection of these objects are identical, but their masses (and densities) are different.

> Two objects similar in size and shape will experience the same dynamic fluid forces, but the more massive object will experience less acceleration.

Summary

Fluid forces affect movements in a variety of sports. Buoyant force affects all aquatic sport participants, and dynamic fluid forces affect participants in aquatic as well as land-based sports where fast velocities are involved. Buoyant force is an upward force equal in magnitude to the weight of water displaced by the immersed object. The dynamic fluid force is resolved into drag and lift components and is proportional to the density of the fluid, the frontal area of the object, and the square of the relative velocity of the fluid. Relative velocity of the fluid thus has the greatest influence on drag and lift forces, because these forces are proportional to the square of the velocity. Ways to reduce drag include adopting a streamlined shape, keeping the surface of the object smooth (in most cases), and reducing the frontal area. Lift can be controlled by the shape and orientation of an object. Spin on a ball can also cause lift, thus making ball games much more interesting.

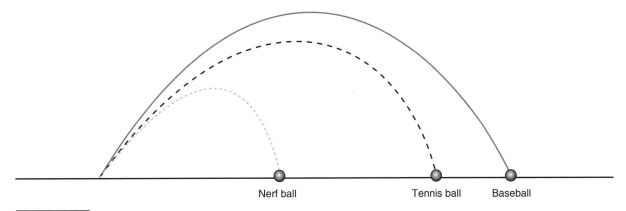

Nerf ball   Tennis ball   Baseball

Figure 8.16 The trajectories of three balls with different masses, demonstrating that the effect of fluid forces is influenced by the mass of the object.

KEY TERMS

Bernoulli's principle (p. 229)
buoyant force (p. 218)
center of pressure (p. 232)
density (p. 220)
drag force (p. 222)

dynamic fluid force (p. 232)
form drag (p. 224)
laminar flow (p. 225)
lift force (p. 228)
Magnus effect (p. 231)

pressure (p. 219)
specific gravity (p. 220)
surface drag (p. 224)
turbulent flow (p. 225)
viscosity (p. 225)

REVIEW QUESTIONS

1. Why is it difficult or impossible for you to float on your back in a perfect horizontal position with your legs together and fully extended at the hips and knees?

2. What drag force is reduced by the presence of dimples on a golf ball?

3. On a human-powered vehicle such as a bicycle, the rider's body creates most of the drag force unless the rider is enclosed or a fairing is used. If you were designing a human-powered vehicle, which rider position would you choose to maximize drag force reduction—upright but crouched (as on a racing bicycle), reclining (as on a recumbent bicycle), or prone (with head and arms forward)?

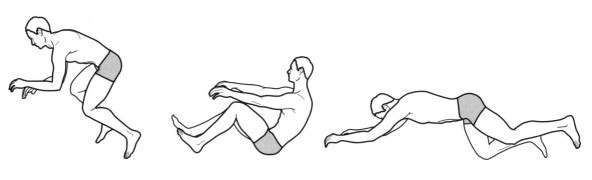

Crouched   Reclining   Prone

4. Why are bugs (those that cannot fly) often able to walk away from falls that would be lethal to humans?

5. How do the stitches on baseballs affect their flight?

6. How does the spin of a tennis ball affect its flight?

7. Why is it beneficial to impart topspin to the tennis ball in a tennis drive shot?

8. Why is it beneficial to impart backspin to the golf ball in a golf drive off the tee?

9. In which sporting event does air resistance have the largest effect: 100 m sprinting, 10,000 m running, or 500 m speed skating?

10. How is the optimal angle of projection in the discus throw affected by fluid forces on the discus during its flight?

11. Rank the following items according to the effect air resistance has on each of them: shot, discus, baseball, Nerf ball, basketball, beach ball, tennis ball, Frisbee.

12. Explain how it is possible for a sailboat to have a velocity component into the wind.

13. Why do swimmers shave their bodies?

14. Why are bicycle helmets tapered posteriorly?

15. What is the purpose of drafting in bicycle racing?

16. When a boat remains still, it sits lower in the water than when it is moving through the water; that is, the boat displaces more water when it is still than when it is moving. Why?

PROBLEMS

1. When Grant fully exhales, he has a body density of 1.02 g/cm³ (1020 kg/m³). During a triathlon swim, Grant wears a wetsuit and his density thus decreases to 0.99 g/cm³ (990 kg/m³). The water he swims in has a density of 1.00 g/cm³ (1000 kg/m³).

   a. If he fully exhales, can Grant float in the water if he doesn't wear a wetsuit?

   b. If he fully exhales, can Grant float in the water if he does wear a wetsuit?

2. K-1 whitewater kayak has a volume of 225 liters (0.225 m³) and a mass of 10 kg. Sean wonders if the kayak is big enough for him to use. Sean's mass is 70 kg. What percent of the kayak will be submerged when Sean sits in it? Assume that the water density is 1.00 g/cm³ (1000 kg/m³).

3. Marathon swimmer DJ is swimming in the Atlantic Ocean. She is moving north with a velocity of 2 m/s relative to the water. The ocean current is also moving north but at 1 m/s. What is DJ's absolute or true velocity?

4. Lily is sprinting in a 100 m dash. At the 80 m mark her velocity is 9.5 m/s. She is running into a 3 m/s headwind. What is the velocity of the air relative to Lily?

5. Carter is running a 400 m race around a 400 m track. On the backstretch his velocity is 8 m/s, but he is running into a 2 m/s headwind. How large is the drag force that acts on Carter? Assume that the density of the air is 1.2 kg/m³, that Carter's cross-sectional area is 0.5 m², and that the coefficient of drag for Carter is 1.1.

6. During the 400 m race in question 5, Carter is now in the homestretch. His velocity is still 8 m/s, but he now has a 2 m/s tailwind. How large is the drag force acting on Carter now? Assume that the density of the air is 1.2 kg/m³, that Carter's cross-sectional area is 0.5 m², and that the coefficient of drag for Carter is 1.1.

7. Amy and Josh are coasting on their bicycles down a 10° slope at 15 m/s through still air. The mass of Amy and her bicycle is 60 kg. The mass of Josh and his bicycle is 90 kg. The cross-sectional area of Amy and her bicycle is 0.45, while the cross-sectional area of Josh and his bicycle is 0.60. The drag coefficient for both cyclists is 0.70. Other than gravity and air resistance, the external forces acting on the two bicycle and rider systems are the same. Which cyclist is more affected by air resistance?

8. When Paul rides his bicycle at 5 m/s, the drag force acting on him and his bicycle is 6 N. If he speeds up to 10 m/s, how large will the drag force become?

Internal Biomechanics

Internal Forces and Their Effects on the Body and Its Movement

The movements of the body as a whole are determined by the external forces that act on the body, but what about the internal forces that hold the body together? How do these forces affect the body and its movement? These questions are addressed in part II, most of which is devoted to the "bio-" part of biomechanics. Part II includes four chapters, chapters 9 through 12. The first chapter in this part, chapter 9, concerns the mechanics of biological materials. The next two chapters, chapters 10 and 11, deal with the structural mechanics of the musculoskeletal system. Chapter 12 presents a brief overview of the control of the neuromuscular system. ■

Mechanics of Biological Materials
Stresses and Strains on the Body

objectives

When you finish this chapter, you should be able to do the following:

- Define mechanical stress
- Define strain
- Define tension (tensile strain and tensile stress)
- Define compression (compressive strain and compressive stress)
- Define shear (shear strain and shear stress)
- Identify and describe bending loads
- Describe the stress patterns that develop under a bending load
- Identify and describe torsion loads
- Describe the general stress–strain relationship for an elastic material
- Define elastic modulus (Young's modulus)
- Describe elastic and plastic behavior
- Define the various descriptors of material strength: yield strength, ultimate strength, and failure strength
- Define stiff, ductile, brittle, and pliant
- Define toughness
- Understand the material properties of bone
- Understand the material properties of tendons and ligaments
- Understand the material properties of muscle

An American football running back is trying to gain enough yardage for a first down when he is hit from the side by a linebacker. A loud pop is heard as the running back's knee bends sideways. Ouch! The force from the linebacker's hit was directed medially against the lateral part of the knee, and the medial collateral ligament of the knee snapped. Why wasn't the knee able to sustain this load without injury? Bones break, ligaments tear, cartilage rips, and muscles pull. These failures occur when the loads imposed cause stresses that exceed the strength of the material. This chapter explains the mechanical behavior of biological materials and how they are able to withstand the loads they encounter.

The external forces that act on

the body affect the movements of the entire body. These forces also impose loads on the body that affect the internal structures of the body—cartilage, tendons, ligaments, bones, and muscle. An understanding of the mechanical properties of these internal structures is important for preventing injury and evaluating the causes of injury.

> The external forces that act on the body impose loads that affect the internal structures of the body.

Stress

In part I, we considered the body as a system of linked rigid segments and thus used rigid-body mechanics to study the effects of the forces acting on the body. In part II, we will discard our assumption of rigidity and consider the segments to be deformable bodies. The external forces that act on the body are resisted by internal forces and cause deformation of the body. The amount of deformation produced is related to the stress caused by the forces and the material that is loaded. Try self-experiment 9.1 to see how forces can cause deformation.

Self-Experiment 9.1

Find a couple of rubber bands. Take one rubber band and pull it until it is twice as long as it was. The rubber band stretches or deforms because of the force exerted. Now take another rubber band and pull on it as well as the first one. If you don't have two rubber bands, double the first one by folding it in half. Pull with about the same force you needed to double the length of the single rubber band. Do the two rubber bands (or the doubled-over rubber band) stretch as much as the single rubber band did under the same force?

The deformation or stretch of the two rubber bands (or the one doubled-over rubber band) was not as large as that of the single rubber band unless you exerted twice as much force. The stress in the single rubber band was twice as large as the stress in the double rubber band.

Mechanical **stress** is the internal force divided by the cross-sectional area of the surface on which the internal force acts. Stress may vary within an object and is associated with a specific internal surface. The three principal stresses are tension, compression, and shear. Mathematically, stress is represented by the Greek letter σ (sigma) and is defined as

$$\sigma = \frac{F}{A} \tag{9.1}$$

where

σ = stress,

F = internal force, and

A = cross-sectional area of the internal surface.

Force is measured in newtons, and area is measured in units of length squared (m^2), so the unit for stress is N/m^2, which is a derived unit also called a pascal (for the French physicist, philosopher, and mathematician Blaise Pascal). The abbreviation for pascal is Pa.

> Mechanical stress is the internal force divided by the cross-sectional area of the surface on which the internal force acts.

Before we go further, a clear explanation of the internal force, F, and the cross-sectional area, A, in equation 9.1 is necessary. Consider the single rubber band in self-experiment 9.1. If it is held still in a stretched position, it is in static equilibrium. A free-body diagram of the rubber band is shown in figure 9.1a.

If we ignore the weight of the single rubber band in self-experiment 9.1, the pulling force, P_1, at the left end must equal the pulling force, P_2, at the right end. Let's call the magnitude of this force P. Now imagine a plane that cuts at right angles through the rubber band at A–A. The free-body diagram of the rubber band to the left of this plane is shown in figure 9.1b. The free-body diagram of the rubber band to the right of the imaginary cut plane is shown in figure 9.1c. The piece of rubber band shown in figure 9.1b is not in static equilibrium, as shown. The pulling force, P, that acts on the left end of the rubber band must be countered by some force, equal in magnitude to

P, that acts to the right. That force is shown in figure 9.1d. This is the force that is used to define stress in equation 9.1. What agent is responsible for this force?

The bonds between the molecules of the rubber band to the right and left of the plane of the imaginary cut produce the pulling force, P, that acts at the cut. This force does not act at a single point, as depicted in figure 9.1d, however. This force actually represents the resultant of all the individual intermolecular bond forces that act across the surface of the imaginary cut and hold the rubber band together (see figure 9.1e). The cross-sectional area of the rubber band at this imaginary cut plane (let's call this the analysis plane) is the area, A, used to define the stress at this plane in equation 9.1.

Tension

Tension is one of the two axial stresses (also called normal or longitudinal stresses), and one of the three principal stresses. The stress described and defined earlier is an example of tensile stress. **Tensile stress** is the axial or normal stress that occurs at the analysis plane as a result of a force or load that tends to pull apart the molecules bonding the object together at that plane. Tensile stress acts perpendicular or normal to the analysis plane and is thus called a normal stress or axial stress.

> ⟳ Tensile stress is the axial or normal stress that occurs at the analysis plane as a result of a force or load that tends to pull apart the molecules bonding the object together at that plane.

Many anatomical structures (long bones, muscles, tendons, ligaments) are longer in one dimension than in the other two dimensions. We typically analyze stress in such structures by considering analysis planes that cut through the structure perpendicular to its longest dimension. The stresses that act perpendicular to the analysis planes (and thus along the long axis of the structure) are called axial, normal, or longitudinal stresses. Consider your humerus (the bone between your elbow and shoulder). If forces pulled on both ends of your humerus, which might be the case when you hang from a chin-up bar, the humerus would be loaded in axial tension. This is an example of axial loading, a loading situation in which the forces act in the direction of the long axis of the bone. A free-body diagram of this situation is shown in figure 9.2a.

If an analysis plane were cut perpendicular to the long axis of the bone, as shown in figure 9.2b, we could determine the stress in the bone at that location, as we did with the rubber band. Divide the load acting across the plane by the cross-sectional area of the bone at the

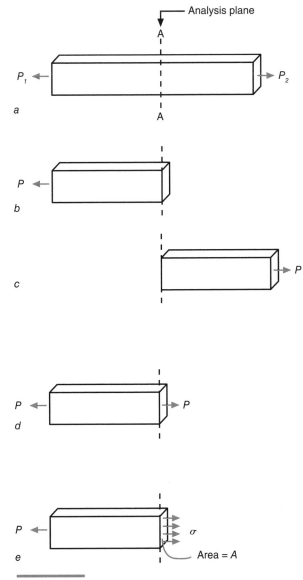

Figure 9.1 Analysis of a stretched rubber band.

surface of the plane. Because the force tends to pull the bone apart and because the stress acts away from the analysis plane, the stress is a tensile stress.

The rubber band had the same cross-sectional area along its length, so no matter where you placed the analy-

sis plane, the stress was the same. The cross-sectional area of the humerus changes as you move proximally to distally along its length. The stress varies with these changes as well. Where the cross-sectional area is larger, the stress is smaller and the bone is stronger. Where the

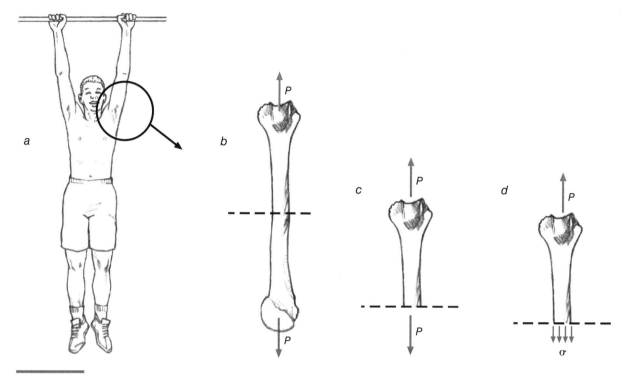

Figure 9.2 The humerus is loaded axially in tension when you do a chin-up.

SAMPLE PROBLEM 9.1

A sample of biological material is loading in a material-testing machine. The cross-sectional area of the specimen is 1 cm². The machine applies a tensile load to the specimen until it fails. The maximum tensile force applied was 700,000 N. What was the maximum stress in the material when it carried this load?

Solution:

Step 1: Identify the known quantities.

$A = 1 \text{ cm}^2 = 0.0001 \text{ m}^2$

$F = 700,000 \text{ N}$

Step 2: Identify the unknown variable to solve for.

$\sigma = ?$

Step 3: Search for an equation with the known and unknown variables.
$\sigma = F/A$

Step 4: Substitute the known quantities and solve for the unknown variable.

$\sigma = (700,000 \text{ N})/(0.0001 \text{ m}^2)$

$\sigma = 7,000,000,000 \text{ Pa} = 7 \text{ GPa}$

(The prefix G before Pa means giga-, or 1 billion. So the stress is 7 billion pascal. Other prefixes used in the SI system are shown in table A.3 in appendix A.)

cross-sectional area is smaller, the stress is larger and the bone is weaker.

When an object or material is axially loaded in tension with forces pulling on either end, tensile stress is produced within the object, and the object also tends to deform by stretching or elongating in the direction of the external loads. For most materials, this elongation is directly proportional to the magnitude of the stress. The increase in deformation of the rubber band was easily seen as it stretched farther with an increasing load (and increasing tensile stress). In the human body, very large tensile loads may sprain or rupture ligaments and tendons, tear muscles and cartilage, and fracture bones.

> ⊃ **When an object or material is axially loaded in tension, the object tends to deform by stretching or elongating in the direction of the external loads.**

Compression

Compression or compressive stress is the axial stress that results when a load tends to push or squash the molecules of a material more tightly together at the analysis plane. Your femur and tibia are under compression when you are standing, as a result of your body weight pushing down on the proximal end and the reaction force from below pushing up on the distal end.

Let's take a look at the humerus again, but this time when it is axially loaded in compression. During a push-up, forces push on each end of the humerus, and it is axially loaded in compression (see figure 9.3).

> ⊃ **Compression or compressive stress is the axial stress that results when a load tends to push or squash the molecules of a material more tightly together at the analysis plane.**

As before, an analysis plane is cut through the humerus. A free-body diagram is drawn of the humerus on either side of the analysis plane, and the stress is determined by dividing the internal force acting at the cut surface by the cross-sectional area of the humerus at the analysis plane. In this case, because the force acting at the analysis plane is pushing toward the surface of the plane, the resulting stress is compressive.

When an object is axially loaded in compression with forces pushing on either end, compressive stress is produced within the object, and the object tends to deform by shortening in the direction of these external forces. If you squeeze a rubber ball, you can easily see the deformation

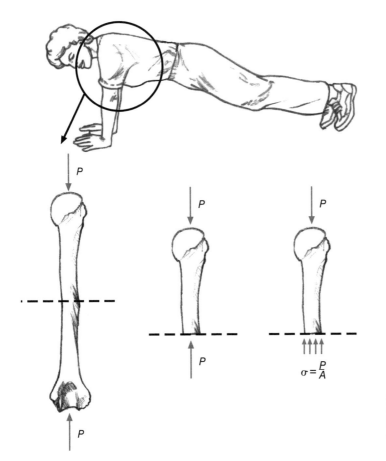

Figure 9.3 The humerus is loaded axially in compression when you do a push-up.

of the ball. The ball is under compression. In the human body, large compressive loads may cause bruising of soft tissues and crushing fractures of bones.

> When an object is axially loaded in compression, the object tends to deform by shortening in the direction of these external forces.

Shear

The third of the three principal stresses is **shear**. Compression and tension are axial stresses that act perpendicular to the analysis plane as a result of forces acting perpendicular to this plane. Shear stress is a transverse stress that acts parallel to the analysis plane as a result of forces acting parallel to this plane. These forces tend to slide the molecules of the object past each other. Try self-experiment 9.2 for an example of shear stress.

> Shear stress is a transverse stress that acts parallel to the analysis plane as a result of forces acting parallel to this plane.

Self-Experiment 9.2

Grab a pencil (or pen) with both hands and hold it near the middle between your thumbs and index fingers as shown in figure 9.4a. Hold the pencil still while pulling back with your left index finger and pushing forward with your right thumb. A free-body diagram of the pencil is shown in figure 9.4b.

Consider an analysis plane cut through the pencil just between the two forces created by the right thumb and left index finger. A free-body diagram of the left half of the pencil is shown in figure 9.4c. Since the left half of the pencil is in static equilibrium, the force pulling back on the pencil must be counteracted by a force pushing forward to maintain force equilibrium. No other external forces act on the left half of the pencil, so the counteracting force must be an internal force acting at the analysis plane. Figure 9.4d shows this force, which arises from the molecular bonds between the two surfaces of the pencil at the analysis plane. This is the shear force from which shear stress is computed.

Shear stress is equal to the force at the analysis plane (which acts parallel to the analysis plane) divided by the cross-sectional area of the object at the analysis plane, or

$$\tau = \frac{F}{A} \tag{9.2}$$

where

τ = shear stress,

F = shear force, and

A = cross-sectional area at the analysis plane.

Axial stresses (tension and compression) are represented by the symbol σ (sigma), but transverse stress (shear) is represented by its own symbol, τ (tau).

Scissors are an example of a tool that creates large shear forces, thus creating large shear stresses in the material worked on. Scissors are also referred to as "shears."

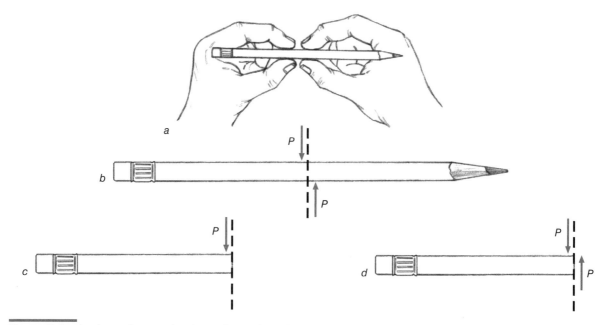

Figure 9.4 Analysis of a pencil withstanding a shear load.

In the human body, shear loads cause blisters of the skin, and large shear loads acting on the extremities may cause joint dislocations or shear fractures of bones. A shear load causes a deformation of the object as well, but rather than stretching or squashing it, shear causes a change in the orientation of the sides of the object, or a skewing. The three different types of stress are illustrated in figure 9.5.

Mechanical Loads

The external forces acting on an object impose a mechanical load on the object. Simple tensile, compressive, or shear loads were described in the preceding examples. For simple tensile or compressive loads, the external forces act along the same line producing a uniaxial load; the analysis plane is chosen perpendicular to the long axis of the loaded object; and the stresses present at the analysis plane are of only one type—tension alone or compression alone. When you stretched the rubber band, this was an example of a uniaxial tensile load, and the stress produced at the analysis plane was tension alone. This stress was also uniform across the analysis plane (i.e., at any point within the analysis plane, the tensile stress was the same

as at any other point within the analysis plane). Not all mechanical loads are uniaxial. The specific arrangements of external forces produce different types of loads. The number, direction, and location of the external forces acting on an object, as well as the shape of the object itself, define the type of load imposed. Bending, torsion, and combined loads are examples of the more complex loads that may be imposed on an object.

Bending

Simple uniaxial loads produce only one type of stress, which is uniform across the analysis plane. Most loading situations are not so simple: Multiple stresses occur at the analysis plane, or the stress may vary across the analysis plane. A **bending load** is an example of a load that produces different stresses at the analysis plane. Try self-experiment 9.3. This self-experiment and figure 9.6 illustrate how bending loads produce tensile and compressive stresses.

Self-Experiment 9.3

Take a pencil (or pen) and grip it by the ends with your right and left hands so that your index fingers are closest to the pencil ends and your thumbs are closer to the middle. Bend the pencil by pushing away with your thumbs and pulling toward yourself with your fingers. The forces produced by the index finger and thumb of each hand produce equal but opposite force couples or torques on the pencil and cause the pencil to bend slightly (and possibly break if the forces you exert are large enough). What stress does this load impose on the pencil? Are parts of the pencil stretched? Are other parts of the pencil compressed? A free-body diagram of the pencil is shown in figure 9.6a.

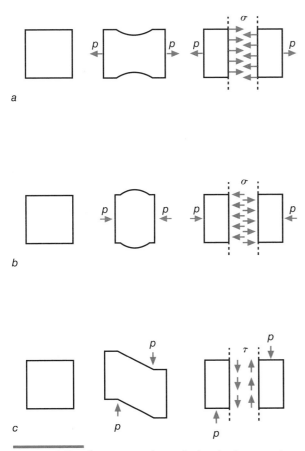

Figure 9.5 Illustration of simple loads that produce the three mechanical stresses: tension (a), compression (b), and shear (c).

The stresses in the pencil in self-experiment 9.3 may be analyzed with an analysis plane cutting through the pencil between the two thumbs. The free-body diagram of the pencil to the left side of the analysis plane is shown in figure 9.6b. The forces acting on this pencil piece are in equilibrium (if the pulling force from the index finger is equal to the pushing force from the thumb), but the torques created by these forces are not in equilibrium. The two forces acting on the left piece of the pencil create a force couple, and the torque caused by this force couple would tend to rotate and angularly accelerate this pencil piece counterclockwise. Some other forces must act on the pencil to create a torque to counter the force couple. The only place where forces could act to create a counter torque is at the analysis plane. To create a torque, though, any forces acting at this analysis plane must be paired as force couples in order to maintain force equilibrium while still creating a torque. Figure 9.6c shows how forces

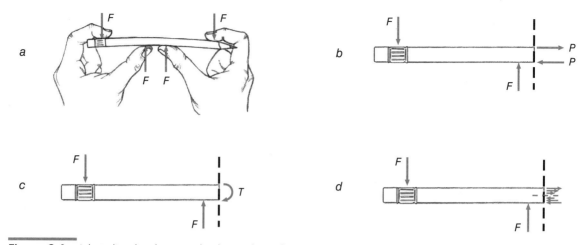

Figure 9.6 A bending load creates both tensile and compressive stresses.

could act as a force couple at the analysis plane and create a torque to counter the unbalanced torque. Such a force couple would be created by the molecules near the top of the right pencil piece pulling on the molecules at the top of the left pencil piece and by the molecules near the bottom of the right pencil piece pushing on the molecules near the bottom of the left pencil piece. The forces shown acting perpendicular to the analysis plane in figure 9.6c represent the net forces created by the pulling and pushing of the molecules at the analysis plane. These forces are distributed among the molecules across the area of the analysis plane as tensile stress on the upper half of the plane and compressive stress on the lower half of the plane as shown in figure 9.6d. The stresses become larger as the distance away from the center line (neutral axis) of the pencil increases.

To maintain force equilibrium in the pencil in self-experiment 9.3, the net tensile force acting at the analysis plane must equal the net compressive force acting at the analysis plane. Because these net forces act some distance away from each other, they create the force couple that causes a torque counter to the torque created by your thumb and index finger. The moment arm (the distance between these internal forces) of the internal force couple is obviously limited by the depth of the object withstanding the bending load. For the pencil, this moment arm is small, so the internal forces (and thus the stresses) must be large to create a large enough countering torque. Recall the equation for torque we learned in chapter 5:

$$T = F \times r \tag{9.3}$$

where

T = torque (or moment of force),

F = force, and

r = moment arm (or perpendicular distance).

An object with greater depth (and more cross-sectional area farther from its neutral axis) is able to withstand greater bending loads because it has a larger moment arm. Under similar bending loads, the stresses in such an object will be smaller. Structural members used in construction are designed with this in mind. In houses, 2 × 6, 2 × 8, or 2 × 10 boards are used as beams. Which dimension is the depth of the beam? The longer dimension is aligned in the plane of the bending loads to maximize the moment arm. I beams used in steel construction are an even better example, because the beam depth is great and the area of the cross section is largest near the top and bottom. Figure 9.7 illustrates the effect of an object's depth on its ability to withstand bending loads.

> An object with greater depth (and more cross-sectional area farther from its neutral axis) is able to withstand greater bending loads because it has a larger moment arm.

The primary structural members of the human body are long bones. Beams used in building construction are designed for specific loading situations with forces acting to bend them in a specific plane. Long bones must contend with a wide variety of loading situations with forces acting to bend them in many different directions. Thus the cross sections of long bones do not have definite deeper and narrower dimensions. Instead, long bones are more like pipes, round in cross section but most dense near the perimeter, where the most stress occurs, and least dense (or hollow) near the central (or neutral) axis, where the least stress occurs.

An object such as a beam that is subjected to a bending load will deform by curving. The tension side of the beam will elongate, whereas the compression side will compact and shorten, thus causing the beam to bend. Figure 9.8

shows a variety of other bending load configurations and the deformations of the loaded beams.

Most bones in the human body are constantly under some bending load as a result of the forces exerted on them by gravity, muscles, tendons, ligaments, and other bones. The neck of the femur is a good example of a cantilever beam subjected to a bending load; a force is exerted on the head of the femur, and the shaft of the femur provides the rigid support for the neck of the femur. Tensile stress develops in the upper part of the neck,

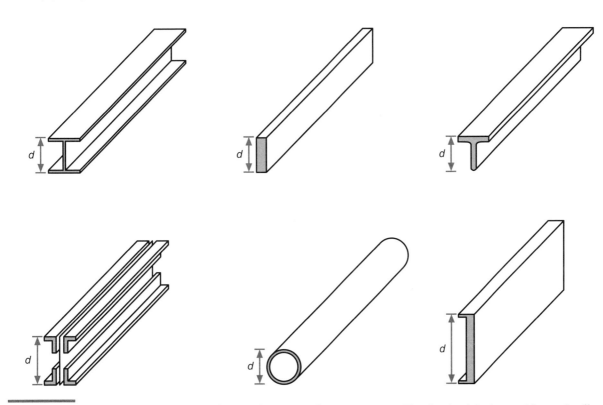

Figure 9.7 Cross sections of a variety of beam shapes used in construction. The depth of the beam (d) greatly affects its ability to withstand bending loads.

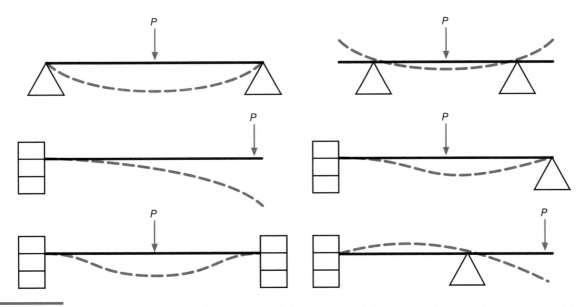

Figure 9.8 Various beam support configurations and the resulting deflections under a single concentrated load. The triangle supports represent pinned or hinged connections that allow the beam to rotate at the support point. The blocks represent rigid connections that do not allow the beam to rotate.

whereas compressive stress develops in the lower part of the neck. Figure 9.9 illustrates these stresses.

> An object such as a beam that is subjected to a bending load will deform by curving. The tension side of the beam will elongate, whereas the compression side will compact and shorten, thus causing the beam to bend.

When the foot is bearing weight, it is an example of an anatomical beam in which the load is distributed among several structures. Figure 9.10 illustrates the foot during weight bearing. It behaves as a simple beam. Supports are at the heel and metatarsophalangeal joints, and the load is applied from the tibia to the talus. In this situation, the bones of the foot carry the compressive stress, while the plantar fascia and dorsal muscles carry the tensile stress.

Torsion

A **torsion load** is another type of load that produces something other than uniaxial stresses. Torsion loading occurs when torques act about the long axis of the object at each end. Take a pencil and hold it with one hand at each end. Now try to twist the pencil in one direction with your right hand and in the opposite direction with your left hand. Figure 9.11a shows a three-dimensional free-body diagram of the whole pencil.

A free-body diagram of the left half of the pencil with an analysis plane cutting through it is shown in figure 9.11b. This pencil piece is not in equilibrium. A single torque acts on the pencil, which would cause it to spin about its longitudinal axis. Another torque must be present to counteract this torque. The countering torque is created by the shear force between the molecules of the pencil at the right and left sides of the analysis plane. Figure 9.11c shows a free-body diagram of the left half of the pencil with this internal torque included. Figure 9.11d shows the shear stresses that produce this internal torque at the analysis plane. The stresses become larger with increasing distance from the central axis of the pencil (the axis of rotation for the torques). The shear stress is thus not uniform across the analysis plane.

The moment arm between the internal shear forces on either side of the longitudinal axis is limited by the diameter of the object under torsion. For the pencil, this moment arm is small, so the internal forces (and stresses) must be large to create a large enough countering torque (refer to the equation for torque, equation 5.1). If the stresses become too large, the pencil will break. An object with a larger diameter is able to withstand greater torsional loads since the shear stresses are smaller as a result of the larger diameter. Figure 9.12 illustrates this graphically.

This is similar to the situation with bending loads, except that increases in both width and depth (increases in diameter) are desirable for increasing torsional strength. Thus structural members or tools designed to resist torsional loads are usually circular in cross section, for example driveshafts, axles (although axles carry bending loads as well), and screwdriver shafts. This is another reason for the circular cross section of bones. Torsion loading of bones is common in a variety of situations. For example, when you step onto your foot and then pivot, torque is produced around the longitudinal axis of your tibia, and it becomes torsionally loaded.

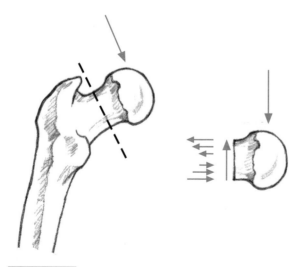

Figure 9.9 The neck of the femur acts as a cantilever beam and develops tensile stress in its upper part and compressive stress in its lower part.

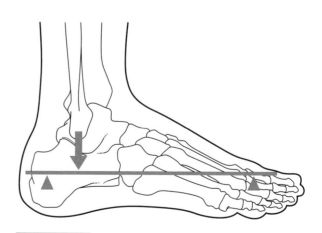

Figure 9.10 The foot as an example of a simple beam.

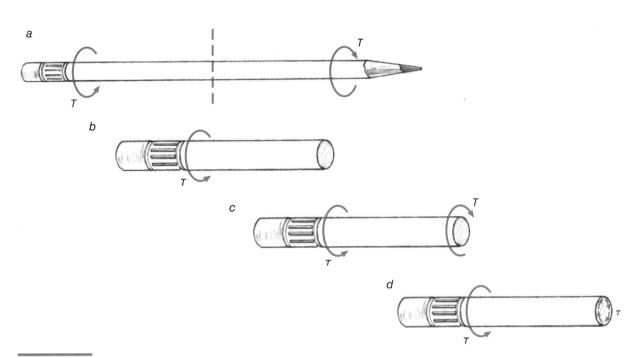

Figure 9.11   A pencil withstanding a torsional load is subjected to shear stress.

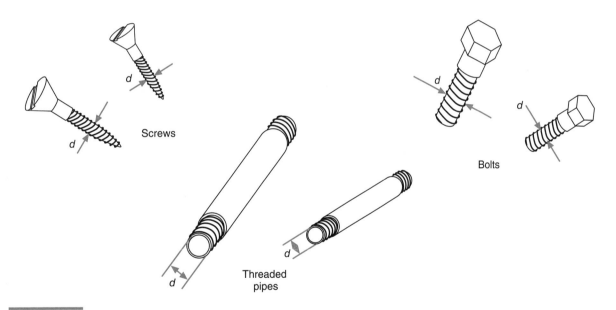

Screws

Bolts

Threaded
pipes

Figure 9.12   The diameter (d) of an object's cross section greatly affects its ability to withstand torsional loads.

Combined Loads

Because of their construction, certain anatomical structures, such as muscles, tendons, and ligaments, behave like ropes or cables and effectively carry only one type of load: uniaxial tension. Bones and cartilage, on the other hand, may be loaded in a variety of ways, from uniaxial tension, compression, or simple shear loads that produce uniform stress to bending and torsion loads that produce more complex stress patterns. Bones and cartilage often encounter a combination of these loading configurations. This type of load is called a **combined load**.

Analysis of the stresses produced by combined loads is complex but basically involves simplifying the loading into several fundamental load configurations. The stresses produced under each of these loading configurations are then added to determine the actual stresses produced by the combined load. Consider the shaft of the femur during weight bearing, as shown in figure 9.13a. If we ignore the muscle and ligament forces that act on it, the

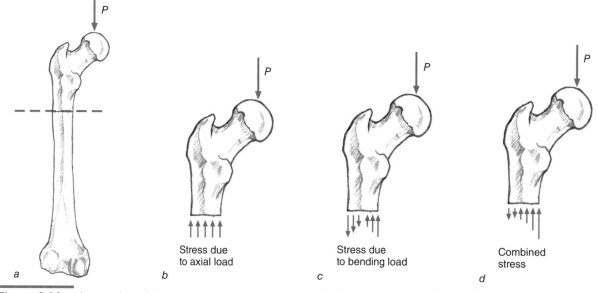

Figure 9.13   The combined stresses due to the compressive load and bending load on the femur.

femur basically acts as a column (to support the weight of the body) and as a beam (because the body's weight is eccentric to the shaft of the femur), so it carries a bending as well as a compressive load. The stresses in the shaft of the femur resulting from the axial compressive column load are shown in figure 9.13b, and the stresses resulting from the bending load are shown in figure 9.13c. The combined stresses are shown in figure 9.13d.

Strain

Objects deform when they are subjected to external forces. These deformations may be large or small, depending on the nature of the material and the stresses involved. Strain is the quantification of the deformation of a material. Linear strain occurs as a result of a change in the object's length. Linear strain is produced by compressive or tensile stresses. Shear strain occurs as a result of a change in the orientation of the object's molecules.

⟳   **Strain is the quantification of the deformation of a material.**

Linear Strain

When the loading of an object causes tensile or compressive stress within the object, some change in length accompanies this stress. This deformation can be measured in absolute terms that describe the change in length of the object as a result of the loading—for example, the rubber band stretched 10 cm or the intervertebral disc compressed 2 mm. The deformation may also be described in relative terms as a proportion of the change

in length (the difference between the undeformed and the deformed lengths) divided by the original length (the undeformed length). This relative deformation measure is called linear strain. Mathematically, linear strain is defined as

$$\varepsilon = \frac{\text{change in length}}{\text{original length}}$$

$$\varepsilon = \frac{\ell - \ell_o}{\ell_o} \tag{9.4}$$

where

ε = linear strain,

ℓ = stretched length,

ℓ_o = original, undeformed length, and

$\ell - \ell_o$ = change in length.

Try self-experiment 9.4.

Self-Experiment 9.4

Take a long rubber band and lay it, unstretched, alongside a ruler. Use a pen to make a mark on it about 1 cm from the left end. Align this mark with zero on the ruler and then make another mark on the rubber band 5 cm from the first mark. Make smaller marks at every 5 mm interval between the zero and 5 cm marks. Now, hook the left end of the rubber band over the zero end of the ruler and pull the slack out of the rubber band. Measure the length of the unstretched rubber band by noting where the right end of the rubber band lines up with the ruler. Now, pull on the right end of the rubber band and stretch it until it's twice its original length. How far apart are the original zero and 5 cm marks on the stretched rubber band? How far apart are any two adjacent 5 mm marks? The end marks are about twice as far apart or 10 cm apart, and the 5 mm marks are now spaced 10 mm apart.

> ## Linear strain occurs as a result of a change in the object's length.

Let's determine the strain in the stretched rubber band in self-experiment 9.4. The distance between the end marks on the rubber band was 5 cm in the unstretched condition and 10 cm after stretching. The absolute deformation of this 5 cm section of the rubber band was

$$10 \text{ cm} - 5 \text{ cm} = 5 \text{ cm.}$$

The relative deformation or strain would be

$$\varepsilon = \frac{\ell - \ell_o}{\ell_o}$$

$$\varepsilon = \frac{10 \text{ cm} - 5 \text{ cm}}{5 \text{ cm}}$$

$$\varepsilon = 1 \text{ cm/cm}$$

or

$$\varepsilon = 100\%$$

Now look at any 5 mm section of the rubber band. Each of them has stretched to 10 mm long, so the strain in every 5 mm section is

$$\varepsilon = \frac{\ell - \ell_o}{\ell_o}$$

$$\varepsilon = \frac{10 \text{ mm} - 5 \text{ mm}}{5 \text{ mm}}$$

$$\varepsilon = 1 \text{ mm/mm}$$

or

$$\varepsilon = 100\%$$

The strain between any two adjacent 5 mm marks on the rubber band is also 100%. This makes sense; because the tensile stress is uniform throughout the rubber band, it follows that the strain would be also.

You may have noticed that we have reported strain in centimeters per centimeter, millimeters per millimeter, and as a percentage. Actually, strain is a dimensionless quantity because it is a ratio of length to length. So, centimeters per centimeter is the same as millimeters

SAMPLE PROBLEM 9.2

A sample of biological material is loaded into a material-testing machine. The material is 2 cm long in its unloaded state. A 6000 N tensile force is applied to the material, and it stretches to a length of 2.0004 cm as a result of this force. What is the strain in the specimen when it is stretched this much?

Solution:

Step 1: Identify the known quantities.

$\ell = 2.0004$ cm

$\ell_o = 2.0$ cm

Step 2: Identify the unknown variable to solve for.

$\varepsilon = ?$

Step 3: Search for an equation with the known and unknown variables.

$$\varepsilon = \frac{\ell - \ell_o}{\ell_o}$$

Step 4: Substitute the known quantities and solve for the unknown variable.

$$\varepsilon = \frac{\ell - \ell_o}{\ell_o} = \frac{2.0004 - 2.000}{2.000}$$

$$\varepsilon = 0.0002 = 0.02\%$$

per millimeter is the same as inches per inch, because the units cancel out. Usually, strain is reported as a percentage, which seems to make more sense. In that case, the ratio is multiplied by 100% to get the percentage of strain. In the preceding example, 1 cm/cm is the same as 1 mm/mm because each equals 100%. Most biological materials are not as elastic as a rubber band, and rupture or failure will result at strains much, much less than 100%.

Shear Strain

Linear strain occurs with a change in length as a result of molecules being pulled apart or pushed together. Shear strain occurs with a change in orientation of adjacent molecules as a result of these molecules slipping past each other. Figure 9.14 graphically illustrates shear strain.

> Shear strain occurs with a change in orientation of adjacent molecules as a result of these molecules slipping past each other.

Shear strain is measured as follows. Imagine a line perpendicular to the analysis plane through the object at the location of interest. In two dimensions, this line and the line of the analysis plane form a right angle when the object is undeformed. But, when a shear load is applied to the object, this angle changes. The change in the angle (θ) is the measure of shear strain in the direction of interest. Shear strain is abbreviated with the Greek letter λ (lambda) and is measured in radians.

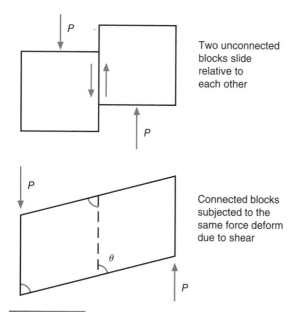

Two unconnected blocks slide relative to each other

Connected blocks subjected to the same force deform due to shear

Figure 9.14 Illustration of deformation caused by shear. The change in the angle (θ) indicates the shear strain.

Poisson's Ratio

Consider the rubber band again. As you stretched it, what happened to the width of the rubber band? Its width became narrower as it was stretched in length. Consider a rubber ball loaded in compression. The diameter of the ball shortens in the direction of the compressive load, but what about the diameter in the lateral direction? When the rubber ball is compressed, it becomes wider in the lateral direction while shortening in the direction of the compressive load. This is called the Poisson effect, after S.D. Poisson, a French scientist who investigated this phenomenon in the 1820s. A specific ratio of strain in the axial direction to strain in the transverse direction exists for each different type of material. This ratio is called **Poisson's ratio**. Values of Poisson's ratio can be as low as 0.1 and as high as 0.5, but for most materials they are between 0.25 and 0.35.

The behavior of intervertebral discs provides a good example of the Poisson effect. During the day, your intervertebral discs are loaded in compression (unless you are lying down all day). This compressive load shortens the vertical dimension of the discs, but laterally, the discs bulge out. Under extremely large compressive loads, a disc may bulge out too much and rupture.

Mechanical Properties of Materials: The Stress–Strain Relationship

How much stress can an intervertebral disc withstand before it ruptures? How far can a bone bend? How much energy can a ligament absorb before it breaks? How far can a tendon stretch before the stretch becomes permanent? These questions all relate to the manner in which materials respond to forces. The relationship between the stress and strain in a material may help explain the behavior of a material under load.

Elastic Behavior

Let's consider the rubber band once more. The more you stretch it, the greater the force you have to pull with. If we could measure this force and the corresponding elongation of the rubber band, we would probably get a plot similar to that shown in figure 9.15.

Because the rubber band was stretched by a uniaxial load, the stress and strain in the rubber band are uniform. The plot shown in figure 9.15 could be represented by a similar plot of stress versus strain, as shown in figure 9.16. The load–deformation plot shown in figure 9.15 was specific to the rubber band being stretched, but the stress–strain curve shown in figure 9.16 is a characteristic of the material the rubber band is made of.

ity (bulk modulus for compression or shear modulus for shear), is defined as

$$E = \frac{\Delta\sigma}{\Delta\varepsilon} \tag{9.5}$$

where

E = elastic modulus,

$\Delta\sigma$ = change in stress, and

$\Delta\varepsilon$ = change in strain.

Since there is zero strain when there is zero stress, the stress–strain curve passes through zero, and equation 9.5 can be expressed as

$$E = \frac{\sigma}{\varepsilon} \tag{9.6}$$

When we compute the modulus of elasticity, strain is expressed in the unitless ratio—not as a percentage. A material that is more stiff has a steeper slope of its stress–strain curve, and thus a larger elastic modulus, than a material that is more pliant. Steel is stiffer than rubber. Bones are stiffer than ligaments or tendons. Figure 9.17 shows the stress–strain curves for materials with different stiffnesses.

> The ratio of stress to strain (which is shown graphically as the slope of the stress–strain curve) is called the elastic modulus of a material.

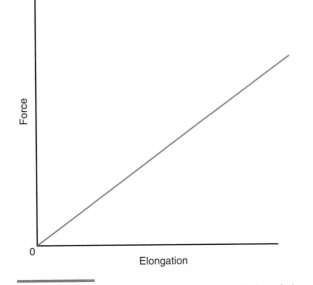

Figure 9.15 A plot of tensile force applied and the corresponding elongation of a rubber band.

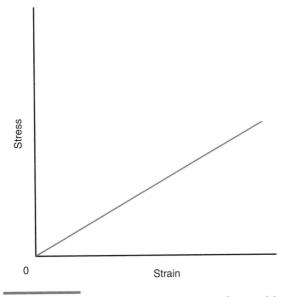

Figure 9.16 A plot of stress and strain for a rubber band's material.

The stress–strain curve shown in figure 9.16 is an example of the stress–strain curve for a linearly **elastic** material. Rubber is elastic (versus plastic). It stretches under a tensile load but returns to its original shape when the load is removed. This property is called elasticity. Rubber is linearly elastic because, as the stress increases, the strain increases a proportional amount. The ratio of stress to strain (which is shown graphically as the slope of the stress–strain curve) is called the **elastic modulus** of a material. Mathematically, the elastic modulus, which is also called **Young's modulus** or the **modulus of elastic-**

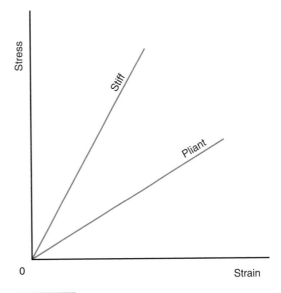

Figure 9.17 Stress–strain curves for stiff versus pliant materials.

Plastic Behavior

The behavior of a material under load may change as the load increases. Under small loads, the object may be elastic; if the load is removed, the object returns to its original shape and dimensions. But if the load exceeds a certain size, some permanent deformation of the object may occur. This is called **plastic** behavior. Try self-experiment 9.5 for an illustration of elastic and plastic behaviors.

Self-Experiment 9.5

Find a paper clip and pull lightly on one of its wire ends so that it tends to unbend one of the curves in the wire. If you pull with only a small force and then let go, the wire springs back to its original shape. This loading occurs in the elastic region of the metal's stress–strain curve. But if you pull with a larger force, the wire does not spring back to its original paper clip shape. This loading occurs in the plastic region of the metal's stress–strain curve. In this region, the stress is no longer proportional to

the strain, and the material will not return to its original shape when the stress is removed. An example of the stress–strain curve for the metal of the paper clip is shown in figure 9.18.

The point on the stress–strain curve where further stress will cause permanent deformation is called the yield point or the **elastic limit**. It often coincides with the **proportional limit**, the end of the linear elastic range of the curve. Below this load, the material behaves elastically, and above this load it behaves plastically. Most materials exhibit some elastic and some plastic behavior.

Material Strength and Mechanical Failure

Physiologically, muscle strength is defined as the ability of a muscle to produce force. Mechanically, the strength of a material has to do with the maximum stress (or strain) the material is able to withstand before failure. The key word here is *failure*. Failure may be defined as

SAMPLE PROBLEM 9.3

A material is subjected to a tensile load of 80,000 N (80 kN). Its cross-sectional area is 1 cm². The elastic modulus for this material is 70 GPa. What strain results from this tensile load?

Solution:

Step 1: Identify the known quantities.

$F = 80,000$ N

$A = 1$ cm² $= 0.0001$ m²

$E = 70$ GPa

Step 2: Identify the unknown variable to solve for.

$\varepsilon = ?$

Step 3: Search for an equation with the known and unknown variables.

$$E = \frac{\sigma}{\varepsilon}$$

$$\sigma = \frac{F}{A}$$

Step 4: Substitute the known quantities and solve for the unknown variable.

$\sigma = (80,000$ N$) / (0.0001$ m²$) = 800,000,000 = 800$ MPa

$E = 70$ GPa $= (800$ MPa$) / \varepsilon$

$\varepsilon = (800$ MPa$) / (70$ GPa$) = (800,000,000$ Pa$) / (70,000,000,000$ Pa$)$

$\varepsilon = 0.0114 = 1.14\%$

the inability to perform a function. Thus there are several quantifications for the strength of a material, depending on which function of the material is of interest. These are illustrated in the stress–strain curve shown in figure 9.19.

> Mechanically, the strength of a material has to do with the maximum stress (or strain) the material is able to withstand before failure.

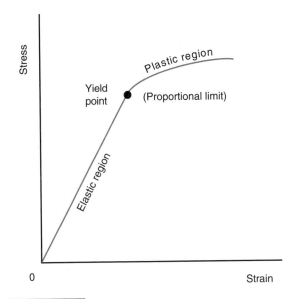

Figure 9.18 Elastic region, plastic region, and yield point on a stress–strain curve for the metal of a paper clip.

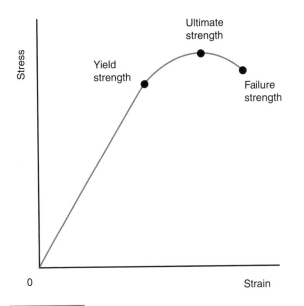

Figure 9.19 Material measures of strength shown on a stress–strain curve.

The stress at the elastic limit of a material's stress–strain curve is the **yield strength** of the material. Although no breakage or rupture of the material occurs, stresses beyond the yield strength will cause permanent changes in the dimensions of the material. Beyond this point, the material fails in the sense that it is unable to regain its shape. A severe strain of the ankle ligaments may cause a permanent lengthening, and thus loosening, of these ligaments.

The **ultimate strength** of a material is the maximum stress the material is capable of withstanding. This is represented by stress corresponding to the highest point on the stress–strain curve shown in figure 9.19. If the function of interest is the ability of the material to withstand large loads, the ultimate stress determines how much load may be carried. The load is the measure of the material's strength in this case.

The rupture or **failure strength** of a material is the stress where failure actually occurs. Failure in this sense means breakage or rupture. Failure strength is represented by stress corresponding to the endpoint of the stress–strain curve in figure 9.19. Failure strength (rupture strength or fracture strength) usually has the same value as ultimate strength.

Material strength may also be expressed as the failure strain. **Failure strain** is the strain exhibited by a material when breakage occurs. It is the strain corresponding to the endpoint of the stress–strain curve in figure 9.19. Materials with large failure strains are ductile materials, whereas materials with small failure strains are brittle materials. Glass is brittle, and rubber is ductile. Bones become more brittle with age (primarily because of disuse). Figure 9.20 illustrates these material properties.

Perhaps the best way to characterize the strength of a material is by its **toughness**. Mechanically, toughness is the ability of a material to absorb energy. Put another way, a material is tougher if more energy is required to break it. An estimation of the toughness of a material is given by the area under the stress–strain curve, as shown in figure 9.21.

> Mechanically, toughness is the ability of a material to absorb energy.

Recall the work–energy relationship we discussed in chapter 4. It is the basis for the measurement of material toughness. Something that is able to exert great force, but only over a very short displacement, cannot do as much work or cause as great a change in energy as something that is able to exert less force over a much longer displacement. A hard, brittle material is able to withstand great stress but only small strain, so its energy-absorbing ability is less than that of a softer but more ductile mate-

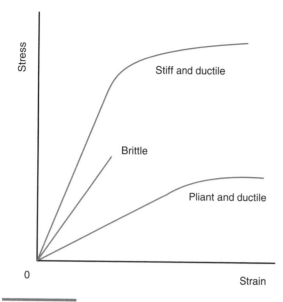

Figure 9.20 Material characteristics illustrated by their stress–strain curves.

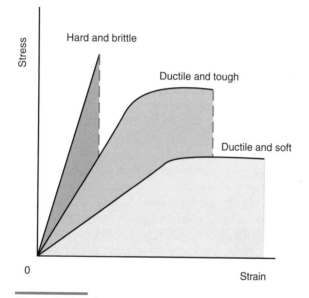

Figure 9.21 The toughness of different materials is indicated by the area under the material's stress–strain curve.

rial. Dry bone is an example of a hard, brittle material. Although it may be able to withstand large stresses, it is relatively easy to break because it cannot deform easily. Moist, live bone, on the other hand, deforms more easily and is less brittle. It is much more difficult to break than dry bone because it is tougher; more work must be done to fracture it.

Mechanical Properties of the Musculoskeletal System

Muscle tissue and connective tissue form the structural units of the musculoskeletal system. The specific connective tissues important to the structure of the musculoskeletal system are bone, cartilage, ligament, and tendon. Muscles may be thought of as the active elements of the musculoskeletal system, whereas connective tissues are the passive elements.

All connective tissues are composed of living cells as well as extracellular components consisting of collagen, elastin, ground substance, minerals, and water. The mechanical properties of the various tissues are determined in part by the proportions of these components and their arrangement. Collagen, which is a fibrous protein, is the most abundant substance in all connective tissue. Molecules of collagen align together to form collagen fibrils that bind together to form collagen fibers. Collagen is thus very stiff (failure strain of 8% to 10%) and has high tensile strength. On the other hand, collagen is unable to resist compression because its long fibers are not supported laterally. It collapses or buckles like a rope. Elastin is also fibrous, but unlike collagen, it is pliant and very extensible (failure strain as high as 160%). The ground substance consists of carbohydrates and proteins that, combined with water, form a gel-like matrix for the collagen and elastin fibers.

The composition of connective tissues (and most biological tissues) causes them to be anisotropic, unlike many synthetic materials, which are isotropic. **Isotropic** materials have the same mechanical properties in every direction. **Anisotropic** materials have different mechanical properties depending on the direction of the load. For example, the ultimate tensile strength of a tendon is very high if the tensile load is in the direction of the tendon fibers, but its ultimate tensile strength is low if the tensile load is perpendicular to these fibers (see figure 9.22).

Age and activity affect the mechanical properties of all connective tissues. Bone, cartilage, tendon, and ligament strengths increase with regular cycles of loading and unloading. Usually this strength increase is due to an increase in size of the tissue cross section, but the stiffness and ultimate strength of these tissues may also increase—an indication that size alone is not responsible for the strength increase. The exact threshold of the loads and the number of loading cycles required to stimulate these strength gains are unclear. Inactivity and immobilization result in decreased strength of these tissues and a shortening of the ligaments and tendons.

⊃ **Age and activity affect the mechanical properties of all connective tissues.**

All of these connective tissues show an increase in ultimate strength with age until the third decade of life, after which strength decreases. Bones become more brittle and less tough with increasing age. Tendons and ligaments, on the other hand, become less stiff.

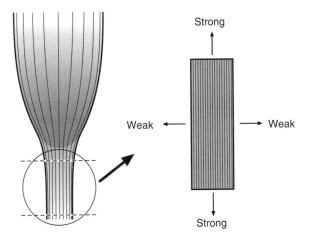

Figure 9.22 Anisotropic behavior of a tendon. The tendon is stronger if the tensile load is aligned with the tendon fibers rather than transverse to the fibers.

Bone

Bones carry almost all of the compressive loads experienced by the body, and they are capable of resisting large shear and tensile loads as well. By weight, 30% to 35% of bone is collagen, 1% to 2% is ground substance, 45% is mineral, and about 20% is water. Bone is the strongest and stiffest material of the musculoskeletal system. The stress–strain curve for a piece of **cortical bone** removed from a long bone and loaded axially is shown in figure 9.23. Cortical or **compact bone** is found in the dense and hard outer layers of bone. **Cancellous (trabecular or spongy) bone** is the less dense, porous bone that is spongy in appearance and found deep to cortical bone near the ends of long bones.

The elastic region of the stress–strain curve is not quite linear but slightly curved. Values for the elastic modulus for bone thus have some inherent error associated with them. In the plastic region of the curve, the loading is still able to increase past the elastic limit. Values for the elastic modulus, yield strength, ultimate strength, failure strength, and failure strain of cortical bone vary from bone to bone as well as from location to location on an individual bone. In addition, values of these properties change with the age of the bone. The porosity of the bone primarily determines its strength and stiffness.

Another factor that affects the mechanical strength and stiffness of bone is the rate of loading. Bone is stronger and stiffer if a load is applied quickly but is weaker and less stiff if a load is applied slowly. Figure 9.24 illustrates

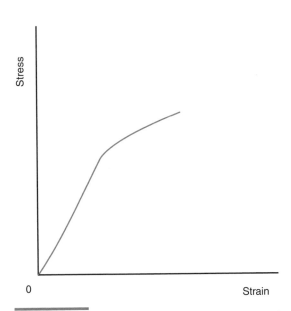

Figure 9.23 General shape of stress–strain curve for cortical bone.

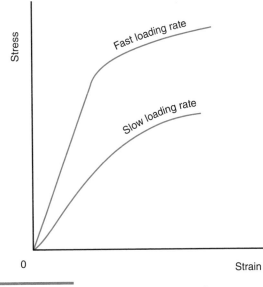

Figure 9.24 Effect of loading rate on the stress–strain relationship of cortical bone.

this rate dependency. Thus a load applied slowly across a joint may result in an avulsion fracture (the ligament pulls bone out of its attachment site), whereas the same load applied quickly across the joint will rupture the ligament.

Bones are strongest in compression and weakest in shear. The high mineral content of bone, primarily calcium and phosphate, accounts for its high compressive strength. The collagen fibers intertwined with these mineral salts give bone its high tensile strength.

> **Bones are strongest in compression and weakest in shear.**

Cartilage

There are three different types of cartilage: hyaline cartilage, fibrous cartilage, and elastic cartilage. **Hyaline cartilage**, also called articular cartilage, is the cartilage that covers the ends of long bones at joints. **Fibrous cartilage** (fibrocartilage) is found within some joint cavities (the menisci of the knee), in the intervertebral discs (the annulus fibrosus), at the edges of some joint cavities, and at the insertions of tendons and ligaments into bone. **Elastic cartilage** is found in the external ear and in several other organs that are not part of the musculoskeletal system. Cartilage is able to withstand compressive, tensile, and shear loads.

By weight, hyaline cartilage consists of 10% to 30% collagen, 3% to 10% ground substance, and 60% to 80% water. Superficially, the collagen fibers in hyaline cartilage are arranged in layers parallel to the articular surface. Deeper, the collagen fibers seem to be arranged randomly. At the deepest layer, where the hyaline cartilage joins with bone, the collagen fibers are perpendicular to the articular surface and stick into the bone like fingers. This arrangement fuses the articular cartilage to the bone. Articular cartilage (as well as fibrous and elastic cartilage) has no nerve or blood supply, so it must be very thin to allow the diffusion of nutrients to cells for normal metabolism. Human articular cartilage is usually only 1 to 3 mm thick.

Articular cartilage transmits the compressive loads from bone to bone at joints. But how does a material composed primarily of collagen and water resist compression loads, when collagen can withstand only small compressive stresses? Imagine a balloon filled with water. As you push down on the balloon, its sides bulge out and stretch. Because water doesn't compress, it pushes out on the sides of the balloon. The sides of the balloon are thus under tension. If the balloon is made of a material capable of resisting large tensile stress, you can push on the balloon with more force. Articular cartilage is similar. Collagen's tensile strength holds the cartilage together under compressive loads due to

the arrangement of the collagen fibers near the exterior surface of the cartilage.

Unlike a balloon, however, articular cartilage is not watertight, so some fluid is exuded (squeezed out) when it is loaded in compression. This behavior causes **creep** and stress relaxation effects. Articular cartilage loaded under a constant compressive stress will not experience a corresponding constant strain. The strain will increase with time as fluid is exuded from the cartilage until it reaches a point where no more fluid is exuded. This increase in strain under a constant stress is called creep. Creep rate is how quickly the cartilage reaches a constant strain. Creep rate depends on the magnitude of the compressive stress, the thickness of the articular cartilage, and the permeability of the cartilage. The time it takes to reach constant strain may range from 4 to 16 h for human articular cartilage. Figure 9.25 illustrates creep.

What happens to the fluid exuded by articular cartilage? It may assist with lubricating the articular surfaces, and it is reabsorbed by the cartilage as the compressive stresses are reduced.

Stress relaxation is another effect caused by fluid squeezing out of articular cartilage. If articular cartilage is loaded so that it experiences a constant strain, it will not experience a corresponding constant stress. The initial strain will cause an increase in stress, but the stress will then decrease (or relax) to some lower value. Self-experiment 9.6 illustrates stress relaxation.

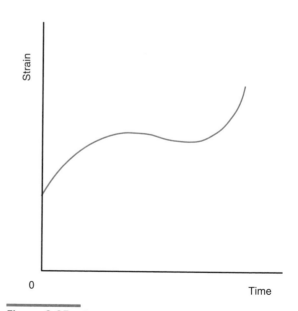

Figure 9.25 Creep in articular cartilage under constant compressive stress.

Self-Experiment 9.6

Fill a small balloon with water and place it on a table or other flat surface. When the balloon first makes contact with the table, only a small area of the balloon makes contact and carries the load. This part of the balloon is under large pressure because of the small area of contact. But this part of the balloon "gives" so that more of the balloon is in contact with the table and pressure is reduced. Because the balloon deforms and conforms to the shape of the supporting surface, pressure on the balloon is reduced. Articular cartilage behaves in a similar way, but, while the balloon deforms quickly, articular cartilage deforms more slowly. Thus large stresses may be present at first, and these stresses decrease as the rest of the cartilage deforms.

Imagine the surfaces of the articular cartilages. There may be only a small area of contact between these surfaces in the unloaded condition. As load is applied, this small area of contact is compressed and fluid is squeezed to other areas of the cartilage, which then expand and contact the other cartilage surface to carry some of the load. This increase in contact area results in a decrease or relaxation of stress at the initial contact point. This process is illustrated with the plot of stress versus time in figure 9.26.

The elastic modulus of articular cartilage is much smaller than the elastic modulus of bone, but articular cartilage transmits compressive loads from one bone to another. The difference in stiffness of these materials accounts for a slight damping effect of articular cartilage.

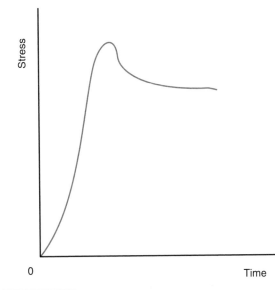

Figure 9.26 Stress relaxation in articular cartilage under constant compressive strain.

But because articular cartilage is so thin, this shock-absorbing effect is limited.

Tendons and Ligaments

Tendons and ligaments are similar in composition and structure, so we will consider them together. By weight, tendons and ligaments consist of approximately 70% water, 25% collagen, and 5% ground substance and elastin. Ligaments have more elastin than tendons. Other than water, collagen is the primary component of tendons and ligaments, but this is also true of cartilage. The major difference between cartilage and tendons or ligaments is in the arrangement of their collagen fibers. The groups of collagen fibers in tendons are bound together in parallel (tendon) or nearly parallel (ligament) bundles that lie along the functional axis of the tendon. The parallel arrangement produces a structure that is very stiff and high in tensile strength but has little resistance to compression or shear. The slight difference in the arrangement of the bundles of collagen fibers and the slightly larger elastin component in ligaments make them less stiff and slightly weaker than tendons. But because the collagen fibers of ligaments are not as well aligned as those of tendons, ligaments can carry loads that are nonaxial. Figure 9.27 illustrates the fiber arrangements in tendons and ligaments.

Collagen fibers have a wavy or crimped appearance. These crimps or waves in the collagen fibers may account for the unusual behavior of ligaments and tendons under uniaxial tensile loads and the stress–strain curves that result. When a tendon is pulled with low tensile force, it stretches easily and behaves like a fairly elastic material. But beyond a certain load, the tendon is very stiff. The low

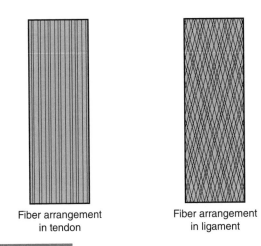

Fiber arrangement in tendon

Fiber arrangement in ligament

Figure 9.27 Parallel arrangement of collagen fibers in tendon, and nearly parallel arrangement of collagen fibers in ligament.

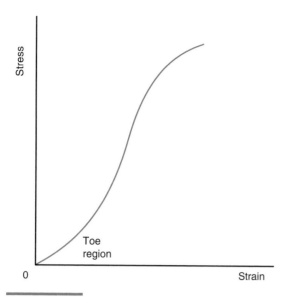

Figure 9.28 General shape of stress–strain curve for a tendon.

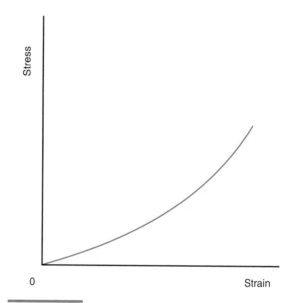

Figure 9.29 General shape of stress–strain curve for passive muscle.

stresses that accompany the initial strains are a result of the straightening of the crimps or waves in the collagen. Once these have been straightened, the actual stretching of the collagen begins, and larger stresses are required to produce the same change in strain. The stress–strain curve for a tendon is shown in figure 9.28. The tensile stress–strain curves for ligaments and cartilage have shapes similar to those of tendons.

The first part of this curve is called the toe region and corresponds to the straightening of the crimps in the collagen. Past the toe region, the behavior is fairly linear; an increase in stress results in a proportional increase in strain. The modulus of elasticity for tendons and ligaments is the modulus that corresponds to this linear region of the stress–strain curve.

Tendons and ligaments also exhibit creep and stress relaxation behavior similar to cartilage as shown in figures 9.25 and 9.26. The creep response of these tissues is relevant to stretching exercises. Stretching exercises to increase flexibility are more effective if you move into the stretch position slowly and then hold this position by keeping the tissues involved under constant stress. The strain in the tendons and ligaments, and thus their lengths, will slowly increase. If the stretch is done very quickly, using swinging or bouncing movements, the strain in the tendons and ligaments is not as large, and the change in their lengths is not significant.

Muscle

Unlike connective tissues, muscle tissue is capable of actively contracting to produce tension within itself and the structures to which it attaches. The active contractile

components of muscle thus determine the stiffness of the muscle at any instant. Muscle stiffness varies as a function of the number of active contractile elements. We will examine the structure and function of muscle more thoroughly in chapter 11. We can learn some things about the mechanical properties of muscle if we consider it in its relaxed or passive state.

If a passive muscle is slowly stretched, there will be some resistance to this stretching and thus some stress developed. This resistance to stretching originates from two different sources: the contractile elements and the connective tissue sheaths that surround the muscle fibers and are connected to the tendons. The stiffness of the passive contractile component caused by the filaments of the contractile elements sliding past each other is very low, so the resistance to this movement is small as well. Stretching of the muscle past the point where the contractile filaments no longer overlap is resisted by the connective tissue component of the muscle. This connective tissue behaves like tendon. A stress–strain curve for passive muscle may look something like that shown in figure 9.29.

⊃ **Muscle stiffness varies as a function of the number of active contractile elements.**

The toe region of this curve is much longer due to the very low stiffness of the passive contractile elements. Stiffness increases when the contractile filaments no longer overlap and the connective tissue begins to strain. The failure strain for muscle is much larger than that of tendon or ligament due to the ability of the muscle's contractile filaments to slide past each other as the muscle increases its length by 50%. Only after the muscle has stretched this much will the connective tissue begin to be strained. The total strain this connective tissue can withstand is similar to that of tendon or ligament: 8% to 15%. The failure strain for muscle is thus 50% plus 8% to 15%, or a total strain of 58% to 65%. Ultimate strength is much lower than that of tendon because a cross section includes mostly the contractile elements of the muscle (which are composed of contractile filaments not bound to each other in the passive state).

Summary

Knowledge of stress and strain and the mechanical properties of musculoskeletal tissues will assist in preventing injury, analyzing injury mechanisms, and evaluating rehabilitation or training exercises. The external forces acting on the body are ultimately carried by the bones, cartilage, ligaments, tendons, and muscles of the musculoskeletal system. These external loads cause stresses and strains. Stress is internal force per cross-sectional area. Strain is the ratio of change in length (or deformation) to the undeformed length. Compressive stress is caused by loads that push molecules together, resulting in compressive strain as the object shortens. Tensile stress is caused by loads that tend to pull molecules apart, resulting in tensile strain as the object lengthens. Shear stress is caused by loads that tend to slide molecules past each other, resulting in shear strain as the object's shape distorts.

Uniaxial compression and shear are examples of loads that result in uniform stress. Bending results in tensile stress on one side of a beam and compressive stress on the other. Torsion results in shear stresses that vary in magnitude across a section. Combined loads may result in compressive, tensile, and shear stresses all in the same cross section.

The strength of a material may be characterized by its stiffness (the ratio of stress to strain during elastic deformation), yield stress (the stress it can withstand without plastic deformation), ultimate stress (the largest stress it can withstand before it breaks), failure strain (the strain it can withstand before it breaks), or toughness (the energy it can absorb before it breaks). These properties may all be estimated from a stress–strain curve of a material.

Bone is the strongest of the connective tissues as determined by its stiffness, ultimate strength, yield strength, and toughness. Its strength is dependent on the rate of loading, however. Ligaments, tendons, and cartilage have similarly shaped stress–strain curves due to their collagenous composition. Under low stresses, these materials are pliant, but as the stresses increase past a certain threshold, they become much stiffer. The mechanical properties of muscle are not as easily examined because of its contractile ability. The ultimate stress of muscle is less than that of tendon, ligament, or bone, whereas its failure strain is much greater.

KEY TERMS

anisotropic (p. 256)
bending load (p. 245)
cancellous bone (p. 257)
combined load (p. 249)
compact bone (p. 257)
compression (p. 243)
cortical bone (p. 257)
creep (p. 258)
elastic (p. 253)
elastic cartilage (p. 258)
elastic limit (p. 254)

elastic modulus (p. 253)
failure strain (p. 255)
failure strength (p. 255)
fibrous cartilage (p. 258)
hyaline cartilage (p. 258)
isotropic (p. 256)
modulus of elasticity (p. 253)
plastic (p. 254)
Poisson's ratio (p. 252)
proportional limit (p. 254)

shear (p. 244)
spongy bone (p. 257)
stress (p. 240)
tensile stress (p. 241)
torsion load (p. 248)
toughness (p. 255)
trabecular bone (p. 257)
ultimate strength (p. 255)
yield strength (p. 255)
Young's modulus (p. 253)

REVIEW QUESTIONS

1. When an athletic trainer pulls laterally on your ankle while pushing medially on your knee, a bending load is placed on your lower extremity. Which side of your knee undergoes compressive stress, and which side of your knee undergoes tensile stress as a result of this manipulation?

2. A snowboarder catches an edge and falls. Her board twists in one direction as her body twists in the opposite direction. The torsional load places large shear stresses on her tibia and knee. If the load on the tibia is axial torsion only, where on a cross section of the tibia will the shear stress be greatest?

3. The shafts of two prosthetic legs have exactly the same surface area in cross section, but the shaft of leg A has a diameter 2% smaller than the shaft of leg B. The two shafts are made from the same material.

   a. Under a uniaxial tensile load, which shaft is stronger?

   b. Under a uniaxial compressive load, which shaft is stronger?

   c. Under a torsional load, which shaft is stronger?

   d. Under a bending load, which shaft is stronger?

4. Which can withstand greater tensile stress—tendon or ligament?

5. Which can withstand greater tensile stress—bone or ligament?

6. Which can withstand greater tensile strain—bone or ligament?

7. Which is stiffer—bone or ligament?

8. Which more ductile—live bone or dry bone?

9. What type of stress is bone strongest in resisting? What type of stress is bone weakest in resisting?

10. Describe the various measures of strength for biological materials.

PROBLEMS

1. A 1 cm long section of the patellar ligament stretches to 1.001 cm when it is subjected to a tensile force of 10,000 N. What is the strain in this segment of ligament?

2. The section of the patellar ligament in the previous question is 50 mm² in cross section. What is the stress in this section of ligament as a result of the 10,000 N tensile force?

3. The modulus of elasticity (for compression) for a section of compact bone in the femur is 12 GPa (12×10^9 Pa). If this bone is subjected to a compression stress of 60 MPa (60×10^6 Pa), what strain results from this compression?

4. The modulus of elasticity (for tension) of a tendon is 1.2 GPa (1.2×10^9 Pa). The tendon is subjected to a strain of 4%.

   a. What is the stress in the tendon when its strain is 4%?

   b. If the unloaded tendon is 2 cm long, what is its deformed length as a result of this 4% strain?

5. The Achilles tendon is subjected to a large tension stress that results in a strain of 6%. If the unloaded tendon is 10 cm long, how much does it elongate as a result of this strain?

6. The yield strength of a material is 10 MPa. The yield strain for this material is 0.10%. What is the modulus of elasticity for this material?

7. The modulus of elasticity for a prosthetic material is 20 GPa. A 3 cm long sample of this material is circular in cross section with a radius of 1 cm. This sample is stretched 3.003 cm. What tensile force was applied to the material to create this stretch?

8. Compact bone from the tibia has an ultimate tensile stress of 150 MPa and a tensile strain at failure of 1.5%. The anterior cruciate ligament has an ultimate tensile stress of 35 MPa and a strain at failure of 30%. Which is probably mechanically tougher, the bone or the ligament?

chapter 10

The Skeletal System
The Rigid Framework of the Body

objectives

When you finish this chapter, you should be able to do the following:

- Identify the parts of the skeletal system

- Describe the functions of the skeletal system

- Describe the anatomical features of bones

- Classify bones as long, short, flat, irregular, or sesamoid

- Describe the growth process of long bones

- Describe the structural and functional classification systems for joints

- Classify synovial joints as gliding, hinge, pivot, ellipsoidal, saddle, or ball and socket

- Describe the anatomical features of synovial joints

- Describe the functions of articular cartilage

- Describe the functions of synovial fluid

- Identify the factors that contribute to the stability of synovial joints

- Identify the factors that contribute the flexibility of synovial joints

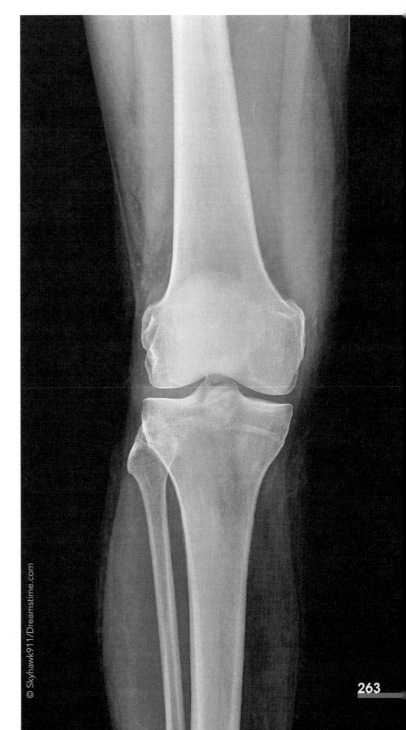

© Skyhawk911/Dreamstime.com

Imagine what life would be like without a skeleton. What would you look like? How would you move? Would you be able to move? If you were in water, you might be functional, but few terrestrial creatures exist without some sort of skeleton. Earthworms and slugs don't have skeletons. They can't stand up, and they move very slowly and inefficiently. Your skeleton allows you to stand up, and its bones and the joints between these bones allow you to move rapidly. This chapter presents these and other functions of the skeletal system and joints.

The skeletal system includes the

bones of the skeleton, the joints where these bones meet, and the cartilage and ligaments associated with the joint structures. The skeleton is sometimes partitioned into the axial skeleton and the appendicular skeleton, as shown in figure 10.1. The **axial skeleton** includes the bones of the trunk, head, and neck: the skull, vertebral column, and rib cage. The **appendicular skeleton** includes the bones of the extremities along with the shoulder girdle (clavicle and scapula) and pelvic bones (ilium, ischium, and pubis). An adult has 206 bones: 126 in the appendicular skeleton, 74 in the axial skeleton, and 6 auditory ossicles (bones of the inner ear).

Mechanically, the skeletal system may be thought of as an arrangement of rigid links connected to each other at joints to allow specific movements. Muscles that attach

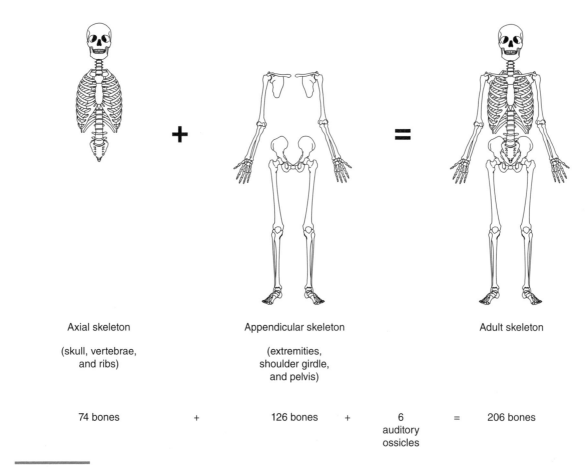

Axial skeleton

(skull, vertebrae, and ribs)

Appendicular skeleton

(extremities, shoulder girdle, and pelvis)

Adult skeleton

74 bones   +   126 bones   +   6 auditory ossicles   =   206 bones

Figure 10.1   The axial and appendicular skeletons together form the adult human skeleton.

to the bones provide the forces that may cause changes in the positions of the bones relative to each other. This is the primary mechanical function of the skeletal system. The importance of the skeletal system in sport is that it provides the long levers that enable movement. The rigid framework of the skeleton also supports the soft tissues and organs of the body. Various bony structures of the skeleton provide protection to vital organs as well. The skull and vertebrae encase and protect the central nervous system (the brain and spinal cord). The ribs and sternum protect the heart, lungs, major blood vessels, liver, and spleen. The pelvis protects the uterus in women and the bladder. The skeletal system also has metabolic functions. The red marrow in some bones produces red blood cells, white blood cells, and platelets. In addition, bones store calcium and phosphorus. The functions of the skeletal system are summarized in table 10.1.

Table 10.1 Functions of the Skeletal System

Mechanical functions	Physiological functions
Support	Hemopoesis (blood cell
Protection	production)
Movement	Mineral storage

⊃ Mechanically, the skeletal system may be thought of as an arrangement of rigid links connected to each other at joints to allow specific movements.

Bones and joints are the basic components of the skeletal system. This chapter presents the structure and development of these components.

Bones

Bones, which account for about 16% of an adult's total body weight, are composed of two different types of bony tissue: cortical or compact bone and cancellous or spongy bone. As their names imply, cortical (compact) bone is more dense and compact, and cancellous (spongy) bone is more porous and spongy in appearance. As discussed in the previous chapter, the composition of bone makes it the strongest and stiffest of all body tissues. It has great compressive, tensile, and shear strength. These qualities of bone make the skeleton very well suited to its function as a supporting structure.

Bones are living tissues and much different from the dried bones or plastic models of the human skeleton used in anatomy classes and laboratories. Bones are innervated and supplied with blood. As living tissue, bone is adapt-

able. It responds to the stresses imposed on it by growing thicker in the areas of local stress or becoming more dense. The density of bone decreases if it is not stressed regularly. Astronauts returning from extended flights in space thus suffer from loss of bone density, as does anyone who does not maintain a minimal level of activity.

Bone Anatomy and Classifications

Bones come in a variety of shapes and sizes. The mechanical stresses imposed on a bone and its function determine its form. Bones are classified according to their shape. Long bones generally have long, hollow shafts with knobby ends. They are designed for large movements. Long bones occur in the extremities and include the humerus, radius, ulna, femur, tibia, fibula, metacarpals, metatarsals, phalanges, and clavicle. Short bones are small, solid, and blocklike. These bones are well suited to transferring forces and shock absorption, but they are not very mobile. The wrist and ankle bones (the carpals and tarsals) are all short bones. As their name implies, flat bones have flat surfaces and are thinner in one dimension. These bones are designed for protection. The ribs, skull, scapula, sternum, and pelvic bones are flat bones. Irregular bones are those that fit into none of the other categories. These bones are designed for support, protection, and leverage. The vertebrae (including the sacrum and coccyx) and the facial bones are examples of irregular bones. Some bones are also described as **sesamoid bones**. These bones develop within tendons, often in an effort to decrease stress or increase leverage. The patella is an example of a sesamoid bone.

⊃ The mechanical stresses imposed on a bone and its function determine its form.

Long bones are the bones most involved in motion. Knowledge of the structure of long bones is thus appropriate for sport biomechanics. Figure 10.2 shows a typical long bone and its structure. A thin, fibrous membrane called the periosteum covers the external surface of the bone except at articular surfaces. The shaft or body of a long bone is a hollow tube with walls of cortical bone. The hollow core of long bones is the medullary or marrow cavity, which contains the red marrow responsible for blood cell production. In adults, most of the red blood cell production takes place in the flat bones, and yellow marrow occupies the medullary cavity of long bones. Yellow marrow is composed primarily of fat cells. As noted in the previous chapter, the long, hollow, cylindrical shape of long bones makes them lightweight, yet still quite strong in resisting bending loads.

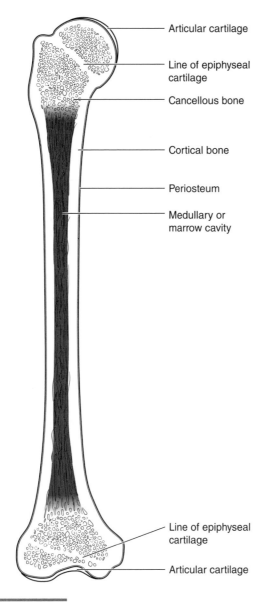

Figure 10.2 Longitudinal section of a typical long bone.

The ends of long bones are typically the location of articulations with other bones. Articular cartilage covers the joint surfaces at the ends of long bones. Again, the walls of the ends of long bones are composed of cortical bone, but the interior spaces are filled with cancellous bone rather than marrow.

Specific terms are used to refer to the surface irregularities of bones. Some of these terms are listed and defined next and are shown graphically in figure 10.3.

- Condyle—a rounded projection that articulates with another bone
- Epicondyle—a rounded projection near the ends of a long bone but lateral to the axis and not necessarily a part of the articulation

- Facet—a small, smooth, and usually flat articular surface
- Foramen—a hole, usually for nerves or vessels to pass through
- Fossa—a hollow depression or pit
- Fovea—a smaller hollow depression or pit
- Head—the spherical articular end of a long bone
- Line—a raised line or small ridge
- Neck—the part of the bone that joins the head to the shaft
- Notch—an indentation on the border or edge of a bone
- Process—a projecting part of bone
- Spine—a sharp projecting part of bone
- Trochanter—a large, knobby projection
- Tubercle—a small, knobby projection
- Tuberosity—a knobby projection

Growth and Development of Long Bones

Long bones develop via endochondral ossification: Cartilage is replaced by bone. In a human fetus, a cartilage model of the skeleton forms. Endochondral ossification of this cartilage skeleton begins prior to birth. At birth, the shaft of a long bone has ossified, but the ends are still composed of cartilage. In the metacarpals, metatarsals, and phalanges of the hands and feet, only one end of these bones remains cartilaginous at birth. The cartilage ends of long bones ossify soon after birth, except for the cartilage that separates the ends from the rest of the bone. This cartilage is called the **epiphyseal cartilage** (epiphyseal plate, epiphyseal disc, growth plate), and the separate end of the bone is called the **epiphysis**. The remaining part of the bone on the other side of the epiphyseal cartilage is called the **diaphysis**. Because a single long bone in children may actually consist of two or three separate bones, children have more bones than adults. Illustrations of the long bones of an adult and child are shown in figure 10.4.

⟳ Long bones develop via endochondral ossification: Cartilage is replaced by bone.

The epiphyseal cartilage is responsible for the growth in length of long bones, as shown in figure 10.5. As this cartilage grows, the cartilage nearest the diaphysis ossifies. If the rate of these processes is equal, longitudinal bone growth occurs. If the rate of ossification exceeds the rate of cartilage growth, however, the entire epiphyseal

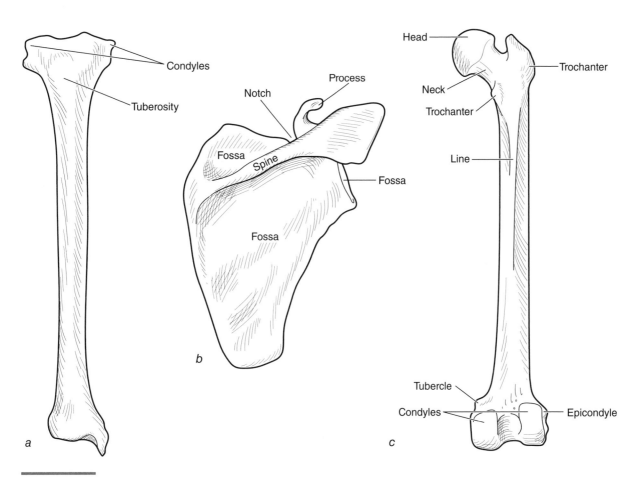

Figure 10.3 Surface irregularities of the femur (a), the scapula (b), and the humerus (c).

cartilage ossifies, joining the diaphysis with the epiphysis, and longitudinal growth stops. Epiphyseal plate closure, as this is called, occurs naturally at specific ages for each bone. Most epiphyses ossify during puberty, but some may not close until after age 25.

Growth at the epiphyseal plates is affected by the stresses occurring there. A certain amount of mechanical stress is necessary to stimulate growth, but too much may cause premature closure. Certain hormones also affect growth at the epiphyseal plate. Too little growth hormone retards growth, whereas too much may extend growth past the normal age of closure. Sex hormones increase the rate of cartilage replacement and thus lead to closure of the epiphyseal plates. This is the primary reason for the closure of most growth plates during puberty.

Growth in the diameter of long bones occurs where the periosteum (the membrane covering the surface of the bone) interfaces with bone. New bone is deposited there to increase the thickness of the walls and the diameter of the bone. At the same time, however, bone is absorbed at the inner surface of the wall, and the central cavity of the bone is enlarged. Growth in the diameter of bones ceases with closure of the epiphyseal plates; however, throughout life, bone continues to adapt to changes in mechanical stress with increases or decreases in wall thickness and density.

Joints

A **joint** or articulation is any place where two bones meet or join. Joints have a variety of functions. Their primary function is to join bones together while controlling the motion allowed between them. Joints can provide rigid or highly mobile connections between bones, depending on their individual functions. In addition to joining bones together, another joint function is to transfer forces between bones. These two competing functions, force transferal and motion control, lead to interesting structural designs of joints.

⟳ A joint or articulation is any place where two bones meet or join. The joint's primary function is to join bones together while controlling the motion allowed between them.

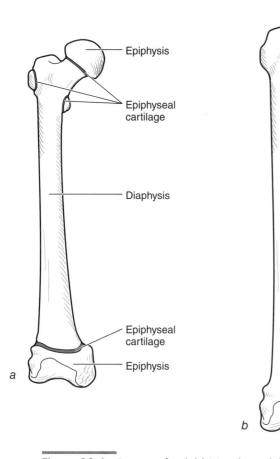

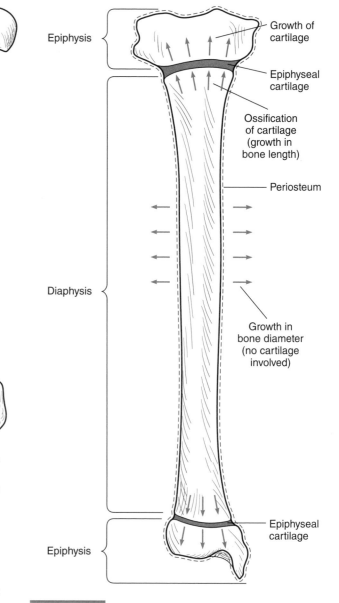

Figure 10.4 Femurs of a child (a) and an adult (b). The child's femur consists of several bones connected via epiphyseal cartilage, whereas the adult's femur consists of one bone.

Figure 10.5 Growth of a long bone.

Joint Classifications

Joints have been categorized in a variety of ways, but most classification schemes are based on joint structure or function (mobility). Structurally, joints can be classified into three general groups. These groups may be further subdivided into subgroups or types of joints. The structural classifications of joints are fibrous (sutures and syndesmoses), cartilaginous (synchrondoses), and synovial. Bones connected by fibrous connective tissue form a **fibrous joint**. These joints are typically (although not necessarily) rigid. The sutures of the skull are examples of fibrous joints. Bones connected by cartilaginous tissue form a **cartilaginous joint**. This type of joint may be rigid or may allow slight movement. The pubic symphysis between the left and right pubic bones of the pelvis is an example of a cartilaginous joint. The joint between the diaphysis and epiphysis in an immature skeleton is another example of a cartilaginous joint. Bones connected by ligaments and separated by a joint cavity form

synovial joints. Synovial joints are highly mobile. Their distinguishing characteristic is a joint cavity that encloses the space between the joints. Most of the joints of the appendicular skeleton are synovial joints.

Functionally, joints can be classified by how much movement they allow. The functional classifications for joints are synarthrodial (immovable), amphiarthrodial (slightly movable), and diarthrodial (freely movable). Some functional classification systems group immovable and slightly movable joints together as synarthroses. Fibrous and cartilaginous joints are classified as synarthrodial and amphiarthrodial, respectively, in the structural classification scheme, whereas synovial joints are classified as diarthrodial joints. The synovial joints

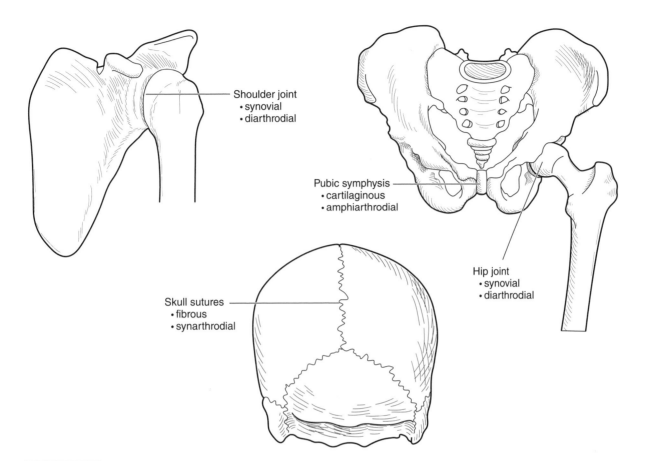

Shoulder joint
• synovial
• diarthrodial

Pubic symphysis
• cartilaginous
• amphiarthrodial

Hip joint
• synovial
• diarthrodial

Skull sutures
• fibrous
• synarthrodial

Figure 10.6 Joints as classified by structure and function.

are of most interest to us because these are the joints where movement occurs. Examples of joints classified by function and by structure are shown in figure 10.6.

Synovial (or diarthrodial) joints are subclassified into six different types according to the movements allowed and the structure of the joint: gliding, hinge, pivot, ellipsoidal, saddle, and ball and socket. The shapes of the articulating surfaces often determine the movements allowed and thus the joint type. Because few joints are exactly like the idealized representations in figure 10.7, joints are classified based on which joint type they most resemble.

Gliding joints are also called irregular, plane, or arthrodial joints. The articulations are flat and small, and planar sliding movements are allowed at these joints. The intercarpal (wrist), intertarsal (ankle), and acromioclavicular (shoulder girdle) joints are examples of gliding joints.

Hinge joints are uniaxial and allow only one degree of freedom of movement (only one number, say the angle between the two bones of this joint, would be needed to fully describe the orientation of the bones with respect to each other). Hinge joints are also called ginglymus joints. The pair of articulating surfaces in a hinge joint approximate a round cylinder (oriented perpendicular to

the long axis of the bone) that fits into a matching shallow trough. The movements allowed at a hinge joint are flexion and the return movement of extension (or plantar flexion and dorsiflexion at the ankle). The humeroulnar (elbow), tibiofemoral (knee), talotibial and talofibular (ankle), and interphalangeal (finger and toe) joints are all examples of hinge joints.

Pivot joints are also uniaxial, allowing only one degree of freedom of movement. These joints may also be called trochoid or screw joints. The articulating surfaces of a pivot joint may approximate a pin inserting into a hole or a cylinder (aligned with the long axis of the bone) that fits into a shallow trough. Rotation about a longitudinal axis is allowed at a pivot joint. The proximal radioulnar joint (between the bones of the forearm) and the atlantoaxial joint (between the first and second cervical vertebrae) are examples of pivot joints. The rotary movements of the proximal radioulnar joint are called supination and pronation. The rotary movements of the atlantoaxial joint are called rotation to the right or left.

Ellipsoidal joints are biaxial and allow two degrees of freedom of movement. These joints may also be called condyloid or ovoid joints. The articulating surfaces of an ellipsoidal joint approximate the shape of an ellipse

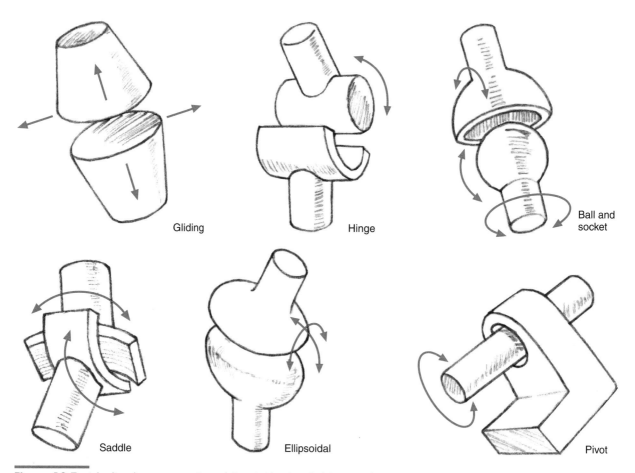

Figure 10.7 Idealized representation of the six diarthrodial (synovial) joint types.

(or egg) that fits into a matching oval depression. These joints have also been described as oval ball-and-socket joints. The movements allowed at an ellipsoidal joint are flexion and extension; abduction and adduction; and circumduction, a combination of these movements. The radiocarpal (wrist), metacarpophalangeal (fingers), metatarsophalangeal (toes), and occipitoatlantal (head and neck) joints are all examples of ellipsoidal joints.

Saddle joints are also biaxial and allow two degrees of freedom of movement. These joints are also called sellar joints. The articulating surfaces of a saddle joint look like a pair of saddles turned 90° to each other. Cup the fingers of each hand, then put your hands together, rotating one of them 90°. This approximates a saddle joint. The U-joints in the driveshaft of a car are like saddle joints. A saddle joint also allows flexion and extension, abduction and adduction, and circumduction. The first carpometacarpal joint (at the base of the thumb) is an example of a saddle joint.

Ball-and-socket joints are triaxial and allow three degrees of freedom of movement. These joints are also called enarthrodial, spheroidal, or cotyloidal. The articu-

lating surfaces of these joints look like a ball and a socket. Ball-and-socket joints are the most freely movable of the synovial joints. They allow flexion and extension, abduction and adduction, and internal and external rotation. The glenohumeral (shoulder) and hip joints are examples of ball-and-socket joints. The sternoclavicular joint (between the shoulder girdle and the axial skeleton) is classified as a ball-and-socket or gliding joint.

Structure of Synovial Joints

The distinguishing characteristic of a synovial joint is the joint cavity formed by the **articular capsule**. The articular capsule is a sleeve of ligamentous tissue that surrounds the joint. It attaches to the bones on either side of the joint, connecting them together. The articulating ends of the bones are covered with a thin layer of hyaline cartilage called the **articular cartilage**. The other exposed bony surfaces within the articular capsule and the inner surface of the articular capsule are lined with a **synovial membrane**. The synovial membrane and the articular cartilages thus seal the joint cavity. A synovial joint can be compared to sealed bearings, which aren't exposed to the elements and thus rarely need to be regreased. But unlike sealed bearings, a synovial joint is self-lubricating. The synovial membrane secretes synovial fluid, which fills the joint cavity. In some synovial joints, such as the knee or sternoclavicular joint, a disc or partial disc of fibrocartilage separates the articulating surfaces of the bones. Figure 10.8 shows the various structures of a synovial joint.

⟩ **The distinguishing characteristic of a synovial joint is the joint cavity formed by the articular capsule. The articular capsule is a sleeve of ligamentous tissue that surrounds the joint.**

The synovial fluid excreted by the synovial membrane is a viscous fluid similar in appearance to egg whites (*syn-*, "like," and *ovum,* "egg"). One function of the synovial fluid is to lubricate the joint and reduce friction. Under pressure, the articular cartilage also secretes synovial fluid that lubricates the joint. Another function of the synovial fluid is to nourish the articular cartilage, because articular cartilage has no nerve or blood supply. The synovial fluid also cleanses the joint cavity and imparts some hydrostatic shock-absorbing properties to the joint.

The articular cartilage is the bearing surface between moving bones. It improves the bone-to-bone fit at the joint, increasing the joint's stability and reducing pres-

sure when the joint is loaded. It also reduces friction and prevents wear. Because hyaline cartilage is more resilient than bone, the articular cartilage also provides some shock absorption at the joint. In joints that have discs or partial discs of fibrocartilage, the discs perform similar functions: improving bone-to-bone fit and absorbing shock.

⟩ **The articular cartilage is the bearing surface between moving bones.**

Although the term *cavity* implies that a space of some sort exists between the ends of the articulating bones in a synovial joint, this space is quite small, and in normal situations the articular cartilages actually contact each other. If a joint is injured, however, the swelling caused by increased fluid volume may greatly enlarge the cavity.

Stability of Synovial Joints

"Stable" was defined in chapter 5 as "not easily moved." Joint stability refers to a joint's resistance to movement in planes other than those defined by the degrees of freedom of movement for the joint, or to movement of the articulating surfaces away from each other through shear dislocation (sliding laterally) or traction dislocation (pulling apart). Stability of a hinge joint thus refers to its ability to resist abduction and adduction, internal and external rotation, or dislocation. Joint flexibility refers to the range of motion possible in the planes of motion defined by the degrees of freedom of movement for the joint and how easily these motions can occur. Flexibility of a hinge joint thus refers to its range of motion in flexion and extension and the ease of these movements.

⟩ **Joint stability refers to the joint's resistance to movement in planes other than those defined by the degrees of freedom of movement for the joint, or to movement of the articulating surfaces away from each other through shear dislocation (sliding laterally) or traction dislocation (pulling apart).**

The reciprocal convex and concave shapes of the articulating ends of bones in a synovial joint are primarily responsible for determining the planes of motion allowed at the joint. Compressive and shear forces develop between matching parts of articulating bones to resist any shear dislocations or rotations in planes other than those for which the joint was designed, as shown in figure 10.9. The tighter the bone-to-bone fit and the

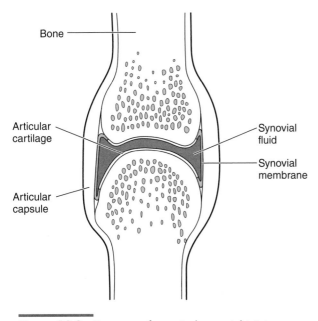

Bone

Articular cartilage

Articular capsule

Synovial fluid

Synovial membrane

Figure 10.8 Features of a typical synovial joint.

deeper the matching convex and concave surfaces of the joint, the more stable the joint. The hip is more stable than the shoulder because the hip socket (the acetabulum) is much deeper than the shoulder socket (the glenoid fossa). Articular cartilage and fibrocartilage discs increase bone-to-bone fit and also assist with joint stability. The menisci of the knee give the condyles of the femur a deeper depression to sit in than the relatively flat tibial plateau.

Whereas the articulating bones resist compressive and shear forces at joints, ligaments provide the tensile forces to resist traction dislocations when something tends to pull the bones apart at a joint. Their tensile strength is also required when a bending load acts on the joint in a plane different from the plane of motion of the joint. In this case, the joint must act like a beam, with compression stress on one side and tensile stress on the other. The articular surfaces of the bones on one side of the joint resist the compression component of the bending load, whereas the ligaments on the opposing side resist the tensile component of the bending load. The articular capsule itself provides some of this tensile strength, but most joints have ligaments as well to strengthen the joint so it can resist dislocating torques or traction forces. The ligaments may occur as thickened bands in the articular capsule itself or as external ligaments separate from the articular capsule. The locations of ligaments relative to the joint they protect thus determine the stability of the joint to bending loads that may cause dislocating abduction, adduction, or rotation. Figure 10.10 illustrates how the medial collateral ligament of the knee carries a tensile load to prevent the knee from abducting when a medial force acts on the knee.

⟳ **Whereas the articulating bones resist compressive and shear forces at joints, ligaments provide the tensile forces to resist traction dislocations when something tends to pull the bones apart at a joint.**

Tendons and the muscles attached to them also resist tensile forces and thus contribute to joint stability in a manner similar to ligaments. The line of pull of most muscles is such that a component of the force generated during a contraction tends to pull the bones of a joint more tightly together, thus affording resistance to any traction-dislocating forces that tend to pull the bones apart. This action of muscles and ligaments is shown in figure 10.11a. In certain joint positions, muscles may actually create forces having components that tend to dislocate rather than stabilize a joint. This action is shown in figure 10.11b. The muscles, tendons, and tendinous sheets (aponeuroses) that cross joints also provide some lateral support to the joints they cross.

Another factor that contributes slightly to the stability of synovial joints is the pressure within the joint cavity.

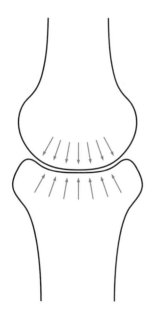

Figure 10.9 Tight bone-to-bone fit allows for compressive and shear force development to prevent dislocations.

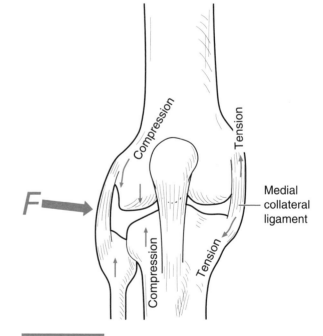

Figure 10.10 Tensile loading of the medial collateral ligament prevents medial dislocation of the knee joint when it is struck from the lateral side.

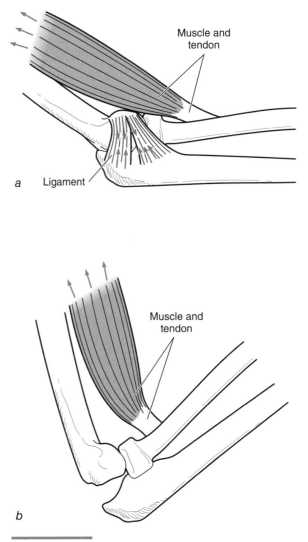

a — Muscle and tendon

Ligament

b — Muscle and tendon

Figure 10.11 Tensile forces from muscles may contribute to joint stability (a) or instability (b).

This internal pressure is less than the pressure external to the joint cavity, so a suction force is created. The articulating bones are thus sucked together as long as the joint cavity is sealed and intact.

Flexibility of Synovial Joints

Joint flexibility refers to the range of motion in the planes in which the joint is designed to move. Range of motion of a limb is limited by some of the same factors that contribute to joint stability: the bones, ligaments, and muscles. Range of motion is limited by other factors as well.

> Joint flexibility refers to the range of motion in the planes in which the joint is designed to move.

Range of motion is limited by the extensibility of the muscles that cross a joint. Muscles that cross more than one joint (multiple-joint muscles) may not be able to stretch far enough to allow full range of motion at a joint if the positions at each joint the muscle crosses stretch the muscle. The hamstring muscles are an example of a multiple-joint muscle since these muscles cross both the hip and knee joints. Try self-experiment 10.1 to see how range of motion at the hip joint is affected by the hamstring muscles.

Self-Experiment 10.1

Lie on your back on the floor. Flex your right hip as far as you can while keeping your right knee extended as shown in figure 10.12a. Now allow your knee to flex and see how much farther you can flex your hip (see figure 10.12b). In the first case, your range of hip flexion was

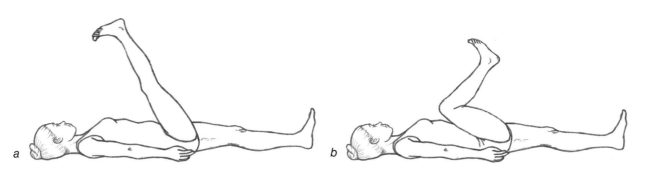

a   b

Figure 10.12 Muscle extensibility affects joint range of motion, as illustrated in this test of hip joint flexibility. In (a), the hamstring muscles limit the range of motion of hip flexion. In (b), tension in the hamstring muscles is released by flexing of the knee, resulting in greater range of hip flexion.

limited by your hamstring muscles. These muscles are two-joint muscles that cross posterior to the hip joint and the knee joint. They are stretched if the knee extends and stretched farther if the hip then flexes (as in the first case). The tension in the hamstring muscles limited your hip flexion when the knee was extended. When the knee flexed, the tension in these muscles was relieved, and you were able to flex the hip farther.

Most muscles that cross only one joint (single-joint muscles) are extensible enough that they do not limit range of motion at that joint. With the exception of the rectus femoris, the quadriceps muscles are all single-joint muscles. Flex your hip and knee at the same time. The quadriceps muscles are not stretched to their limit at the knee's range-of-flexion limit. Other factors limit the range of knee flexion.

Ligaments limit range of motion also. Ligaments that do not align radially with respect to an axis of the joint become looser during one joint action and tighter as the reverse joint action progresses, until the ligament stops the movement. At ball-and-socket and pivot joints, twisting of the ligaments limits the range of rotation. The articular capsule itself also limits range of motion.

The shapes of the articulating bones may also limit range of motion. Elbow extension stops when the olecranon process of the ulna becomes constrained by the olecranon fossa of the humerus, or vice versa. The lateral malleolus of the fibula limits eversion of the ankle when it contacts the calcaneus.

The bulkiness of soft tissues or clothing around a joint may also limit range of motion. Someone with large elbow flexor muscles (biceps brachii and brachialis) may not have as much range of elbow flexion as someone with less developed muscles. A child bundled up in a bulky coat for outdoor winter weather provides an extreme example of how clothing may limit range of motion.

Mobility, or the ease of movement through a range of motion, is affected by friction within the joint and by the inertia and tension in the muscles around the joint, especially those that oppose the joint motion in question (antagonist muscles). Any wearing or damage to the articular cartilage increases joint friction and thus reduces mobility. Wear and tear of the articular fibrocartilage discs reduces mobility as well in joints with these structures. Damage to the articular cartilage or bone within the joint cavity may produce loose particles that decrease mobility. Damage to the synovial membrane and its capacity for producing synovial fluid also results in loss of joint mobility. People suffering from arthritic diseases often have one or more of these problems.

Summary

The skeletal system provides the body with a framework of rigid links that enables movement. The 200-plus bones of the skeleton provide support and protection for internal organs. The bones also function as red blood cell producers and as a storage site for minerals.

The bones of the skeleton are joined together at joints that may allow a great range of movement, slight movement, or none at all. Joints may be classified according to their function or structure. Diarthrodial or synovial joints are the joints of interest to us because they allow free movement. The six types of synovial joints are gliding, hinge, pivot, ellipsoidal, saddle, and ball and socket. All synovial joints have a joint cavity lined with a synovial membrane that secretes synovial fluid. The bearing surfaces of the bones in a synovial joint are coated with a thin layer of articular cartilage that is nourished and lubricated by the synovial fluid. A ligamentous sleeve surrounds synovial joints and connects the bones to form the articular capsule.

Joint stability is affected by bone-to-bone fit, articular cartilage and fibrocartilage discs, ligaments, muscles and tendons, and negative pressure (suction). Joint flexibility or range of motion is affected by muscle extensibility, ligaments, bone-to-bone contact, and the bulk of soft tissue and clothing. Joint mobility or ease of range of motion is affected by antagonist muscle tension and friction within the joint cavity.

KEY TERMS

appendicular skeleton (p. 264)
articular capsule (p. 270)
articular cartilage (p. 270)
axial skeleton (p. 264)
cartilaginous joint (p. 268)

diaphysis (p. 266)
epiphyseal cartilage (p. 266)
epiphysis (p. 266)
fibrous joint (p. 268)

joint (p. 267)
sesamoid bone (p. 265)
synovial joints (p. 268)
synovial membrane (p. 270)

REVIEW QUESTIONS

1. What are the mechanical functions of the skeletal system?
2. What are the metabolic functions of the skeletal system?
3. What type of bone is
   a. the femur?
   b. the third thoracic vertebra?
   c. the calcaneus?
   d. the first metacarpal?
   e. the patella?
4. What component of bone is primarily responsible for resisting compression?
5. At what age do most epiphyseal plates ossify?
6. What are the six types of synovial joints?
7. Give an example of a uniaxial joint.
8. Give an example of a biaxial joint.
9. Give an example of a triaxial joint.
10. What type of synovial joint is the glenohumeral joint?
11. What are the functions of synovial fluid?
12. What keeps the synovial fluid in the joint cavity?
13. What are the functions of articular cartilage?
14. What factors influence the stability of synovial joints?
15. In general, which are more stable—the joints of the upper extremities or the joints of the lower extremities?

chapter 11

The Muscular System
The Motors of the Body

objectives

When you finish this chapter, you should be able to do the following:

- Describe the three types of muscle tissue: smooth, skeletal, and cardiac

- Discuss the functions of skeletal muscle

- Describe the microstructure of skeletal muscle

- Describe the macrostructure of skeletal muscle

- Discuss the three different types of muscular actions: concentric, eccentric, and isometric

- Describe the roles that muscles can assume

- Discuss the factors that influence the force developed during muscular activity

- Understand the relationship between power output and velocity of contraction of a muscle

You are watching a bodybuilding show as the athletes pose to music. You are struck by the number and sizes of the muscles displayed and by how well defined and large they are. You have always thought of muscles as the active force producers within the body, but now you realize what artists have known for centuries: Muscles give our bodies the shape and form that make it aesthetically pleasing. What is the anatomy of muscles that allows them to take on such variety of shapes? How do muscles produce force and what factors affect this force production? This chapter attempts to answer these and other questions concerning the structure and function of muscles.

Muscles are the active components of the musculoskeletal system. Although the bones and joints of the skeleton form the framework of the body, this framework would collapse without the active force generation of muscles providing stiffness to the joints. Muscles are the motors of the musculoskeletal system that allow the levers of the skeleton to move or change position.

The distinguishing characteristic of muscle, which gives it the capacity to move or stiffen joints, is its ability to actively shorten and produce tension. This characteristic is not unique to the muscles of the musculoskeletal system alone, however. All muscle tissue has the ability to contract. Three different types of muscle tissue are present in the human body: smooth (visceral) muscle, cardiac muscle, and striated (skeletal) muscle. The walls of vessels (veins and arteries) and hollow organs (stomach, intestines, uterus, bladder, and so on) are smooth muscle. This type of muscle is smooth in appearance, as its name implies, and is greatly extensible. Cardiac muscle is irregularly striped (striated) in appearance. The walls of the heart are cardiac muscle. Cardiac and smooth muscle are innervated by the involuntary (autonomic) nervous system. Skeletal muscle has regularly spaced parallel stripes that give it its striated appearance. Unlike cardiac or smooth muscle, skeletal muscle cells are multinucleated due to their length. Skeletal muscle contractions are largely voluntary, and thus skeletal muscle is controlled by the somatic (voluntary) nervous system.

⟩ **The distinguishing characteristic of muscle is its ability to actively shorten and produce tension.**

The Structure of Skeletal Muscle

Skeletal muscles serve a variety of functions, including movement and posture maintenance, heat production, protection, and pressure alteration to aid circulation. More than 75% of the energy used during a muscular contraction is released as heat. Muscles are the body's furnace. Muscles may serve as shock absorbers to protect the body. The walls of the abdomen and chest are covered with muscles that protect the underlying organs. A final function of skeletal muscle is pressure alteration. This is primarily a function of cardiac or smooth muscle, but the contraction of skeletal muscles may also alter pressure in veins, thus assisting with the venous return of blood. Because skeletal muscle attaches to bone and is responsible for controlling joint movements, this is the function that is most interesting in sport and exercise. The more than 400 skeletal muscles in the human body are of various shapes and sizes, but their common functions lead to some general similarities in their structure.

Microstructure of Skeletal Muscles

A single muscle cell is a **muscle fiber**. A muscle fiber is a long, threadlike structure 10 to 100 µm (10 to 100 millionths of a meter) in diameter and up to 30 cm long. Covering the muscle fiber is a thin cell membrane called the **sarcolemma**. External to the sarcolemma is the **endomysium**, the connective tissue sheath that encases each muscle fiber and anchors it to other muscle fibers and connective tissue and eventually to the tendon.

Within each muscle fiber are hundreds of smaller (1 µm in diameter) threadlike structures lying parallel to each other and running the full length of the fiber. These are the myofibrils, whose number may vary from less than 100 to more than 1000, depending on the size of the muscle fiber. Transverse light and dark bands appear across each myofibril and align with the same bands on adjacent myofibrils. These bands of light and dark repeat every 2.5 µm and give skeletal muscle its striated appearance. Within a single myofibril, the repeating unit of the myofibril between stripes is called a sarcomere.

The **sarcomere** is the basic contractile unit of muscle. Thick (myosin) and thin (actin) protein filaments, or myofilaments, overlap within the sarcomere. The thin actin filaments are free at one end, where they overlap with the thick myosin filaments; at the other end, they anchor to adjacent sarcomeres in series with each other at the transverse Z line or Z band (for *Zwischenscheibe*, or "between disc"). A sarcomere is thus the part of a myofibril from Z line to Z line. The region that includes only actin filaments and the Z band (or the region where the actin filaments do not overlap with the myosin filaments) is called the I band (for "isotropic"—named for its refraction of one wavelength of light) and appears as a light band. The darker band or region that includes the full length of the myosin filaments along with the region of overlap with the actin filaments is called the A band (for "anisotropic"—named for its refraction of more than one wavelength of light). The region within the A band where the actin and myosin do not overlap is called the H band or H zone (for *Hellerscheibe*, or "clear disc"). At the middle of the H band is the M band or M line (for *Mittelscheibe*, or "middle disc"), the transverse band that connects adjacent myosin filaments with each other. The ends of the myosin filaments have projections that give them a brushlike appearance. These projections are the cross bridges that link to the actin filaments and create the active contractile force during a muscular contraction. Figure 11.1 illustrates conceptually the structure

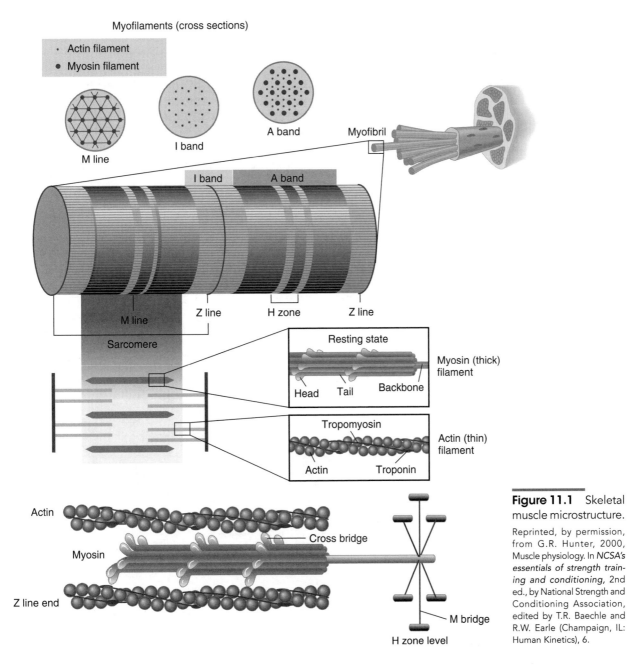

Figure 11.1 Skeletal muscle microstructure.

Reprinted, by permission, from G.R. Hunter, 2000, Muscle physiology. In *NCSA's essentials of strength training and conditioning*, 2nd ed., by National Strength and Conditioning Association, edited by T.R. Baechle and R.W. Earle (Champaign, IL: Human Kinetics), 6.

of a muscle fiber, its myofibrils, and a sarcomere. A transverse or cross section through a myofibril shows that each myosin filament is surrounded by six actin filaments in the overlap zone, whereas each actin filament is surrounded by only three myosin filaments.

> The sarcomere is the basic contractile unit of muscle.

Macrostructure of Skeletal Muscles

Muscle fibers are bundled together in groups of 100 or more to form a **fascicle** (or fasciculus). Each fascicle is encased by a connective tissue sheath called the **perimysium**. The endomysium of each muscle fiber is bound to this perimysium. Several fascicles are then bound together within a connective tissue sheath called the **epimysium** to form a whole muscle, as shown in figure 11.2.

Myofilaments run the entire length of a muscle fiber, and the ends of the myofilaments attach to the endomysium. In long muscles, individual muscle fibers are not long enough to run the entire length of the muscle and may not be as long as the fascicle in which they are bound. The endomysium of a muscle fiber thus attaches to the endomysium of the next muscle fiber in series (aligned with it end to end) or to the perimysium of the fascicle. The perimysium of a fascicle then attaches to the perimysium of the next fascicle in series or to the epimysium of the entire muscle. Thus the connective tissue

sheaths of the muscle become bound together at the ends of the muscle. This connective tissue extends beyond the region of the muscle containing the contractile elements. It weaves itself into cords or sheets of connective tissue that connect the muscle to bone. These cords are called **tendons**, and the sheets are called **aponeuroses**. The force of a muscular contraction is thus transmitted from the endomysium of the muscle fiber to the perimysium and epimysium and then to the tendon, which is a continuation of these connective tissues.

The insertion or attachment of a tendon to bone is similar to the attachment of a ligament to bone. Because tendons (and ligaments) are less stiff than bone, there must be a gradual increase in the stiffness of the tendon (or ligament) from the muscle end to the bone end. At the site of insertion, the tendon gradually contains more ground substance so that it becomes fibrocartilage. This fibrocartilage part of the tendon then becomes more mineralized, and thus bony, as the tendon attaches to the periosteum of the bone and blends into the bone itself. This attachment and transition at the tendon–bone interface is shown in figure 11.3.

Because muscles attach to movable limbs, the muscles and their attached tendons move when the muscles shorten or lengthen. Movements of the belly of the muscle may create friction between it and adjacent muscles. Loose connective tissue between adjacent muscles functions to reduce this friction. Where tendons or muscles would rub against bones or ligaments, bursae may be present to reduce friction or prevent damage to the muscle

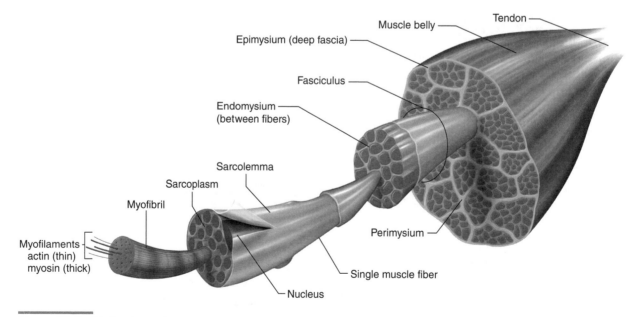

Figure 11.2 Skeletal muscle macrostructure.

Reprinted, by permission, from G.R. Hunter and R.T. Harris, 2008, Structure and function of the muscular, Neuromuscular, cardiovascular, and respiratory systems. In *Essentials of strength training and conditioning*, 3rd ed., by National Strength and Conditioning Association, edited by T.R. Baechle and R.W. Earle (Champaign, IL: Human Kinetics), 6.

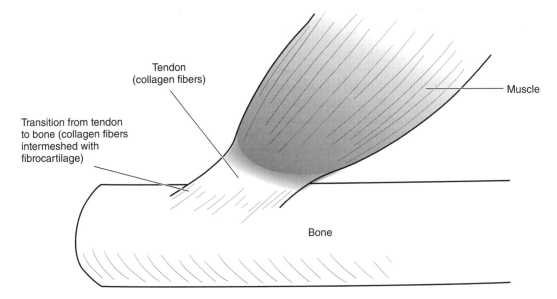

Figure 11.3 The insertion of tendon to bone, illustrating the transition from collagen fibers to fibrocartilage to bone.

or tendon. A bursa is a sac filled with synovial fluid that pads and lubricates soft tissues on bones. Tendon sheaths are similar to bursae in structure and function, but they wrap completely around a tendon.

Typically, muscles have two ends that are attached by tendons to bones on either side of a joint. When a muscle contracts, it pulls with equal force on each attachment, and each bone will tend to move. Anatomically, the origin of a muscle is its more proximal attachment, and the insertion of a muscle is its more distal attachment. Mechanically, the origin of a muscle attaches to the bone that moves less, and the insertion attaches to the bone that moves more. The anatomical insertion of the triceps brachii is always the olecranon process of the ulna, the more distal attachment of this muscle. Mechanically, the origin and insertion of the triceps brachii depend on the movement. The olecranon process is the mechanical insertion of the triceps brachii during a bench press; but during a push-up, the olecranon process is the mechanical origin of the triceps. The anatomical terminology is most common because it is understood even if no movement is referred to.

> **When a muscle contracts, it pulls with equal force on each attachment.**

A muscle's line of pull refers to the direction of the resultant force produced at an attachment. The direction of the resultant force is along a line from the origin to the insertion of the muscle. This is true for muscles that have single points of origin and single points of insertion. Some muscles, such as the triceps brachii, have more than one insertion or origin; other muscles, such as the trapezius,

have broad origins or insertions. In these situations, no single line of pull is apparent, and the line of pull of these muscles depends on which fibers are active.

> **A muscle's line of pull refers to the direction of the resultant force produced at an attachment.**

As alluded to in the previous paragraph, the arrangement of fibers in a muscle may affect its function. Muscles may have more than one head or anatomical origin (e.g., biceps, triceps, quadriceps). The fibers in these muscles align in several different directions. Long muscles whose fibers all align parallel to the line of pull of the muscle are called longitudinal, strap, or fusiform muscles. Muscles that have shorter fibers not aligned with the line of pull are called pennate (for feather) muscles. The fibers of these muscles insert at an angle to a tendon and thus resemble the barbs of a feather. Pennate, or penniform, muscles may be unipennate, bipennate, or multipennate. In general, pennate muscles are able to exert greater forces than similar-sized longitudinal muscles, but the longitudinal muscles are able to shorten over a greater distance. Figure 11.4 shows examples of the various shapes and forms of muscles.

Muscle Action

The distinguishing characteristic of muscle is its ability to contract. The development of tension within a muscle causes it to pull on its attachments. This action of a muscle is usually referred to as a muscular contraction; however, this use of the word *contraction* is confusing because it

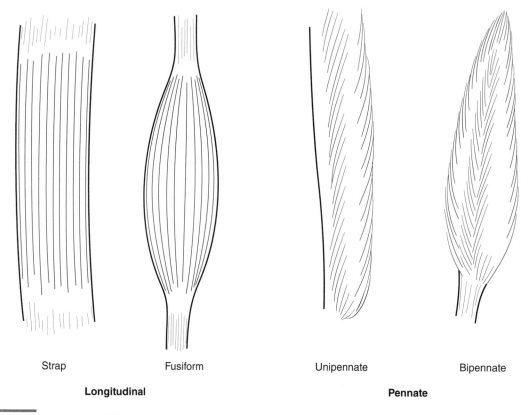

Strap   Fusiform   Unipennate   Bipennate

Longitudinal   **Pennate**

Figure 11.4 Examples of different fiber arrangements in skeletal muscles.

implies that the muscle shortens in length during this activity. But a muscle may be contracting and not changing length at all or may even be lengthening. A more accurate word to denote muscle "contraction" is muscle *action*. When a muscle is active, it develops tension and pulls on its attachments. The active muscle may shorten, stay the same length, or lengthen.

Types of Muscular Actions

When a muscle is active and its attachments draw closer together, the muscle is active concentrically (or contracting concentrically). The muscle action is a concentric action (or a **concentric contraction**). This occurs when the torque created by the active muscle on either limb at a joint is in the same direction as the rotation of the limb. Concentric muscle action occurs when

- the flexor muscles are active and flexion occurs,
- the extensor muscles are active and extension occurs,
- the abductor muscles are active and abduction occurs,
- the adductor muscles are active and adduction occurs,

- the internal rotator muscles are active and internal rotation occurs, or
- the external rotator muscles are active and external rotation occurs.

Because the force and torque created by the muscle are in the same direction as the movement of the muscle's points of attachment or the rotation of the limbs, positive mechanical work is done by the muscle. Recall from chapter 4 that work done causes a change in energy. Positive work results in an increase in kinetic or potential energy or both. If the muscle force is the only force acting on the limb (other than gravitational forces), then a concentric muscle action results in an increase in the limb's kinetic or potential energy. Generally, if a limb's velocity is speeding up or its position is being elevated, concentric muscular activity is likely to be occurring. During the up phase of a push-up, potential energy is increased because the body is being elevated. Elbow extension occurs, and the elbow extensor muscle (the triceps brachii) is active concentrically. Figure 11.5 illustrates examples of concentric muscle activity.

⟳ **When a muscle is active and its attachments draw closer together, the muscle is active concentrically.**

Figure 11.5 Examples of phases of movements that involve concentric muscular activity.

When a muscle is active and its attachments are drawn farther apart, the muscle is active eccentrically (or contracting eccentrically). The muscle action is an eccentric action (or an **eccentric contraction**). This occurs when the torque created by the active muscle on either limb at a joint is in the direction opposite the rotation of the limb. Eccentric muscle action occurs when

- the flexor muscles are active and extension occurs,
- the extensor muscles are active and flexion occurs,
- the abductor muscles are active and adduction occurs,
- the adductor muscles are active and abduction occurs,
- the internal rotator muscles are active and external rotation occurs, or
- the external rotator muscles are active and internal rotation occurs.

Because the force and torque created by the muscle are in the direction opposite the movement of the muscle's points of attachment or the rotation of the limbs, negative mechanical work is done by the muscle. Negative work results in a decrease in kinetic or potential energy or both. If the muscle force is the only force acting on the limb (other than gravitational forces), an eccentric muscle action results in a decrease in the limb's kinetic or potential energy. Generally, if a limb's velocity is slowing down or its position is being lowered, eccentric muscular activity is likely to be occurring. During the down phase of a push-up, potential energy is decreased because the body is being lowered. Elbow flexion occurs, and the elbow extensor muscle (the triceps brachii) is active eccentrically. Figure 11.6 illustrates examples of eccentric muscle activity.

⟳ **When a muscle is active and its attachments are drawn farther apart, the muscle is active eccentrically.**

Figure 11.6 Examples of phases of movements that involve eccentric muscular activity.

When a muscle is active and its attachments do not move relative to each other, the muscle is active isometrically (or contracting isometrically). The muscle action is an isometric action (or an **isometric contraction**). No rotation of the limb occurs, so no mechanical work is done and no change in kinetic or potential energy occurs. Because the isometrically active muscle exerts a torque about the joint, some other opposing torque must be exerted about the joint to keep it in static equilibrium and not moving. This opposing torque may be provided by the opposing muscles on the other side of the joint or by some external force such as gravity. Generally, if no movement occurs at a joint but other forces act on the limbs at that joint, an isometric muscular activity is likely to be occurring. If you push yourself halfway up during a push-up and hold that position, no movement is occurring, but the elbow extensor muscle is active to prevent gravity from flexing your elbows and pulling you back down to the ground. The elbow extensor muscle (the triceps brachii) is active isometrically.

> When a muscle is active and its attachments do not move relative to each other, the muscle is active isometrically.

Roles of Muscles

A variety of terms are used to refer to the roles of muscles either relative to a joint action or relative to another muscle. An **agonist** muscle is capable of creating a torque in the same direction as the joint action referred to. Thus muscles that are active concentrically are agonists to the action occurring at the joint they cross. The biceps brachii is an agonist for elbow flexion; the triceps brachii is an agonist for elbow extension; the quadriceps femoris is an agonist for knee extension; the gastrocnemius is an agonist for ankle plantar flexion; and so on. These muscles may also be referred to as prime movers or protagonists. The terms *agonist, prime mover,* and *protagonist* are undefined unless a joint action is referred to in relation to them. The statement, "The biceps brachii is an agonist," is incomplete.

An **antagonist** is a muscle capable of creating a torque opposite the joint action referred to or opposite the other muscle referred to. The biceps brachii is an antagonist for elbow extension; the triceps brachii is an antagonist for elbow flexion; the quadriceps femoris is an antagonist for knee flexion; the gastrocnemius is an antagonist for ankle dorsiflexion; and so on. In the preceding examples, the term *antagonist* is used in reference to a joint action, and thus muscles active eccentrically are antagonists to the action occurring at the joint they cross. *Antagonist* may also be used in reference to another muscle. The biceps brachii is an antagonist to the triceps brachii; the quadriceps femoris is an antagonist to the hamstrings; the gastrocnemius is an antagonist to the tibialis anterior; and so on. In these cases, the torque that the antagonist creates opposes the torque created by the muscle referred to.

Other terms have been used to denote other roles that muscles play, including *synergist, neutralizer, stabilizer, fixator,* and *supporter.* These terms are all used in reference to another muscle. The terms **stabilizer**, *fixator,* and *supporter* refer to muscles that are active isometrically to keep a limb from moving when the reference muscle contracts. When a muscle is active, it tends to move both bones to which it is attached when movement of only one limb may be desired. Other external forces may cause unwanted movement of limbs as well. The isometric action of the stabilizing (fixating, supporting) muscles keeps the limb from moving. During the down phase of a push-up, the shoulder girdle tends to adduct due to the weight of the trunk and the upward reaction force acting on the shoulder joint. The serratus anterior acts as a stabilizer to prevent this adduction by acting isometrically.

A **neutralizer** is a muscle that creates a torque to oppose an undesired action of another muscle. The torques created by many muscles have components in several planes; thus, when these muscles are active, they create torques about more than one axis of a joint. For example, when the biceps brachii is active, it creates a flexion torque at the elbow as well as a supinating torque. If supination is the desired action, the triceps would act isometrically to neutralize the undesired flexion torque of the biceps.

The word **synergy** is used to refer to mutually beneficial actions of two or more things. In terms of muscle function, a synergist muscle has been described as a muscle that assists in producing the desired action of an agonist muscle. Neutralizers and stabilizers could thus be described as synergists. A synergist could also be another muscle whose torque adds to the torque of an agonist. The term *synergist* is not very exclusive, and its use should be avoided.

Contraction Mechanics

The action responsible for the contraction of a muscle occurs within the sarcomere. In response to a stimulus from the motor neuron that innervates a muscle, the muscle becomes active, and the cross bridges of the thick myosin filament attach, pull, release, and reattach to specific sites on the thin actin filament. The cross bridges of the myosin filament tug on the adjacent actin filament and pull themselves past it, similar to the way you might pull on (or climb up) a rope hand over hand. The myosin filaments thus slide past the actin filaments, and the muscle shortens (in a concentric contraction). An eccentric contraction is like lowering a bucket on a rope by letting it drop a short distance, then regripping it, then letting it drop, and so on. The cross bridges of the myosin filament attach and tug, but another force pulls them off that attachment site; and as the actin filament slides away, the myosin cross bridges quickly reattach to another site. The force developed during a contraction is the sum of the pulling forces that each myosin cross bridge exerts on the actin filament. The more cross bridges attached to actin filaments, the larger the contraction force.

A single stimulus from the motor neuron (the nerve cell that innervates the fiber) results in a twitch response of the fiber. The cross bridges attach briefly and then release, with muscle tension rising and then falling as shown in figure 11.7*a*. The duration of tension in the muscle is short. A repeated series of stimuli received from the motor neuron results in a repeated series of twitch responses of the muscle fiber if the time between each successive stimulus is long enough. With increased frequency of stimulus (and less time between stimuli), there will still be tension in the fiber when the next stimulus occurs. The subsequent tension in the fiber will be greater. If the frequency of stimuli is rapid enough, a tetanic response of the fiber results, as shown in figure 11.7*b*. The cross bridges attach, release, and reattach, with increasing tension developed until a maximum value is reached. The maximum tension achieved in a tetanic response is much greater than that achieved in a twitch response. Continued stimuli keep the tension in the muscle high until fatigue occurs.

> A single stimulus from the motor neuron results in a twitch response of the fiber.

Muscle Contraction Force

The maximum force a muscle is capable of developing depends on several factors concerning the state of the muscle. If all the fibers of a muscle are stimulated to

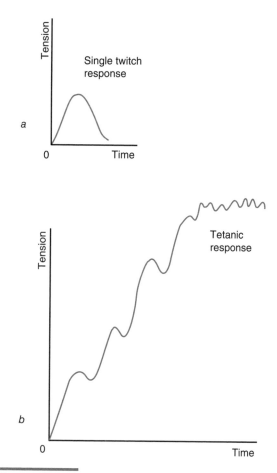

Figure 11.7 Tension developed in a muscle fiber as a result of a single stimulus (*a*) is smaller than that developed as a result of repeated stimulation (*b*) if the stimulation frequency is great enough.

contract, these factors include the physiological cross-sectional area of the muscle, the length of the muscle, and the velocity of muscle shortening, as well as other factors.

Physiological Cross-Sectional Area

The tensile force produced by an active muscle would appear to be determined by the number of myosin cross bridges attached to actin myofilaments throughout the muscle. But this is not exactly the case. Let's consider a single muscle fiber. As described earlier, a muscle fiber consists of a bundle of myofibrils arranged parallel to each other (see figure 11.1). A number of sarcomeres lined up end to end in series with each other make up each myofibril, and within each sarcomere are the force-generating myofilaments. If we increased the number of sarcomeres lined up end to end in series with each other, we would increase the number of cross bridges and thus increase the number of possible attachments of

the cross bridges to the actin myofilaments. The length of the myofibril would increase; and if we did this to each myofibril, the fiber length would increase as well as the length of the resting muscle. But would the muscle be capable of producing any more force?

Series Arrangement of Fibers and Myofibrils

The arrangement of muscle fibers in a muscle or of myofibrils in a muscle fiber affects the behavior of the muscle or muscle fiber. Fibers and myofibrils arranged and attached end to end are described as in a series with each other. Fibers and myofibrils arranged alongside each other are described as parallel to each other. Self-experiment 11.1 illustrates the behavior of fibers in series.

Self-Experiment 11.1

You'll need three rubber bands and two paper clips for this experiment. Loop one rubber band over the index finger of your right hand and over the index finger of your left hand as shown in figure 11.8. Now pull until the rubber band stretches to twice its length. Unhook the rubber band from your left hand and link a second rubber band in series to this one using a paper clip. Now loop the second rubber band over your left index finger and pull. Pull until the two rubber bands double their length. Unhook the second rubber band from your left hand and hook the third rubber band to this one and then over your left index finger. Pull again until the three rubber bands double their length. After adding one or two rubber bands in series with the first rubber band, did you have to pull with more force to double the length of the rubber bands? No! The force you had to pull with to double the length was about the same. Muscles behave similarly—longer muscles can stretch and shorten over greater lengths than short muscles, but they are not any stronger even though they may have more myofilaments. Strength gains don't occur in weight training as a result of athletes growing longer muscles; instead, their muscles grow wider.

Parallel Arrangement of Fibers and Myofibrils

Now suppose that, rather than adding sarcomeres in series to a myofibril, we added them in parallel by increasing the number of myofibrils, or instead of adding sarcomeres, we increased the number of myofilaments within each sarcomere of a myofibril. In these cases, we have increased the number of possible attachments of the cross bridges to the actin myofilaments (as was the case before), but rather than increasing the length of the myofibril or muscle, we have increased its diameter and cross-sectional area. Would the muscle be capable of

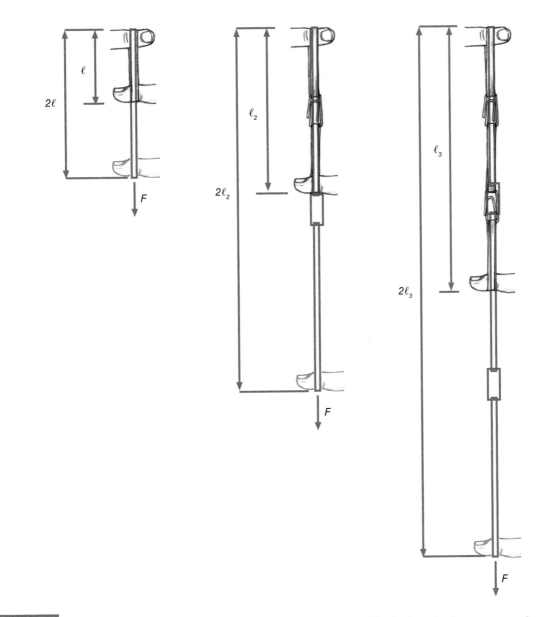

Figure 11.8 Stretching rubber bands in series. The force required to double the length of one, two, or three rubber bands in series is the same in all three cases.

producing any more force in these cases? Self-experiment 11.2 illustrates the behavior of fibers in parallel.

Self-Experiment 11.2

Again, let's use rubber bands for our model of muscle. Loop one rubber band over your right and left index fingers as in self-experiment 11.1. Pull until the rubber band doubles in length. Now loop a second rubber band over the index fingers of your right and left hands so that both rubber bands are looped over your fingers as shown in figure 11.9. Pull again until the rubber bands double in length. Now loop the third rubber band over the index fingers of your right and left hands so that three rubber bands are looped over your fingers. Pull again until the rubber bands double in length. After adding one or two rubber bands in parallel with the first rubber band, did you have to pull with more force to double the length of the rubber bands? Yes!

Muscles behave similarly to the rubber bands in self-experiment 11.2. Increasing the number of fibers side by side and parallel to each other increases the strength of the muscle. The cross-sectional area of the muscle perpendicular to the muscle fibers and line of pull of a muscle gives an indication of the maximal tensile force that a longitudinal muscle can produce.

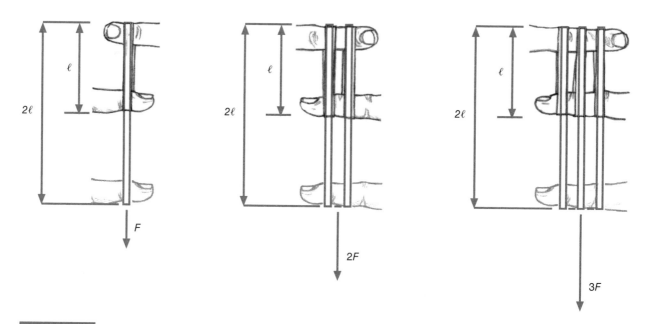

Figure 11.9 Stretching rubber bands in parallel. The force required to double the length of two rubber bands in parallel is twice the force required to stretch one rubber band the same distance. For three rubber bands, the force required is three times as large.

Longitudinal Versus Pennate Muscles

Human muscle can produce a tensile force of approximately 30 N/cm² of cross-sectional area of active muscle during an isometric contraction, or the tensile stress produced by muscle during isometric contractions is 30 N/cm² if all the muscle fibers in that cross section are active.

$$F_m = A_m \sigma_m \qquad (11.1)$$

where

F_m = maximal isometric contraction force,

A_m = cross-sectional area of muscle, and

σ_m = maximum stress developed by isometric muscle contraction.

Thus a longitudinal muscle with a diameter of 3 cm (and a cross-sectional area of 7 cm²) can produce a maximal isometric contraction of

$$F_m = (7 \text{ cm}^2)(30 \text{ N/cm}^2)$$

$$F_m = 210 \text{ N}$$

Growth in the size of muscles produces increases in the strength of the muscle as well. Gains in size and strength are thus commonly observed by novice weightlifters.

The maximum tensile force a pennate muscle can produce cannot be estimated by the cross-sectional area perpendicular to the muscle fibers and line of pull of a muscle because the muscle fibers and the line of pull of a pennate muscle are not in the same direction. A cross section taken perpendicular to the line of pull of a pennate muscle would not include all the fibers of the muscle (see figure 11.10). A cross section taken perpendicular to the muscle fibers may not include all the fibers either.

The solution is to take several cross sections perpendicular to the fibers so that all the parallel fibers are included. This area gives an indication of how much total tension the fibers can produce, but this tensile force is not in the same direction as the muscle's line of pull. The component of this total tensile force parallel to the line of pull must be determined. We compute this component by multiplying the total area by the cosine of the angle of pinnation (this product gives the physiological cross-sectional area of the muscle) and then by the maximum tensile stress of the muscle, or

$$F_m = (A_m \cos \theta)\sigma_m \qquad (11.2)$$

where

F_m = maximal isometric contraction force,

$A_m \cos \theta$ = physiological cross-sectional area of muscle,

θ = angle of pinnation (the angle of the muscle fibers with the line of pull), and

σ_m = maximum stress developed by isometric muscle contraction.

$$F_m = (11.54 \text{ cm}^2 + 11.54 \text{ cm}^2)\,(\cos 30°)\,(30 \text{ N/cm}^2)$$

$$F_m = (23.08 \text{ cm}^2) \times (0.866)\,(30 \text{ N/cm}^2)$$

$$F_m = 600 \text{ N}$$

In this example, the same volume (and mass) of muscle tissue from a unipennate muscle is able to produce 2.0 times as much tensile force during an isometric contraction as tissue from a longitudinal muscle. The shorter fibers of the pennate muscle and their orientation relative to the angle of pull, however, limit the distance over which a pennate muscle can shorten. A comparison of minimum contraction lengths for the longitudinal and unipennate muscles used in the previous example is shown in figure 11.12.

Muscle Length

The physiological cross-sectional area of active muscle gives an indication of the maximum tensile force a muscle is capable of producing, but this maximum is dependent on the length of the muscle during the contraction. The mechanics of contraction within the sarcomere determine the relationship between the length of a muscle and the maximum force of contraction.

> The physiological cross-sectional area of active muscle gives an indication of the maximum tensile force a muscle is capable of producing.

Active Tension

The attachment of the myosin cross bridges to the actin filament within a sarcomere is the basis for the active tensile force developed during a muscular contraction. The range of lengths through which the actin and myosin myofilaments can overlap determines the range of lengths over which a whole muscle can actively develop tension, as shown in figure 11.13. At maximum tension, the overlap between the actin and myosin myofilaments is maximal. Let's call this the sarcomere's resting length, or ℓ_o (see figure 11.13a). The sarcomere can shorten beyond this point, but then opposing actin filaments begin to overlap; and because the myosin cross bridges cannot attach to these filaments, the tension decreases. Shortening of the sarcomere can still continue until the actin and myosin myofilaments become jammed against the opposite Z disk. At this point, no tension is produced in the muscle and the sarcomere is a little more than half its resting length, or about 60% of ℓ_o (see figure 11.13b). A whole muscle can thus produce tension and shorten to approximately half its resting length (not including the length of its tendons). If the sarcomere is stretched longer than

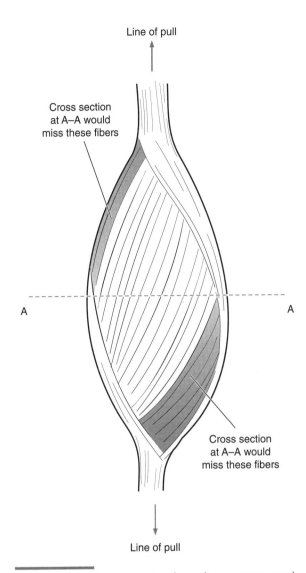

Line of pull

Cross section at A–A would miss these fibers

A   A

Cross section at A–A would miss these fibers

Line of pull

Figure 11.10 A cross section through a pennate muscle and perpendicular to its line of pull may not include all its fibers.

For example, let's compare the tension-producing capabilities of two equal volumes of muscle tissue, one from a longitudinal muscle and one from a unipennate muscle, as shown in figure 11.11.

Using equation 11.1, the maximum isometric tensile force developed in the longitudinal muscle is

$$F_m = A_m \sigma_m$$

$$F_m = (5 \text{ cm} \times 2 \text{ cm}) \times (30 \text{ N/cm}^2)$$

$$F_m = 300 \text{ N}$$

And, using equation 11.2, the maximum isometric tensile force developed in the unipennate muscle is

$$F_m = (A_m \cos \theta)\sigma_m$$

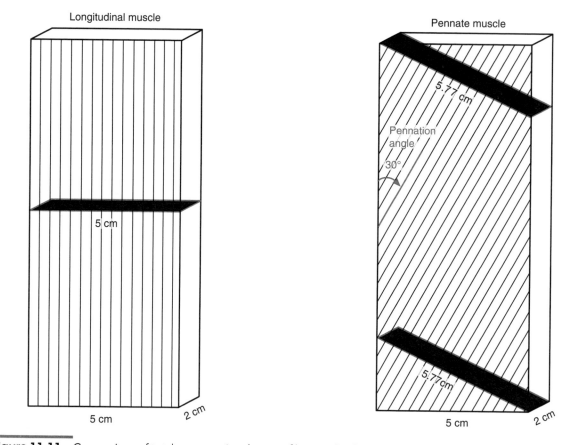

Figure 11.11 Comparison of total cross-sectional areas of longitudinal and pennate muscles of the same volume. The cross-sectional area, A, of the longitudinal muscle is 5 cm × 2 cm = 10 cm². The cross-sectional area of the pennate muscle is $\Sigma A = A_1 + A_2 = (5.77 \text{ cm} \times 2 \text{ cm}) + (5.77 \text{ cm} \times 2 \text{ cm}) = 11.54 \text{ cm}^2 + 11.54 \text{ cm}^2 = 23.08 \text{ cm}^2$.

its resting length, fewer myosin cross bridges can attach to the actin myofilaments, and the tension decreases. Active tension can be developed at longer lengths until the actin and myosin no longer overlap. At this point, no active tension is produced and the sarcomere is a little more than 1.5 times its resting length, or 160% of ℓ_o (see figure 11.13c). A whole muscle can thus actively produce tension at lengths 60% to 160% of its resting length.

Passive Tension

Passive tension can be developed in the sarcomere and within the whole muscle by stretching of the connective tissue structures: the sarcolemma, endomysium, perimysium, epimysium, and tendon. The passive stretching of these tissues allows a muscle to be stretched and develop tension beyond 160% of its resting length. The tension developed in a whole muscle thus depends on the tension produced by active contraction of the contractile elements (the myofilaments) plus the passive tension developed when the muscle is stretched beyond its resting length.

The relationship between muscle length and tension is shown in figure 11.14. Maximum tension can be developed in a whole muscle when it is just a little longer than (about 120% of) its resting length. If the muscle is able to stretch beyond 160% of its resting length, the maximum tension is reached when the muscle is stretched to its maximum length.

⟳ **Maximum tension can be developed in a whole muscle when it is just a little longer than (about 120% of) its resting length.**

Single-Versus Multiple-Joint Muscles

Most single-joint muscles are limited by the range of motion of the joint they cross to operating well within 60% to 160% of their resting lengths. Their maximum tension is developed at about 120% of their resting length because they cannot be stretched beyond 160% of their length and into the right side of the curve shown in figure

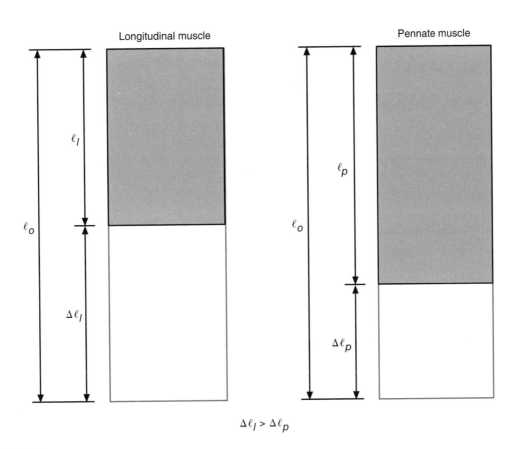

Figure 11.12 The minimum contraction length of a longitudinal muscle (ℓ_l) is shorter than that of a pennate muscle (ℓ_p) if their resting lengths (ℓ_o) are the same. The longitudinal muscle can create tension through a greater range.

11.14. You are stronger in some positions than in others, partly because single-joint muscles are able to create the greatest tension at a specific length.

Multiple-joint muscles are not so constrained to operate within 60% to 160% of their resting length. The arrangements of multiple-joint muscles usually do not allow shortening below 60% of their resting length, but they may be stretched beyond 160% of their resting length. They are able to operate in the right side of the curve shown in figure 11.14. Their maximum tension is developed at lengths beyond 160% of their resting length due to passive stretching of the connective tissue structures. Self-experiment 11.3 illustrates how passive tension is produced in stretched muscles.

Self-Experiment 11.3

Consider the muscles that flex or extend your fingers. The finger flexor muscles are located in your anterior forearm, while the finger extensor muscles are located in your posterior forearm. Their tendons cross the wrist joint, the carpometacarpal joints, the metacarpophalangeal joints, and the interphalangeal joints. Flex your fingers and grip a pencil as tightly as you can. Now flex your wrist as far as you can. You may notice that you cannot flex your wrist as far with your fingers flexed as you can without your fingers flexed. You may also notice that your grip on the pencil weakened as you flexed your wrist (try to pull the pencil out in both positions). If you push on your hand and cause it to flex farther, the pencil may even fall out of your grip. The finger flexor tendons shortened at each joint they crossed, and thus the finger flexor muscles were unable to produce much tension. The finger extensor tendons were lengthened at each joint they crossed and stretched the extensor muscles beyond 160% of their resting length. Passive tension created by the stretching of the connective tissue resulted in an extension of the fingers when you pushed your hand into further wrist flexion.

The hamstring muscles provide another good example of the practical implications of the length–tension relationship for muscles. These muscles on the posterior of

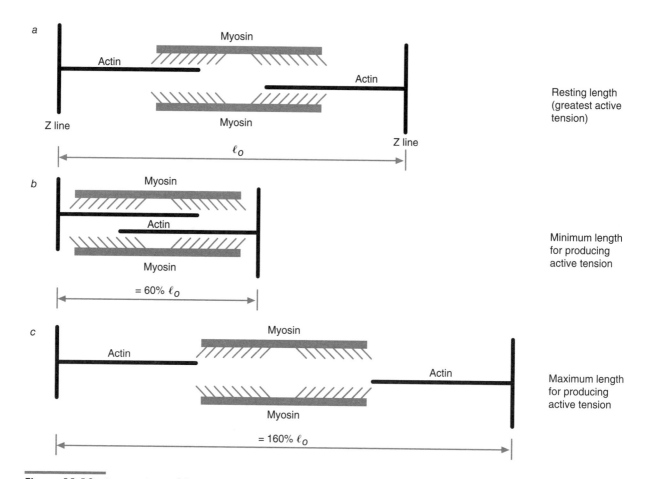

Figure 11.13 Interactions of the actin and myosin filaments with each other and the Z disks determine the magnitude of active tension developed in a sarcomere. The greatest active tension is at resting length (*a*); (*b*) shows the minimum length for producing active tension; and (*c*) shows the maximum length for producing active tension.

your thigh create a flexor torque about the knee and an extensor torque about the hip. If you have ever done hamstring curl exercises using a flat bench with weights and a pulley system, you may have noticed that your hamstring muscles are very weak in the fully flexed position shown in figure 11.15*a*. This is because your hamstrings are as short as you can get them. They are close to the minimum length at which they can develop tension—they are on the left end of the curve shown in figure 11.14. The position you are in, with hips extended and knees flexed, may shorten the hamstrings to less than 60% of their resting length. The hamstring curl exercise performed on a flat bench works the muscle only in the 60% to 100% range of its resting length. To improve this exercise, you could modify your position so that your hips are slightly flexed rather than extended. This would lengthen the hamstrings a little and thus keep them active in a stronger range of lengths. Rather than using a flat bench, a bench with a hump in it, as shown in figure 11.15*b*, would allow you to keep your hips flexed while doing the hamstring curl.

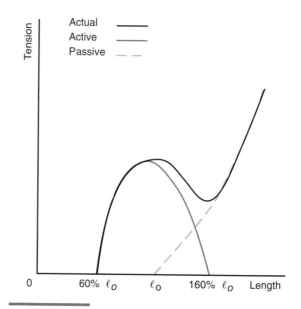

Figure 11.14 The relationship between muscle length and tension.

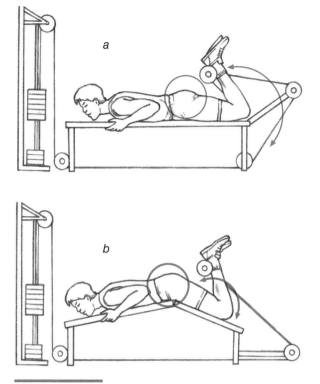

Figure 11.15 Machines for exercising hamstring muscles. In (a), muscle length is very short, and its ability to create tension is limited. In (b), muscle length is longer due to flexion at the hip, and its ability to create tension is greater.

Contraction Velocity

The maximum tensile force a muscle is capable of producing is dependent on the velocity of shortening of the muscle as well as its length. Just as the mechanics of contraction within the sarcomere determine the relationship between the length of a muscle and its maximum force of contraction, these mechanics also determine the relationship between the velocity of a muscle's contraction and its maximum force of contraction.

> **The maximum tensile force a muscle is capable of producing is dependent on the velocity of shortening of the muscle as well as its length.**

Activity Within the Sarcomere

Remember that the active tension produced during a muscular contraction is a result of the attachment of the cross bridges of the myosin filament to the actin filament within the sarcomere. As the muscle shortens during a contraction, the cross bridges attach to the actin myofilament, pull it toward them, release it, and then reattach

to it farther along its length. Thus there are three steps in the cycle for the cross bridges: attach, pull, release. The force developed within a sarcomere is proportional to the number of cross bridges attached and pulling on the actin filaments. When a muscle is shortening, some cross bridges are in the release step of the process and are not contributing at all to the tension development. If the velocity of shortening is slow, only a small percentage of the cross bridges are in the release step, and each cross bridge spends a small proportion of the contraction time in that step, so a large amount of tension is developed. If the velocity of shortening is faster, more of the cross bridges are in the release step, and each cross bridge spends a larger proportion of the contraction time in that step, so less tension is developed. Thus, in a sarcomere, greater tension is developed at slower velocities of shortening. The same is true in a whole muscle. It is easy to lift lightweight objects quickly, but as things get heavier, you cannot move them as quickly. You can lift 10 lb with one hand very quickly; you cannot lift 50 lb with one hand as quickly. This is partly due to the inertia of the object, but it is also due to your inability to create large forces if your muscles must shorten quickly. This relationship between contraction velocity and muscle tension is shown in figure 11.16.

Concentric, Eccentric, and Isometric Activity

You may notice that negative velocities of shortening are shown on the graph in figure 11.16. These represent eccentric contractions. A muscle contracting eccentrically or isometrically is capable of producing more force than a muscle contracting concentrically. If you have ever done any weight training, you'll find that this makes sense. Imagine that you are in the weight room trying to determine your maximum lift in the bench press. A graphic representation of these lifts on the velocity–tension curve is shown in figure 11.17. You're lying on the bench with your partner spotting you as you try to lift 150 lb. You lower the barbell to your chest and then lift it quickly. The pectoral muscles contract eccentrically during the lowering phase and concentrically during the raising phase. Your partner adds two 25 lb plates to the bar to make the total weight 200 lb. You lower the barbell to your chest and then lift it upward, but this time you can only move the bar slowly. The speed of the muscular contraction is now limited by the size of the force your muscles must develop. You cannot lift 200 lb as quickly as you lifted 150 lb.

> **A muscle contracting eccentrically or isometrically is capable of producing more force than a muscle contracting concentrically.**

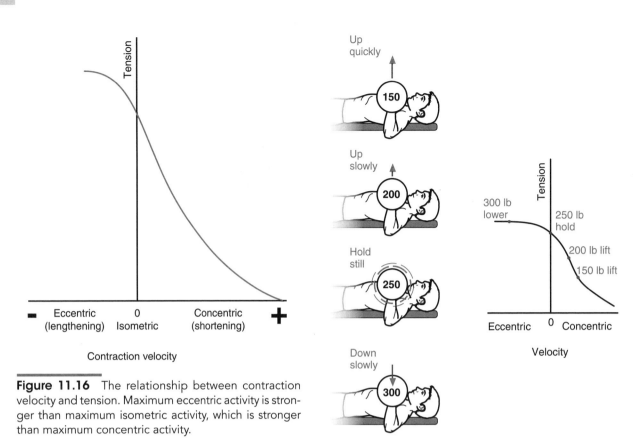

Figure 11.16 The relationship between contraction velocity and tension. Maximum eccentric activity is stronger than maximum isometric activity, which is stronger than maximum concentric activity.

Figure 11.17 Graphic representation of four bench press lifts on the velocity–tension curve.

Your partner adds two more 25 lb plates to the bar, so the total weight is now 250 lb. You slowly lower the barbell to your chest, but you cannot lift it off your chest (accelerate it upward) without help from your spotter. You and your spotter lift the barbell halfway up; then the spotter lets you handle the weight. You are unable to keep it moving upward (constant velocity with zero acceleration), but you are able to hold the barbell at that position. The pectoral muscles contracting isometrically were able to produce a force large enough to hold the 250 lb statically, but they were unable to produce the same force concentrically to lift the 250 lb. The tension developed in the isometric contraction was greater than the maximum tension that could be developed in the concentric contraction.

Your spotter now adds another 50 lb to the barbell to make a total of 300 lb. You are able to lower the barbell slowly (at constant velocity) to your chest, but you are not able to lift it (accelerate it upward), even with a slight assist from your partner. Your spotting partner helps you lift it halfway, then releases and lets you try again. You cannot even hold it still this time, and the barbell begins to lower slowly (at constant velocity) toward your chest before your spotter helps you complete the lift. The pectoral muscles contracting eccentrically were able to produce a force large enough (300 lb) to lower the 300 lb slowly, but they were unable to produce the same force isometrically to hold the 300 lb still or concentrically to lift the 300 lb. The tension developed in the eccentric contraction was greater than the maximum tension that could be developed in the isometric or concentric contraction.

Other Factors

Several other factors may affect the maximum force produced by an active muscle. Temperature is one of these factors. Increasing the temperature of a muscle slightly increases its ability to produce tension. Muscle temperature can be raised externally through rubbing or heating or internally through engagement in warm-up exercises before an activity. The next sections review additional factors.

Prestretch

Prestretching a muscle just before a concentric contraction may also affect the force of the concentric contraction. The less time between the stretching of the muscle and the subsequent concentric contraction, the greater the force of the contraction. The mechanical bases for this effect are not well understood. The slack in the connective tissue and other noncontractile elements in series with the contractile elements (the myofilaments) may be reduced by prestretching, so that the contraction force is immediately transmitted to the attachments. There is also a neuromuscular basis for this behavior, which is discussed in the next chapter. In any case, you can demonstrate the effect of prestretch to yourself while performing self-experiment 11.4.

Self-Experiment 11.4

Try to jump up as high as possible using the following two techniques. First, flex your knees and hips slightly as shown in figure 11.18*a,* and hold this position for a second before extending your legs and jumping up as high as you can. Now try it again, but this time start in an upright position (see figure 11.18*b*). Quickly flex your knees and hips and then extend them and jump up as soon as you reach the start position of the first jump. You probably jumped higher using the second technique, and the forces exerted by your knee and hip extensors were larger. The muscles were stretched quickly immediately before contracting concentrically in the second technique. The concentric contraction was more forceful as a result of the prestretch.

Stimulus Duration

As discussed in the previous section under "Contraction Mechanics," the frequency of muscle stimulation affects its tension. Maximum tension does not develop within a muscle fiber for a short time (between 0.001 and 0.300 s,

depending on the muscle type) after it is stimulated. This rate of tension development affects the maximum tension developed in a muscle if a large force is needed in a short time. Thus muscle contractions of very short duration are weaker than contractions lasting longer than 0.001 to 0.300 s.

Fatigue

The maximum tension a muscle can develop is also affected by fatigue. Continuous stimulation of a muscle results in an eventual decline in the tension it produces. The demand of the contracting muscle for adenosine triphosphate (ATP) eventually exceeds the supply of ATP to the muscle, resulting in diminishing force production by the muscle. You can demonstrate this to yourself even if the muscle does not produce a maximal contraction. Take the largest weight you can lift with your extended arm (perhaps several books) and hold it at arm's length away from you. How long can you hold the weight in this position? Eventually, the tension your muscles produce weakens, and you are unable to hold up the weight.

Fiber Type

Fatigue and the rate of tension development within a muscle are both affected by the type of muscle fiber. Skeletal muscle fibers differ in terms of their fatigue resistance and rate of tension development. Fibers are classified into three types according to these differences: type I, or slow-twitch oxidative (SO); type IIA, or fast-twitch oxidative-glycolytic (FOG); and type IIB, or fast-twitch glycolytic (FG). **Type I muscle fibers (SO)**

Figure 11.18 Two different standing vertical jump techniques. The second technique uses the prestretch of the muscles involved in the jumping action to produce a higher jump.

have a high density of mitochondria and are thus highly aerobic and fatigue resistant. These fibers have a slow rate of tension development and are smaller in diameter; thus their maximum tension is lower. **Type IIB muscle fibers** (FG) are glycogen rich and oxygen poor. These fibers have a high anaerobic capacity and a low aerobic capacity, so they fatigue quickly, but their rate of tension development is fast. Type IIB fibers are larger in diameter than type I fibers, so they can produce greater tension, but not for long durations. **Type IIA muscle fibers** (FOG) have characteristics of type I and type IIB fibers. They have relatively high aerobic and anaerobic capacities, so they develop tension quickly and can maintain it for long durations.

The muscles of the average person contain about 50% to 55% type I fibers, 30% to 35% type IIA fibers, and 15% type IIB fibers, but great variations exist. The more successful athletes engaged in endurance activities have a higher than normal percentage of type I fibers, whereas those in activities requiring short bursts of power (sprinters, jumpers, throwers, weightlifters, and so on) have a higher than normal percentage of type II fibers.

Torque Produced

A final consideration in the discussion of maximum force produced by muscle concerns the effectiveness of the force produced. Muscles produce forces that pull on bones and create torques at joints. These torques are used to rotate the limbs and move external loads. The effectiveness of the muscle force thus refers to the effectiveness of the force in producing torque. Because torque is force times moment arm ($T = F \times r$), the torque produced by a force depends on the length of the moment arm. The moment arm of a muscle about a joint depends on the point of attachment of the muscle on each limb and the line of action of the muscle force. The maximum length of the moment arm is determined by the distance between the axis of the joint and the muscle's point of attachment closest to the joint. The moment arm of the muscle will equal this distance when the muscle's line of pull is perpendicular to the axis of the bone at this closest attachment. When the limbs are in the position that places the muscle's line of pull at 90° to the bony lever, the torque produced by the muscle is maximum. At all other positions, both the moment arm and the torque are smaller if the muscle force doesn't change.

Once again, torque is force times moment arm ($T = F \times r$). The angle of a joint affects the moment arm of the muscles crossing the joint (r) as well as the relative length of the muscles crossing the joint. The relative length of the muscle affects the maximal tensile force it can develop (F). So, torque (T) production by a muscle is sensitive to the angle of the joint or joints it crosses since joint angle affects both torque variables, force (F) and moment arm (r).

Muscle Power

Power is the rate of doing work. In chapter 4, we learned that power could also be expressed as the product of force times velocity. The power output of a muscle is thus the tensile force produced by the muscle times the velocity of shortening of the muscle. Figure 11.16 showed the relationship between the force and velocity of a contraction. From this graph, we can obtain the relationship between muscle power and velocity of contraction by multiplying the velocity values by their corresponding force values to determine muscle power. Figure 11.19 illustrates this relationship.

⟩ **The power output of a muscle is the tensile force produced by the muscle times the velocity of shortening of the muscle.**

Bicyclists use this relationship when choosing a gear ratio. In a low gear, they exert small forces on the pedals but spin the crank fast to get the same output as in a high gear, when they exert large forces but spin the crank very slowly. According to the velocity–power curve for muscle, cyclists should choose a gear that allows them to spin the crank at a moderate speed, thus keeping the

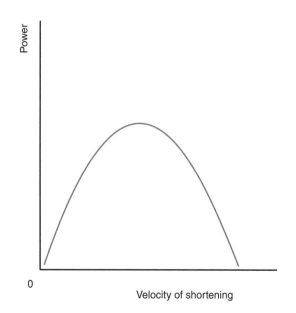

Figure 11.19 The relationship between velocity of shortening and power produced in a muscle.

muscle's velocity of shortening near the range of values that produces the maximum power output.

Summary

Skeletal muscles are the motors and active force generators of the musculoskeletal system. The forces they produce move the limbs of the body and impart stiffness to the joints. Muscles also produce heat, provide protection, and alter pressure.

Skeletal muscle is composed of numerous muscle cells or fibers arranged in series and in parallel and supported by a complex of connective tissue sheaths. The sarcomere is the fundamental structure of a muscle fiber. Within the sarcomere, the attachments of cross bridges of myosin filaments to actin filaments generate the active tension of a muscular contraction. Fiber arrangements in whole muscles may be longitudinal or pennate. Longitudinal muscles have more muscle fibers in series with each other and may shorten over a greater distance. Pennate muscles have more fibers parallel to each other and may produce greater forces.

A muscular contraction or action may be concentric, isometric, or eccentric. Concentric action occurs when the muscle shortens while it develops tension. Positive mechanical work is done by a concentric muscle action. Isometric action occurs when the muscle does not change length while it develops tension. No mechanical work is done by an isometric muscle action. Eccentric action occurs when the muscle lengthens while it develops tension. Negative mechanical work is done by an eccentric muscle action.

Terms such as *agonist, antagonist, neutralizer,* and *stabilizer* are used to refer to muscles when they function in a specific way relative to another muscle or to a joint action.

A number of variables determine the force produced by an active muscle. If all the fibers of a muscle are stimulated, the maximum force developed by a muscular action depends on the physiological cross-sectional area of the muscle, the length of the muscle relative to its resting length, the velocity of contraction, the temperature, and the stimulus rate. The arrangement of a muscle relative to the joint or joints it crosses affects the torque produced and thus determines the effectiveness of the muscular force. Muscle power is determined by the velocity and force of contraction. Maximum power output occurs at a specific contraction velocity.

KEY TERMS

agonist (p. 284)	fascicle (p. 280)	stabilizer (p. 285)
antagonist (p. 285)	isometric contraction (p. 284)	synergy (p. 285)
aponeuroses (p. 280)	muscle fiber (p. 278)	tendons (p. 280)
concentric contraction (p. 282)	neutralizer (p. 285)	type I muscle fibers (p. 295)
eccentric contraction (p. 283)	perimysium (p. 280)	type IIA muscle fibers (p. 296)
endomysium (p. 278)	sarcolemma (p. 278)	type IIB muscle fibers (p. 296)
epimysium (p. 280)	sarcomere (p. 279)	

REVIEW QUESTIONS

1. What are the functions of skeletal muscle?

2. What is the basic contractile unit of a muscle?

3. What is the sarcolemma?

4. Describe how the force developed between the actin and myosin filaments of a sarcomere is transferred to the tendon of the muscle.

5. During what phase of a baseball throw are the arm muscles' actions concentric? Eccentric?

6. During what phase of a vertical jump are the hip muscles' actions concentric? Eccentric?

7. What muscle is an antagonist to elbow extension?

8. What muscle is an agonist to hip flexion?

9. What muscle is a neutralizer for the supination torque produced during concentric activity of the biceps?

10. Discuss the advantages and disadvantages of longitudinal versus pennate fiber arrangements in muscle.

11. Which is stronger—a maximal eccentric contraction or a maximal concentric contraction?

12. Which is stronger—a fast concentric contraction or a slow concentric contraction?

13. In which activity are you able to produce greater forces in your hip muscles: jumping up or landing from a jump? Why?

14. In which activity are you able to produce greater forces in your shoulder muscles: throwing a baseball or putting a shot? Why?

15. How does joint angle affect the torque production capabilities of a muscle crossing that joint?

16. Discuss the advantages and disadvantages of muscle attachments close to joints.

chapter 12

The Nervous System
Control of the Musculoskeletal System

objectives

When you finish this chapter, you should be able to do the following:

- List the elements of the nervous system

- Describe the different parts of a neuron

- List the three types of neurons

- Define motor unit

- Understand the two strategies (recruitment and summation) used by the central nervous system to control muscle force

- Define proprioceptor and list different types of proprioceptors

- Describe the function of the muscle spindles

- Describe the stretch reflex

- Describe the function of the Golgi tendon organs

- Describe the Golgi tendon organ response

- Describe the proprioceptors of the vestibular system

- Describe the righting and tonic neck reflexes

- Define exteroceptor and list different types of exteroceptors

- Describe reflexes initiated by exteroceptors

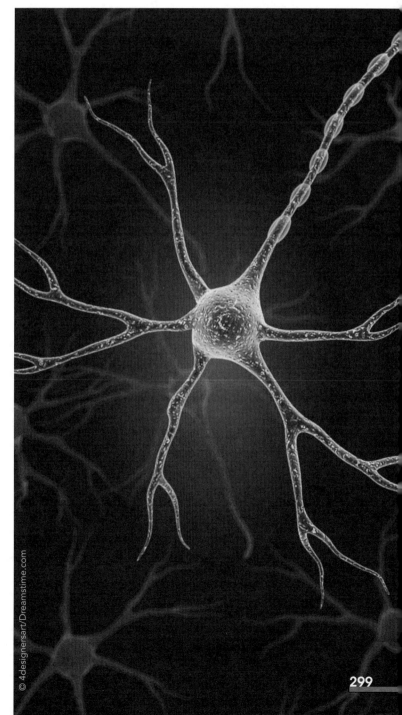

© 4designersart/Dreamstime.com

Close your eyes and have someone drop a book into your outstretched hand. Were you able to catch it? How did you know when the book hit your hand? How were you able to detect the weight of the book? How were you able to regulate the amount of force you exerted on the book so that you were able to stop it and catch it? More generally, how does the human body detect external forces acting on it? How does it detect changes in limb positions at joints or changes in orientation of the whole body? How does the body control the contraction force of a muscle? These questions all deal with some aspect of musculoskeletal system control, which is the topic of this chapter.

Control of the musculoskeletal system
is managed by the nervous system. The nervous system collects information from external and internal stimuli, processes this information, and then initiates and controls the musculoskeletal system's response to the stimuli. A basic understanding of the nervous system and how it controls movement is thus necessary in our study of the biomechanics of human movement. A thorough treatment of the nervous system and its control of muscular function is provided in other books, which you are encouraged to read. The purpose of this chapter is to present a simple, brief overview of the nervous system and how it operates to control the musculoskeletal system.

The Nervous System and the Neuron

The nervous system is organized into the central nervous system and the peripheral nervous system. The brain and spinal cord form the **central nervous system**. Each of these elements is protected by bony structures: the skull and the vertebral column. The brain is the central processor of the nervous system, and the spinal cord transmits signals to or from the peripheral nervous system to the brain. All the nervous tissue that lies outside the skull and the vertebral column forms the **peripheral nervous system**. The 12 cranial nerves and 31 pairs of spinal nerves are part of the peripheral nervous system. These nerves are actually bundles of nerve fibers that may include both sensory nerves, which detect information about the external environment and the internal state of the body, and motor nerves, which send stimuli to muscles.

⟳ The nervous system is organized into the central nervous system and the peripheral nervous system.

The nervous system is sometimes divided functionally into the somatic nervous system and the autonomic nervous system. The **somatic nervous system** is involved in conscious sensations and actions and is also referred to as the voluntary nervous system. The **autonomic nervous system** is involved with unconscious sensations and actions and is also referred to as the involuntary nervous system. It includes sympathetic and parasympathetic nerves. The autonomic nervous system controls and regulates the functions of most internal organs, whereas the somatic nervous system controls and regulates movement.

The fundamental unit of the nervous system is the neuron or nerve cell. A neuron has a cell body, which contains the cell nucleus and other metabolic structures. Numerous branched hairlike projections from the cell body are the dendrites. The axon, or what is usually thought of as a nerve fiber, is a single, long, threadlike projection that usually extends from the cell body in the opposite direction of the dendrites. The axon often branches into many short fibers at its distal end. Figure 12.1 shows an illustration of two neurons.

⟳ The fundamental unit of the nervous system is the neuron or nerve cell.

Basically, there are three types of neurons: (1) sensory, or afferent, neurons; (2) motor, or efferent, neurons; and (3) interneurons, or connector neurons. The sensory and motor neurons are situated primarily in the peripheral nervous system, whereas the interneurons are situated within the central nervous system. The **sensory neurons** are responsible for sensations. They receive stimuli from our external or internal environment and send this information back toward the central nervous system, where they interface with **interneurons** and occasionally motor neurons. The cell bodies of sensory neurons lie close to but outside the spinal cord. **Motor neurons** receive stimuli from interneurons or sensory neurons and send

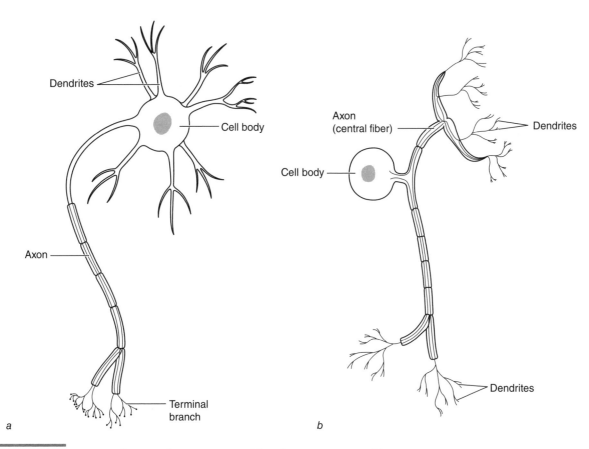

Figure 12.1 Features of a typical motor neuron (a) and sensory neuron (b).

signals to muscles, away from the central nervous system. The cell bodies of motor neurons are located within the spinal cord. Because the cell bodies of the sensory and motor neurons are situated close to the spinal cord whereas the terminal endings of their axons or dendrites may be located in the distal ends of the extremities, nerve fibers (axons and fibers to peripheral dendrites) may be more than 1 m long.

> There are three types of neurons: (1) sensory, or afferent, neurons; (2) motor, or efferent, neurons; and (3) interneurons or connector neurons.

The cell membrane of a neuron is an excitable membrane whose electrical potential (voltage across the membrane) can change as the result of stimulation. Normally, neurons may be stimulated at their cell body and at their numerous dendrite endings. A stimulus can be excitatory (facilitating) or inhibitory, depending on whether the stimulus results in an increase or decrease in the electrical potential of the cell membrane. If the net effect of all the excitatory and inhibitory stimuli is an increase in the membrane potential above a threshold level at the base

of the axon (where it leaves the cell body), an action potential is generated within the neuron at the base of the axon. If the net effect of all the excitatory and inhibitory stimuli does not increase the membrane potential above the threshold level, no action potential is generated. The action potential or neural impulse is a depolarization of the cell membrane at a specific site. The action potential is propagated along the membrane, away from the cell body, and down the axon to its terminal endings. At the terminal endings, the axon synapses with the dendrites of other neurons if it is a sensory neuron or interneuron, or it synapses with muscle fibers if it is a motor neuron. The neural impulse is then transmitted by chemical means across the synapse and presents a stimulus to the next membrane. Figure 12.2 illustrates how neurons communicate with each other and the direction of neural impulses.

The Motor Unit

The fundamental unit of the neuromuscular system is the **motor unit**, which is composed of a single motor neuron and all the muscle fibers with which it synapses. The number of muscle fibers innervated by a single motor neuron may be smaller than 20 or larger than 1000. This number represents the number of branches at the end of an axon, and the larger motor neurons usually have larger threshold potentials. The ratio of muscle fibers to motor neurons in a whole muscle indicates the degree of control a person has over a muscle contraction (see figure 12.3).

The smaller the number of muscle fibers per motor unit, the more precise the control of the muscle. The larger the number of muscle fibers per motor neuron, the coarser the control of the muscle.

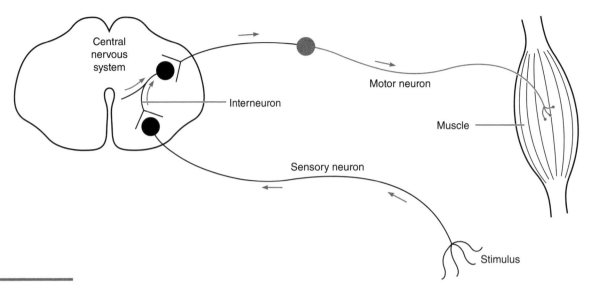

Figure 12.2 Schematic diagram of the pathways of a neural impulse from external stimulus to muscle action.

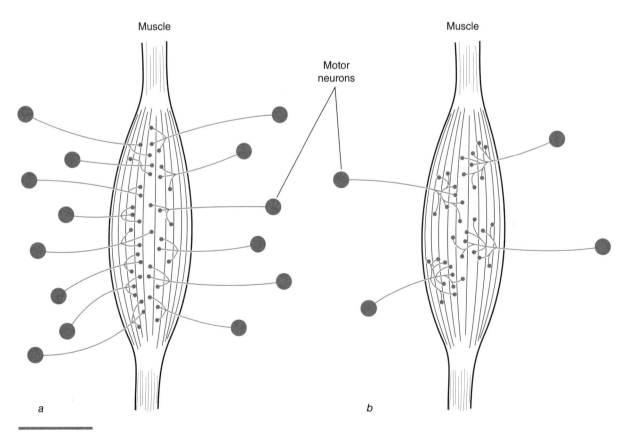

Figure 12.3 Idealized example of a muscle with many motor neurons and fine control (a) versus one with fewer motor neurons and coarse control (b).

⊃ The fundamental unit of the neuro-muscular system is the motor unit, which is composed of a single motor neuron and all the muscle fibers with which it synapses.

The muscle fibers of a single motor unit are not grouped together within the whole muscle but are dispersed throughout a local area of the muscle. Thus adjacent muscle fibers are usually parts of different motor units.

The neural action potential propagated in the axon of a motor neuron is transmitted to the synapses of each axon motor end plate with each muscle fiber of the motor unit. The threshold potential of the membrane (sarcolemma) of an individual muscle fiber is usually small relative to the chemical stimuli provided by the neural action potential at the synapses, so a neural action potential almost always generates a muscle action potential. All the muscle fibers of a motor unit thus contract together as a result of an action potential in the motor neuron.

One method of controlling the size of the force produced in a whole-muscle contraction is to control the number of active motor units. If only a small force is desired, only a small number of motor units are recruited, and only a small number of muscle fibers contract to produce the tension. If a larger force is desired, more motor units are recruited, and a larger number of muscle fibers contract to produce more tension.

Another method of controlling the size of the force produced in a whole-muscle contraction is to control the rate of stimulation. A single action potential generates a single twitch response or contraction in a muscle, as shown in figure 12.4. If another action potential is received by the muscle fiber before the tension that was developed in the first twitch response has subsided, the tension developed in the second twitch response sums with the tension of the first twitch response. If the time between action potentials is short enough, a tetanic contraction occurs, and the maximum tension developed is much larger than that of a single twitch. The two methods (recruitment and summation) of controlling muscle force occur simultaneously.

It appears that a certain pattern of motor unit recruitment is followed as the desired tension in a muscle increases. The first motor units recruited are the small motor units, and the stimulation rate is low. These motor units have the least number of muscle fibers and a larger percentage of slow-twitch (type I) fibers. As greater tension is desired, larger motor units with more fast-twitch (type II) fibers are recruited. At maximal tension, all the motor units have been recruited, and the stimulation rate is high.

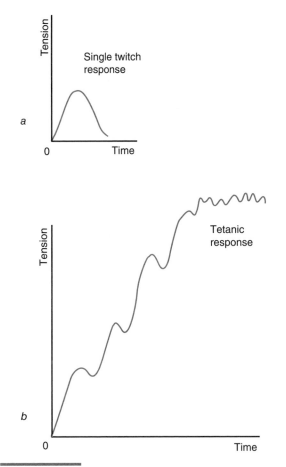

Figure 12.4 Tension developed in a muscle fiber as a result of a single stimulus (a) is smaller than that developed as a result of repeated stimulation (b) if the stimulation frequency is great enough.

During a maximal voluntary contraction of a muscle, it may be that not all of that muscle's motor units are recruited. Larger contraction forces may still be elicited from the muscle by electrical stimulation or other means. This result indicates that some of the strength gains associated with a weight training program (especially the initial strength gains) may be due to training of the nervous system and its improved ability to recruit more motor units. You may have experienced this yourself if you have ever done any weight training. Some strength gains occur without any weight gain or increase in muscle mass.

Whether or not a motor unit generates an action potential (and thus causes its muscle fibers to contract) depends on the stimuli it receives from neurons with which it synapses. The dendrites of a motor neuron synapse with thousands of other neurons. Some of these synapses may be inhibitory, whereas others may be excitatory. Volitional control of a motor unit occurs only through the synapses between the motor neuron and the neurons whose fibers descend from the higher-order structures in the central nervous system. The other synapses transmit neural impulses that may augment or detract from the desired volitional response. Through feedback from the proprioceptors, the proper intensity of the volitional neural impulse occurs at the motor neuron synapses with the neurons whose fibers descend from the central nervous system. This process of constant feedback and control determines the behavior of a motor unit. The following section describes some of the receptors and the reflex actions they elicit through excitation or inhibition of the motor neurons with which they synapse.

Receptors and Reflexes

Sensory neurons or afferents receive stimuli from specialized receptors. **Exteroceptors** respond to stimuli that come from sources external to the body. These include the receptors for the five senses: sight, hearing, taste, smell, and touch. **Interoceptors** respond to stimuli from sources within the body. These include the receptors associated with the internal organs, or visceroceptors, and the receptors associated with the musculoskeletal system, or **proprioceptors**.

Some receptors initiate reflexes. Reflexes are involuntary responses that result from sensory input and little if any input from the higher-order structures in the central nervous system. Reflexes may have some protective function during development, and a few continue to have a protective function in adults. A simple spinal reflex may involve only two neurons: a sensory neuron that detects a stimulus, and a motor neuron that synapses with this sensory neuron and is thus activated by it. A sensory stimulus to this spinal reflex arc causes the sensory neuron to excite the motor neuron and elicit a muscular contraction. Most reflexes are much more complex and involve more than two neurons, but these rudimentary spinal reflex arcs may serve to enhance or inhibit certain movements.

> **Reflexes are involuntary responses that result from sensory input and little if any input from the higher-order structures in the central nervous system.**

Proprioceptors and Proprioceptive Reflexes

Proprioceptors may be thought of as the sensory organs that monitor the status of the musculoskeletal system. Receptors are located within joint capsules to give feedback about the joint position (flower-spray or Ruffini endings) or rapid changes in joint position (pacinian

corpuscles). Receptors are located within muscles and tendons to give feedback about increases in muscle length (muscle spindles) or muscle tension (Golgi tendon organ). The vestibular system of the inner ear also has specialized receptors that give feedback about the head's position and changes in its position.

The Muscle Spindle and Stretch Reflex

The **muscle spindle** is a proprioceptor that detects the stretch of a muscle or its relative changes in length. Numerous muscle spindles are present in a muscle. Each muscle spindle includes some short muscle fibers that are arranged in parallel to the other fibers of the muscle. These muscle spindle fibers have several sensory and motor neurons connected to them. When the whole muscle is stretched, the muscle spindle is also stretched. The muscle spindle's sensory neurons are stimulated by this stretch. Slow stretches result in a slow rate of stimulation, whereas faster stretches result in a faster rate of stimulation. Once a muscle has changed its length, the motor neuron to the muscle spindle itself is activated to reset the tension in the muscle fibers of the spindle. The muscle spindle can thus respond to increases in muscle length, no matter what the length of the muscle is.

> **The muscle spindle is a proprioceptor that detects the stretch of a muscle or its relative changes in length.**

The sensory neurons of the muscle spindle synapse with the motor neurons that innervate the whole muscle. If the stimulation of these sensory neurons is great enough, the motor neurons are also stimulated, and contraction of the stretched muscle is facilitated. This reflex is called the stretch reflex. The response is stronger if the increase in length occurs quickly rather than slowly.

You may have experienced an embarrassing demonstration of the stretch reflex if you have ever fallen asleep during a lecture. As you begin to doze off, your neck extensor muscles relax, and your head begins to fall forward. Before your chin reaches your chest, the stretch reflex is evoked by the quick stretch of the neck extensor muscles. These muscles then contract strongly and pull your head up. This action is usually vigorous enough to startle you out of your dreams and back to the lecture. Try self-experiment 12.1 for another demonstration of the stretch reflex.

Self-Experiment 12.1

Physicians test the stretch reflex when they strike your patellar ligament. The resulting stretch of the quadriceps muscle elicits a stretch reflex and this muscle contracts, causing extension at the knee. You can try this yourself. Sit on the edge of a table or in a chair that's high enough that your legs hang freely and do not touch the floor. Relax your muscles, especially your quadriceps. Now gently but quickly strike your patellar ligament (on the anterior part of your leg just below your patella but above your tibial tuberosity) with the side of your hand or the spine of a book. Try it several times or have someone else do it if you're unsuccessful. Can you elicit this reflex?

Athletes also use the stretch reflex when they incorporate a backswing or a windup or any sort of prestretch before a movement. The quicker this prestretch or backward movement, the larger the contraction of muscles creating the torque for the forward movement. The windup of a baseball pitcher during a pitch, the backswing of a tennis player during a forehand stroke, and the backswing of a golfer during a golf drive are all examples of athletes using the stretch reflex to enhance performance.

The sensory fibers from the muscle spindle also synapse with an interneuron that synapses with the motor neurons of the antagonist muscle. This synapse is an inhibitory synapse, so the activation of these motor neurons is inhibited. Thus the stretch reflex also inhibits contraction of the antagonist muscle. This effect is called reciprocal inhibition.

The response of the stretch reflex to slow stretching of a muscle is used to control posture and limb position involuntarily. When you sway forward while standing, the posterior posture muscles are slowly stretched. This stretch initiates the stretch reflex, and those posterior muscles contract, stop the forward sway, and pull you back so you remain in balance. The response of the muscles to a slow stretch is not as large as the response to a fast stretch.

The stretch reflex functions to protect the joints crossed by a muscle. A quick lengthening of a muscle occurs when limb positions at a joint change quickly. Such changes may lead to dislocation at the joint unless the limb movements are slowed down. The stretch reflex causes the stretching muscle to contract eccentrically and slow the movement. During the follow-through of many throwing or striking movements, the stretch reflex may be evoked to slow down the throwing or striking limbs.

⤴ **The stretch reflex causes the stretching muscle to contract eccentrically and slow the movement.**

The stretch reflex also influences flexibility exercises. Muscles are stretched most effectively if they are relaxed. A quick stretch of a muscle elicits the stretch reflex and causes it to contract. The slower the stretch, the slower the rate of stimulation of the muscle spindle sensory fibers,

and the smaller the stretch reflex response. Slow, static stretching is thus more effective.

The Golgi Tendon Organ and Tendon Reflex

The **Golgi tendon organ** is another proprioceptor associated with muscle function. A Golgi tendon organ is typically located within the tendon, close to the muscle, and is in series with the muscle. The sensory fibers from this organ are stimulated by tension within the tendon, whether caused by stretching or by contraction of the muscle. The greater the tension, the greater the stimulation. These sensory fibers from the Golgi tendon organ synapse with the motor neurons of the muscle; but rather than transmitting excitatory impulses, the Golgi tendon sensory fibers transmit inhibitory impulses across the synapses. The contraction of the tensed muscle is thus inhibited by the response of the Golgi tendon organ. If the tension is great enough, the Golgi tendon organ will completely inhibit contraction of the tensed muscle and cause it to relax. This reflex is sometimes referred to as the tendon reflex.

⤴ **The contraction of a tensed muscle is inhibited by the response of the Golgi tendon organ.**

The tendon reflex protects the muscle from rupturing or tearing by turning off the active development of tension if the tensile stress within the muscle is too great. This reflex works in opposition to the stretch reflex. The stretch reflex response is usually greater unless the tension within the muscle is very great. An example of the tendon reflex is occasionally seen when the leg of a high jumper or triple jumper buckles or collapses due to the extreme forces in the knee extensor muscles at takeoff.

Some of the strength gains observed in resistance training may occur due to the increased ability of the central nervous system to provide enough excitatory stimulation to the motor neurons to override the inhibitory stimulation by the Golgi tendon organs.

The Vestibular System and Its Related Reflexes

The vestibular system is composed of the sense organs of balance. Each inner ear contains three proprioceptors: the semicircular canals, the utricle, and the saccule. These proprioceptors are bony tunnels filled with fluid called **endolymph**. The walls of these tunnels are lined with sensory hair cells that are surrounded by a gelatinous substance. There are three semicircular canals arranged in three mutually perpendicular planes roughly corresponding to the sagittal, frontal, and transverse planes. The endolymph fluid in these canals moves relative to the head when the head is accelerated. This movement bends the sensory hair cells, thus giving feedback about changes in the motion of the head or its acceleration. The utricle and saccule are bulbous lumps in the bony tunnels that contain otoliths, tiny calcium carbonate stones that are embedded in the gelatinous substance along with the hair cells. The otoliths bend the hair cells according to the direction in which they are pulled by gravity, thus giving feedback about the position of the head relative to the gravitational force.

> The vestibular system, or the labyrinths of the inner ear, is composed of the sense organs of balance.

The vestibular system and the proprioceptors associated with the joints of the neck give rise to several rudimentary reflexes associated with the head and neck positions. In adults, these reflexes are normally difficult to observe due to overriding volitional control. Righting reflexes occur when the vestibular system detects a nonupright position of the head. Reflex actions of the limbs and the trunk and neck musculature are attempts to correct the head position and maintain upright posture. Figure 12.5 shows examples of righting reflex responses.

Reflexes resulting from the stimulus of the proprioceptors of the neck are called **tonic neck reflexes**. These proprioceptors communicate with the motor neurons of the upper extremity muscles. Neck flexion initiates a tonic neck reflex response that facilitates contraction of the muscles that cause pulling actions of the arms, whereas neck extension facilitates contraction of the muscles that cause pushing actions of the arms. Rotating the head

to the right initiates a tonic neck reflex response that facilitates contraction of the elbow extensors and shoulder abductors on the right side and the elbow flexors and shoulder adductors on the left side. Figure 12.6 illustrates the positions facilitated by the tonic neck reflexes.

Exteroceptors and Exteroceptive Reflexes

The exteroceptors that significantly influence the movement of the body include the receptors for sight, hearing, touch, and pain. The receptors for touch are **pacinian corpuscles,** which are also the receptors for change in joint position. These receptors are sensitive to pressure, but only when changes in pressure occur. Large changes in pressure result in a large response from the pacinian corpuscles, but only during the change.

The pacinian corpuscles beneath the skin on the soles of your feet or the palms of your hands are responsible for the extensor thrust reflex. The sensory fibers from the pacinian corpuscles synapse with the motor neurons of the extensor muscles of the limb, thus facilitating contraction of these muscles and extension of the limb. The large change in pressure occurring when you land on your feet after a jump initiates the extensor thrust reflex. The extensor muscles of the leg contract eccentrically and stop you from falling. This contraction is facilitated by the extensor thrust reflex.

Cutaneous pain receptors are also involved in reflex actions of the musculoskeletal system that function to protect the body. The flexor reflex, or withdrawal reflex, occurs when pain is sensed at some distal location on a limb, as illustrated in figure 12.7. The response of this

Figure 12.5 An example of the righting reflex. As a person trips and falls forward, this reflex causes his neck and back to extend and his arms to extend at the elbows and flex at the shoulder.

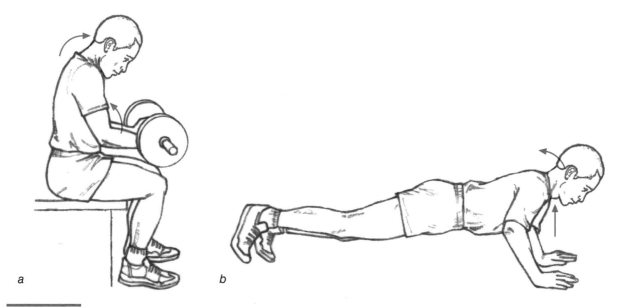

Figure 12.6 Use of tonic neck reflexes to facilitate specific actions: *(a)* Neck flexion facilitates contraction of the elbow flexors, and *(b)* neck extension facilitates contraction of the elbow extensors.

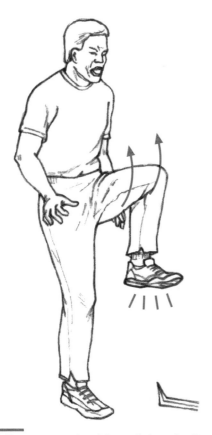

Figure 12.7 An example of the withdrawal reflex.

reflex is withdrawal or flexion of the affected limb. Placing your hand on a sharp tack or a hot surface will initiate the flexor reflex. The hand will be pulled away from the source of pain.

The crossed extensor reflex is an example of the combined effects of the flexor reflex and the extensor thrust reflex. This reflex is initiated by pain in the supporting limb. The response is flexion of that limb and contraction of the extensor muscles of the opposite limb. When you walk barefoot and step on a sharp stone, you quickly shift your weight to the opposite leg by withdrawing the foot and contracting the extensor muscles of the opposite leg so that it can support your weight.

Summary

The nervous system is structurally organized into the central nervous system and the peripheral nervous system. Nerves are bundles of nerve fibers (axons or dendrites) from individual nerve cells. A neuron is a nerve cell. There are three types of neurons: sensory or afferent neurons, motor or efferent neurons, and interneurons. Sensory neurons detect the status or change in status of the external environment, via exteroceptors, and of the internal environment, via interoceptors, and deliver this information encoded as neural impulses to the central nervous system. Motor neurons receive stimuli from the central nervous system and from sensory neurons and send neural impulses to muscle fibers. Interneurons transmit neural impulses between neurons.

The motor unit is a motor neuron and all of the muscle fibers it innervates. Motor units may have as few as 20 or as many as 1000 muscle fibers. Muscles with very few fibers per motor unit have precise control. The contraction force of a muscle may be controlled by the recruitment of motor units and the rate of stimulation of the individual motor units.

Receptors are the special sensory organs that detect the status or change in status in the body's environment. Reflexes are involuntary actions that result from sensory input. Proprioceptors detect the status or change in status of the musculoskeletal system. Muscle spindles detect increases in muscle length. The stretch reflex is initiated by a lengthening of the muscle spindle and results in facilitation of the contraction of the muscle being stretched. Golgi tendon organs detect tension in tendons. The tendon reflex is initiated by tension in the Golgi tendon organ and results in inhibition of the contraction of the muscle under tension. The vestibular system and neck joint proprioceptors detect position and changes in position of the head and also of the head relative to the neck. Righting reflexes facilitate the maintenance of an upright body and head posture. Tonic neck reflexes facilitate various upper extremity positions when the neck is flexed, extended, or rotated to either side. Pacinian corpuscles and Ruffini endings detect changes in pressure and absolute pressure. The extensor thrust reflex, flexor withdrawal reflex, and crossed extensor reflex are initiated by these receptors.

Volitional control of movement requires proprioceptive feedback to determine intensity of the neural impulses to the motor neurons. Most reflexes can be suppressed through volitional control if the reflex detracts from the desired action. Sometimes reflex responses enhance the desired action, and in these situations, it may be beneficial to elicit a more intense reflex response.

KEY TERMS

autonomic nervous
  system (p. 300)
central nervous system (p. 300)
endolymph (p. 305)
exteroceptors (p. 304)
Golgi tendon organ (p. 305)

interneurons (p. 300)
interoceptors (p. 304)
motor neurons (p. 300)
motor unit (p. 301)
muscle spindle (p. 304)
pacinian corpuscles (p. 306)

peripheral nervous system (p. 300)
proprioceptors (p. 304)
sensory neurons (p. 300)
somatic nervous system (p. 300)
tonic neck reflexes (p. 306)

REVIEW QUESTIONS

1. Is a motor neuron part of the central nervous system or part of the peripheral nervous system?

2. In which direction does an action potential travel along a motor neuron axon: toward the cell body or away from the cell body?

3. Describe the two methods employed by the central nervous system to increase the active tension in a muscle.

4. What proprioceptor detects increase in muscle length?

5. What proprioceptor detects muscle tension?

6. What proprioceptors are positioned parallel to muscle fibers?

7. What proprioceptors are positioned in series with the muscle fibers?

8. Give three examples of activities in which a preparatory movement evokes a stretch reflex to enhance the desired motion.

9. Which reflex protects joint structures?

10. Which reflex protects muscles?

11. What position of your head and neck would facilitate a biceps curl exercise?

12. What stimulus does a pacinian corpuscle detect?

13. A stretching routine that uses the actions facilitated by certain proprioceptors is called PNF (proprioceptive neuromuscular facilitation) stretching. Describe how muscle spindles and Golgi tendon organs may enhance or detract from the effectiveness of a stretching exercise.

part III

Applying Biomechanical Principles

The previous chapters of this book have presented information about how the body as a whole responds to external forces acting upon it and about how the various components of the musculoskeletal system create internal forces in response to those external forces. Several examples of applications of mechanical principles are also presented in these chapters. The purpose of part III is to introduce you to several methods that may be used to conduct qualitative biomechanical analyses of activities. Recall from the introduction that the goal of biomechanics is performance improvement and that a secondary goal is injury prevention and rehabilitation. Further recall that biomechanics can be used to improve performance through improved technique, training, or equipment. Part III describes qualitative biomechanical analysis methods directed toward these goals. Chapter 13 presents an overview of qualitative biomechanical analysis for improving technique, chapter 14 presents an overview of qualitative biomechanical analysis for improving training, chapter 15 overviews qualitative biomechanical analysis for diagnosing and preventing injury, and chapter 16 concludes the book with an overview of technology used to measure biomechanical variables. ■

chapter 13

Qualitative Biomechanical Analysis to Improve Technique

objectives

When you finish this chapter, you should be able to do the following:

- Understand the difference between a qualitative and quantitative biomechanical analysis
- List the steps involved in a qualitative biomechanical analysis for technique improvement
- Understand the process of developing a theoretical mechanically based cause-and-effect model of a skill
- Describe the important guidelines for observing a performance
- Understand how to identify and evaluate errors in technique
- Understand how to give instruction to a student or athlete that will help the person correct errors in technique

© MaxiSports/Dreamstime.com

You are watching a high jumper attempt to set a personal record. Some aspects of the jumper's technique don't seem so odd now that you have some knowledge of biomechanics. His arm action at takeoff increases his vertical velocity at the instant of takeoff. His back layout puts his center of gravity much closer to the bar, so the bar can be higher. But can you explain the basis for the other actions in the jumper's technique? This chapter may help you by outlining a structured approach to a qualitative biomechanical analysis of technique.

Analysis involves breaking something into smaller parts and then examining those parts. A qualitative analysis involves breaking something into smaller parts and then examining those parts without measuring or quantifying their characteristics. A **qualitative biomechanical analysis** of a movement or sport skill is thus breaking down the movement into its basic elements and then qualitatively examining those elements from a biomechanical perspective. Conducting a qualitative biomechanical analysis requires you to use all the material you have learned from this book, but not in a haphazard way. Applying the mechanical and anatomical knowledge presented in the previous chapters requires an organized approach if the resulting qualitative biomechanical analysis is to be successful. The approach may differ depending on the goal of the analysis. Is the goal to improve technique? To improve training? To prevent injury? To improve equipment? This chapter presents the framework for conducting a qualitative biomechanical analysis with the goal of technique improvement. Subsequent chapters focus on training improvements and injury prevention.

Types of Biomechanical Analysis

Before getting into the details of a qualitative biomechanical analysis to improve technique, we should first learn more about the difference between qualitative and quantitative biomechanical analyses. The adjectives *qualitative* and *quantitative* describe how the characteristics of the performance are observed and analyzed by the coach, teacher, or clinician. If the performance or any of its aspects is quantified or measured (described with numbers), the resulting analysis based on these measurements is a **quantitative biomechanical analysis**. If the performance or any of its aspects is evaluated using only the senses of the observer, the resulting analysis is a qualitative biomechanical analysis. This and subsequent chapters focus on qualitative biomechanical analysis methods.

Qualitative Biomechanical Analysis

Teachers and coaches often perform qualitative biomechanical analyses, but they rarely perform any quantitative biomechanical analyses. They observe their athletes and students performing and describe the mechanical characteristics of the performance subjectively. Comparative descriptors (faster, slower, higher, lower, shorter, longer, larger, smaller, and so on) may be used to denote these characteristics. The sense of sight, or visual observation, is the basis for most qualitative analyses.

How a coach or teacher observes a performance affects the subsequent qualitative biomechanical analysis. The details of how and from what vantage point to observe performances are discussed later in this chapter.

Quantitative Biomechanical Analysis

Comprehensive quantitative biomechanical analyses are usually limited to performances by elite athletes; however, teachers and coaches may make some performance measurements and thus do limited quantitative biomechanical analyses. A stopwatch and a tape measure may be used to measure and thus quantify many biomechanical parameters. Counting steps and timing how long it takes to take that many steps gives the coach a measure of step rate. Measuring a specific distance and timing how long it takes to move that distance gives a measure of speed. If assistants record where each footfall lands, step length can be measured. These types of measurement allow the coach or teacher to do limited quantitative biomechanical analyses, but taking such measurements prevents the coach or teacher from observing the whole performance.

A comprehensive quantitative biomechanical analysis requires specialized and expensive equipment for recording and measuring the biomechanical variables of interest. Biomechanists or trained technicians, rather than teachers or coaches, usually conduct comprehensive biomechanical analyses. The measuring tools used in quantitative biomechanical analyses are reviewed in chapter 16.

The cost of completing a comprehensive quantitative biomechanical analysis of a performance is high, so these types of analyses are usually done only for elite performers. The other reason comprehensive quantitative biomechanical analyses are usually limited to elite performers relates to the observer's ability to detect errors. As the level of performance increases, the magnitude of the errors in performance decreases. Errors made by novices are large and easy to detect visually using qualitative biomechanical analysis techniques. With improved performance, the errors decrease in size and become more difficult to detect; and at the elite level, a comprehensive quantitative biomechanical analysis may be necessary to detect them.

Steps of a Qualitative Biomechanical Analysis

A variety of procedures for conducting qualitative biomechanical analyses exist (Arend and Higgins 1976; Brown 1982; Hay 1984; Hay and Reid 1988; Knudson and Morrison 2002; McPherson 1988; Norman 1977). The method presented here is the one recommended by the author. It is not novel, but it includes procedures common to existing methods and provides a systematic way of biomechanically analyzing human movements. A qualitative biomechanical analysis to improve technique involves four steps:

1. Description. Develop a theoretical model of the most effective technique and describe what it would look like. Determine what you want to see when you observe your students or athletes.

2. Observation. Observe the performance of your student or athlete to determine what that person's technique actually looks like.

3. Evaluation. Compare the ideal technique to the observed performance. Identify and evaluate the errors.

4. Instruction. Educate the student or athlete by providing feedback and the instruction necessary to correct those errors.

The following sections describe and illustrate the four steps of a qualitative biomechanical analysis.

Describing the Ideal Technique

Why undertake a biomechanical analysis? As a coach, teacher, therapist, or other human movement professional, when evaluating a student's or athlete's performance, you should be able to distinguish between what is important and what is unimportant, what is correct and what is incorrect, what is possible and what is impossible, what is effective and what is ineffective, what is safe and what is unsafe, and so on. The process of making these distinctions is part of a biomechanical analysis. As a student or athlete, you probably experienced this when your teacher or coach watched you perform and then provided you with feedback to correct deficiencies in your performance. Biomechanical or movement analysis is often thought of as just that—the process of observing (or measuring) the performance of a skill, identifying faults in the performance, and providing feedback to the performer to help correct those faults. Indeed, these steps are part of a biomechanical analysis. But what you didn't see the teacher or coach do is the most important part of a biomechanical analysis. The reason you didn't see this part of the analysis is that it doesn't occur in the gym or on the playing field. It occurs before any observations and corrections are made.

When coaches or teachers come before their athletes or students, they already have an image of the ideal technique for performing a skill. This ideal becomes the standard against which observed performances are compared. Proposing and conceptually developing the ideally effective technique for a skill is the first and most important step of a biomechanical analysis. It also requires the most use of biomechanical knowledge.

Fundamental Knowledge of the Skill

Before performing any evaluation of a sport skill or other human movement activity, you should have some familiarity with the skill. This is absolutely necessary if you are going to propose the ideally effective technique for that skill. At the very least, if you are analyzing a sport skill, you should know the rules for the activity.

Rules impose constraints. If you don't know what those constraints are, you may suggest improvements in technique or equipment that are considered unfair or cheating and illegal according to the rules. Some rules are obvious. For example, in baseball, your analysis of pitching techniques indicates that the pitcher could increase the velocity of the pitch by taking several running steps toward home plate before releasing the ball. But this technique would violate the rule that the pitcher's back foot must be on the pitching rubber before the release of the ball.

Let's look at an example in which equipment is changed. Suppose your analysis of goaltending techniques in ice hockey indicates that the goalie's equipment needs improving, and you design a stick with a much longer and wider blade (figure 13.1). Such a stick would obviously make the goalie more effective, but it would put the other team's offensive players at a great disadvantage. The dimensions of a goalie's stick in ice hockey are thus specified by the rules.

Other rules of competition are not always as well known or obvious. The sport of triathlon is a race that includes swimming, bicycling, and running. An analysis of the cycling portion of this sport may indicate that a competitor could ride more effectively by drafting behind the bicycle of another athlete, thus reducing the aerodynamic drag forces. In fact, this is a common tactic in bicycle racing, but it is not allowed in a triathlon according to the rules of that sport.

Knowledge of the techniques and equipment traditionally used in a sport may also be valuable for the biomechanical analysis, but such knowledge is not absolutely essential. The sport skill techniques developed and taught by coaches and teachers are generally effective. Usually, their effectiveness has been proven through the trial-and-error method of coaching. You might use these techniques as the basis for proposing the most effective technique, but to do so, you must know what the traditional techniques are.

Where can you find information about sport techniques? Much of this information can be found on the internet or in coaching journals or magazines, textbooks, and videos. You can also learn by talking with coaches and athletes, watching successful athletes perform, or trying to perform the skill yourself. While gathering this information, be curious but skeptical. Learn to "question authority," as the bumper sticker says. Why is the skill performed that way? Should everyone swing the arms in that manner? What purpose does that type of follow-through serve? Why are the hips turned that way during the preparatory phase? Why, why, why? Realize also that completely different ways of performing certain aspects of the traditional technique may be described (or actually demonstrated by elite athletes). That is why you

Figure 13.1 A large goalie stick may be much more effective in ice hockey than a smaller one, but it gives the goalie an unfair advantage and thus the rules don't allow it.

must propose and develop the most effective technique yourself, based on your knowledge and understanding of the underlying biomechanics of the movement.

Purpose or Goal of the Skill

Now that you have some fundamental knowledge about the skill, the next step is to identify the purpose or goal of the skill and interpret that purpose or goal in mechanical terms if possible. The purpose of a skill is its desired outcome, the measure of performance or success, or the performance criterion. What is the athlete striving to achieve during performance of the skill? The purpose of some sport or movement skills is easy to define; for others, it's much more difficult. Some sport skills have only one purpose; others have more than one. The complexity of this task depends on the skill being analyzed.

Generally, it is easier to identify the purpose of those skills whose outcomes are determined objectively. The outcomes of all track and field skills are determined objectively by answers to the questions how fast, how far, or how high. Similarly, the purpose of any racing activity is to complete the course (whatever it may be) in the least time. The purposes of other skills may not be as easy to identify or define mechanically. What is the purpose of a tennis serve (see figure 13.2), a football block, a triple axel in ice skating, or a reverse one-and-a-half in a layout position in diving?

The purpose of the tennis serve may be to score an ace by putting the ball past the receiver or to gain an advantage by putting the receiver in a poor position to return the ball effectively. In either case, the speed and accuracy of the serve are two parameters that may be used to define the purpose mechanically. The purpose of the block in American football may be to prevent the defensive end from reaching the quarterback and tackling him. Mechanically, this purpose can be defined as decreasing the stability of the opponent and decreasing his velocity in the direction of the quarterback. The purpose of the triple axel in ice skating may be to impress the judges. Mechanically, the purpose is to jump as high as possible, complete three spins in the air, and land on one skate without falling. The purpose of the dive is to score as many points as possible with the judges. Mechanically, the purpose is to jump as high as possible with just enough angular momentum to complete one and a half revolutions in a layout position, then enter the water in a position perpendicular to the surface with minimal splashing.

Although identifying the purpose of these activities wasn't too difficult, defining that purpose in terms of mechanical parameters may be. Most skills have goals that can be defined mechanically, however. Think of the sports or movement activities you are interested in and try to define the goals of the specific skills used in those sports. Then try to translate those goals into mechanical

Figure 13.2 What is the goal of a tennis serve?

terms. The more precisely you can identify the purpose of a skill, the better your analysis of that skill will be.

Characteristics of the Most Effective Technique

The next step in the biomechanical analysis is to identify the characteristics of the most effective technique. This step definitely requires your biomechanical knowledge and some common sense. Two different methods may be used to accomplish this task. The first involves an assessment of traditional techniques to determine the characteristics of the most effective technique. The second involves development of a deterministic (cause-and-effect) model that establishes the characteristics of the most effective technique.

The first method of determining the characteristics of the most effective technique is based on the assumption that the traditional technique probably exhibits many of these characteristics. The task here is to determine which aspects of the traditional technique are important and which aren't. What matters?

This question can be answered in two ways. First, observe as many performances of the skill by elite performers as possible. Try to identify the actions and positions that are common to all the athletes. Then, from the start of the skill to its end, assess each action or position by asking how it contributes mechanically to achieving the purpose of the skill. Do the actions or positions contribute to the achievement of the goal, or are they detrimental? Those actions or positions that contribute to the achievement of the goal should be included as characteristics of the most effective technique. Those actions or positions that are detrimental to the achievement of the goal should definitely not be included. And those actions or positions that neither contribute to nor detract from the goal are perhaps manifestations of style and also should not be included.

The second way of answering the question "What matters?" is to assess the actions or positions common to the techniques illustrated in most of the coaching and teaching materials you have reviewed. Follow the same procedure as before by asking how each action or position mechanically contributes to the purpose of the skill.

In both of these assessments, it may be easier to break down the purpose of the skill into parts. Then, during the assessment of specific actions or positions, ask how the action or position contributes to achieving a specific part or subgoal of the purpose. For example, the purpose of high jumping is to clear a crossbar set as high as possible. The performance measure is defined mechanically as crossbar height cleared. This can be subdivided into parts. The height of the crossbar cleared (h_b) is determined by the height of the jumper's center of gravity at takeoff (h_{to}), the vertical displacement of the center of gravity from takeoff to maximum height (h_f), and the difference between the maximum height of the jumper's center of gravity and the crossbar height (Δh). Crossbar height cleared can thus be expressed as an equation:

$$h_b = h_{to} + h_f - \Delta h \qquad (13.1)$$

Various aspects of the traditional high jumping technique may then be assessed according to whether or not they contribute to any of these subgoals.

High jumpers swing their arms forward and upward during the takeoff phase of a high jump, but some jumpers "block" their arms at shoulder height before leaving the ground, and others continue to move the arms upward above shoulder height until they have left the ground. Are these differences in arm actions just manifestations of style, or is one technique better than the other? The positions of a jumper's arms affect the takeoff height, h_{to} in equation 13.1. The jumper's center of gravity should be as high as possible at the instant of takeoff. The technique of continuing to move the arms above shoulder height before takeoff moves the center of gravity slightly higher, thus contributing to achievement of a high center-of-gravity height at takeoff. The movements of the arms during the takeoff phase contribute to flight height, h_f in equation 13.1. The jumper becomes a projectile once takeoff has occurred. The vertical displacement of a projectile (flight height, h_f in this case) is determined by its vertical velocity at takeoff. The vertical velocity of the jumper's center of gravity at the instant of takeoff is determined by the vertical velocities of the jumper's body parts. The arms have a vertical velocity at takeoff in the technique of continuing to move the arms above shoulder height before takeoff. They contribute to the vertical velocity of the jumper's center of gravity, thus contributing to achieving a high flight height. In the other technique, the arms stop contributing to the vertical velocity of the center of gravity because they stop their upward swing before takeoff. This aspect of technique is detrimental to high jumping performance.

The second method of determining the characteristics of the most effective technique is to develop the mechanically based foundation for achievement of the skill's purpose. Through this process the characteristics of effective techniques are revealed. This method uses deterministic, cause-and-effect relationships based on mechanical principles. The mechanical purpose of the goal or skill is broken down into subgoals, as in the high jump example. These subgoals are then further reduced by identifying the mechanical bases for each of them. The mechanical bases of the subgoals define the characteristics of the most effective technique.

Let's look at the high jump again. We've defined the mechanical purpose of the skill as maximizing the height of the crossbar cleared, and we've defined this height as h_b. This mechanical goal was then broken down into three subgoals: maximize the height of the jumper's center of gravity at takeoff (h_{to}), maximize the vertical displacement of the center of gravity from takeoff to maximum height (h_f), and minimize the difference between the maximum height of the jumper's center of gravity and the crossbar height (Δh). To define the characteristics of the most effective technique, we must investigate each of these subgoals.

The jumper's center-of-gravity height at the instant of takeoff (h_{to}) is affected by the position of the jumper at the instant of takeoff. The rules of high jumping indicate that the jump must be made from one foot, so the problem is to determine what body position maximizes the center-of-gravity height during standing on one foot. Standing up

is obviously better than lying down. An upright standing position with full extension of the takeoff leg is better than one in which the jumper is leaning in any direction. Raising both arms as high as possible raises the center of gravity. Raising the free leg as high as possible also raises the center of gravity. These are thus characteristics of an effective technique.

The center of gravity is higher at takeoff if the jumper is tall, which is why most elite high jumpers are tall. The first man to high jump 8 ft (2.4 m) was Javier Sotomayor, who is 6 ft 4.75 in. (1.96 m) tall. The world record holder in the women's high jump is Stefka Kostadinova, who is 5 ft 11 in. (1.80 m) tall. Both high jumpers are well above average height for their sex.

The second subgoal in the high jump is to maximize the vertical displacement of the center of gravity from takeoff to maximum height (h_f). The jumper becomes a projectile as soon as the takeoff foot leaves the ground, so the vertical displacement of the jumper's center of gravity after takeoff is determined by the vertical velocity of the jumper's center of gravity at the instant of takeoff. To make h_f as large as possible, the jumper needs to make the vertical velocity of the center of gravity at the instant of takeoff (v_{to}) as high as possible. What causes a high velocity? Recall from chapter 3 and equation 3.30 that impulse (average net force times the duration of force application) causes a change in momentum (mass times velocity):

$$\Sigma \bar{F} \Delta t = m(v_f - v_i)$$

This equation can be rewritten as equation 13.2, which describes the conditions in the high jump:

$$(\bar{R} - W)\Delta t = m(v_{to} - v_{td}) \tag{13.2}$$

where

$\Sigma \bar{F} = \bar{R} - W$ = the average net vertical force acting on the jumper,

$\bar{R}$ = the average vertical reaction force exerted by the ground on the takeoff foot,

W = the weight of the jumper,

Δt = the duration of force application or the duration of takeoff foot contact with the ground,

m = the mass of the jumper,

$v_f = v_{to}$ = the final velocity or vertical velocity of the center of gravity at the instant of takeoff, and

$v_i = v_{td}$ = the initial velocity or vertical velocity of the center of gravity at the instant of touchdown at the start of the takeoff phase.

If equation 13.2 is rewritten and solved for v_{to}, the parameter we want to maximize, it appears as

$$v_{to} = \frac{(\bar{R} - W)\Delta t}{m} + v_{td} \qquad (13.3)$$

To maximize vertical takeoff velocity, the jumper should maximize the average vertical reaction force acting on the takeoff foot $(\bar{R})$, the duration of takeoff foot contact (Δt), and the vertical velocity at touchdown (v_{td}). The jumper's mass (m) or weight (W) should also be minimized (i.e., the jumper should be fairly lean). Javier Sotomayor weighs 180 lb (his mass is 82 kg). Stefka Kostadinova weighs 132 lb (her mass is 60 kg).

To maximize the average vertical reaction force acting on the jumper's takeoff foot, the jumper's run-up velocity should be fast. An increase in run-up velocity will increase the size of the reaction force acting beneath the takeoff foot at the end of the run-up. The jumper should also strive to accelerate the arms and free leg upward during the takeoff phase. These actions will cause the takeoff leg to push down on the ground with greater force.

To maximize the duration of takeoff foot contact (and force production) during the takeoff phase, the jumper's center of gravity should be low at the start of the phase when the takeoff foot touches down and high at the end of the phase just before the takeoff foot leaves the ground. The takeoff phase is prolonged by the swinging of body parts through a large range of motion vertically. Thus, when the takeoff foot first touches down, the arms should be extended and in a low position; the free leg should also be in a low position, and the jumper should be leaning backward and away from the bar. These actions will put the jumper's center of gravity in a low position.

Finally, to maximize the vertical velocity of the jumper's center of gravity at touchdown, the last step should be taken quickly. Actually, the jumper's center of gravity is unlikely to have an upward vertical velocity when the takeoff foot hits the ground. It will probably have a downward vertical velocity (or negative velocity, because we have been describing upward as positive). Thus, the task is to minimize the downward vertical velocity of the center of gravity at the instant the takeoff foot touches down.

The third subgoal in an effective high jump technique is to minimize the difference between the maximum height of the jumper's center of gravity and the crossbar height (Δh), or the bar clearance. To achieve this goal, the jumper's center of gravity needs to be as close to the crossbar as possible, or even below it, when the center of gravity reaches maximum height. But the jumper doesn't want to touch the crossbar or dislodge it. If the jumper goes over the bar feetfirst in an upright position, the difference (Δh) is large. If the jumper goes over the bar in a lying-down position, Δh is much smaller. If the jumper arches or drapes the body over the bar, Δh is even smaller.

Longer-limbed individuals may be at an advantage here because they can drape their limbs farther below the crossbar while their trunk is above it, making Δh smaller or even negative in value.

From the previous mechanical assessment of the purpose of high jumping, the following characteristics of an effective technique are proposed:

1. Approach the bar with a fast run-up speed.
2. Make the last step of the run-up quick by putting the takeoff foot down early.
3. Lower the center of gravity during the last step, and especially at the start of the takeoff phase, by positioning the arms and free leg low and leaning back and away from the bar.
4. Use the takeoff leg to push down against the ground with as much force as possible by accelerating the arms and free leg upward for as long as possible.
5. Raise the center of gravity as high as possible at the instant of takeoff by achieving an upright trunk position, fully extending the takeoff leg, and reaching high upward with both arms and the free leg.
6. Clear the crossbar with the body in a draped position to minimize the distance between the crossbar and the jumper's center of gravity.

Your model of the most effective technique for high jumping would include these characteristics. You are now ready for the next step of the biomechanical analysis.

Observing the Performance

The next step of a qualitative biomechanical analysis to improve technique is to observe the actual performance. Visual observation is the primary basis for most qualitative biomechanical analyses, but simply watching a performance is not enough. Your observation must be planned in advance, with consideration of the following: Who will you observe? What conditions will the subject perform under? Where will the subject perform? Where will you observe the performance? What will you look for?

Who?

Who you observe, especially when considering skill level, may affect the manner in which you observe and conduct your analysis. As skill level increases, the size of observable errors in technique decreases, as does the variability in performance. Novice performers of a skill exhibit errors that are easily recognized but that may not be repeated from one performance to the next. Their performances may be so variable that a detailed analysis would be painstaking and futile. Instead, your observations of these students or athletes should be directed at identifying their gross errors in each performance.

More care must be taken in observing and analyzing experienced, highly skilled athletes. Their errors will be minor, so as the observer, you must pay more attention to the details of their technique. In addition, their performances are probably less variable, so your observation of these athletes should be directed at identifying errors that are repeated from performance to performance.

What Conditions?

Ideally, the environment in which you watch your students perform should be carefully controlled, if possible. Try to duplicate as closely as possible the environment in which the performances normally occur. An actual competition may be the best environment for an athlete because it is one in which the athlete normally performs; however, it may not be best for the observer. The best locations for viewing the performances may not be available to you in a competition, or your view of the performances may be obstructed by officials or other athletes.

A class or team practice session is not ideal either. The activities of other students or athletes may distract you (especially if you are responsible for their supervision) and the performer, thus affecting both your ability to observe and the athlete's ability to perform. It may be difficult to simulate the normal performance environment in a class or practice session as well.

Creating the ideal conditions for observing a performance is difficult, if not impossible, so you'll have to compromise. In any case, do your best to simulate the normal performance environment and minimize distractions. For novices or beginners, a class or practice session may be the normal performance environment.

Where to Observe?

Your vantage point when observing the performance is important because it affects what parts of the performance will be visible to you. First, determine if the activity has a principal plane of movement. Do most of the movements occur in the same plane? If so, your line of sight should be perpendicular to this plane. Running, long jumping, gymnastics vaulting, gymnastics high bar routines, basketball free throw shooting, and backstroke swimming are all examples of activities whose movements occur primarily in the sagittal plane. To observe these movements, position yourself to the side so that you view the sagittal plane. You can get a sagittal view from either the left or right side, so choose the position that provides the least obstructed view of the subject or body parts of interest. You must also decide how close you should be to view the action. A more distant view is best for evaluating the quality of the performance as a whole. A closer position is desirable if you are evaluating only certain aspects of the performance. In any case, you should be far enough away that you can see the movements of interest in your field of view without having to move your eyes too quickly to track them.

Even if the activity occurs primarily in one plane, several viewing positions may be necessary if the performance involves multiple skills. But because you are only one person, multiple viewing positions require multiple trials or performances by the student or athlete (unless you have several assistants equipped with video cameras). For example, the three separate parts of a long jump performance—the approach run, takeoff, and landing—may require three separate vantage points, as shown in figure 13.3.

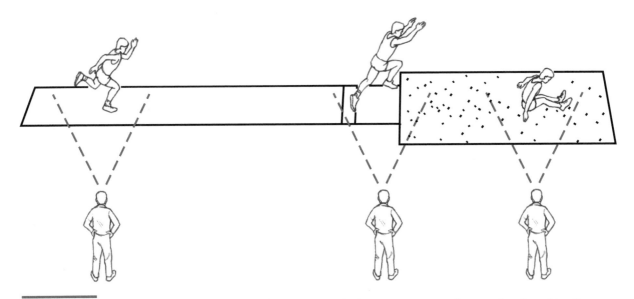

Figure 13.3 Three different viewing positions for observing the long jump approach run, takeoff, and landing.

Most activities have multiple planes of motion and usually require multiple viewing positions. Your initial position for observing these activities is determined by the part of the skill that is most important, which should have been identified by your model. Determine the primary plane of movement for this part of the skill. Your viewing direction should be perpendicular to this plane. Subsequent observations may then be from various vantage points that give better views of the other important movements in the skill. Two viewing positions for observing a high jumper are shown in figure 13.4. In this example, the primary viewing position gives a good view of the approach and takeoff, whereas the secondary viewing position gives a good view of the bar clearance.

One last note about viewing position. If possible, videotape the performance from each of the viewing positions. Preferably, have an assistant (another teacher or coach, or another student or athlete) videotape the performance. With video recordings of performances, you can view the same trials repeatedly without having fatigue affect your subjects (or your evaluations of them). In addition, the video record is useful when you are giving students or athletes feedback about their performance.

What to Look For?

Now you've set up the ideal conditions for the performance, you're in the best position to view the performance, and you've got an assistant to videotape the performance. How do you observe the performance?

What do you look for? Generally, you will have to view several performances or several replays of the videotaped performance. During the first trial (or video replay), get a general view of how the student or athlete performs. (If the subject you are observing is a novice or beginner, this may be all it takes to identify technique errors.) During subsequent trials (or video replays), the movements or actions you have previously identified as characteristics of the most effective technique should guide your observation. The following are some of the distinguishing features of these actions:

1. The position of the body or body segments at specific instants (usually at the start and end of some force-producing phase). For high jumping, is the jumper leaning back and in a low position at the start of the takeoff phase (when the takeoff foot first touches the ground), and is the jumper upright with the arms and free leg as high as possible at the end of the takeoff phase (when the takeoff foot leaves the ground)?

2. The duration and range of motion of the body and its segments during specific phases of the skill, especially the force-producing phases. For high jumping, are the duration of the takeoff phase and vertical range of motion of the arms and free leg during this phase as long and as large as possible?

3. The velocity and acceleration (along with their directions) of body segments during specific

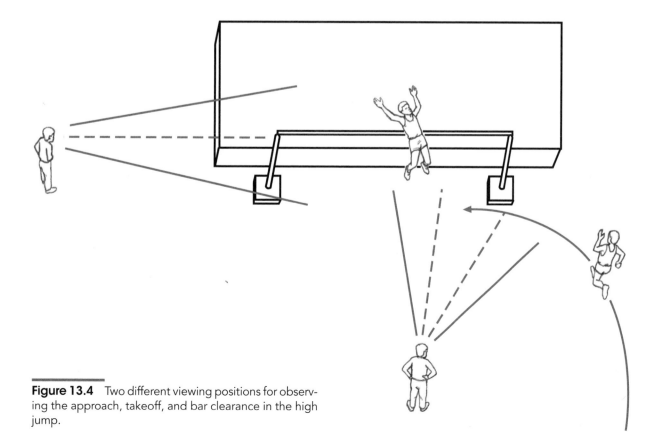

Figure 13.4 Two different viewing positions for observing the approach, takeoff, and bar clearance in the high jump.

phases of the skill. For the high jump, are the vertical velocities of the arms and free leg as fast as possible during the takeoff phase, especially at the end of the phase?

4. The timing of body segment motions relative to each other. For the high jump, do the arms and free leg swing upward together during the takeoff phase?

These are the details you should examine and compare to your model of the most effective technique.

The position of the body and segments at specific instants indicates the direction of force application as well as the range of motion through the phase. Examples of instants when positions should be examined include the instant of release in throwing activities (American football passing, baseball pitching, shot putting, javelin throwing, basketball free throw shooting, and so on); the instant of takeoff in jumping activities (running, hurdling, high jumping, long jumping, volleyball spiking, diving, figure skating, basketball shot blocking, and so on); and the instant of contact in striking activities (kicking in football or soccer, serving or hitting in all racket sports, punching in boxing, spiking in volleyball, shooting slap shots in hockey, and so on). These are examples of instants that mark the end of a movement phase. The instants that mark the beginning of these movement phases should also be examined, such as the start of initial forward movement of the ball or implement in throwing and striking activities and the instant of takeoff foot contact in jumping activities.

The duration and range of body motions during the force-producing phases should be examined, including propulsion (speeding up) and follow-through (slowing down) phases. These phases of motion can also be described as phases of positive and negative work. What is the duration of the propulsion phase during a throwing, striking, or jumping activity, and what is the corresponding displacement of the ball or implement or body during this phase? What is the duration of the follow-through phase during a throwing or striking activity, and what is the corresponding range of motion of the throwing or striking limb during this phase?

The velocities and accelerations of specific body segments during specific phases of a movement should also be examined. Often the velocity of a limb's distal end determines the outcome of the performance (especially in throwing and striking activities). Rather than just assessing the velocity of the limb qualitatively, it is more useful to assess the acceleration of the limb. Once the forward motion is initiated, does the throwing or striking limb or implement continuously accelerate throughout the movement, or are there pauses or periods in which the limb or implement ceases to speed up?

Finally, the timing or coordination of the body segment movements must be examined. In many throwing and striking activities, the movement is initiated by the larger limbs (the legs) and proceeds in sequence to the smaller throwing or striking limb. This sequence of limb movements from largest to smallest increases the work done by increasing the displacement of the throwing or striking limb, as well as the magnitude of the average force applied to the limb.

Use Other Senses

Although visual observation is the basis for most qualitative analyses, other senses may be employed as well. Tasting or smelling a performance wouldn't reveal much about its mechanics (if a performance could be tasted or smelled), but listening might. Many activities have specific rhythms that indicate something about the timing of the movements involved. Touch is another sense that may be used in a qualitative analysis. How do you touch a performance? A coach or teacher spotting an athlete in gymnastics may touch the athlete during the routine. The amount of force the coach or teacher must exert on the athlete to correct the stunt can indicate that errors have occurred in the stunt. Finally, you may rely on an athlete's kinesthesia. Ask the athlete what a performance felt like. Did she pull hard? Did she slip at takeoff? Was the landing hard or soft?

Evaluating the Performance

Once you have described the ideal technique and observed the actual technique, the next step in the biomechanical analysis process is to evaluate the observed technique. This evaluation has two steps. First, errors or deficiencies in the actual performance are identified. These errors are then diagnosed to determine the magnitude of their effect on the performance.

Identify Errors

Errors or deficiencies in the performance should be identified while it is being observed (or immediately after). Do this by comparing the actual performance to the characteristics of the most effective technique. For each element of the performance you examine, ask yourself, How does this movement, position, or timing of movements differ from the most effective technique?

Once a difference has been identified, you must decide whether the difference adversely affects the performance. Differences may indicate style variations in the technique, or they may be actual errors. You should carefully examine discrepancies between the observed performance and the most effective technique by questioning how the action or position being evaluated contributes to (or detracts from) the desired goal of the performance.

Sometimes the differences between the observed and the proposed techniques result from modifications the athlete has made to accommodate morphological constraints.

The morphological constraints of the performer should be identified. In other words, how does the performer's anthropometry and strength constrain, limit, or modify his ability to match the proposed technique? Some constraints cannot be changed and are thus true constraints on the performance of the student or athlete. Other constraints can be modified through training. In the high jump example, height was identified as an advantage. Taller jumpers can have higher centers of gravity at takeoff. The height to which a jumper can raise her center of gravity at the instant of takeoff is constrained by her height. The height of the jumper cannot be changed once growth has ceased after puberty. Body weight was identified as another anthropometric characteristic that affects high jumpers. A lean jumper may have an advantage over a jumper with a higher percentage of body fat. This morphological constraint can be changed with training. Strength of the takeoff leg affects the high jump performance as well. A stronger jumper may be able to run faster and get into a lower position at the start of the takeoff phase, whereas a weaker jumper has to sacrifice some speed or not get into such a low position at takeoff. Strength is another morphological constraint that can be modified with proper training. Physical training is discussed further in chapter 14.

Evaluate Errors

Once the errors or deficiencies in the performance have been identified, they should be evaluated to determine the focus of correction efforts. During this evaluation, consider the causes of the errors as well as their effects. There are several concerns here:

1. Does the error expose the performer to the danger of injury? Such errors should be corrected immediately. For sports involving high risk, ensure that risks are minimized, especially with beginners, by using trained spotters and all appropriate safety equipment such as safety harnesses, crash pads, and helmets.

2. Who are your clients? Are they students learning a new skill, or are they athletes refining a skill they have practiced for years? Are they 5-year-olds or 25-year-olds? For students learning a new skill, focus your attention on errors in the basic elements of the ideal technique.

3. How easy is it to correct the error? Is the error caused by a strength deficiency that may require training to correct, or is it due to an incorrect starting position? Will it take months, weeks, days, or only one practice session to correct? How much time do you have to correct the error before the next competition or before the most important competition?

4. Is the error a result of another error that occurred earlier in the performance? If so, the earlier error should be the focus of correction efforts.

5. How great an effect does the error have on the performance? Would correcting it dramatically improve performance, or would the change be hardly noticeable?

6. Is the error or deficiency due to poorly designed or inappropriate equipment? Can it be alleviated with improved or redesigned equipment?

After evaluating each error in light of these concerns, you must decide what errors to correct and in what order. In all cases, errors that endanger the athlete should be corrected first. Then the errors should be ranked in three categories: (1) from major to minor in terms of their effect on the performance; (2) from earliest to latest in terms of their chronology during the performance; and (3) from easiest to most difficult in terms of the time and effort it will take to correct them. Errors ranked near the top of each category should be attended to first. Those ranked near the bottom should be ignored or attended to last. Specific circumstances may dictate correction of errors with discrepant rankings from category to category. Errors observed in a baseball pitcher's delivery two days before he starts the championship game should not be corrected until after this game and the completion of the season, except for small errors that are easily corrected. On the other hand, the same errors observed during preseason practices should all be corrected, because the corrections are likely to be made during the season and before the most important games at the end of the season.

Instructing the Performer

The final step in a qualitative biomechanical analysis is to instruct the performer by correcting the errors or deficiencies identified and ranked in the previous step. This instruction is what most people think of as teaching or coaching. Because the fundamentals of effective instruction come from the fields of motor learning and sport psychology rather than biomechanics, only a brief overview of this step is presented here.

To correct errors in performance, you must do three things. First, you must clearly communicate to the athlete or student what he did incorrectly (what his error was). Second, you must clearly communicate to the athlete or student what you want him to do (what the ideal technique looks like). Finally, you must devise means for the performer to correct the errors.

Communicate With the Performer

A large part of effective instruction involves effective communication. You must effectively communicate to your athlete or student what her performance looked like (what error in technique you want to correct) and what the ideal technique looks and feels like. You can accomplish this by describing verbally what the athlete or student did and what you want her to do. Or you (or someone else) can demonstrate what the student or athlete did and what you want her to do. Alternatively, you can show the performer photos or a videotape of the performance along with photos or a videotape of an athlete or student performing the skill in the desired way. The most effective method involves all three modes: verbal description, demonstration, and photographic or videotape records.

Keep your descriptions and instructions simple, and work on correcting only one error at a time. Keep the student or athlete focused on that one element of the performance during each exercise or performance.

Your attitude during instruction may affect its effectiveness. Be positive. Praise the athlete or student for the positive aspects of the performance. Realize that during the correction process, some decline in performance may result initially, especially with highly skilled performers. Improvements may not occur immediately. This is frustrating for the athlete, so you must be positive and patient as the coach or teacher.

Correct the Error

Once the performer understands what error he made in his performance and what the performance should look and feel like without the error, you must devise methods to assist the student or athlete in correcting the error. Certain guidelines may be helpful here.

The correction process involves development of a full teaching progression for the skill along with the drills involved in this progression. First, break the skill into parts. Devise drills that duplicate the movements and forces present in each part of the skill. Then use the drills to correct errors in discrete parts of the skill. Have the athlete practice the drills specific to the error slowly at first, then faster, until the speed is the same as that of the actual performance. Extend the drills to link the next part (or the previous part) of the skill to the movements. When the error is consistently absent from the linked drills, the athlete should perform the entire skill (at a slower speed at first, if possible). Throughout the correction process, the speed of the movements should be manipulated. Begin with drills that emphasize positions. Then follow with drills in which initial and final positions are emphasized. Slow movement from one position to the next emphasizes the sequence of movements. Speed up the movement as the student or athlete becomes more proficient.

Repeat the Analysis

Once the instruction phase is completed and the errors are corrected, it may be worthwhile to repeat the entire qualitative biomechanical analysis procedure. This would include revising the description of the most effective technique based on any morphological constraints you have learned about your student or athlete. The observation, evaluation, and instruction phases would also be repeated. The process is thus a loop that eventually results in improved performances.

Sample Analyses

The previous pages described the process for conducting a qualitative biomechanical analysis to improve technique, and high jumping was used as an example to illustrate some of the procedures. In the following pages, models of the most effective techniques for other sport skills are developed for use in qualitative biomechanical analyses. The examples include a throwing activity (a fastball pitch), a striking activity (a tennis forehand drive), and a locomotor activity (sprint running).

Fastball Pitch in Baseball

The pitcher is the most important athlete on a baseball team. An effective pitcher can single-handedly shut down the offense of an opposing team. A fastball pitch is one of the basic pitches in a pitcher's repertoire. What are the best techniques to use for pitching a fastball? Let's do the first step of a qualitative biomechanical analysis to find out.

Theoretical Model of the Ideal Fastball Pitch

Recall that the first step in a qualitative biomechanical analysis is to describe the ideal technique by developing a theoretical model of the technique based on mechanics and cause-and-effect relationships. The first step in developing a model of a fastball pitch is to acquire the necessary fundamental knowledge of baseball pitching. Let's assume we've done this already and get right to developing the model.

The basis of the theoretical model is the purpose or goal of the skill. What is the purpose of a fastball pitch? What is the pitcher trying to achieve with this pitch? The pitcher is trying to pitch a strike, either by having the batter swing at the ball and miss or by putting the ball in the strike zone without the batter's swinging at it.

Now let's identify the factors that determine this goal. In the case of a fastball, the difficulty in hitting a pitch depends on how much time the batter has to react to the pitch and where it is placed in the strike zone. So a strike is more likely to occur if the time is short and the path of the ball places it in the strike zone. These are the two mechanical factors that affect the ultimate goal of pitching a strike with a fastball.

The time it takes for the ball to reach the plate is affected by the horizontal velocity of the ball at the instant it leaves the pitcher's hand, the horizontal distance to the plate from this point of release, and the horizontal component of the forces exerted on the ball by air resistance during its flight (primarily drag forces). A pitch with a faster horizontal velocity, a shorter horizontal distance to travel, and smaller drag forces acting on it will take less time to travel from the pitcher's hand to home plate.

Horizontal velocity of the ball at release is the most important factor in determining the time it takes a fastball pitch to reach the batter. Most of the characteristics of a pitcher's technique are directed at maximizing this factor. At the start of a pitch, the ball is stationary in the pitcher's hand or glove; at the end of the pitching motion, when the pitcher releases it, the ball is moving very fast. During the pitching motion, the ball's kinetic energy increases. Remember from chapter 4 that kinetic energy is energy due to the motion of an object; or, quantitatively, kinetic energy is one-half the object's mass times its velocity squared. This increase in kinetic energy is a result of the work done on the ball by the pitcher. Because work is a product of force and displacement, the horizontal velocity of the ball at release is determined by the average horizontal force exerted on the ball by the pitcher during the pitching motion and the horizontal displacement of the ball while this force is exerted on it. A faster pitch results if the average horizontal force acting on the ball and the horizontal displacement of the ball are greater during the pitching motion. The horizontal displacement of the ball may be greater if the pitcher is taller and has longer limbs.

> ⟳ Horizontal velocity of the ball at release is the most important factor in determining the time it takes a fastball pitch to reach the batter.

The other two factors that affect the time it takes for the ball to reach the plate are the horizontal distance to the plate from the point of release and the horizontal component of the forces exerted on the ball by air resistance during its flight. The horizontal distance from the point of release to home plate is limited by the distance from the pitching rubber to home plate. This distance is 60 ft 6 in. (18.4 m). The pitcher's foot must remain in contact with the pitching rubber until the ball is released, but the horizontal distance from the point of release to home plate can be less than 60 ft 6 in. (18.4 m), depending on the pitcher's anthropometry and position at release. The flight time of the ball will be shorter if this distance is shorter.

Besides the time it takes for the ball to reach the batter, the path of the ball also affects whether or not the pitch is a strike. The path of the ball after release is determined by its height at release, its horizontal and vertical velocities at release, and air resistance. The height at release is determined by the pitcher's position and anthropometry. Air resistance is affected by the air density, the relative velocity of the ball with respect to the air, the roughness of the ball's surface, and the spin on the ball (the ball's angular velocity at release). The spin on the ball at release is determined by the torque exerted on the ball by the pitcher's fingers during the pitching action and the time during which this torque acts. The factors affecting horizontal velocity at release were discussed previously. Vertical velocity at release is affected by similar factors—the average vertical force exerted on the ball and the vertical displacement of the ball during the pitching action. The theoretical model of the factors that mechanically affect the outcome of a fastball pitch is shown in figure 13.5.

Application of the Theoretical Model

In developing the ideal model of the most effective fastball pitching technique, certain mechanical factors were identified as desirable. Now let's examine those factors and determine what actions by the pitcher produce these desirable features. From our mechanical assessment of the purpose of a fastball pitch, the following factors were identified as important:

1. Maximize the horizontal velocity of the ball (and pitching hand) at the instant of ball release. The way to achieve this is to maximize the horizontal force exerted on the ball while maximizing the horizontal displacement of the ball during the pitching action. The drag force due to air resistance also affects this slightly.

2. Minimize the horizontal distance the ball has to travel to get from the point of release to the plate. The pitcher achieves this by releasing the ball as far forward as possible. The pitcher's height and limb lengths also affect this factor.

3. Release the ball such that its path after release intersects the strike zone. The horizontal and vertical velocities of the ball at release affect this factor, as do air resistance and the height of release.

Maximum horizontal velocity of the ball is the most important factor in determining the success of a fastball pitch. This velocity is maximized through maximization of the horizontal force exerted on the ball and the horizontal displacement of the ball during the pitching motion. The pitcher can lengthen the horizontal displacement of the ball during the pitching motion by turning

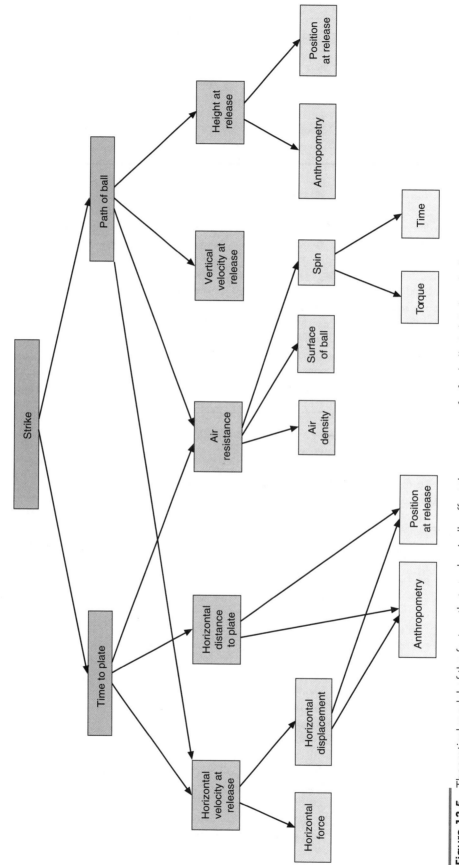

Figure 13.5 Theoretical model of the factors that mechanically affect the outcome of a fastball pitch in baseball.

324

the body (rotating the pelvis and trunk to the right for a right-handed pitcher) so that the pitching arm is away from the batter, then reaching the pitching hand as far back as possible. These actions constitute the pitcher's windup, which places the ball in a starting position from which it is then moved forward (see figure 13.6). A taller, longer-limbed pitcher can put the ball in a starting position farther from the plate. The release position of the pitch should be as far forward of this starting position as possible. Pitchers accomplish this by stepping forward onto the left foot, rotating the pelvis and trunk to the left, flexing the left hip, and extending the throwing arm forward. Again, a taller, longer-limbed pitcher has the advantage of being able to put the ball in a release position closer to the plate. The release position is constrained by the rules, however, which require that the pitcher keep one foot on the pitching rubber before releasing the ball. If the pitch is to be a strike, the direction of the ball's velocity at release is restricted, and this restriction also places constraints on the release position.

The finger positions on the ball also affect the horizontal displacement of the ball during the pitching action. Contact can be maintained slightly longer if the ball is held by the fingers rather than in the palm of the hand.

Further, the middle finger (the longest finger) should lie across the top of the ball, aligned with the direction of the throw (see figure 13.7). Slower pitches result if the ball is held deeper in the hand because the ball is released sooner, resulting in less horizontal displacement of the ball during the pitching action.

The average horizontal force exerted against the baseball during the pitching motion is maximized through use of a kinetic chain with movement proceeding in sequence from proximal segments (those closest to contact with the ground) to distal segments. The forces are generated and transmitted from proximal to distal and from larger to smaller segments as well. A kinetic chain is similar to a whip. The muscles of the lower extremity and trunk generate large forces initially as the pitcher steps forward. These forces are transmitted to the upper extremity and ultimately to the baseball through the bones, ligaments, and stretched muscles and tendons of the throwing arm. The pelvis rotates forward and the hip flexes as the left foot touches down. This stretches the trunk muscles, which then contract more forcefully due to the stretch reflex and cause the trunk to rotate. The muscles crossing the shoulder joint (especially the internal rotators and extensors) are then stretched maximally and contract forcefully to produce the rapid internal rotation and extension of the throwing arm that occur just before release.

Minimizing the horizontal distance from the point of release to the plate also reduces the time it takes for the fastball pitch to reach the plate. How small this distance is depends on the pitcher's position at the instant of ball release and the pitcher's anthropometry. If the pitcher steps forward onto his opposite foot while moving his pitching hand forward, this distance can be reduced by several feet. A taller, longer-limbed pitcher is able to stretch farther forward from the pitching rubber while still keeping his foot on it, thus minimizing this distance. The position of the ball at release affects both the horizontal distance from the point of release to the plate and the

Figure 13.6 The windup or starting position for the forward movement of the ball during a fastball pitch.

Figure 13.7 Finger positions for pitching a fastball.

Figure 13.8 The release position for a fastball pitch.

horizontal displacement of the ball during the pitching action. A release position closer to the plate is beneficial to each of these factors. A good release position is shown in figure 13.8.

The path the ball follows from its release by the pitcher until it crosses the plate is the other major factor affecting the success of a fastball pitch. This path depends on the direction and speed of the ball at release (its horizontal and vertical velocities), the height of release, air resistance, and the spin of the ball (its angular velocity) at release. Air resistance and horizontal velocity also affect the time it takes for the ball to reach the plate. The pitcher can minimize the effect of air resistance by holding the ball so that only two seams lie in the spin plane (see figure 13.7). This also minimizes deviations from the parabolic path that the ball would follow if there were no air resistance. The backspin imparted on the ball causes it to be slightly above this parabolic path due to the Magnus effect. Remember from chapter 8 that the Magnus effect is a ball's deviation from its normal flight path due to the effect of air resistance and the spin of the ball. Pitchers can increase this deviation by holding the ball so that four seams lie in the spin plane. A fastball pitch thrown with such a grip is called a rising fastball.

The direction of the ball at release depends on the pitcher's position at release. If the pitcher is slightly ahead of or behind the normal release position, the direction of the ball will be affected. The stride taken during the pitching action is forward and down the front of the pitching mound. This stride flattens the path followed by the ball and pitching hand before release. This flattening of the ball's path minimizes errors caused by early or late releases.

This analysis has identified some of the fundamental characteristics of the most effective technique for throwing a fastball. Not all aspects of technique were discussed, however. The follow-through (the movements of the pitcher that occur following release of the ball) was not discussed because it does not directly affect the performance criterion. This does not mean that the follow-through is unimportant. The purpose of the follow-through is to slow down the limbs safely and prepare the pitcher for fielding a hit ball. Other aspects of technique may vary and do not necessarily concern the mechanics of the pitch. For instance, it is advantageous for the pitcher to hide the ball for as long as possible to give the batter less time to react to the pitch. Some actions of the pitcher's glove hand are directed at this task. The presence of base runners may also alter the pitcher's technique.

Forehand Drive in Tennis

The forehand drive in tennis is an example of a striking activity in sport. It is also an example of an open skill—a skill performed under conditions that vary and are unknown from one performance to the next. In a forehand drive, some of the conditions that vary are your location on the court, your opponent's location on the court, and the incoming direction and velocity of the ball. Your return may be offensive or defensive, depending on the situation. For this example, let's assume that your position on the court and the incoming velocity and direction of the ball are such that your return is offensive. What are the characteristics of an effective offensive forehand drive? Let's develop a theoretical model of the most effective technique to find out.

Theoretical Model of the Ideal Tennis Forehand Drive

Again, the first step in a qualitative biomechanical analysis is to describe the ideal technique by developing a theoretical model of the technique based on mechanics and cause-and-effect relationships. The first step in developing a model of a forehand drive in tennis is to acquire the necessary fundamental knowledge of tennis. Let's assume we've done this work already and get right to development of the model.

The basis of the theoretical model is the purpose or goal of the skill. What is the purpose of a forehand drive?

the application of our model are the interactions among the factors identified. The incoming velocity of the ball, the angle of the racket face, the velocity and direction of the racket, and the racket characteristics all affect the horizontal and vertical velocities of the ball as well as the spin of the ball. These factors, in turn, affect the racket-to-court time and the path of the ball.

The open nature of the forehand drive and the interactions between the determining factors make it difficult to identify the factors that are controlled exclusively by the player and that are obviously important in determining the success of the skill. One such factor is the horizontal velocity of the racket head at impact. To impart a fast horizontal velocity to the ball after impact, the horizontal velocity of the racket head must be fast at impact. Keep in mind, however, that the horizontal velocity of the ball after impact must be such that it allows the ball to clear the net and land in the court. Players achieve a fast horizontal velocity by exerting a large horizontal force on the racket while maximizing the horizontal displacement of the racket head during stroke execution.

Players can lengthen the horizontal displacement of the racket head during the stroke by turning the body (rotating the pelvis and trunk to the right for a right-handed player) and extending the racket as far back from the oncoming ball as possible. These actions constitute the player's backswing, which places the racket in a starting position from which it is then moved forward (see figure 13.10a). A taller, longer-limbed player can start the racket farther back. The racket's impact with the

ball should be as far forward of the starting position as possible (see figure 13.10b). Again, taller, longer-limbed players have the advantage of being able to hit the ball in a position farther forward. The total horizontal displacement of the racket during the stroke is constrained by how much time the player has to get in position to execute the stroke. If the time is short, preparation for the stroke is limited, and the horizontal displacement of the racket head may not be as long.

The grip position along the racket and the length of the racket also affect the horizontal displacement of the racket head during the stroke. With a longer racket, the racket head can be moved through a longer horizontal displacement during the stroke. The rules limit racket length, but the grip position can be used to maximize the horizontal displacement of the racket within the limits imposed by its length. Gripping farther from the racket head and closer to the heel of the handle (see figure 13.11b) lengthens the horizontal displacement of the racket head by increasing the effective radius of the stroke. Gripping farther up the handle and closer to the racket head (see figure 13.11a) shortens this displacement. There is a tradeoff here, however. The axis of the swing moves farther from the racket head as the grip moves down the handle, increasing the racket's moment of inertia and making it harder to control.

The average horizontal force exerted against the racket during the forehand stroke is maximized through use of a kinetic chain, with movement proceeding in sequence from proximal to distal segments (meaning from the

Figure 13.10 The starting (a) and ending (b) positions of a forehand drive in tennis.

Figure 13.11 Grip positions on a tennis racket affect the displacement of the racket head during a stroke.

ground up), as in the baseball pitch. The forces are generated and transmitted from proximal to distal and from larger to smaller segments as well. For example, a right-handed player initially pivots and shifts the center of gravity toward the rear and over the right (back) foot. The muscles of the lower extremity and trunk then generate large forces as the player shifts the center of gravity forward over the left foot. Depending on the time available and the position of the incoming ball, the player may also step forward rather than just shifting her weight forward. The forces generated by the lower extremities and trunk are transmitted to the upper extremity and ultimately to the racket through the bones, ligaments, and stretched muscles and tendons of the swinging arm. The pelvis and trunk rotate forward as the center of gravity moves forward over the left foot. This movement stretches the trunk muscles, which then contract more forcefully due to the stretch reflex and cause the trunk to rotate. The muscles crossing the shoulder joint (especially the horizontal adductors) are then stretched and contract forcefully to produce the arm and racket movements that occur just before ball impact.

⟳ **The horizontal forces on the racket are generated and transmitted from proximal to distal and from larger to smaller segments.**

Another important factor is racket orientation (the angle of the racket face), because it affects the velocity

and direction of the ball after impact. The angle of the racket face also affects the spin on the ball, which affects the path taken by the ball. There is no one best racket face angle, because it is influenced by the preimpact velocity and direction of the ball and the racket head. These factors all interact to determine the postimpact velocity and direction of the ball. Let's analyze a simple situation first. Assume that the racket is stationary and that the ball hits the racket face at an angle. If the impact between the ball and the racket were perfectly elastic (a coefficient of restitution between the ball and racket of 1.0, meaning that the velocity of the ball with respect to the racket face before the collision is equal to the velocity of the ball with respect to the racket face after the collision), the angle of incidence (the preimpact angle of the ball with the racket face) would equal the angle of reflection (the postimpact angle of the ball with the racket face) (see figure 13.12a). In reality, the collision is not perfectly elastic (the coefficient of restitution is less than 1.0), so the postimpact angle is less. This means that if you hold your racket still and let the incoming ball strike it, the racket face should be facing a little more than halfway toward the incoming direction of the ball from the direction you want the ball to go (see figure 13.12b). If the racket is swung and the racket head has some velocity perpendicular to the racket face before impact, the postimpact angle will be equal to or greater than the preimpact angle (see figure 13.12c). Thus, as you swing the racket faster and faster, you should aim the racket face closer and closer to the direction you want the ball to go. You would never aim the racket face

exactly in the direction you want the ball to go, however, unless that is the direction the ball came from.

Spin on the ball influences the path it takes due to differing air pressures on the sides of the ball (remember the Magnus effect from chapter 8). Topspin causes the ball to follow a slightly lower than normal trajectory; backspin causes it to follow a slightly higher than normal trajectory; and clockwise or counterclockwise spin causes it to veer slightly sideways. Spin also affects the bounce of the ball due to friction between the ball and the court.

Spin is imparted to the ball by friction between the racket strings and the ball when it strikes the racket and rolls or slides across the strings. This friction force creates a torque about the center of the ball, which creates the spin. The friction force also acts to change the linear direction of the ball. No spin is imparted to the ball if the velocity of the incoming ball and the velocity of the racket head are exactly opposite in direction and if the racket face is perpendicular to those directions. The angle between the direction of the racket face and the direction of the relative velocity of the ball with respect to the racket head determines the amount of spin imparted to the ball (see figure 13.13). The greater the angle, the more spin imparted to the ball. A swing path from low to high

imparts topspin to the ball *(b)*, and a swing path from high to low generally imparts backspin to the ball *(c)*. A swing path from left to right imparts counterclockwise spin *(d)*, whereas a swing path from right to left imparts clockwise spin *(e)*.

Ideally, the forehand drive should have a fast horizontal velocity, so the racket face should be directed close to the desired direction of the ball. This fast horizontal velocity increases the chances that the ball will go past the baseline, so some topspin must be imparted to the ball to keep it in the court. In a good forehand drive, the racket head undergoes some vertical displacement along with its horizontal displacement.

This has been an incomplete analysis of the factors affecting the performance of the forehand drive in tennis. Because the forehand drive is an open skill, it can be performed effectively in many ways based on the numerous combinations of factors the performer is unable to control. This reveals one of the shortcomings of the proposed qualitative biomechanical analysis procedure. Developing a model of the most effective technique is more difficult for open skills. A qualitative biomechanical analysis of a player's forehand drive technique should include observation and consideration of the various conditions under which the stroke is executed.

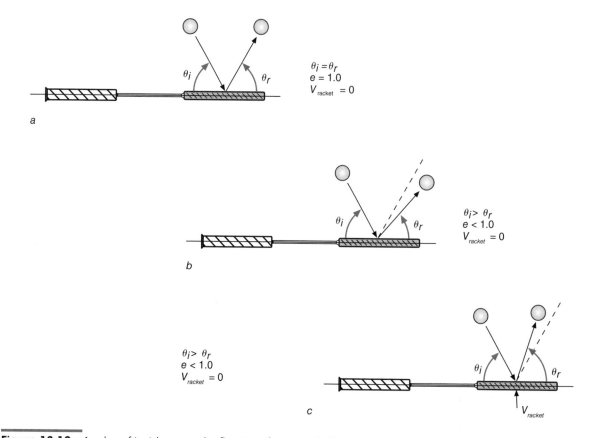

Figure 13.12 Angles of incidence and reflection of a tennis ball on a stationary *(a, b)* and a moving *(c)* racket.

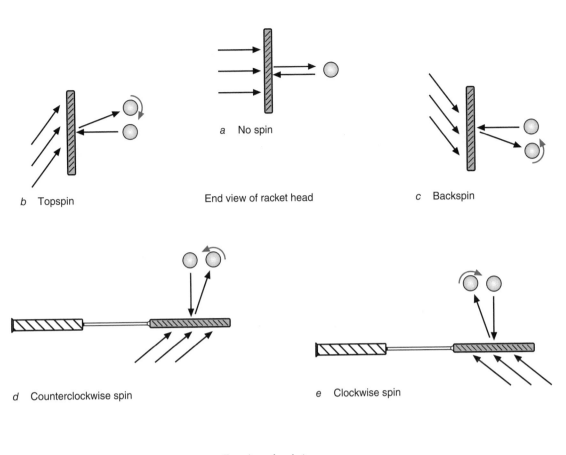

a No spin

b Topspin

End view of racket head

c Backspin

d Counterclockwise spin

e Clockwise spin

Top view of racket

Figure 13.13 Directions of the ball velocity, racket head velocity, and racket face determine ball spin.

Sprint Running

Running is a basic human movement. How fast a person can run or sprint may determine the level of success in many sports. The skill of sprinting is the last skill we will analyze. Unlike the previous sport skills, sprinting is a cyclic skill—the movements occur in cycles that are repeated over and over again. For sprinting, one stride represents one cycle. A stride begins and ends with the body parts in the same relative positions. If left foot touchdown is identified as the start of a sprinting stride, the end of the stride or cycle occurs at the next instant the left foot touches down. For our sample analysis, let's consider a 100 m dash in track and field. This skill has several parts: the start, the acceleration phase, and the velocity maintenance phase. Our analysis focuses on the last phase, the velocity maintenance phase, in which the sprinter attempts to maintain maximum velocity.

Theoretical Model of Sprinting

As stated previously, the first step in a qualitative biomechanical analysis is to describe the ideal technique by developing a theoretical model of the technique based on mechanics and cause-and-effect relationships. The first step in developing a model of sprinting is to acquire the necessary fundamental knowledge of sprinting. Let's assume we've done this work already and get right to developing the model.

The basis of the theoretical model is the purpose or goal of the skill. What is the purpose of sprinting in a 100 m dash? What is the runner trying to achieve? The runner is trying to win the race and get to the finish line first—in the least amount of time—so the performance criterion may be considered the time taken to run 100 m. The time it takes to run 100 m is a function of the average horizontal velocity of the sprinter during the race and the 100 m distance. Thus, the performance criterion for the velocity maintenance phase of the 100 m dash is the average horizontal velocity of the sprinter during this phase.

A faster sprinter has a faster average horizontal velocity during each step of the velocity maintenance phase of the 100 m dash. The average velocity of the sprinter is determined by the sprinter's average step length and step rate. Longer steps and more steps per second result in faster sprinting. How are longer steps and a faster step rate accomplished?

⊃ **The average velocity of the sprinter is determined by the sprinter's average step length and step rate.**

Step length is the sum of the sprinter's horizontal displacement during the support phase of a step (when the sprinter is in contact with the ground) and the sprinter's horizontal displacement during the flight phase of a step (when the sprinter is not in contact with the ground). Greater displacements during each of these phases result in a longer step length. The horizontal displacement during the support phase can be further divided into the touchdown distance (how far behind the foot the sprinter's center of gravity is at touchdown) and the takeoff distance (how far ahead of the foot the sprinter's center of gravity is at takeoff). Each of these distances depends on the sprinter's position at these instants and the sprinter's anthropometry. A taller, longer-legged sprinter is able to achieve longer touchdown and takeoff distances.

The horizontal displacement during the flight phase of a step is determined by the horizontal velocity of the sprinter at takeoff and the duration of the phase. The faster the horizontal velocity and the longer the duration of the flight phase, the greater the horizontal displacement during the phase. The horizontal velocity at takeoff is, in turn, determined by the horizontal velocity at touchdown and the work done by the friction force acting on the takeoff foot during the previous support phase. This work could be subdivided into the negative work done during the braking part of the support phase (when the friction force slows down the sprinter) and the positive work done during the propulsive part of the support phase (when the friction force speeds up the sprinter). The negative and positive work done are determined by the average horizontal forces and horizontal displacements occurring during each period. A sprinter who minimizes the negative work done and maximizes the positive work done will accelerate forward. During that part of a 100 m dash in which the sprinter maintains the same velocity from step to step, the negative and positive work done during a typical step are equal.

The duration of the flight phase (or flight time) also affects the horizontal displacement during the phase. The flight time depends on the vertical velocity of the sprinter at takeoff, the height of the sprinter's center of gravity (CG) at takeoff, the height of the sprinter's center of gravity at touchdown, and the average drag force due to air resistance. The vertical velocity at takeoff is, in turn, determined by the vertical velocity at touchdown and the work done by the normal reaction force acting on the takeoff foot during the previous support phase. This work could be subdivided into the negative work done when the upward reaction force slows the downward velocity of the sprinter at the beginning of the support phase and the positive work done when the upward reaction force speeds the upward velocity of the sprinter during the second half of the support phase. The negative and positive work done are determined by the average normal reaction forces and vertical displacements occurring during each period. A sprinter must do equal amounts of positive and negative work to maintain the same horizontal velocity from step to step.

The flight time during a step is also affected by the sprinter's center-of-gravity height at the beginning and end of the flight phase (i.e., at takeoff and touchdown). If the takeoff height is greater than the touchdown height, the time of flight is extended. If the takeoff height is less than the touchdown height, the flight time is shortened.

Average step length and average step rate determine a sprinter's average horizontal velocity. The step length determinants were discussed earlier. Step rate is determined by the time per step. The shorter the time per step, the faster the step rate. The time per step can be subdivided into support time and flight time. The flight time determinants were also described previously, because flight time is one of the factors affecting horizontal displacement during the flight phase. A greater displacement during the flight phase results if the flight time is longer; however, the time per step is shorter and the step rate is faster if the flight phase is shorter. The theoretical model of the factors that mechanically affect sprinting is shown in figure 13.14.

Application of the Theoretical Model

Sprinting is a closed skill, so it should be easier to interpret and apply the model of the most effective technique to sprinting than it was for an open skill such as the tennis forehand drive. However, because sprinting is cyclic by nature, some of the input factors in the model are outcomes of a previous cycle (a previous step), and thus our model becomes sort of an endless loop. For instance, the horizontal displacement during the flight phase is determined by the flight time and the horizontal velocity at takeoff. Horizontal velocity at takeoff is, in turn, determined by the work done during the previous support phase and the horizontal velocity at touchdown. This horizontal velocity is a performance measure of the previous step. Despite this shortcoming of our model, let's examine it and see how it can be applied.

Before applying the model, let's look at how the instantaneous horizontal velocity of a sprinter's center of gravity varies during a single step. Figure 13.15 shows a plot of the horizontal velocity of a sprinter during one step. Notice how the velocity decreases at the start of the support phase and increases during the second half of that phase. In which part of the step is the sprinter's average horizontal velocity faster? The sprinter is obviously faster during the flight phase. The horizontal velocity during the

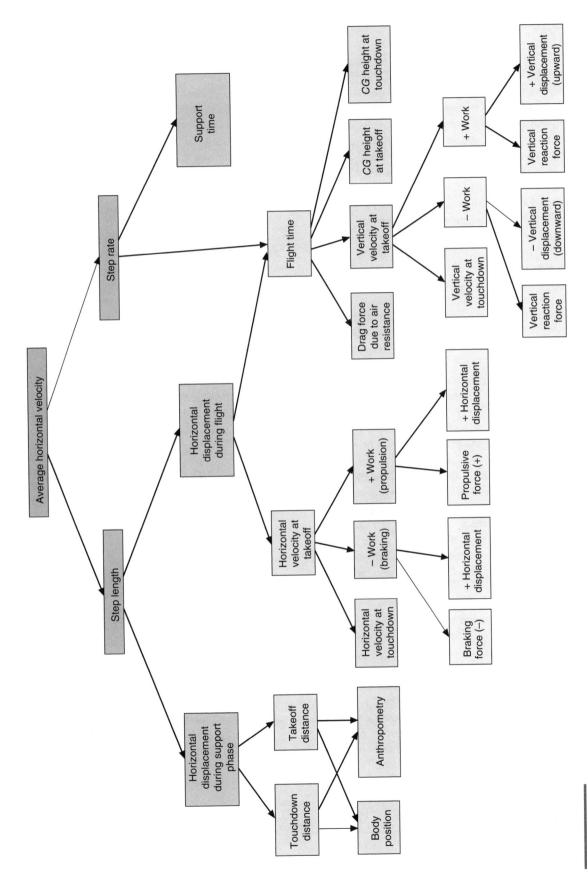

Figure 13.14 Theoretical model of sprinting.

flight phase is always faster than the horizontal velocity during the support phase. When the sprinter's foot strikes the ground, friction between the ground and the sprinter's foot creates a braking force, and the sprinter slows down. In the second half of the support phase, the sprinter speeds up to regain the velocity lost during the braking phase. Thus, the average horizontal velocity during the support phase of a sprinting step is slower than the horizontal velocities at the beginning or end of the phase, which are equal to the average velocities during the preceding or succeeding flight phases. A sprinter should thus minimize the support time and maximize flight time. Let's keep this in mind as we analyze our model.

In our sprinting model, the factors affecting the step length during the support phase are touchdown and takeoff distance. Although a longer touchdown distance

would appear to be beneficial, it increases the braking forces as well as lengthening the support phase. Therefore, touchdown distance should be kept short by having the foot strike the ground almost directly beneath the sprinter. No detrimental effect is associated with takeoff distance, so it should be maximized. By fully extending the takeoff leg and rotating the pelvis toward the side of takeoff, a sprinter can add another 3 or 4 cm to this distance (see figure 13.16).

The factors affecting step length during the flight phase are flight time and horizontal velocity at takeoff. Both factors are affected by the work (positive and negative) done during the previous support phase. The negative work done to slow the vertical velocity of the sprinter at touchdown should have a large average vertical reaction force and a short vertical displacement. The hip, knee, and ankle should flex enough to reduce the reaction force and the force required in the muscles to values below injury threshold. Too much flexion and downward vertical displacement increases the time spent in the support phase. Another benefit of minimizing the length of this negative work phase is that the larger load and quicker stretching of the lower extremity muscles evokes a stronger stretch reflex. The muscles will contract more forcefully during the propulsive or positive work phase. The larger vertical reaction force also causes a larger frictional force between

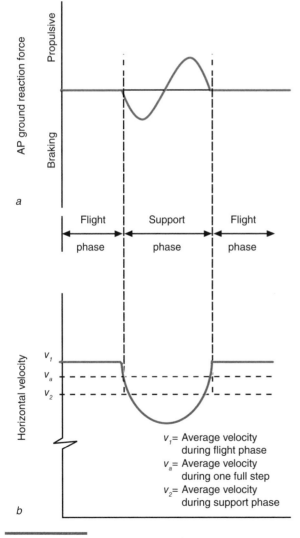

Figure 13.15 Theoretical plot of the anteroposterior ground reaction force acting on a sprinter (a) and the horizontal velocity of the sprinter (b).

Figure 13.16 A sprinter's takeoff distance (d_{to}) is the horizontal distance between the takeoff foot toe and the sprinter's center of gravity (cg).

the foot and the ground, increasing the negative work done to slow the runner's forward velocity. Driving the free leg and arms upward during the second half of the support phase increases the vertical reaction force as well as the vertical displacement of the sprinter. These actions, along with full extension of the support leg, increase the positive work done to speed up the vertical velocity of the sprinter during the support phase.

The negative work done to slow the horizontal velocity of the sprinter at the beginning of the support phase should be minimized by having the sprinter's foot touch down almost directly beneath the center of gravity. The positive work done to speed up the runner's horizontal velocity should be maximized by thrusting the swing leg forward vigorously, but the knee should be flexed to minimize the leg's moment of inertia about the hip joint so that the movement is quicker. The arms should swing in opposition to counter the angular momentum generated by the swinging leg. Some elbow flexion is desirable to reduce the arm's moment of inertia about the shoulder joint and thus increase its movement speed. Again, full extension of the support leg as well as rotation and tilt of the pelvis is desirable to increase the horizontal displacement of the runner during the positive work phase.

During the support phase, especially the propulsive portion, the sprinter's trunk should be fairly rigid to keep the propulsive forces from being dissipated and absorbed when the trunk and vertebral column flex. This is the phase in which positive work is done, and the more powerful the sprinter, the faster the sprinter will run. If the upward- and forward-thrusting forces from the legs do not cause upward and forward displacement of the trunk but rather a deflection or collapsing of the trunk and vertebral column, little or no positive work is done, and power is not transmitted from the lower extremity to the trunk.

A maximum sprint represents a fine balance between step rate and step length. Both need to be as large as possible, but overemphasizing one will adversely affect the other. Too long a step decreases step rate by increasing the landing distance and the braking forces at the start of the support phase. These conditions increase the length of the support phase, thus slowing the step rate and reducing the velocity. On the other hand, an increased step rate requires a shorter flight time and support time. If stride length remains the same but the support time is shorter, the sprinter has to be more powerful to absorb the energy of the downward motion when the foot first strikes the ground and then produce the same energy during the propulsive part of the phase. If the support time is too short, less energy will be absorbed or produced, and the subsequent takeoff velocity will be slower, resulting in a shorter step length. This analysis reveals the primary physical requirement of sprinting—a sprinter must be powerful. The sprinter's muscles must be able to develop large forces (both eccentrically and concentrically) at high contraction velocities.

Summary

The biomechanics presented in the previous chapters of this book can be applied systematically to analyze human movement. One system for conducting a qualitative biomechanical analysis to improve technique includes the following steps:

1. Develop a theoretical model of the most effective technique.
2. Observe the actual performance.
3. Evaluate the performance by comparing it to the most effective technique.
4. Instruct the performer by providing feedback about discrepancies between the actual performance and the most effective technique.

This biomechanical analysis may be qualitative or quantitative. A qualitative biomechanical analysis relies on subjective observations of the performance, whereas a quantitative biomechanical analysis uses actual measurements to quantify certain mechanical parameters of the performance. Practitioners (coaches, teachers, clinicians) use this or some similar procedure to evaluate the movement or performance of their students, athletes, or clients.

KEY TERMS

qualitative biomechanical analysis (p. 312)
quantitative biomechanical analysis (p. 312)

REVIEW QUESTIONS

1. What steps are involved in a qualitative biomechanical analysis for technique improvement?

2. What is the difference between a qualitative and a quantitative biomechanical analysis?

3. Identify the performance criteria for the following activities:
   a. A 110 m hurdle race
   b. A basketball pass
   c. A football punt
   d. A discus throw
   e. Batting in baseball
   f. A triple toe loop in figure skating
   g. A hockey slap shot
   h. A caber toss in Scottish Highland Games

4. Describe where you should position yourself to observe the following:
   a. The takeoff phase of a long jump
   b. The clearance technique of the trail leg in a 110 m hurdle race
   c. The start of a 50 m freestyle swimming race
   d. A field goal kick in football
   e. The arrow release in archery
   f. A back squat
   g. A wide-grip bench press

5. How are errors that are detected in a technique prioritized for correction?

6. Develop a theoretical model for the technique used in a human movement skill of your choice.

chapter 14

Qualitative Biomechanical Analysis to Improve Training

cbjectives

When you finish this chapter, you should be able to do the following:

- Distinguish between technical training and physical training

- Understand how the results of a qualitative analysis to improve technique may also be used to improve technical training

- Evaluate a technical exercise or drill and determine whether it is appropriate

- Describe the steps involved in a qualitative anatomical analysis

- Identify the active muscle groups during any phase of a human movement and whether their contraction is concentric, eccentric, or isometric

© Aspenphoto/Dreamstime.com

You are watching the men's javelin competition in the Summer Olympic Games on television. The winning throw flies well over 300 ft (91 m). You wonder how the thrower strengthens his arm to generate the forces necessary to throw the javelin so far. But, obviously, the thrower uses more than just his arm muscles. What other muscle groups must be well developed to throw the javelin so far? What kinds of exercises and drills are effective for training a javelin thrower? This chapter may help you answer these questions by providing you with the basis for completing a qualitative biomechanical analysis to improve training.

In the introduction, we learned

that the primary goal of biomechanics is performance improvement and that a secondary goal is injury prevention and rehabilitation. In the previous chapter, we discussed how biomechanics can be used to improve technique and thus improve performance. But what if the performer is not strong or powerful enough to execute the most effective technique? Just as an athlete's technical errors may limit performance, deficiencies in the athlete's physical capabilities may limit performance or prevent execution of the most effective technique. In this chapter, we will learn how biomechanics can be used to improve training and thus improve performance. Specifically, we will learn how to conduct a qualitative anatomical analysis of a performance or exercise to identify the active muscle groups.

Biomechanics and Training

A fundamental principle of training is the principle of specificity. Exercises and drills must be specific to the sport or activity. If the exercises and drills you use in training closely match the aspects of the skill you are training for, you will have greater gains in performance. For some sports or activities, the type of training that is specific to the sport or activity is obvious. If you want to run a marathon, your training should consist primarily of running long distances. In other cases, the training exercises that are specific to the sport or activity may not be as obvious. For example, the specific strengths needed to be successful in pole vaulting are not obvious to the casual observer. Biomechanics can contribute to the improvement of training by identifying the specific technical or physical requirements necessary to perform a skill well. Before learning more about how to identify the specific technical or physical requirements of a skill, let's consider the various types of training.

⟳ If the exercises and drills you use in training closely match the aspects of the skill you are training for, you will have greater gains in performance.

Technical Training

Practice and training time for specific sport or movement activities is usually devoted to improving technique (**technical training**) or physical condition (physical training). The proportion of time allocated to technical training depends in part on the technical difficulty of the activity. Technical training may involve performing the actual skill or performing drills that mimic specific aspects of the skill.

Biomechanics can contribute to improvements in technical training in several ways. First, a qualitative biomechanical analysis of an actual performance can identify deficiencies in technique. We discussed the procedure for completing this type of analysis in chapter 13. If the athlete is unable to correct the deficiency after being made aware of it, you prescribe drills and exercises that are specific to the aspect of the skill that is deficient.

Another way biomechanics can contribute to improving technical training is by identifying exercises and drills that closely simulate the specific technical aspects of the skill. Examine the drills and exercises used in the technical training for an activity you are familiar with. What is the purpose of the drill or exercise? What aspect of the skill is it specific to? Are the joint positions, velocities, and ranges of motion of the exercise similar to those of the skill? Are the muscle forces and contraction velocities similar? Are the external forces similar? These are questions you should ask yourself when evaluating drills or exercises for your students, clients, or athletes. The greater the fidelity between the exercise and the actual skill, the more specific the exercise is to that skill and the greater its potential for improving the performance of the skill. If few similarities exist between the exercise and the skill, the exercise is inappropriate for developing that skill. The muscle forces, contraction velocities, and magnitudes of the external forces are difficult to assess qualitatively, but you can view the joint positions and ranges of motion and from these assess joint angular velocities. Quantitative biomechanical analyses of exercises and drills for specific sport skills would be more valuable, but few have been reported in the biomechanical literature.

Physical Training

Physical training is the other part of training. Whereas technical training is directed primarily at correcting or improving aspects of technique, **physical training** is directed at altering performance limitations due to the physical condition of the performer. Be aware that there is usually some overlap between the two types of training. Technical training may have some effect on the performer's physical condition, and physical training may have some effect on the performer's technical proficiency. In any case, physical training is directed at improving the components of physical fitness, including

1. muscular strength,
2. muscular power,
3. muscular endurance,
4. flexibility,
5. cardiovascular fitness, and
6. body composition.

Again, biomechanics is directed at improving the specificity of physical training. A biomechanical analysis of an activity can identify the specific muscle groups whose strength, power, endurance, or flexibility limits performance. Specific exercises can then be chosen that strengthen or stretch these specific muscle groups. Likewise, a biomechanical analysis of an exercise can identify if the muscles used in the exercise are those used in the sport or activity.

In this text, the type of qualitative biomechanical analysis used to identify the active muscle groups during each phase of a movement is called a qualitative anatomical analysis. The next section of this chapter describes the steps involved in a qualitative anatomical analysis of a movement. The remainder of the chapter presents sample qualitative anatomical analyses.

Qualitative Anatomical Analysis Method

The purpose of a **qualitative anatomical analysis** is to determine the predominant muscular activity during specific phases of a performance and to identify instants when large stresses may occur due to large muscle forces or extremes in joint ranges of motion. The teacher or coach may complete such an analysis on a student or on an elite performer who demonstrates effective technique. The analysis of the elite performance identifies which muscles are involved in performance of the most effective technique, whereas the analysis of the student identifies the muscles used in performing the specific technique. In either case, the methods used to identify the muscles involved are the same.

Let's think for a minute. Do you already know of any methods for identifying what muscles are active during a sport or movement? One way is to touch and feel the performer's superficial muscles during the movement. If the muscle is firm and rigid, it is actively developing force. If the muscle is soft and flabby, it is inactive. This method works only if the activity involves static positions or slow movements (and if the performer is comfortable with being touched). Certain weightlifting exercises and gymnastics routines are examples of movements that may be appropriate for this type of analysis. Obviously, this method is impractical for analyzing dynamic activities such as throwing or running. The method is also intrusive and affects how the subject performs.

Another qualitative method for determining which muscles are active during an activity is to perform the activity vigorously and then wait a day or two to see which muscles become sore. This works only when the person hasn't been practicing the activity regularly. It may also identify only those muscles that experience large eccentric activity during the performance, because it is eccentric muscular action that is associated with muscle soreness.

Quantitatively, a researcher can wire the performer's muscles with electrodes and use electromyography to monitor the electrical activity of the muscles. The EMG recordings of a muscle indicate whether or not the muscle was active. Further analysis of the EMG signal will provide a general idea of the magnitude of the muscular action. This type of quantitative analysis is expensive, time-consuming, and unavailable to most teachers and coaches. In addition, the electrodes and associated wiring are likely to cause the athlete's performance to differ from a typical performance.

Another quantitative method involves recording the movement on film or videotape and then digitizing the film or video to get a complete kinematic description of the movement. Force platforms or transducers can also be used to measure any external contact forces acting on the subject. Anthropometric measurements are then taken from the subject to estimate body segment masses and moments of inertia. Free-body diagrams of each segment are then used to identify the resultant joint forces and torques that cause acceleration of the segments. Equations of motion based on Newton's second law are solved to determine these forces and torques. The computed resultant joint torques are related to the muscles crossing those joints and indicate which muscles are active. This type of mechanical analysis, called an inverse dynamic analysis, is also one that few coaches or teachers have the equipment, time, or expertise necessary to complete.

None of these methods are practical or reasonable for coaches or teachers to use. Muscle activity cannot be directly observed; however, we can get a general idea of which muscles are active based on the principles used in an inverse dynamic analysis. This type of qualitative anatomical analysis may be practical for coaches and teachers, and it provides reasonable results for most activities. The following are step-by-step procedures for completing a qualitative anatomical analysis:

1. Divide the activity into temporal phases.

2. Identify the joints involved and the movements occurring at those joints.

3. Determine the type of muscular contraction (concentric, eccentric, or isometric) and identify the predominant active muscle group at each joint.

4. Identify instances when rapid joint angular accelerations (rapid speeding up or slowing down of joint motions) occur and where impacts occur.

5. Identify any extremes in joint ranges of motion.

The results of a qualitative anatomical analysis can then be used to determine appropriate strength or flexibility exercises for the muscle groups identified. The five steps of a qualitative anatomical analysis are described and illustrated next.

Temporal Phases

The first step of a qualitative anatomical analysis is to break down the performance into specific phases or motions. For the simplest movements, such as slow movements or those that involve raising and lowering, the end of one phase and the beginning of the next may simply be designated by the change in direction of movement. For example, let's look at a bench press. The bench press involves two phases: the down phase, in which the lifter lowers the barbell to the chest, and the up phase, in which the lifter raises the barbell.

In contrast, faster, more complex movements may require breakdown into more phases than are obvious. The phases of a baseball pitch include the windup, delivery, and follow-through, but breaking the pitch into only these three phases may not provide enough detail about the joint motions to allow determination of the muscle activity that occurs. For such high-speed movements, it is better to make a video recording of the performance. The motion should then be examined frame by frame rather than phase by phase. One can make comparisons with elite performers by examining videotapes or sequence photos of such performers. Sequence photos of elite performances are often found in textbooks or coaching journals on the sport in question. These publications, as well as those of the National Strength and Conditioning Association, may also include results of qualitative anatomical analyses or indications of what muscle groups are important in the execution of the skill being examined.

Joint Motions

Once the performance has been broken down into specific phases or recorded on video, the next step is to identify which body segments and joints to examine. Which segments and joints are involved in the performance of the skill? Which segments move, and which joints are involved in the movements? For skills involving gross movement of the whole body, most of the major joints are involved. For example, running involves the ankle, knee, hip, shoulder, and elbow joints on either side of the body. Other skills, especially strength training exercises, may involve movements of only a few segments and joints. Let's go back to the example of the bench press exercise. Which segments move, and which joints are involved in this exercise? The upper arm and forearm are the moving segments, and the elbow and shoulder are the joints primarily involved in the movement.

Once the segments and joints involved in the activity have been identified, the next step is to identify the motions occurring at each joint during each phase of the activity. Is flexion or extension occurring at the joint? Abduction or adduction? Internal or external rotation? Is more than one motion occurring at the joint? Here you are identifying the motion or change in position of the segments that occur at the joint during the phase or between phases. How did the adjoining segments move relative to each other? Joint motion differs from joint position, and the two must be distinguished from each other. For example, during the down phase in the bench press, flexion (the joint motion) occurs at the elbow joint, and the elbow is in positions of flexion. During the up phase, extension (the joint motion) occurs at the elbow joint, but the elbow is in positions of flexion until the very end of the up phase. At the shoulder joint, horizontal extension (or horizontal abduction) occurs during the down phase, and horizontal flexion (or horizontal adduction) occurs during the up phase. In a qualitative anatomical analysis, we are interested in joint motion, not joint position.

Muscle Contractions and Active Muscle Groups

The next step in a qualitative anatomical analysis is to determine what types of muscle contractions occur within the active muscle groups crossing each joint and to identify these active muscle groups. Recall from chapter 11 that concentric muscular contractions produce positive work that results in an increase in mechanical energy; eccentric muscular contractions produce negative work that results in a decrease in mechanical energy; and isometric muscular contractions produce zero work that results in no change in mechanical energy. If no external contact forces act on the body, the easiest way to identify the type of muscular contraction is to identify only the change in mechanical energy that occurs during the phase of motion under examination. Was something lifted (increasing potential energy) or lowered (decreasing potential energy)? Was something sped up (increasing kinetic energy) or slowed down (decreasing kinetic energy)? Was something stretched or deformed (increasing strain energy) or slowly unstretched and reformed (decreasing strain energy)? If the total mechanical energy (the sum of potential, kinetic, and strain energies) increases, the contraction of the active muscle group is concentric. If the total mechanical energy decreases, the contraction of the active muscle group is eccentric. If the total mechanical energy does not change, the contraction of the active muscle group is isometric or the muscles are inactive.

This general method usually gives correct results, even when external contact forces act on the body. In some cases, however, the direction of the force that the body exerts in reaction to the external force and the direction of displacement of the object that the body exerts a force against must be examined. If the directions are the same, the force does positive work; the muscles contract concentrically. If the directions are opposite, the force does negative work; the muscles contract eccentrically. If no displacement occurs, no work is done; the muscles contract isometrically or are inactive.

Identifying the active muscle groups is simple once the type of muscular contraction has been determined. If the contraction has been identified as concentric, the active muscles are those that create torque in the same direction as the observed joint motion. For instance, if the observed joint motion is flexion and the contraction is concentric, the flexor muscles are active at that joint (the opposing extensor muscles might be active as well, but the resultant torque is a flexor torque, so the flexor muscles must produce greater torque). If the muscle contraction is eccentric, the active muscles are those that create torque opposing the observed joint motion. For example, if the observed joint motion is flexion and the contraction is eccentric, the extensor muscles are active at that joint (the opposing flexor muscles might be active as well, but the resultant torque is an extensor torque, so the extensor muscles must produce greater torque).

> ## Identifying the active muscle groups is simple once the type of muscular contraction has been determined.

Let's look at the example of the bench press again. During the down phase, what happens to the mechanical energy of the barbell? Its potential energy decreases, as do the potential energies of the arm and forearm segments of the lifter, because they are all lowered. The kinetic energies of these segments increase, however, because they change from no movement to movement downward. The downward acceleration of these segments is not as fast as the acceleration due to gravity that would occur if the decrease in potential energy exactly equaled the increase in kinetic energy—that is, if you just let the barbell drop to your chest. Potential energy decreases more than kinetic energy increases, so there is a net decrease in the total mechanical energy. This indicates that the elbow and shoulder muscles act eccentrically during this phase. Because elbow flexion occurs and the contraction is eccentric, the active elbow muscle group is the elbow extensors. At the shoulder, horizontal extension occurs and the contraction is eccentric, so the active muscle group must be the shoulder horizontal flexors.

During the up phase of the bench press, the potential and kinetic energies of the barbell and arm segments increase, indicating that the elbow and shoulder muscles contract concentrically. Because elbow extension occurs during this phase, the elbow extensors are again the active muscle group. Likewise, because shoulder horizontal flexion occurs during this phase, the shoulder horizontal flexors are again the active muscle group. The pattern of muscular activity observed in the bench press is typical for most movements and exercises that primarily involve changes in potential energy (lifting and lowering). The same muscle groups are active throughout the movement or exercise. The muscle groups contract eccentrically during the lowering phase and then concentrically during the lifting phase.

Generally, if something is raised slowly, concentric muscular contractions are responsible. If something is lowered slowly, eccentric muscular contractions are responsible. If something is held still, isometric muscular contractions are responsible.

Analysis of the forces and displacements during the bench press yields the same result. During the down phase, the lifter exerts an upward force on the barbell, but the barbell's displacement is downward. The force and displacement are opposite in direction, so the work done by the force is negative. The elbow and shoulder muscles contract eccentrically during the down phase. During the up phase, the lifter continues to exert an upward force on the barbell, but in this phase, the barbell's displacement is also upward. The force and displacement are in the same direction, so the work done by the force is positive. The elbow and shoulder muscles contract concentrically during the up phase.

How is the active muscle group determined when the muscle contraction at a joint has been identified as isometric? In this case, we can determine the active muscle group by imagining what joint motion would occur if all the muscles around the joint relaxed. If movement would occur, the active muscle group (the muscle group that is active isometrically) would be the muscle group that produces a torque opposing this movement. In the bench press, for example, suppose you pause during the up phase and hold the barbell halfway up. No movement occurs at the elbow joint, so the muscle contraction is isometric or the muscles are inactive. If the muscles were inactive, the barbell would crash to your chest, and your elbow would flex. The active muscle group must be the extensor muscles that prevent elbow flexion.

Rapid Joint Angular Accelerations and Impacts

After identification of the active muscle groups, the next task in a qualitative anatomical analysis is to identify any instances when rapid joint angular accelerations and impacts occur. This task identifies the active muscle groups that must produce the largest forces as well as the positions of the limbs when these large forces must be produced. These are the muscle groups whose strength will be tested and the muscle groups that must be trained. This step is more important when we are analyzing fast movements, but we can still use the bench press as our example.

At the beginning of the down phase, the barbell accelerates in the downward direction, and elbow flexion and shoulder horizontal extension occur. Because the force exerted by the lifter is upward at this instant, it is reduced to create this downward acceleration. Less torque is created by the active muscle groups at this instant. At the end of the down phase, the downward movement of the barbell is slowed, so its acceleration is upward. Elbow flexion and shoulder horizontal extension also slow down. The force exerted by the lifter is upward at this instant, so it must be increased to cause the upward acceleration of the barbell. More torque must be created by the active muscle groups at this instant to cause this acceleration. The same is true at the start of the up phase, when the barbell is accelerated upward again. At the end of the up phase, the barbell is slowed and its acceleration is downward. The force exerted by the lifter is upward at this instant, but it is reduced to create this downward acceleration. Thus, in the bench press, the active muscle groups are most stressed at the end of the down phase and the beginning of the up phase.

Extreme Joint Ranges of Motion

The last step in a qualitative anatomical analysis is to identify any extremes in joint ranges of motion. The purpose of this step is to identify those muscles and soft tissues that may be stretched and possibly injured. Flexibility exercises may be appropriate for these muscle groups. As with the previous step, this step is more important when we are analyzing fast movements.

In the bench press, the elbow is at full extension at the start of the down phase and at the end of the up phase. This is not an unusual position for the elbow joint, however. At the end of the down phase and the beginning of the up phase, the elbow is fully flexed, and the shoulder may be close to the limit of its horizontal extension. The muscular torques are greatest at this position as well, so specific flexibility exercises to stretch the elbow extensors and shoulder horizontal flexors may be appropriate.

Charting the Analysis

Keeping track of what has been identified in each step of a qualitative anatomical analysis is difficult without a record of what you have identified in the previous steps. This job can be easier if you make a chart or table of the results of each step of the analysis. The chart might include columns for the phases of the movement (or the frames of video), the joints analyzed, the joint motions, the muscle contractions, the active muscle groups, the rapid joint angular accelerations or impacts, and the extreme joint ranges of motion. Table 14.1 is an example of such a chart representing our analysis of a bench press.

Sample Analyses

The steps in conducting a qualitative anatomical analysis to improve training were described in the previous pages using a simple exercise, the bench press, as an example to illustrate the procedures. A qualitative anatomical analysis of a skill may be simple or complex, depending on the

Table 14.1 Sample Qualitative Anatomical Analysis of a Wide-Grip Bench Press

Joint	Phase of motion	Joint motion	Muscle contraction	Active muscle group	Rapid acceleration or impact	Extreme range of motion
Elbow	Down	Flexion	Eccentric	Extensors	At end of phase	Full flexion at end of phase
	Up	Extension	Concentric	Extensors	At beginning of phase	
Shoulder	Down	Horizontal extension	Eccentric	Horizontal flexors	At end of phase	Full horizontal extension at end of phase
	Up	Horizontal flexion	Concentric	Horizontal flexors	At beginning of phase	

activity being analyzed. Activities that primarily involve changes in potential energy (such as the bench press and most weightlifting exercises) are generally easier to analyze than activities that include faster movements and involve changes in kinetic energy. This section provides sample qualitative anatomical analyses of some of these more difficult-to-analyze activities. The examples include a jumping activity (a vertical jump), a striking activity (a football punt), a locomotor activity (sprint running), and a throwing activity (a javelin throw). The analyses progress in difficulty from simplest (vertical jump) to most complex (the javelin throw).

Vertical Jump

A standing vertical jump is only slightly more difficult to analyze than the bench press, because it still involves primarily changes in potential energy. The vertical jump can be broken down into three phases, as shown in figure 14.1: a preparatory (or down) phase, a propulsive (or up) phase, and a flight phase. The first two phases occur while the jumper is on the ground. Because the height reached by the jumper's center of gravity is determined by what the jumper does on the ground, let's analyze only the two ground phases.

The movements that occur during the vertical jump primarily involve movements about the ankle, knee, hip, and shoulder joints. Some movement occurs around the elbow joints, but it doesn't appear to be as important as that at the other joints, so let's examine just these four joints. Let's assume the jumping action is symmetrical, so the left and right sides move together.

During the preparatory phase, the ankles dorsiflex, the knees flex, the hips flex, and the shoulders hyperextend. During the propulsive phase, the opposite motions occur at each joint: The ankles plantar flex, the knees extend, the hips extend, and the shoulders flex.

During the preparatory phase, the body is lowered, so its potential energy decreases. The segment immediately above the ankle joint is lowered (its motion is downward relative to the ankle joint), so its potential energy relative to the ankle joint decreases. The contraction of the active ankle joint muscle group is eccentric. The ankle joint motion is dorsiflexion, but the muscle contraction is eccentric, so the ankle plantar flexors are the active muscle group. The segment immediately above the knee joint (the thigh) is also lowered relative to the knee joint, so its potential energy relative to the knee joint decreases. The contraction of the active knee joint muscle group is eccentric. The knee extensors are the active muscle group.

Figure 14.1 Breakdown of a standing vertical jump into three phases for analysis.

The segment immediately above the hip joint (the trunk) is also lowered relative to the hip joint, so its potential energy relative to the hip joint decreases. The contraction of the active hip joint muscle group is eccentric, and the hip extensors are the active muscle group. The segment distal to the shoulder joint (the arm) is raised (the arm moves upward relative to the shoulder joint), so its potential energy relative to the shoulder joint increases. The action of the active shoulder joint muscle group is concentric. Hyperextension occurs at the shoulder joint, so the shoulder extensors are the active muscle group.

During the propulsive phase, the potential and kinetic energies of all the body segments increase. The contraction of the active muscles at each of the joints is concentric. The ankle plantar flexes, so the plantar flexors are active; the knee extends, so the knee extensors are active; the hip extends, so the hip extensors are active. Closer examination of the shoulder joint reveals that the arm initially moves downward relative to the shoulder joint and then moves upward. This slight decrease in potential energy is much smaller than the segment's large increase in kinetic energy, so positive work is done, and the shoulder flexors contract concentrically.

At the end of the preparatory phase and the beginning of the propulsive phase, the body is rapidly accelerated upward. The joints all experience accelerations at these instances as well. Strength is required in the ankle plantar flexors, knee and hip extensors, and shoulder flexors to perform a vertical jump well. (Further examination of the flight and landing phases would reveal an impact at landing that would also stress all these muscles except the shoulder flexors.) Strength (and power) training exercises specific to these muscle groups may be appropriate.

The only extremes in joint ranges of motion observed occur at the shoulder joint. The hyperextension of the shoulder joint that occurs during the preparatory phase may be limited by the shoulder flexors. Flexibility exercises for the shoulder flexor muscles may be appropriate.

Table 14.2 is a completed chart of the qualitative anatomical analysis of the preflight phases of a vertical jump.

Football Punt

The American football punt is an example of an activity in which the work done by the muscles primarily causes changes in kinetic energy. Thus, it is more difficult to analyze than the previous examples. Because it involves quick movements, a frame-by-frame analysis is required. Sequence drawings of a football punter are shown in figure 14.2.

Joint Motions

The most important segments in football punting are the legs, so we should analyze the hip, knee, and ankle joints of both legs. The activity occurs primarily in the sagittal plane, so for simplicity, let's confine the analysis to

Table 14.2 Sample Qualitative Anatomical Analysis of the Preflight Phases of a Standing Vertical Jump

Joint	Phase of motion	Joint motion	Muscle contraction	Active muscle group	Rapid acceleration or impact	Extreme range of motion
Ankle	Down	Dorsiflexion	Eccentric	Plantar flexors	At end of phase	
	Up	Plantar flexion	Concentric	Plantar flexors	At beginning of phase	
Knee	Down	Flexion	Eccentric	Extensors	At end of phase	
	Up	Extension	Concentric	Extensors	At beginning of phase	
Hip	Down	Flexion	Eccentric	Extensors	At end of phase	
	Up	Extension	Concentric	Extensors	At beginning of phase	
Shoulder	Down	Hyperextension	Concentric	Extensors	At end of phase	Full hyper-extension
	Up	Flexion	Concentric	Flexors	At beginning of phase	

movements occurring in this plane. Look at figure 14.2. We'll begin with the right hip joint, which hyperextends from frames 1 to 2 as the punter steps forward with his left leg and flexes from frames 2 to 5 during the kicking action and follow-through.

The left hip actually extends slightly from frames 1 to 2, even though the punter is stepping forward onto this leg. This extension occurs primarily because the trunk was leaning forward in frame 1 and is upright in frame 2. This extension continues from frames 2 to 3. From frames 3 to 4, slight flexion appears to occur at the left hip joint. The position of the right leg in frame 4 prevents a clear view of the left thigh, so the action at the left hip cannot be determined with certainty. From frames 4 to 5, the hip joint appears to maintain the same position, so no joint motion occurs.

From frames 1 to 2, the right knee extends slightly as it pushes off. It then flexes from frames 2 to 3. It appears to maintain position from frames 3 to 4, so there is no joint motion. From frames 4 to 5, the right knee extends rapidly as the ball is kicked.

From frames 1 to 3, the left knee extends. It then flexes from frames 3 to 4. It appears to maintain the same flexed position in frame 5, so no joint motion occurs from frames 4 to 5.

From frames 1 to 2, the right ankle plantar flexes as it pushes off. It appears to maintain the same plantar-flexed position in frames 2 to 5, so no joint motion occurs from frames 2 to 5.

The left ankle plantar flexes from frames 1 to 3, then dorsiflexes from frames 3 to 4 and plantar flexes from frames 4 to 5.

Muscle Contractions and Active Muscle Groups

Now that we've identified the joint motions occurring during the football punt, the next step is to determine the muscle contractions and identify the active muscle groups. Again, look at figure 14.2. At the right hip, hyperextension occurs from frames 1 to 2. This appears to be the end of an extension phase of the right hip, so the muscle contraction is concentric at the beginning of this phase (as the extension speeds up and kinetic energy increases) and eccentric at the end (as the extension slows down and kinetic energy decreases). The hip extensors contract concentrically in the beginning, and the hip

Figure 14.2 Sequence drawings of a football punt executed by a skilled performer.

flexors contract eccentrically to slow the extension at the end. From frames 2 to 5, the right hip flexes. This flexion is initiated and continued by concentric contraction of the hip flexors from frames 2 to 4 (as flexion speeds up and the kinetic energy of the right leg increases). Concentric contraction of the hip flexors continues during the initial period from frames 4 to 5 (as the flexion continues to speed up), but just after foot contact with the ball, this flexion slows down and the right leg's kinetic energy decreases. Eccentric contraction of the hip extensors causes this slowing of hip flexion.

From frames 1 to 3, the left hip extends due to a concentric contraction of the hip extensors. Although difficult to determine, the extension of the hip appears to speed up from frames 1 to 3, indicating an increase in kinetic energy. From frames 3 to 4, the left hip flexes slightly. In these frames, the left foot is in contact with the ground. The ground reaction force pushing upward and slightly backward is what causes the hip to flex. Also, the kinetic and potential energies of the body decrease when the foot strikes the ground. The muscle contraction is thus eccentric, and the hip extensors are the active muscle group. From frames 4 to 5, the hip does not change position, so any muscle contraction that occurs is isometric. If no muscles were active at the left hip joint during this phase, the hip would flex, so the hip extensors must contract isometrically to maintain the joint angle.

From frames 1 to 2, the right knee extends. This extension phase is short, with the extension speeding up at the beginning and slowing down at the end, so the muscle contraction is concentric initially and eccentric finally (corresponding to an increase in kinetic energy followed by a decrease in kinetic energy). Because knee extension occurs, the active muscle groups are the knee extensors initially and the knee flexors finally. From frames 2 to 3, the right knee flexes. Again, this flexion phase is short and involves an initial speeding up of flexion followed by a slowing down. Thus, the muscle contraction is concentric at the beginning of this phase and eccentric at the end. The knee flexors contract at the beginning of the phase, and the knee extensors contract at the end. From frames 3 to 4, no motion occurs at the right knee, so any muscle contraction that occurs is isometric. If all the knee muscles were relaxed at this point, the knee would extend, so the flexors must be the active muscle group. From frames 4 to 5, the right knee continues its extension, speeding up until ball contact and then slowing down. The increase and then decrease in kinetic energy indicate that the muscle contractions are concentric and then eccentric. The knee extensors contract concentrically, followed by an eccentric contraction of the knee flexors.

From frames 1 to 3, the left knee extends. The muscle contraction is concentric from frames 1 to 2 because the extension speeds up and the leg's kinetic energy increases.

The active muscle group is the knee extensors. From frames 2 to 3, the extension slows down, and the leg's kinetic energy decreases, but the foot has contacted the ground. The reaction force from the ground slows down the extension. The muscle contraction is still concentric, and the knee extensors are still the active muscle group. From frames 3 to 4, the left knee flexes due to the ground reaction force. The potential and kinetic energies of the left leg decrease, thus muscle contraction is eccentric. The knee extensors are still the active muscle group. From frames 4 to 5, no joint motion occurs at the left knee, so any muscle activity occurring is isometric. If no muscles were active at the left knee joint during this phase, the knee would flex, so the knee extensors must be contracting isometrically.

The right ankle joint plantar flexes from frames 1 to 2. The potential and kinetic energies of the segments above the joint increase, so the muscle contraction is concentric and the plantar flexors are the active muscle group. No change in position occurs in frames 2 to 5, so any muscle activity occurring at the right ankle joint is isometric. From frames 2 to 4, the ankle plantar flexors are probably active to hold the plantar-flexed position. From frames 4 to 5, the ankle dorsiflexors must be active to brace the ankle and foot for impact with the ball.

The left ankle joint plantar flexes from frames 1 to 3. To initiate this action, the muscle contraction is concentric from frames 1 to 2, and the plantar flexors are the active muscle group. As the heel strikes the ground, the ground reaction force accelerates this plantar flexion. To slow down this plantar flexion, the muscle contraction is eccentric from frames 2 to 3, so the dorsiflexors are the active muscle group. From frames 3 to 4, the left ankle dorsiflexes. This dorsiflexion is also caused by the ground reaction force, now acting through the forefoot. The potential and kinetic energies of the left leg decrease, which means the muscle contraction is eccentric and the plantar flexors are the active muscle group. From frames 4 to 5, the left ankle plantar flexes, and the kinetic and potential energies of the body increase. The muscle contraction is concentric, and the plantar flexors continue to be the active muscle group.

Rapid Joint Angular Accelerations and Impacts

The next step in analyzing the football punt is to identify any instances of rapid joint angular acceleration and any impacts. Still looking at figure 14.2, we focus our attention on the kicking leg here. From frames 4 to 5, the right hip rapidly accelerates in flexion and then quickly slows during the follow-through. The right knee rapidly accelerates in extension and then quickly slows during the follow-through. During this period, the foot's impact with the ball also occurs. The right hip and knee extensors and flexors must be specifically strengthened to produce the torques necessary to cause these accelerations. In addition, the right dorsiflexors must be strong to maintain foot position during impact with the ball. Another impact occurs when the left foot strikes the ground between frames 2 and 3. The left knee and hip extensors must be strong to stabilize the body while the right leg swings. Strength and power training exercises are indicated for the hip extensors and flexors as well as the knee extensors and flexors. In addition, strength training exercises are indicated for the right ankle dorsiflexors.

Extreme Joint Ranges of Motion

The obvious extreme joint range of motion occurs at the right hip and right knee joints in frame 5. Here the right hip is almost fully flexed while the right knee is almost fully extended. The hamstring muscle group that is a hip extensor and knee flexor is stretched maximally during this activity. Flexibility exercises are warranted for this muscle group.

Table 14.3 is a completed chart of the qualitative anatomical analysis of the football punt illustrated in figure 14.2.

Table 14.3 Sample Qualitative Anatomical Analysis of a Football Punt

Joint	Frames	Joint motion	Muscle contraction	Active muscle group	Rapid acceleration or impact	Extreme range of motion
Right hip	1-2	Hyperextension	Concentric, then eccentric	Extensors, then flexors		
	2-3	Flexion	Concentric	Flexors		
	3-4	Flexion	Concentric	Flexors		
	4-5	Flexion	Concentric, then eccentric	Flexors, then extensors	Yes	Full flexion
Left hip	1-2	Extension	Concentric	Extensors		
	2-3	Extension	Concentric	Extensors		
	3-4	Flexion	Eccentric	Extensors		
	4-5	No motion	Isometric	Extensors		
Right knee	1-2	Extension	Concentric, then eccentric	Extensors, then flexors		
	2-3	Flexion	Concentric, then eccentric	Flexors, then extensors		
	3-4	No motion	Isometric	Extensors		
	4-5	Extension	Concentric, then eccentric	Extensors, then flexors	Yes	Full extension
Left knee	1-2	Extension	Concentric	Extensors		
	2-3	Extension	Concentric	Extensors		
	3-4	Flexion	Eccentric	Extensors		
	4-5	No motion	Isometric	Extensors		
Right ankle	1-2	Plantar flexion	Concentric	Plantar flexors		
	2-3	No motion	Isometric	Plantar flexors		
	3-4	No motion	Isometric	Plantar flexors		
	4-5	No motion	Isometric	Dorsiflexors	Ball impact	
Left ankle	1-2	Plantar flexion	Concentric	Plantar flexors		
	2-3	Plantar flexion	Eccentric	Dorsiflexors	Ground impact	
	3-4	Dorsiflexion	Eccentric	Plantar flexors		
	4-5	Plantar flexion	Concentric	Plantar flexors		

Sprint Running

Sprinting is another activity in which the work done by the muscles primarily causes changes in kinetic energy. It is more difficult to analyze than the football punt because it involves more segments. Again, because it involves quick movements, a frame-by-frame analysis is required. Sequence photos of a collegiate sprinter are shown in figure 14.3.

Joint Motions

Let's analyze one full sprinting stride, from takeoff of the right foot to the next takeoff of the right foot. The movements of the right and left sides are similar but out of phase with each other, so we'll analyze only the joints on one side—in this case, the left side. Sprinting occurs primarily in the sagittal plane, so we'll simplify our analysis further by considering only the sagittal plane movements. The joints we'll examine are the left hip, knee, ankle, and shoulder.

The left hip flexes from frames 1 to 5, then extends from frames 5 to 7 and hyperextends from frames 7 to 8.

The left knee flexes from frames 1 to 3, then extends from frames 3 to 6. It flexes slightly from frames 6 to 7 and extends again from frames 7 to 8.

No noticeable joint motion occurs at the left ankle from frames 1 to 2. The ankle dorsiflexes from frames 2 to 5 and then plantar flexes slightly from frames 5 to 6. The left ankle dorsiflexes at foot contact with the ground from frames 6 to 7 and then plantar flexes during push-off from frames 7 to 8.

The left shoulder extends from frames 1 to 2 and then hyperextends from frames 2 to 5. It flexes from frames 5 to 8.

Muscle Contractions and Active Muscle Groups

Similar to what happens with the football punt, the muscle contractions in sprinting are primarily due to changes in kinetic energy of the sprinter's segments. From frames 1 to 4, left hip flexion speeds up as the kinetic energy of the leg increases, so the muscle contraction is concentric. The hip flexors are the active muscle group. From frames 4 to 5, the hip flexion slows down as the kinetic energy of the leg decreases, so the muscle contraction is eccentric. The hip extensors are the active muscle group. From frames 5 to 7, hip extension speeds up as the kinetic energy of the leg increases, so the muscle contraction is concentric. The hip extensors are the active muscle group. From frames 7 to 8, kinetic energy increases, but at the end of this phase, hip hyperextension slows down. The muscle contraction is initially concentric and then eccentric. The hip extensors are the active muscle group initially, followed by the hip flexors.

At the left knee, flexion speeds up from frames 1 to 2 and slows down from frames 2 to 3. The muscle contraction is thus concentric from frames 1 to 2 and eccentric from frames 2 to 3. The knee flexors contract concentrically from frames 1 to 2, and the knee extensors contract eccentrically from frames 2 to 3. Knee extension speeds up from frames 3 to 5 but slows down from frames 5 to 6 (and possibly even during the end of the period from frames 4 to 5). The muscle contraction is concentric from frames 3 to 5 and eccentric from frames 5 to 6 (and maybe even eccentric during the end of the period from frames 4 to 5). The knee extensors contract concentrically from frames 4 to 5; then the knee flexors contract eccentrically to slow down the extension from frames 5 to 6. From frames 6 to 7, the knee flexes when the left foot strikes the ground due to the large ground reaction force acting on the left foot. Negative work is done at the knee joint, and the muscle contraction is eccentric. The knee extensors are the active muscle group during this phase. From frames 7 to 8, the left knee extends. Initially, knee extension speeds up, but at the end, it slows down. The knee extensors contract concentrically initially, followed by an eccentric contraction of the knee flexors at the end of the phase.

No noticeable joint motion occurs at the left ankle joint from frames 1 to 2, so any muscle activity that occurs is isometric. The previous motion at this joint was plantar flexion as the sprinter pushed off. An eccentric contraction of the dorsiflexors followed the push-off, so the dorsiflexors are probably still active from frames 1 to 2. From frames 2 to 5, the ankle dorsiflexes. The muscle contraction is concentric from frames 2 to 4 as the dorsiflexion speeds up. The dorsiflexors are the active muscle group. From frames 4 to 5, the muscle contraction is eccentric as the dorsiflexion slows down. The plantar flexors are the active muscle group. From frames 5 to 6, plantar flexion speeds up, so the muscle contraction is concentric, and the plantar flexors are still the active muscle group. From frames 6 to 7, dorsiflexion occurs when the left foot strikes the ground due to the large ground reaction force acting on the left foot. Negative work is done at the ankle joint, and the muscle contraction is eccentric. The plantar flexors continue to be the active muscle group. From frames 7 to 8, the ankle plantar flexes. Initially, the muscle contraction is concentric as the muscles do positive work and kinetic and potential energies are increased. The plantar flexors continue to be the active muscle group. At the end of this phase, the dorsiflexors contract eccentrically to slow plantar flexion.

At the left shoulder, extension or hyperextension occurs from frames 1 to 5. This extension speeds up from frames 1 to 4 and slows down from frames 4 to 5. The muscle contraction is thus concentric from frames 1 to 4 and eccentric from frames 4 to 5. The shoulder extensor

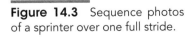

Figure 14.3 Sequence photos of a sprinter over one full stride.

muscles are the active muscle group from frames 1 to 4; then the shoulder flexor muscles contract eccentrically in frames 4 to 5. From frames 5 to 8, the left shoulder flexes. Shoulder flexion speeds up from frames 5 to 7 and slows down at the end of the period, from frames 7 to 8. The muscle contraction is thus concentric starting at frame 5 until just before frame 8, when it becomes eccentric. The shoulder flexors contract concentrically from frames 5 to 7; then the shoulder extensors contract eccentrically.

Rapid Joint Angular Accelerations and Impacts

Several instances of rapid joint angular acceleration occur during the sprinting stride. The hip joint rapidly accelerates in flexion from frames 1 to 3 due to the action of the hip flexors. This flexion slows rapidly from frames 4 to 5 due to the eccentric action of the hip extensors. From frames 7 to 8, hip hyperextension speeds up rapidly while the leg is bearing weight due to a large and powerful action of the hip extensors. The hip hyperextension then slows down before frame 8 due to eccentric action of the hip flexors.

Knee flexion rapidly accelerates from frames 1 to 2 due to the concentric action of the knee flexors. The eccentric action of the knee extensors rapidly slows this knee flexion from frames 2 to 3. The subsequent concentric contraction of the knee extensors causes rapid knee extension from frames 3 to 4. This extension is suddenly slowed by the eccentric action of the knee flexors at the end of the period, from frames 4 to 5 and frames 5 to 6. From frames 4 to 5, the two-joint hamstring muscle group (a knee flexor and hip extensor) must produce a large force as it acts eccentrically to slow both knee extension and hip flexion. Hamstring muscle pulls usually occur at this instant during the sprint stride. Rapid acceleration of the knee in extension also occurs at the start of frame 7 and continues through frame 8 as the leg pushes off due to the strong concentric action of the knee extensors.

The left ankle joint undergoes a rapid dorsiflexion from frames 6 to 7 as the foot's impact with the ground occurs. The strong eccentric action of the plantar flexors prevents the joint from sustaining injury. From frames 7 to 8, the ankle joint rapidly accelerates in plantar flexion while bearing weight due to the strong concentric action of the plantar flexors.

Rapid extensor acceleration of the left shoulder joint occurs from frames 2 to 3 due to concentric action of the shoulder extensors. Rapid slowing of this extension occurs in frames 4 to 5 due to eccentric action of the shoulder flexors. The shoulder flexors then act concentrically to accelerate flexion of the shoulder joint from frames 5 to 6.

Appropriate strength and power training exercises are indicated for the hip flexors and extensors, the knee flexors and extensors (especially the hamstring muscle group), the plantar flexors, and the shoulder flexors and extensors.

Extreme Joint Ranges of Motion

Few extreme joint ranges of motion are demonstrated by the sprinter in figure 14.3. The hip joint reaches a position of extreme hyperextension during push-off in frames 1 and 8, so the hip flexors are stretched. The ankle joint reaches positions of extreme plantar flexion in frames 1 and 8, thus possibly stretching the dorsiflexors. The ankle joint also reaches a position of extreme dorsiflexion at touchdown in frame 7, so the plantar flexors may be stretched. The shoulder joint reaches a position of extreme hyperextension in frame 5, so the shoulder flexors are stretched. Flexibility exercises are indicated for each of these muscle groups.

Table 14.4 is a completed chart of the qualitative anatomical analysis of the sprinting stride illustrated in figure 14.3.

Javelin Throw

As with the football punt and sprinting stride, the javelin throw is an activity in which the work done by the muscles primarily causes changes in kinetic energy. It is more difficult to analyze than the previous examples because it involves many segments and is a multiplanar activity. Once again, it involves quick movements, so a frame-by-frame analysis is required. Sequence drawings of an elite javelin thrower are shown in figure 14.4.

Joint Motions

The joints that seem to be important in javelin throwing include the elbow and shoulder of the throwing arm, the trunk (the intervertebral joints), and the right and left hip joints. Let's begin our analysis with the right elbow joint. From frames 1 through 5, the positions of the limbs at the elbow joint do not change, so no joint motion occurs. The elbow begins to flex from frames 5 to 6 and continues flexing through frame 8. From frames 8 to 10, the elbow then extends during the final delivery and release of the javelin.

The shoulder joint also maintains a fairly static position early in the throwing action, from frames 1 through 6. Some external rotation occurs from frames 6 to 8. Rapid internal rotation along with abduction occurs during the delivery, release, and follow-through from frames 8 to 10. From frames 9 to 10, extension also occurs at the shoulder joint.

The trunk (or intervertebral joints) is inactive from frames 1 to 2. It rotates slightly to the right as the crossover step is made from frames 2 to 3. From frames 3 to 6, the trunk unwinds and rotates to the left. From frames 6

Table 14.4 Sample Qualitative Anatomical Analysis of a Sprinting Stride

Joint	Frames	Joint motion	Muscle contraction	Active muscle group	Rapid acceleration or impact	Extreme range of motion
Left hip	1-2	Flexion	Concentric	Flexors	Yes	Hyperextension
	2-3	Flexion	Concentric	Flexors		
	3-4	Flexion	Concentric	Flexors		
	4-5	Flexion	Eccentric	Extensors	Yes	
	5-6	Extension	Concentric	Extensors		
	6-7	Extension	Concentric	Extensors		
	7-8	Hypertension	Concentric, then eccentric	Extensors, then flexors	Yes	Hyperextension
Left knee	1-2	Flexion	Concentric	Flexors	Yes	
	2-3	Flexion	Eccentric	Extensors	Yes	
	3-4	Extension	Concentric	Extensors	Yes	
	4-5	Extension	Concentric	Extensors	Yes	
	5-6	Extension	Eccentric	Flexors	Yes	
	6-7	Flexion	Eccentric	Extensors		
	7-8	Extension	Concentric, then eccentric	Extensors, then flexors	Yes	
Left ankle	1-2	No motion	Isometric	Dorsiflexors		Plantar flexion
	2-3	Dorsiflexion	Concentric	Dorsiflexors		
	3-4	Dorsiflexion	Concentric	Dorsiflexors		
	4-5	Dorsiflexion	Eccentric	Plantar flexors		
	5-6	Plantarflexion	Concentric	Plantar flexors		
	6-7	Dorsiflexion	Eccentric	Plantar flexors	Ground impact	Dorsiflexion
	7-8	Plantar flexion	Concentric, then eccentric	Plantar flexors, then dorsiflexors	Yes	Plantar flexion
Left shoulder	1-2	Extension	Concentric	Extensors		
	2-3	Hyperextension	Concentric	Extensors	Yes	
	3-4	Hyperextension	Concentric	Extensors		
	4-5	Hyperextension	Eccentric	Flexors	Yes	Hyperextension
	5-6	Flexion	Concentric	Flexors	Yes	
	6-7	Flexion	Concentric	Flexors		
	7-8	Flexion	Concentric, then eccentric	Flexors, then extensors		

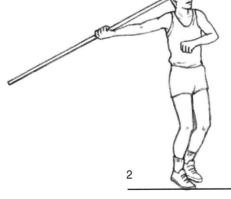

Figure 14.4 Sequence drawings of an elite javelin thrower. *(continued)*

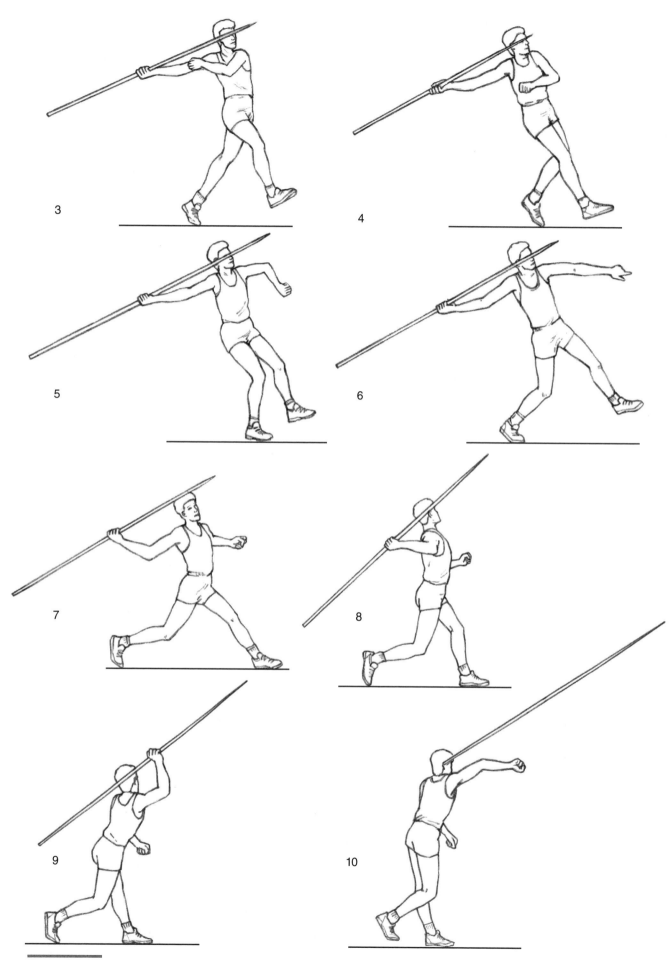

Figure 14.4 *(continued)* Sequence drawings of an elite javelin thrower.

to 7, it again rotates to the right as the left leg is planted. From frames 6 to 7 (and perhaps even earlier), the trunk also extends and hyperextends. From frames 7 to 10, the trunk rotates rapidly to the left and flexes during the delivery and follow-through.

The right hip flexes and adducts from frames 1 to 4. From frames 4 to 5, it flexes even more during the right foot plant, and some abduction occurs. From frames 5 to 6, hip extension begins and abduction continues. From frames 6 to 7, hyperextension occurs along with rapid internal rotation. From frames 7 to 10, the right hip flexes. Internal rotation continues from frames 7 to 8.

The left hip extends and adducts from frames 1 to 3. It then flexes, abducts, and externally rotates from frames 3 to 6. From frames 6 to 8, the left hip internally rotates rapidly and flexes as the left foot is planted. From frames 8 to 10, the left hip internally rotates farther as it extends.

Muscle Contractions and Active Muscle Groups

Now that we've identified the joint motions occurring during the javelin throw, the next step is to determine the type of muscle contractions and identify the active muscle groups. No movement occurs at the elbow joint from frames 1 to 5, so the muscle contraction is isometric or there is no muscle activity at all. The arm is acting as a rope or cable pulling on the javelin. To protect the elbow from hyperextending, the elbow flexor muscles contract isometrically. Elbow flexion speeds up from frames 5 to 7 (kinetic energy increases) and slows down from frames 7 to 8 (kinetic energy decreases). The elbow flexors contract concentrically from frames 5 to 7, and the elbow extensors contract eccentrically from frames 7 to 8. Elbow extension occurs from frames 8 to 10, speeding up from frames 8 to 9 and slowing down from frames 9 to 10. The elbow extensors continue to contract concentrically from frames 8 to 9; then the elbow flexors contract eccentrically from frames 9 to 10.

The shoulder is in abduction from frames 1 to 6 but does not change position, so the muscle contraction is isometric or there is no muscle activity at all. To hold this position, the shoulder abductors must be active isometrically. Some external rotation occurs from frames 6 to 8 due to the rapid rotation of the trunk and the inertia of the javelin and arm. The muscle contraction is eccentric. (Here our energy analysis scheme does not work very well or perhaps is difficult to understand. The body and legs seem to be speeding up, but the arm and javelin don't speed up as much, so there is a loss of energy—or perhaps a storage of elastic energy in the shoulder muscles—across the shoulder joint.) The shoulder internal rotators are the active muscle group. From frames 8 to 10, shoulder abduction and rapid shoulder internal rotation occurs. From frames 9 to 10, shoulder extension also

occurs. Kinetic energy increases greatly as the arm and javelin speed up in these frames, so the muscle contractions are concentric. In frame 10 and beyond (if more of the follow-through were shown), these joint motions continue, but they slow down due to eccentric muscle contractions. The shoulder abductors, internal rotators, and extensors are the active muscle groups until frame 10, when the abductors, external rotators, and flexors become active eccentrically.

The trunk is generally static or rigid from frames 1 to 2, so the normal postural muscles of the trunk contract isometrically. From frames 2 to 3, the trunk rotates to the right as the pelvis rotates to the left. This action is caused by the crossover step of the right leg and the thrust of the left leg, as well as the action of the trunk muscles. The kinetic energy of the pelvis increases initially, then decreases as the pelvic rotation stops, so the muscle contraction is first concentric and then eccentric. The right trunk rotators are the active muscle group initially, followed by the left trunk rotators. From frames 3 to 6, the trunk rotates to the left (as the pelvis rotates to the right). This action is partly due to the planting of the right foot in frame 5, but the increase in energy is caused by the concentric muscle contraction that initiates the movement. The left trunk rotators are the active muscle group. From frames 6 to 7, extension and hyperextension along with rotation to the right occur rapidly as the left leg is planted. At this point, the situation is similar to that with the shoulder joint. The large reaction force from the left leg causes the back to hyperextend as the pelvis rotates to the left, but the rest of the trunk lags behind momentarily. There is a loss of energy within the trunk (or perhaps a storage of elastic energy), so the muscle contraction is eccentric. The trunk flexors and left trunk rotators are the active muscle groups. From frames 7 to 10, the trunk rotates rapidly to the left and flexes. The rotation and flexion speed up from frames 7 to 9, then slow down from frames 9 to 10. The muscle contraction is thus concentric from frames 7 to 9 and eccentric from frames 9 to 10. The trunk flexors and left trunk rotators continue to be the active muscle groups in frames 7 to 9. The trunk extensors and right rotators are the active muscle groups in frames 9 to 10.

The right hip adducts and flexes from frames 1 to 4. These actions speed up from frames 1 to 3, and the muscle contraction is concentric. The hip adductors and flexors are thus the active muscle groups. The adduction and flexion slow down from frames 3 to 4, and the muscle contraction is eccentric. The hip abductors are thus the active muscle group. The right hip continues to flex as it slows down from frames 4 to 5, so the muscle contraction is eccentric. The hip extensors contract eccentrically from frames 3 to 5. From frames 4 to 6, the right hip abducts. This action speeds up throughout these frames, so the muscle contraction is concentric and the hip abductors are the active muscle group. From frames 5 to 7, the right hip extends and then hyperextends. The muscle contraction is concentric from frames 5 to 6, as the extension speeds up, and eccentric from frames 6 to 7 as the hyperextension slows down. The hip extensors contract concentrically from frames 5 to 6; then the hip flexors contract eccentrically from frames 6 to 7. From frames 6 to 8, rapid internal rotation occurs at the right hip. This action speeds up initially and then slows down, so the internal rotators contract concentrically from frames 6 to 7; then the external rotators contract eccentrically from frames 7 to 8. From frames 7 to 10, the right hip slowly flexes. The flexion speeds up slowly, so the hip flexor muscles contract concentrically.

The left hip adducts and extends from frames 1 to 3. These actions increase the body's kinetic energy as the left foot pushes off the ground, so the hip adductors and extensors contract concentrically until the very end of this phase, when these actions slow down and the hip abductors and flexors contract eccentrically. From frames 3 to 6, the left hip flexes, abducts, and externally rotates. These actions speed up from frames 3 to 5, when the hip flexors, abductors, and external rotators contract concentrically. The hip flexion, external rotation, and abduction slow down from frames 5 to 6, when the hip extensors, internal rotators, and abductors contract eccentrically. From frames 6 to 10, the left hip internally rotates as the pelvis rotates to the left. This action speeds up from frames 6 to 9 and slows down from frames 9 to 10. The muscle contraction is thus concentric from frames 6 to 9, and the internal rotators are the active muscle group. The muscle contraction is eccentric from frames 9 to 10, and the external rotators are the active muscle group. From frames 6 to 8, the left hip flexes. The flexion speeds up from frames 6 to 7, when the hip flexor muscles contract concentrically, and slows down from frames 7 to 8, when the hip extensor muscles contract eccentrically. The left hip extends from frames 8 to 10. This extension speeds up throughout this period due to the concentric contraction of the hip extensors.

Rapid Joint Angular Accelerations and Impacts

The most rapid joint angular acceleration in the javelin throw occurs as the right shoulder joint internally rotates, abducts, and extends from frames 8 to 10. The shoulder abductors, internal rotators, and extensors must be strong and powerful to produce these rapidly accelerated joint motions, and the shoulder adductors, external rotators, and flexors must be strong to slow these motions.

The impact of the left foot with the ground in frame 7 places large stresses on the trunk and lower extremity. The trunk musculature, specifically the trunk flexors and left rotators, produces large torques during this phase.

Appropriate strength and power exercises are indicated for all muscles around the shoulder joint as well as the trunk flexors and rotators.

Extreme Joint Ranges of Motion

Extremes in joint ranges of motion are observed in several instances in the javelin throw. The most extreme example is the position of maximal external rotation of the right shoulder joint in frame 8. The shoulder internal rotators are stretched maximally in this position. Other extremes in joint positions include the hyperextension of the trunk and right hip in frame 7. These extremes stretch the trunk and hip flexors. Appropriate flexibility exercises are indicated for all these muscle groups.

Table 14.5 is a completed chart of the qualitative anatomical analysis of the javelin throw illustrated in figure 14.4.

Summary

Biomechanics can be used to improve performance by improving training for that performance. A basic principle of training is specificity. Biomechanics can improve the specificity of training by identifying specific aspects of technique that need to be perfected (as discussed in the previous chapter); by identifying drills and exercises that mimic specific aspects of the technique or that exercise specific muscle groups used during the performance; and by identifying the specific muscles whose strength, power, or flexibility limit the performance. Most of this chapter addressed the last item. A qualitative anatomical analysis of a performance identifies the specific muscles that are active during a movement.

Five steps are involved in a qualitative anatomical analysis:

1. Divide the activity into temporal phases.
2. Identify the joints involved and their motions.
3. Determine the type of muscular contraction (concentric, eccentric, or isometric) and the predominant active muscle group at each joint.
4. Identify instances when rapid joint angular accelerations (rapid speeding up or slowing down of joint motions) occur and where impacts occur.
5. Identify any extremes in joint ranges of motion.

Steps 1, 2, 4, and 5 are self-explanatory and straightforward. Step 3, the heart of the analysis, is more difficult. Muscle contractions are determined through use of a work and energy analysis, because concentric muscle contractions produce positive work and increases in energy and eccentric muscle contractions produce negative work and decreases in energy. Identifying the energy changes that occur at each joint or within the body during a movement usually indicates the type of muscle contraction involved at that joint. We then identify the active muscle groups by considering the joint motion and muscle contraction together. If the muscle contraction is concentric, the active muscles are those that produce torque in the same direction as the joint motion. Thus, if the joint motion is flexion and the muscle contraction is concentric, the active muscles are the flexors. If the muscle contraction is eccentric, the active muscles are those that produce torque in the opposite direction from the joint motion. Thus, if the joint motion is flexion and the muscle contraction is eccentric, the active muscles are the extensors. The muscles that are active during instances when large joint angular accelerations or impacts occur need to be strong, so these are identified as muscle groups to train for strength and power. Likewise, the muscles that are stretched during instances of extreme joint ranges of motion need to be flexible, so these are identified as muscle groups to train for flexibility.

Table 14.5 Sample Qualitative Anatomical Analysis of a Javelin Throw

Joint	Frames	Joint motion	Muscle contraction	Active muscle group	Rapid acceleration or impact	Extreme range of motion
Right elbow	1-2	No motion	Isometric	Flexors		
	2-3	No motion	Isometric	Flexors		
	3-4	No motion	Isometric	Flexors		
	4-5	No motion	Isometric	Flexors		
	5-6	Flexion	Concentric	Flexors		
	6-7	Flexion	Concentric	Flexors		
	7-8	Flexion	Eccentric	Extensors		
	8-9	Extension	Concentric	Extensors		
	9-10	Extension	Eccentric	Flexors		
Right shoulder	1-2	No motion	Isometric	Abductors		
	2-3	No motion	Isometric	Abductors		
	3-4	No motion	Isometric	Abductors		
	4-5	No motion	Isometric	Abductors		
	5-6	No motion	Isometric	Abductors	Yes	
	6-7	External rotation	Eccentric	Internal rotators	Yes	
	7-8	External rotation	Eccentric	Internal rotators	Yes	External rotation
	8-9	Internal rotation	Concentric	Internal rotators		
		Abduction	Concentric	Abductors	Yes	
	9-10	Internal rotation	Concentric, then eccentric	Internal rotators, then external rotators	Yes	
		Abduction	Concentric, then eccentric	Abductors, then adductors		
		Extension	Concentric, then eccentric	Extensors, then flexors		

(continued)

Table 14.5 *(continued)*

Joint	Frames	Joint motion	Muscle contraction	Active muscle group	Rapid acceleration or impact	Extreme range of motion
Trunk (intervertebral joints)	1-2	No motion	Isometric	Postural muscles		
	2-3	Rotation right	Concentric, then eccentric	Right rotators, then left rotators		
	3-4	Rotation left	Concentric	Left rotators		
	4-5	Rotation left	Concentric	Left rotators		
	5-6	Rotation left	Concentric	Left rotators		
	6-7	Rotation right	Eccentric	Left rotators	Ground impact	
		Extension	Eccentric	Flexors	Ground impact	Hyperextension
		Hyperextension	Eccentric	Flexors		
	7-8	Rotation left	Concentric	Left rotators		
		Flexion	Concentric	Flexors		
	8-9	Rotation left	Concentric	Left rotators		
		Flexion	Concentric	Flexors		
	9-10	Rotation left	Eccentric	Right rotators		
		Flexion	Eccentric	Extensors		
Right hip	1-2	Flexion	Concentric	Flexors		
		Adduction	Concentric	Adductors		
	2-3	Flexion	Concentric	Flexors		
		Adduction	Concentric	Adductors		
	3-4	Flexion	Eccentric	Extensors		
		Adduction	Eccentric	Abductors		
	4-5	Flexion	Eccentric	Extensors		
		Abduction	Concentric	Abductors		
	5-6	Extension	Concentric	Extensors		
		Abduction	Concentric	Abductors		
	6-7	Hyperextension	Eccentric	Flexors		Hyperextension
		Internal rotation	Concentric	Internal rotators		
	7-8	Flexion	Concentric	Flexors		
		Internal rotation	Eccentric	External rotators		
	8-9	Flexion	Concentric	Flexors		
	9-10	Flexion	Concentric	Flexors		

Joint	Frames	Joint motion	Muscle contraction	Active muscle group	Rapid acceleration or impact	Extreme range of motion
Left hip	1-2	Extension	Concentric	Extensors		
		Adduction	Concentric	Adductors		
	2-3	Extension	Concentric, then eccentric	Extensors, then flexors		
		Adduction	Concentric, then eccentric	Adductors, then abductors		
	3-4	Flexion	Concentric	Flexors		
		Abduction	Concentric	Abductors		
		External rotation	Concentric	External rotators		
	4-5	Flexion	Concentric	Flexors		
		Abduction	Concentric	Abductors		
		External rotation	Concentric	External rotators		
	5-6	Flexion	Eccentric	Extensors		
		Abduction	Eccentric	Abductors		
		External rotation	Eccentric	Internal rotators		
	6-7	Flexion	Concentric	Flexors	Ground impact	
		Internal rotation	Concentric	Internal rotators	Ground impact	
	7-8	Flexion	Eccentric	Extensors		
		Internal rotation	Concentric	Internal rotators		
	8-9	Extension	Concentric	Extensors		
		Internal rotation	Concentric	Internal rotators		
	9-10	Extension	Concentric	Extensors		
		Internal rotation	Eccentric	External rotators		

KEY TERMS

physical training (p. 341)
qualitative anatomical analysis (p. 341)
technical training (p. 340)

REVIEW QUESTIONS

1. What other qualitative or quantitative methods (besides the qualitative anatomical analysis described in the previous pages) exist for identifying the specific muscle groups that are active during an exercise or movement?

2. What elbow and shoulder joint muscle groups are exercised during a push-up?

3. What lower extremity muscle groups are exercised during a squat?

4. What hip, knee, and ankle joint muscle groups are exercised during the landing phase following a vertical jump?

5. Is the contraction of the muscles in the previous question concentric, eccentric, or isometric?

6. During the follow-through phase of a baseball pitch, what muscle groups are active in the throwing shoulder?

7. Is the contraction of the shoulder muscles in the previous question concentric, eccentric, or isometric?

8. Complete a qualitative anatomical analysis of an underhand-grip pull-up.

9. Complete a qualitative anatomical analysis of a jumping jack exercise.

10. Complete a qualitative anatomical analysis of a standing long jump.

chapter 15

Qualitative Biomechanical Analysis to Understand Injury Development

Steven T. McCaw, PhD, FACSM
Illinois State University

objectives

When you finish this chapter, you should be able to do the following:

- Differentiate between force and pressure or stress

- Explain how the stress continuum relates to tissue adaptation and injury

- Describe the concept of the stress threshold

- Differentiate between intrinsic and extrinsic factors related to injury development

- Identify intrinsic factors that predispose an individual to injury

- Identify extrinsic factors associated with high stress during skill performance

- Explain the concepts of cross-training and within-activity cross-training

- Suggest interventions to decrease the risk of injury during performance of a task

You are at the starting line ready to run a popular 10K race. A friend approaches and explains that he is not registered to run because he recently developed a nagging pain in the lower leg. He had modified his training for the race and hoped to achieve a personal best time. You wonder if the changes in the training program might have led to the injury. This chapter discusses the biomechanical basis of injury and outlines a structured approach to identifying factors related to injury prevention.

Promising careers in sport and well-intentioned fitness programs are sometimes cut short by injury. Competitive runners can suffer a variety of ailments in the lower extremity and low back that interrupt training. A baseball pitcher can suffer continuous shoulder or elbow pain that restricts the ability to throw. For many seeking to improve their fitness level, an injury is often reason enough to abandon a cardiovascular training program. Obviously, preventing injury can benefit both competitive and recreational athletes, but injury prevention requires an understanding of why injury occurs.

Mechanical Stress and Injury

The combination of all forces acting on a body is responsible for the observed motion of the body. Improving performance depends on improving the application of force to the body; that is, by altering the size, direction, line of application, and timing of force production, a person can improve performance.

Forces acting on a body are also responsible for the occurrence of injury. An injury is damage to a tissue that inhibits performance. For example, a baseball striking a batter in the face can fracture the zygomatic bone in the cheek; excessive inversion of the ankle can sprain the ligaments on the lateral surface of the ankle joint; and muscle strains occur when high levels of tension developed by a muscle disrupt the continuity of the connective tissue within the muscle.

Although the simplistic interpretation is that "high levels of force" cause injury, a simple example demonstrates that more than the size of the force itself is responsible. Consider this offer: For $100, would you lie down on a wooden floor and let a 10-year-old boy stand barefoot on your abdomen? Many of you would agree to that (even without knowing how heavy he is). The forces applied to your body would be his body weight pushing down on you, gravity pulling down on you, and the floor pushing back up on you with a magnitude equal to the boy's weight plus yours. The contact between the floor and your back spreads the force over a wide area. The soles of his feet spread his weight over your abdomen. Although the experience would be uncomfortable, for $100 most people would put up with the mild discomfort.

Would you lie down on a wooden floor and let the boy stand on your abdomen wearing golf spikes? Except for the greediest individuals, few would agree to this offer. Why? The forces would be the same if we ignored the slight increase in the boy's weight because of the shoes. The contact area between your back and the floor would be similar. The only change is that the boy's weight would now be concentrated over a smaller surface—the narrow tips of the golf spikes. This scenario promises to be so painful that enduring the discomfort would not be worth $100 to most people. It also clearly demonstrates that more than force magnitude and direction are responsible for injury.

The concept of mechanical stress provides the starting point for understanding all injury and clarifies the scenario described. As we saw in chapter 9, mechanical stress (pressure) refers to the distribution of a force over the body that it acts on. Graphically, figure 15.1 compares the difference in stress on your abdomen if the boy were to stand on you with bare feet and with golf shoes. His body weight (the downward force) remains the same, but the area changes with the different conditions. Mathematically, stress is defined as force/area (read "force over area"). With force measured in newtons and area measured in meters, the unit of stress is N/m^2, also known as a pascal. You may be more familiar with the English system unit, which is pounds per square inch (psi).

Mechanical stress is imposed on the body in different ways. Compressive stress occurs when opposing forces squeeze the body together. For example, when you land from a jump, a compressive force is applied to your leg by the body weight above it acting downward and the ground reaction force acting upward. Each vertebra is continually exposed to compressive stress as the weight of the body positioned above it pushes down and the vertebra

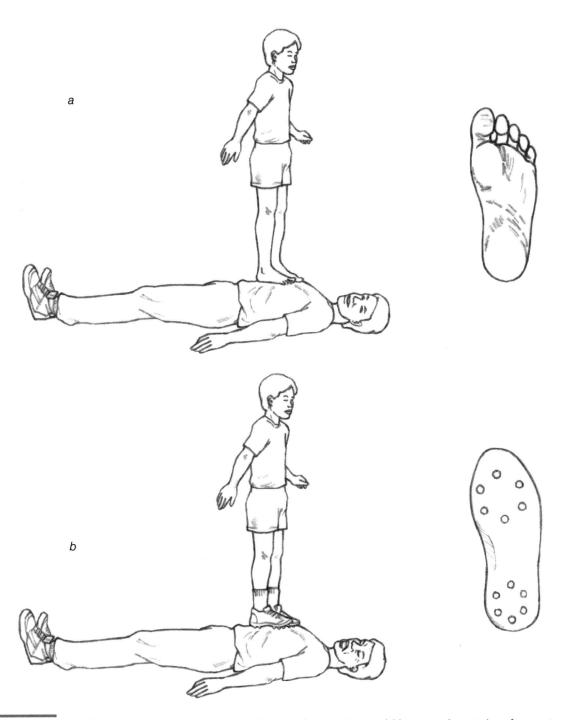

Figure 15.1 The difference in pressure on your abdomen when a 10-year-old boy stands on it barefoot or in golf spikes. His weight is 356 N (approximately 80 lb), and he stands with his weight spread equally on his feet. *(a)* The surface area of the sole of each foot is about 97 cm² (15 in.²). The pressure under the contacting area of each foot is calculated to be $P = F/A$ = 356 N/97 cm² = 3.67 N/cm². *(b)* Each golf shoe has 12 spikes, and each spike has a surface area of about 0.03 cm² (0.004 in.²). The pressure under the contacting area of the shoes is calculated to be 356 N/(12 × 0.03 cm²) = 988.89 N/cm². The pressure under the spikes of a golf shoe is about 269 times the pressure under a bare foot.

below it pushes up. Another form of compressive loading comes from muscle tension applied at a bone. The pull of a muscle can be resolved into two components (see figure 15.2). The rotary component acts perpendicular to the bone and tends to cause rotation of the segment. The parallel component acts along the bone toward the joint crossed by the muscle. This force acts to pull the segment tight against the adjacent bone with which it articulates. The force acts simultaneously with the force acting toward the joint along the adjacent segment, and the two forces compress the joint and stabilize it by squeezing the segments together. Compressive force is associated with a variety of injuries. Chondromalacia patella, characterized by deterioration of the cartilage under the kneecap, is caused by the high compressive component of the quadriceps force during knee motion. Compressive force applied down the vertebral column during a headfirst collision may collapse the vertebrae, causing severe spinal cord injury.

Tensile stress occurs when a force pulls on a body. Sources of tensile stress include loading of the medial collateral ligament of the knee joint when the leg is struck from the lateral side and the loading imposed on a muscle attachment site such as the tibial tuberosity when tension is present in the patellar ligament. Avulsion fractures, which occur when high tensile stress in a tendon or ligament breaks off bone at the site of attachment, are more likely when muscle activity is high or when a joint is pulled to an extreme end of its range of motion. Common sites of avulsion fracture include the lateral malleolus when the ankle joint is extremely inverted and the medial epicondyle of the humerus during the pitching motion. When we calculate tensile stress, area refers to the cross-sectional area of the tissue. For example, the Achilles tendon in a typical adult is somewhat circular, with a diameter of about 0.7 cm (0.28 in.), or a radius of about 0.35 cm (0.14 in.). Using the equation for the area of a circle, $A = \pi r^2$, the cross-sectional area of the tendon is $A = \pi (0.35 \text{ cm})^2 = 0.385 \text{ cm}^2$. If the force developed by the triceps surae muscle group is 2800 N, the calculated tensile stress in the Achilles tendon is $\sigma = F/A = 2800$ N/0.385 cm² = 7273 N/cm², often reported as 72.7 MPa (megapascals, or million pascals).

Shear stress occurs when a force tends to slide two parts of an object across each other. Examples of shear stress in the human body include the loading of knee joint tissues during foot support in running. The momentum of the body mass above the knee joint tends to cause the femur to slide forward over the tibial plateau. The hyaline cartilage of the knee joint and the menisci are subjected to shear stress, and the medial collateral ligament and the anterior and posterior cruciate ligaments are under tensile stress. Commonly, when a knee "blows out," the anterior cruciate ligament, the medial collateral ligament, and the menisci are all injured.

In most situations, the human body is subjected to complex loading—a combination of compressive, tensile, and shear stresses imposed at the same time. For example, consider the forces acting on the calcaneus (heel) when your foot is planted on the ground as you stand in the anatomical position. A compressive force is applied to the calcaneus by the vertical ground reaction force acting upward and your body weight acting downward. Tensile stress is imposed at the insertion point of the Achilles tendon and the plantar fascia, as the soleus is acting to maintain upright balance and the plantar fascia supports the longitudinal arch of the foot. Shear stress is imposed where the calcaneus articulates with the navicular bone (a tarsal). Torsion is present in the calcaneus because of its alignment along the longitudinal axis of the foot. Thus, loading of the calcaneus is a combined load, reflecting the combined action of the multiple forces applied. Think of some other activities and identify the nature of the imposed stresses on various tissues.

Tissue Response to Stress

Most tissues in the human body are remarkable in their ability to adapt to imposed stress. In 1892, German anatomist Julius Wolff summarized the nature of the response to mechanical stress. This summary, known as **Wolff's law**, states that a tissue adapts to the level of stress imposed on it; that is, the level of adaptation in a tissue reflects the level of typical loading. As mechanical stress increases, a tissue gains strength through hypertrophy, or an increase in size. If mechanical stress is removed, a tissue loses strength through atrophy, or a decrease in size. Although Wolff's observations were specific to bone, subsequent study of the body revealed the law to be applicable to other connective tissue such as ligament and tendon.

The study of anatomy offers clear examples of Wolff's law. We learn that the protuberances and landmarks on bones are adaptations to the compressive, tensile, tor-

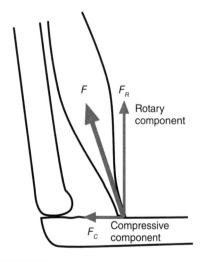

Figure 15.2 The force produced by a muscle can be resolved into two components. The rotary component affects the rotation of the segment, and the compressive component affects joint stability by altering the compressive force at the joint.

sional, and shear forces imposed by the tensile forces of tendons and ligaments and to the compressive forces of gravity and muscle. Structurally, bone adapts by increasing mineralization and aligning trabeculae in the direction of the imposed stress. Muscle tissue adapts to training, or the imposition of an overload stress, by increasing cross-sectional area as individual fibers increase in diameter. The arrangement of collagen and elastin in tendon and ligament similarly reflects the exposure to tensile loading and is affected by the level of training. A notable exception to Wolff's law is neural tissue such as sensory receptors and axons, which do not benefit from applied stress. Although learning is characterized by the adaptation and proliferation of synapses, the nerve tissue itself does not respond to the level of stimulus by hypertrophying or atrophying. In some disease or overuse conditions, one or more nerves may atrophy, but this is technically a loss of myelin rather than a change in the diameter of the nerve itself.

The level of stress on a tissue varies with changes in activity level. The magnitude of the imposed stress ranges from a very low level or no stress to a very high level. For example, the compressive stress on the femur is less when one lies in bed than during aerobic dance. Shear stress on knee cartilage is less when one is standing upright than during downhill skiing. Tensile stress on the skin increases when an object pushing on the skin stretches it. Controlling the level of imposed stress is important in training various tissues and avoiding injury.

Graphically, the level of imposed stress can be viewed as a **stress continuum** (see figure 15.3). The continuum of stress ranges from a low level (pathologic underload zone) to a high level (pathologic overload zone). With an active lifestyle, the level of stress is usually kept within the physiologic loading zone, and the tissue maintains its current status. This can be thought of as a maintenance program, with the tissue getting neither stronger nor weaker. When stress is maintained in the **physiologic loading zone**, muscle maintains the same force-generating capability, bone mineral content stays the same, and tendons and ligaments maintain their ability to withstand tensile stress.

Controlling the level of imposed stress is important in training various tissues and avoiding injury.

In the physiologic training zone, a level of stress is imposed above that to which the tissue has adapted. A stress within this zone exceeds the yield strength of the tissue and causes microdamage (microscopic tissue damage) within the tissue. The larger the imposed stress, the greater the extent of this microdamage. The body's response to the microdamage is to initiate **remodeling**, or rebuilding, of the tissue. The time required for remodeling is related to the extent of damage; the greater the damage, the longer remodeling takes. According to Wolff's law,

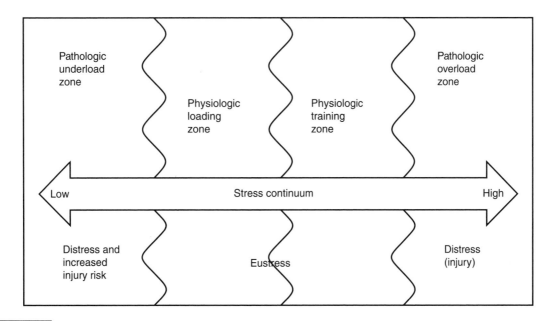

Figure 15.3 The stress continuum demonstrates that the level of stress imposed on a body ranges from low to high. Undesirable effects occur when the level of imposed stress is too low or too high. A training effect occurs if the level of stress is systematically raised above the physiologic loading zone, or the level of stress to which the body has adapted.

the healing of the microdamage leads to hypertrophy, or strengthening, of the tissue. Systematic application of loads in the physiologic training zone is the basis of the overload principle, which is responsible for causing the cellular and structural changes within a tissue called the training effect. Systematic loading means that the imposed tissue loading purposely exceeds the tissue's yield threshold to cause microdamage, but adequate rest time is provided to allow for tissue rebuilding before the next overload session (figure 15.4). Typically, the mineral content of bone will increase and muscle fibers will hypertrophy. The magnitude of the training response is, of course, dependent on other factors, including genetics, diet, rest, and hormonal status.

At opposite ends of the continuum are areas marked distress (*dis* = bad). These areas reflect stress levels that are too low, below the physiologic loading zone, and too high, above the physiologic training zone. Loading of a tissue at these levels leads to undesirable changes that compromise tissue function.

The pathologic underload zone represents continued low loading applied to a tissue such as that occurring during extended periods of inactivity. An inactive, or sedentary, lifestyle may be selected by choice, or it may be imposed by prolonged bed rest or immobilization with a cast. A low level of mechanical stress is also

an unfortunate characteristic of space flight, when the lack of gravity reduces the compressive loading of the skeleton and the amount of muscle tension needed to move body segments and external objects. If the level of imposed stress is not raised out of the underload zone, in accordance with Wolff's law, a detraining effect occurs. An understressed tissue begins to atrophy, or waste away. Muscle decreases in cross-sectional area, bone mineral content decreases, and ligaments and tendons lose their flexibility. You may have observed such tissue changes when a cast is removed and the immobilized limb has noticeably atrophied compared to the uninjured limb. A chronically underloaded tissue becomes weaker and more prone to injury as the yield threshold of the tissue is reduced.

The pathologic overload zone represents a level of loading that causes substantial damage to the tissue. When a single application of a relatively high level of stress is identified as the injury-causing factor, the injury is referred to as an accidental or **traumatic injury**. This type of injury occurs during high-impact collisions between two or more objects, such as another player, the ground, or an obstacle. For example, a medially directed blow to the knee can strain the medial collateral ligament, or a fall onto an outstretched arm can fracture the distal radius. An extended period of recuperation and

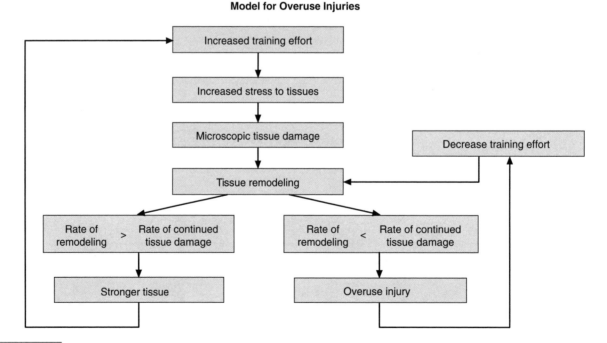

Model for Overuse Injuries

Figure 15.4 Williams' model for the effect of repetitive stress imposed on the body. The imposed stress causes microscopic tissue damage and initiates tissue remodeling. If the rate of remodeling is greater than the rate of tissue damage, a training effect occurs and the tissue gets stronger. If the rate of remodeling is slower than the rate of tissue damage, an overuse injury develops.

Reprinted, by permission, from K.R. Williams, 1993. Biomechanics of distance running. In *Current issues in biomechanics*, edited by M.D. Grabiner (Champaign, IL; Human Kinetics), 21.

rehabilitation is required to return the tissue to a state in which activity can be resumed. Damage to tissue can also be caused by the repeated application of a level of stress that is too low to cause traumatic injury but exceeds the threshold for overuse injury. In the next section we consider overuse injury in more detail.

Mechanism of Overuse Injury

As is evident from clinical records, not all injury is caused by a single, identifiable stress. An **overuse injury** can occur following repeated applications of a stress lower than that required to cause injury in a single application. For example, anterior shin pain may develop due to the repeated foot strikes during a long-distance run. Workers in an office or factory may develop chronic shoulder and neck pain due to prolonged work in an awkward posture involving arm abduction or neck flexion. No single occurrence of a running step or an incident during the work process can be identified as the cause of the pain. Instead, the injury and associated discomfort reflect a combination of the number of repetitions and the magnitude of the stress.

Figure 15.5 graphically presents the theoretical relationship among stress magnitude, number of repetitions, and development of an injury. The vertical axis is the magnitude, or size, of the stress; the horizontal axis is the number of repetitions of the stress, or the frequency of application. The curved line represents the injury threshold and reflects the observed interaction of stress magnitude and frequency with the occurrence of injury.

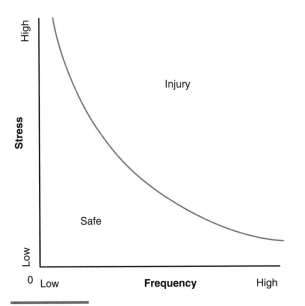

Figure 15.5 Relationship among stress magnitude, frequency of stress application, and injury.

The higher the magnitude of stress imposed, the fewer repetitions needed for an injury to develop.

It appears that what determines whether imposed stress will cause a training effect or lead to injury is the amount of time provided for remodeling to occur. Williams' model of the effect of imposed stress (figure 15.4) shows that with adequate rest, a training effect follows the tissue damage caused when a load falls within the physiologic training zone. However, if the stress is reimposed but adequate time for the tissue to repair is not provided, an overuse injury eventually develops. Without adequate recovery time, stress is imposed on tissue that is already damaged, and the repetitive loading increases the extent of damage. Eventually, the pain and discomfort from the repetitively stressed tissue inhibit performance, and an overuse injury is diagnosed.

Because an overuse injury is caused by the frequency and magnitude of loading combined with inadequate rest time for tissue remodeling, eliminating overuse injury has proven to be quite difficult. Symptoms of an overuse injury are often the first indication that an individual is at risk. Frequently, someone with an overuse injury tries to "work through" the pain, hoping that it is temporary and will disappear on its own. This approach, however, violates the need to treat the injury by providing adequate time away from the stress for the tissue damage to heal.

Once the individual is healed and ready to resume the activity, the probable cause of the overuse injury must be eliminated to prevent recurrence—that is, the frequency of application or the magnitude of the stress must be reduced, or the resistance to injury must be increased through a systematic training program. Without such an intervention, the injury is likely to recur.

Individual Differences in Tissue Threshold

Based on the conceptually simple model of injury just outlined, preventing injury would seem to be relatively easy: Reduce the magnitude and frequency of loading and provide adequate rest for remodeling (the training effect) to occur. Or, in the case of traumatic injury, simply avoid levels of stress beyond those that tissues can withstand. However, injury prevention is complicated by the difficulty in determining the injury threshold and the amount of time to wait before stress is reimposed. Different tissues in the body have different thresholds. For example, the threshold for bone is higher than that for tendon, which is higher than that for ligament, which is higher than that for cartilage. In addition, the injury threshold varies within each tissue according to the direction of stress. For example, bone strength is highest in compression and lowest in shear, with tensile strength falling in between. Establishing a stress threshold that is applicable to all individuals is complicated because the threshold of a tissue such as bone is not the same for everyone. Instead, the injury threshold reflects differences in genetics, training, and tissue adaptation. A level of stress falling within the physiologic loading zone of one individual may fall within the pathologic overload zone of another. Finally, the rate of recovery, or tissue remodeling, also differs among individuals. A recovery period adequate for one person might not be adequate for another. Individual differences in tissue threshold and recovery rate complicate the establishment of training guidelines that will be successful in preventing all injury.

For example, consider the guidelines that have been established to reduce the number of elbow injuries among pitchers. In Little League baseball, a rule limits the number of innings a player can pitch in a single game as well as the number of games per week. Obviously, the first part of the rule is intended to limit the number of times the stress of pitching is imposed, and the second part is intended to ensure that a rest period is provided for recuperation from the imposed stress. Although no such formal rule exists for pitchers in professional baseball, a manager typically limits a starting pitcher to between 100 and 120 pitches per appearance and tries to provide four days of rest between starts. Unfortunately, elbow injuries still occur at both the Little League and professional levels. A reduction in the number of elbow injuries in Little League baseball since the implementation of rules limiting the number of innings and games that a pitcher can pitch, suggests that, for the most part, the guidelines are adequate for reducing the rate of injury. The observed injuries may develop because some players throw many more pitches than expected during the allowed innings, some throw more frequently each week because parents or coaches knowingly violate the rule, and some players may throw hard outside of formal games or practices. However, players who do not violate the rules develop injuries too. Obviously, the guidelines are not adequate for these players, but it is not possible to identify those who will be injured even if the guidelines are followed. Individual differences in tissue threshold and level of imposed stress make it impossible to develop foolproof guidelines for injury prevention.

Intrinsic and Extrinsic Factors Affecting Injury

The observed differences in the types of injury recorded suggest that factors related to both the individual and the task are associated with injury development. Characteristics of the individual are known as intrinsic factors. Characteristics of the task and the environment in which it is performed are known as extrinsic factors. Generally,

intrinsic factors reflect an individual's ability to withstand loading, whereas extrinsic factors reflect the nature of the loading that is imposed on the individual. Examples of intrinsic and extrinsic factors are listed in table 15.1.

Intrinsic factors related to injury include anthropometrics; skeletal structure, such as bone density and joint congruity (alignment); current fitness level, such as muscle strength, endurance, and flexibility; and previous history of injury. These factors are related to an individual's ability to cope with imposed mechanical stress—that is, to how an imposed force creates stress within the individual and how well the tissues are adapted to the level of stress. If you consider the obvious variety of physical shapes among individuals, it will be readily apparent that individual differences in anthropometrics play a potentially large role in protection from or predisposition to injury.

Extrinsic factors related to injury include characteristics of both the task and the environment. Task-related factors include the nature of the task being performed; how a given individual performs the task; the movement patterns involved; and the frequency, speed, and duration of performance. Environmental factors include the type of surface played on, the rules used, the skill level and number of teammates and opponents, the type and condition of protective equipment, the type and condition of implements, and the current weather conditions. These factors primarily affect the magnitude and frequency of the applied stress.

For example, consider how different people walk, run, swim, perform a dance leap, swing a hammer, or throw a ball. Even though there are similar general characteristics for each movement, each person performs a task that has unique individual characteristics. For a pitcher, variable characteristics of the task include type of pitch thrown; finger placement on the ball; length of stride; degree of trunk rotation; and the positions and ranges of motion of the joints of the throwing arm during preparation, execution, and follow-through. For an assembly line employee, extrinsic characteristics include joint posture assumed during the work, rate of assembly, number of hours at the task, percentage of maximum strength required to complete the task, and use of allotted break time. Obviously, differences among individuals in performance characteristics can increase or decrease the risk of injury during performance of a similar task. The risk of injury is increased for assembly workers whose wrists must be in an awkward position when they perform their job if they maintain similar awkward positions during scheduled break periods or time away from work. Altering the posture is a critical component of preventing repetitive overuse injuries in the workplace.

When a given task is performed, the various intrinsic and extrinsic factors interact to set the level of risk for a

Table 15.1 Intrinsic and Extrinsic Factors Related to Injury Development

Intrinsic factors	Influence
Skeletal alignment	Affects the pattern of stress imposed on tissues
Muscle strength	Affects magnitude of loading and shock absorption
Muscle endurance	Affects magnitude of loading and shock absorption
Current level of fatigue	
Joint flexibility	Affects loading pattern of segments
Tissue temperature	
Joint alignment	Affects area of force distribution
Bone mineral density	Affects strength of bone to withstand stress
Diet	
Hormone levels	
Previous injury history (injury status)	Affects tissue threshold
Muscle firing pattern	Affects magnitude of load and pattern of imposed load
Body mass (body weight)	Affects magnitude of imposed load
Body composition	
Psychological factors	Affect pain threshold
Motivation	
Pain tolerance	
Extrinsic factors	**Influence**
Task	
Nature of the task	
Single (discrete) versus repetitive (continuous)	Affects magnitude of loading and recovery
Movement pattern	Affects tissues loaded
Intensity of performance	Affects magnitude of loading
Frequency of performance	Affects recovery time
Environment	
Playing surface	
Slope	Affects magnitude and direction of loading
Hardness	Affects magnitude of loading and friction
Material condition	Affects magnitude of loading and friction
Equipment	
Footwear	
Outsole materials	Affect magnitude of friction
Midsole materials	Affect amount of cushioning
Padding	
Level of participation	Affects magnitude of forces imposed
Recreational versus competitive	
Skill level of opponents	
Rules	Affect the magnitude and frequency of loading, and the pattern of force distribution

particular individual during a specific performance. The nature of this interaction is indicated in figure 15.6. The intrinsic factors taken together set the threshold value for the stress that may cause an injury. The extrinsic factors taken together reflect the potential for the given performance to impose a stress that exceeds the threshold value determined by the intrinsic factors. For example, the strength, flexibility, and anatomical alignment of the vertebral column each represent an intrinsic factor related to the risk of suffering a back injury during the lifting of an object. Strength of the abdominal and back muscles affects the support provided to the column, as well as the size of the force that will be imposed due to the compressive and tensile forces that muscle activity develops. Flexibility of the column reflects the condition of both the annulus fibrosus (fibrous outer covering) and the nucleus pulposus (inner material) of the intervertebral discs and is related to the injury threshold. A person with a weak spot in the fibers of the annulus fibrosus, caused by either a congenital defect or previous loading of the disc, is at increased risk of disc herniation when the back is loaded. Vertebral column alignment (whether scoliotic, kyphotic, or lordotic) affects the pattern of stress distribution over a disc, because the angle between adjacent vertebrae determines the tensile, compressive, and shear loading of the disc. Extrinsic factors for a given lift include the weight and dimensions of the load, the shape of any available grips, and the trunk angle adopted during the performance of the lift. Greater load weight imposes greater stress on the vertebral column. Dimensions of the load determine whether the load can be kept close to the body, affecting the amount of muscle tension required to perform the lift. Grips affect the potential for the load to slip, which can affect the amount of muscle activity required during the lift. Keeping the back "straight," or upright, with the vertebral column in the aligned anatomical position, creates less stress on an intervertebral disc than does flexing forward to a more horizontal position of the trunk. Together, these extrinsic factors influence the magnitude of the stress imposed on the intervertebral discs of the back and determine whether their stress threshold will be exceeded.

Risk of injury to an individual during a task can be considered an interaction between the intrinsic factors characterizing that individual and the extrinsic factors characterizing the specific task and the environment in which it is performed. Multiple extrinsic and intrinsic factors have been identified for various tasks. Consider the risk of running on a road. Most paved roadways have a raised crown in the middle, tapering off to the sides to improve drainage. The slope of the road, an extrinsic factor, can interact with the leg length of the runner, an intrinsic factor, to create, reduce, or increase the leg length inequality (see figure 15.7). Leg length inequality is an intrinsic factor that increases the risk of hip, knee, or low back injury because it shifts the musculoskeletal alignment of the lower back, hip, knee, and ankle. This shift alters the pattern of stress distribution within these joints. A significant leg length inequality (greater than 0.5 cm or 0.25 in.) poses a risk of premature hip or knee and low back degeneration due to the loads imposed during walking or running—loads that represent less risk if there is no leg length inequality. However, the slope of the running surface interacts with the length of the runner's legs to cause, reduce, or increase the difference between the lengths of the runner's right and left legs. When feasible, such as when running on paved bike or walking trails, a runner should switch sides of the path to reduce the "environmental" effect of road slope on leg length inequality. If the extrinsic risk factors are minimized by running on a surface that reduces the leg length difference, or at a pace, in shoes, or on a surface that reduces the load across the malaligned joint, the risk of injury may not be increased.

Conceptually, considering the development of an injury as an interaction between intrinsic and extrinsic factors provides a basis for evaluating the appropriateness of an activity for a particular individual. It also provides a basis for identifying the steps required to reduce the risk of injury. An activity presents a risk if it exposes a tissue to a high level of stress or to repetitive applications of stress, or if it is performed by an individual with a low stress threshold. An intervention to prevent injury must be based on reducing the magnitude of the stress imposed on a tissue, reducing the frequency of the stress,

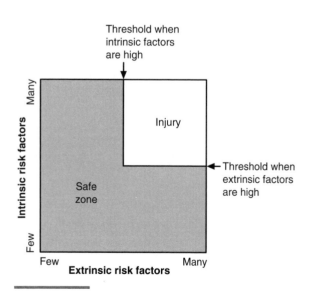

Figure 15.6 The interaction between intrinsic factors and extrinsic factors and the development of injury.

Adapted Messier et al., 1991.

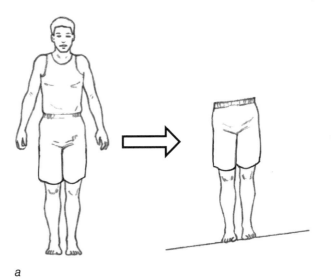

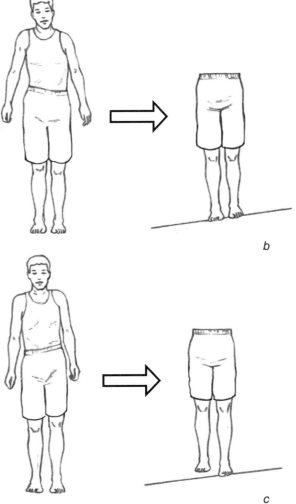

Figure 15.7 The slope of a paved roadway, an extrinsic factor, interacts with the leg length of a runner, an intrinsic factor, to create, reduce, or increase the characteristic of leg length inequality. *(a)* The slope of the road creates a leg length inequality in a runner with equal leg lengths; *(b)* the slope reduces the leg length inequality in a runner with a leg length inequality; *(c)* the slope increases the size of the difference in leg lengths in an individual with a leg length inequality.

or increasing the tissue's ability to withstand stress. If a particular intervention does not affect one of these three factors, it will not provide protection from injury. Let's look in more detail at how intrinsic and extrinsic factors relate to the development of injury in running.

Sample Analysis: Overuse Injuries in Running

Running is a popular pastime for millions of fitness enthusiasts. In addition, running competitively is a sport of choice for many, and running is an integral part of other sports. Although a running program yields considerable cardiovascular training benefit, the development of an overuse injury is relatively common. Clinical data indicate that the knee is the most common site of running-related injury, followed by the lower leg and the foot.

The development and patterns of overuse injury in running are interesting from a biomechanics perspective.

Clinical data reveal that some runners are frequently injured while others are injured rarely. A unilateral pattern of injury development is prevalent in that injury typically develops in only one side of the body despite the cyclical nature of running. These observations raise the question whether some individuals are predisposed to or at greater risk of injury. Finally, the type of running program and the rate of injury, as well as the effect of shoes and surfaces on the development of injury, suggests that the nature of loading is important in development of an injury. Patterns of running injury suggest that the risk of injury is very much related to the interactions among the individual, the task, and the environment, or intrinsic and extrinsic factors.

To understand how an overuse injury develops in running, it's necessary to examine the intrinsic and extrinsic factors that interact to cause injury. As already explained, intrinsic factors are those related to the characteristics of the individual runner, whereas extrinsic factors are related to the movement of running and the nature of the environment in which the runner performs. We'll begin with the general characteristics of the loading imposed during running, taking into account the task and the environment. Then we'll discuss intrinsic factors related to running injury.

General Patterns of Loading During Running

Running is a cyclical activity consisting of alternating periods of single-leg support (stance phase) and no support (the airborne or flight phase). Most runners exhibit a step rate between 50 and 70 steps per minute. In other words, each foot contacts the ground 300 to 900 times per mile. Mechanically, each stance phase begins with contact between the downward-falling runner and the ground. During stance, the downward motion of the runner must be stopped and then reversed to propel the runner upward and forward into the next airborne

phase. With each stance phase, the lower extremity must initially absorb energy as the downward vertical motion of the runner is stopped, briefly support the body during midstance, and finally, generate energy as the runner is pushed back into the air during the propulsion phase. All these actions occur within the 200 to 300 ms that the foot is in contact with the ground.

Most runners exhibit a rearfoot landing with the foot slightly dorsiflexed and inverted at touchdown. Initial contact with the ground occurs along the lateral border of the heel of the shoe. Some runners use a midfoot or forefoot landing, with the foot slightly plantar flexed and inverted at touchdown. Initial ground contact still occurs along the lateral border of the shoe, but closer to the front of the foot.

The pattern of stress distribution on the sole of the foot during the stance phase of running is shown in figure 15.8 for both rearfoot and forefoot strikers. As expected, two main areas of loading are evident in the pattern

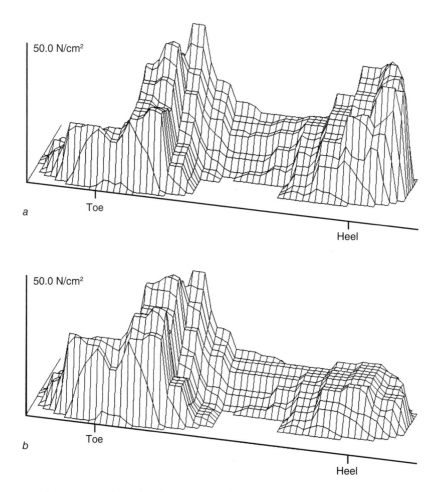

Figure 15.8 Patterns of stress over the sole of the runner's foot. The pattern for a rearfoot striker is shown in (a), and the pattern for a forefoot striker is shown in (b). Note that both runners exhibit a high level of stress in both the front and rear of the shoe.

Reprinted, by permission, from Tom Kernozek, *Patterns of stress over the sole of the runner's foot* (Department of Health Professions, University of Wisconsin-La Crosse).

of rearfoot strikers: both the rearfoot and the forefoot portions of the shoe. The rearfoot is loaded during the initial part of stance, and the forefoot is loaded during the latter part when the foot is plantar flexed for push-off. For forefoot strikers, the forefoot region is the main area loaded during stance.

The majority of runners use the rearfoot striking pattern, so most running shoes are designed for this pattern of loading. A cushioned heel is intended to help absorb and distribute the load during initial contact; and a flexible, cushioned forefoot region helps absorb the load while allowing the shoe to bend during the latter portion of stance. Cushioning the forefoot region of the shoe helps absorb load during both initial contact and the latter portion of stance for forefoot strikers. Cushioning is provided by a combination of materials and shoe construction.

The pattern of observed injury to the foot closely follows the observed pattern of loading on the sole of the foot. A stress fracture of the head of either the first or second metatarsal is a common injury among runners. The magnitude of the stress applied in this area is relatively high for both forefoot and rearfoot strikers. When the number of steps per mile is considered, it is easy to see that the cumulative effect of this loading may lead to a stress fracture in this area of the foot.

> ⟳ **The pattern of observed injury to the foot closely follows the observed pattern of loading on the sole of the foot.**

Chapter 1 introduced the concept of reaction force. Ground reaction force is basically the reaction force applied to the body during contact with the ground. The ground reaction force in gait ignores the actual distributed pattern of the ground contact force over the sole of the foot. The point of application of the single vector representing the ground reaction force applied to the foot is called the center of pressure. Understanding the ground reaction force and the center of pressure simplifies analysis of the effect of the ground contact force on the motion of the runner.

The ground reaction force is applied to the runner's foot beginning with the start of the stance phase, or initial contact with the ground. Figure 15.9 presents patterns of the ground reaction forces typical of rearfoot and forefoot landing styles in running. The ground reaction force consists of three components: the vertical force (acting upward), the anteroposterior force (acting forward or backward on the body), and the mediolateral force (acting from side to side). Measuring the ground reaction force during the support phase of running has long been the focus of research, especially since force platforms became available commercially in the late

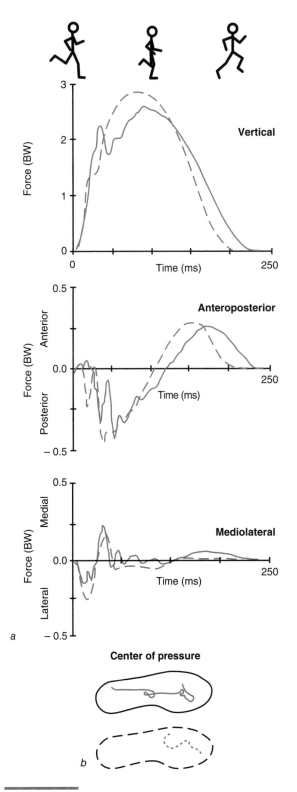

Figure 15.9 (a) Typical patterns of the ground reaction force components for a rearfoot (solid line) and a forefoot (dashed line) striker at 3.6 m/s. (b) The center of pressure pattern for the left foot. VGRF = vertical ground reaction force; APGRF = anteroposterior ground reaction force; MLGRF = mediolateral ground reaction force.

1970s. Although the magnitude of the individual curves varies depending on factors such as running speed, subject weight, shoes, and surface materials, the general pattern and temporal relationship of each curve remain relatively consistent across individuals.

The vertical ground reaction force pushes up on the runner throughout the stance phase. The anteroposterior ground reaction force initially pushes backward as the forward-moving runner contacts the ground and then forward as the runner rotates over the planted foot and pushes back on the ground. The mediolateral ground reaction force tends to push laterally on the runner, although this pattern is much more variable across runners.

The vertical ground reaction force increases in magnitude until approximately midstance, the point of maximum knee flexion. During this time, the vertical ground reaction force acting upward decreases the downward motion of the runner. The maximum value of the vertical ground reaction force increases with faster running speeds, ranging in magnitude from two to five times body weight (BW), depending on running speed. With 300 to 900 foot contacts per mile, the cumulative load to a leg can be measured in tons. During the latter half of the stance phase, the vertical ground reaction force continues to act upward on the runner as the leg is extended. Although decreasing in size during the latter half of stance, the vertical force acts to push the runner up into the next flight phase.

Application of the vertical ground reaction force represents what is known as an **impulsive load**. An impulsive load is an applied force that reaches a relatively high magnitude in a short time. The graph of the vertical force for both forefoot and rearfoot strikers (figure 15.9) shows that within the first 50 ms, the force reaches a magnitude of about 2.0 BW. The force is directed up the body and imposes stress on all tissues of the body. Although the impulsive load is partially absorbed by eccentric muscle action as it passes up the body, traces of it have been recorded at the skull. Tissue responses to this load may be responsible for both the training effect and the injury pattern seen in runners.

Much speculation has focused on the relationship among the magnitude and pattern of ground reaction forces and injury, but research has established no direct relationship. Although the impulsive loading typical of running has been implicated in the development of injury such as osteoarthritis, the observed pattern of running-related injury, the types of injury encountered, and biomechanical models of the runner suggest that factors in addition to ground reaction forces may play a role in overuse injury.

Good running shoes include features for comfort, shock absorption, and rearfoot control. Comfort of a shoe is an obvious selling point, and shock absorption through well-designed cushioning is important for dealing with the impulsive loads applied during stance. Rearfoot control refers to a shoe's capability to allow **rearfoot motion**, the natural sequential pattern of pronation and supination during stance. Pronation is a combination of dorsiflexion, eversion, and abduction. This motion allows the foot to adapt to the running surface and to absorb shock. Supination is the opposite action, a combination of plantar flexion, inversion, and adduction. This motion stabilizes the foot and lets it serve as a rigid lever during the propulsion that occurs in later stance.

Figure 15.10 is a position–time graph of the typical motion pattern exhibited by rearfoot strikers during the support phase of running. For research and clinical purposes, rearfoot motion is measured in the frontal plane as the angle between the shoe and the lower leg or shank. At contact, the foot is slightly supinated (measured as rearfoot inversion). Following contact, the foot initially pronates (measured as eversion) to reach a maximum everted position at approximately midstance before supinating through to toe-off. Typically, the foot everts to an angle of about 5° to 15° past neutral, or straight alignment of the foot and lower leg. Some individuals exhibit excessive pronation, measured as moving more than 18° of eversion past the neutral position. For some individuals, excessive pronation reflects the anatomical structure of the foot; however, the softer midsole materials intended to provide increased cushioning may actually cause the feet of some runners to pronate beyond the typical 15° of eversion.

The ground reaction force components tend to cause pronation of the foot during the initial part of the stance phase but are resisted by eccentric activity in muscles inserting on the foot. Concentric activity of the same

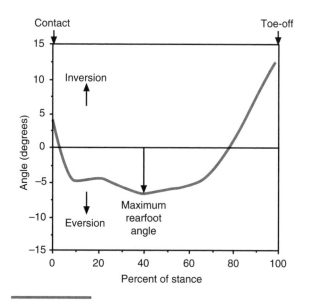

Figure 15.10 Typical pattern of rearfoot motion during the stance phase of running.

muscles is primarily responsible for supination during the latter portion of stance. Figure 15.11 demonstrates how the ground reaction force acts on the foot of a rearfoot striker to cause the observed motion during the initial stance period. In a rearfoot landing, the ground reaction force is applied behind the ankle joint on the lateral portion of the foot. The vertical force pushes upward, the mediolateral force pushes laterally, and the anteroposterior force pushes posteriorly. All three components have a moment arm around the ankle joint. Torque created by both the vertical and mediolateral components tends to cause the foot to evert. In rearfoot strikers, the torque of the anteroposterior force tends to cause the foot to plantar flex. A shoe made higher by the addition of materials to provide more cushioning produces longer moment arms for the ground reaction force components. The greater torque created by these components acting with longer moment arms tends to increase the degree of pronation and plantar flexion. In rearfoot strikers, the anterior and posterior tibialis muscles are active eccentrically to prevent the foot from slapping down and rolling inward too fast. Increased activity in these muscles to resist greater pronation may lead to common injuries reported in runners, such as shinsplints, anterior compartment syndrome, and Achilles tendinitis.

Pronation and supination also affect the magnitude of the stress imposed at the knee joint, the most frequent site of overuse injury in running. During running, the knee joint flexes during the initial portion of stance and extends during the latter portion. Knee flexion during running is controlled by the quadriceps muscle group as it is active eccentrically to absorb energy. Knee extension is caused by shortening of the quadriceps muscles as they release the energy. The tibia, the weight-bearing bone of the lower leg, is rotated internally along its long axis during ankle pronation and externally when the ankle supinates. These actions are caused by forces imposed on the tibia by the motion of the talus bone of the foot as it rotates under the tibia. At its proximal end, the tibia articulates with the femur at the knee joint. Due to the structure of the distal end of the femur, knee flexion and extension also cause the tibia to rotate along its vertical axis. Knee flexion causes internal rotation of the tibia, and knee extension causes external rotation. Ideally, the joint actions of ankle pronation and knee flexion and ankle supination and knee extension should occur simultaneously to avoid placing the tibia in torsion and stressing the knee joint. If the ankle joint overpronates, the synchronization of ankle and knee motion may be disrupted. Knee extension may begin before the ankle reaches maximum pronation. The disruption in synchronization of the coordinated joint action imposes abnormal stress at the knee joint and may alter the muscle activity patterns. The line of pull of the patellar tendon may be altered by misalignment of the femur and tibia, altering the tracking of the patella within the femoral groove. The resulting abnormal stress pattern on the sides and back of the patella may cause injury to the patella or the femur, commonly referred to as patellofemoral pain.

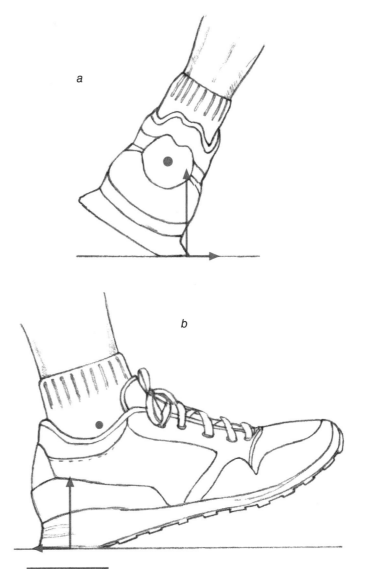

Figure 15.11 Influence of the geometry of loading on muscle activity in the lower leg during running. *(a)* Rear view of the foot–leg system. The vertical ground reaction force vector is applied lateral to the axis of the ankle joint, while the mediolateral force is applied below the axis of the ankle joint. These forces create torques that cause the foot to pronate. Eccentric muscle activity in the tibialis anterior and other foot "everters" is required to control the motion of the foot. *(b)* Side view of the foot–leg system. The vertical ground reaction force vector is applied posterior to the ankle joint, and the posteriorly directed anteroposterior ground reaction force vector is applied below the ankle joint. These forces create torques that cause plantar flexion of the foot. Eccentric muscle activity in the tibialis anterior and other foot "dorsiflexors" is required to control the rate of plantar flexion.

Extrinsic Factors

Extrinsic factors are those characteristics of the environment or the task that are related to the potential for injury. For runners, these factors are the physical conditions of the running area, such as surface texture, and the motions involved in running. This section focuses on task- and environment-specific factors that affect running.

Factors Related to the Task of Running

Variable characteristics of the running task include pace, distance of each run, total number of miles run per week, stride length, vertical motion, and joint position and range of motion during each cycle. Task factors affect the size of the load, the number of times the load is applied to the body, and how the body is aligned to accept the load.

Running pace, or speed, directly influences the size of the ground reaction force components. Faster speeds are associated with higher loading *magnitudes*. The maximum vertical ground reaction force, for example, increases from approximately 2 BW at a slow jog to 6 BW at a fast run. The higher forces associated with faster speeds cause greater torques at the joints. Increased muscle activation is required to control the joint motions incurred by the torques and to create the propulsive ground reaction forces necessary to maintain the faster speed. Thus, a runner must increase running pace slowly over several training sessions to provide time for the neuromusculoskeletal adaptations to occur. Too rapid an increase in training pace is a common training error. Many running overuse injuries are attributable to this "too fast, too soon" syndrome. Setting a more reasonable goal of increasing training pace over a longer period—that is, a smaller increase in pace with each session—helps to avoid overuse injury.

A similar error summarized as "too much, too soon" involves either increasing the distance run too quickly or increasing the number of training sessions per week. The increased number of loading cycles involved in increasing distance or number of sessions causes greater microdamage to the tissue. Additional rest time is required for remodeling to occur after an increase in distance. Without adequate rest, the next training session imposes a load on tissues that have not completed remodeling. Typically, 48 h is recommended between successive training sessions. Extending this period to 72 h is practical advice when a prior training session has involved an increase in mileage. It may also be beneficial to decrease the pace of the subsequent session to reduce the load on the body.

The total number of miles run per week, a combination of miles per session and number of sessions, reflects the total cumulative load applied to the body. Thus, both of the errors just mentioned—"too fast, too soon" and "too much, too soon"—contribute to the risk of injury related to total number of miles run per week. A systematic approach to increasing both pace and distance can reduce the effect of total number of miles per week on injury risk. As running speed is increased, it is wise to limit the number of miles run at the new pace by restricting the time per session. Similarly, an increase in distance should not be accompanied by an increase in pace.

> Controlling running pace and the distance run each session, along with the number of sessions per week, is among the most effective injury prevention techniques.

Understanding the interaction of these three factors relative to the size and repetition of loading provides the best basis for designing a training program. These guidelines apply to runners interested in achieving a greater training effect as well as those preparing for competition.

Extrinsic factors such as stride length, vertical motion, joint position, and the range of motion during each cycle are more closely related to individual running style than are the extrinsic factors discussed earlier. These factors reflect an interaction between the individual's body size and running technique. The question of whether changing these factors is effective or useful is more controversial than are the guidelines related to controlling increases in pace, distance, and total miles per week. Research on experienced runners has indicated that a runner naturally chooses the most efficient stride length for a given pace, one that will minimize energy consumption. The vertical motion of the runner and the observed joint kinematics reflect the efficient gait pattern; however, since the research subjects were experienced runners, whether training with that particular stride length influences energy consumption is unknown. Typically, people learning to run receive little instruction. Most cues provided in a training program relate to racing strategies such as positioning in a crowd or moving up during a race. Most coaches are reluctant to suggest changes in running style to experienced runners. The effect of providing guidelines on running style to inexperienced runners has not been fully investigated and deserves further research.

Factors Related to the Running Environment

Environmental factors related to injury risk during running include the running surface material (asphalt, concrete, wood chips, grass, sand, tartan), the condition of the surface (wet, dry, icy, loose gravel, sand), the material of the running shoes (midsole material, inserts for shock absorption or rearfoot control [orthotics]), the design characteristics of the shoes (special features for shock absorption or rearfoot control), and temperature (hot, warm, cold). Although runners have relatively little control over temperature, other than considering how it might affect pace and distance or even deciding whether to go for a run, they do select the shoes worn and the terrain.

Running surface directly affects the magnitude of the ground reaction forces and the amount of rearfoot control. Harder surfaces such as asphalt or concrete are associated with higher forces than are surfaces such as grass or sand. The higher coefficient of restitution typical of asphalt and concrete (see chapter 3) means that less force is absorbed by the material and greater forces are imposed on the runner, but these firm surfaces provide good traction or grip. Conversely, the lower coefficient of restitution typical of wood chips, grass, and sand results in lower ground reaction forces, but the tendency for the materials to shift can result in reduced traction. Many communities have created running trails made of such softer materials to provide a safer environment for training. However, some runners, especially those who are highly trained, do not train regularly on these surfaces specifically because they feel that the softer surfaces and reduced friction decrease the quality of the workout. In addition, softer materials allow more pronation of the foot, leading to increased muscle pain in the lower extremity due to increased range of motion and muscle activity. Selecting a running surface that provides a balance between ground reaction force magnitude and rearfoot control is an individual preference.

Running shoes are also a matter of personal preference. Shoes marketed today incorporate a variety of features to make them more attractive, comfortable, and safe. The distinction between features is not always clear though. For example, a highly cushioned shoe may feel extremely comfortable when you are walking in the salesroom, giving you the sensation of "walking on air." However, the materials and shoe construction that provide cushioning may not afford the degree of rearfoot control needed when the ground reaction forces typical of running are imposed on the shoe. In addition, a shoe that provides cushioning and adequate rearfoot control on a harder surface may feel too soft and lacking in rearfoot control on a softer surface.

As a guideline for injury prevention, it makes sense to run on a variety of surfaces in a variety of running shoes. This is a form of "within-activity cross-training." Cross-training refers to a training program that incorporates day-to-day rotation among a variety of activities. For example, someone training for cardiovascular fitness might rotate

training sessions among running, cycling, swimming, and aerobics. The idea is that each activity imposes a different pattern of loading on the body. Alternating activities alters the pattern of loading, avoiding the cumulative effect of a consistent loading pattern and reducing the risk of injury. Running on different surfaces and with different shoes has a somewhat similar effect. Although the same general pattern of cyclical movement is involved, the slight differences in load magnitude and loading pattern represent a form of cross-training within the activity. By varying loads from session to session with different shoe and surface combinations, you avoid placing a cumulative load on a tissue that can lead to overuse injury.

In practice, it's a good idea to alter the pace and distance run in accordance with how your body responds to the shoe and the surface. You may find that more pain develops during or following a training session on a particular surface. Respond to this by changing the pace and length of your run. A good rule of thumb, if an overuse injury develops from using a particular shoe or running on a particular surface, is to change the environment. Discard (or return, if possible) a pair of shoes if it causes more pain than usual. When purchasing, select shoes similar to ones in which you have been able to run without pain. Use your own experience to select a particular model or style, and realize that your response to a shoe may differ from advertising hype. Also, shoes wear out with use, and as their condition deteriorates, the shock absorption or rearfoot control they afford decreases. Similar considerations relate to running surfaces. If you experience increased soreness from running on a particular surface, change surfaces. Some people prefer softer surfaces; others prefer harder surfaces. Recognizing that shoes and surfaces interact to create the injury risk, and that the risk is specific to an individual, leads to individual choice and preferences in shoes and surfaces.

Intrinsic Factors

Intrinsic factors relate to the characteristics of the individual performing a task in a given environment. Researchers have attempted to identify individuals for whom running is an inappropriate activity, believing that some people may be predisposed to injury because of individual characteristics affecting the level of stress imposed and the response to stress. In this section we consider a few of these characteristics.

Body Mass

An appropriate starting point is to consider body mass. In mechanics, the mass of a body represents inertia, or resistance to a change in the body's current state of motion. As discussed earlier, during the stance phase the runner's downward motion at contact is slowed and stopped by the vertical ground reaction force, whereas the forward motion is slowed by the anteroposterior ground reaction force. The vertical and anteroposterior forces then act simultaneously to propel the runner upward and forward into the next flight phase. In chapter 3, Newton's second law ($\varepsilon F = ma$) was used to explain that for a given acceleration, a larger force is required to change the motion of a larger mass. If two runners train together at the same running pace, Newton's second law suggests that the heavier runner will be exposed to higher forces; that is, to provide the same acceleration, higher forces must act on the heavier runner as compared to the lighter runner.

This is why ground reaction force data are presented in multiples of body weight. Generally, when performing a task with given kinematic patterns, the more massive individual will be exposed to higher forces. Wolff's law indicates that the body of the more massive runner should be adapted to the higher forces. Typically, higher bone density and more muscle strength are associated with greater body mass, as would be expected according to Wolff's law. Presenting ground reaction force in multiples of body weight is an attempt to scale the size of the imposed load to the mass of the individual to account for assumed adaptations to the higher absolute force. The idea is that an imposed load of 2.5 BW will have a similar effect on individuals despite differences in the absolute size of the imposed force. However, as we pointed out earlier, the stress threshold of various tissues is uncertain. Whether absolute or relative loading is more related to tissue response is still unclear.

Lower Extremity Anatomy

The structure and function of the lower extremity have been implicated in the development of overuse injury in runners. As an example, in this section, we will consider how two common anatomical factors relate to overuse injury.

Foot Structure   Foot structure refers to the type of medial longitudinal arch. The medial longitudinal arch runs along the medial aspect of the foot and is created by the calcaneus, talus, navicular, three cuneiforms, and three medial metatarsals. The arch is supported by muscles and the plantar aponeurosis, the connective tissue band that runs from the calcaneus to the head of the first metatarsal. In running, the arch lengthens during the initial phases of foot contact to help distribute the contact load over the bones of the foot, shortens during midstance to help support the body, then lengthens again to aid in push-off.

Differences in foot structure are associated with differing types of common overuse injury. A lower, more flexible arch allows for absorption of more energy than does a higher, more rigid arch. As a result, in individuals with rigid high arches compared to those with flexible low arches, a greater load is passed on to the tibia and up the leg. This difference in stress absorption is reflected in a different pattern of overuse injury. Individuals with high arches are more prone to stress fractures in the tibia and femur due to reduced absorption associated with less foot motion. Individuals with low arches are more prone to metatarsal stress fractures due to the increased stress imposed on these bones as they contribute to greater load absorption.

People should consider the structure of the foot when selecting running shoes. Individuals with low arches should choose a shoe that provides rearfoot control. This will help prevent the overpronation (>15°) that may occur due to increased foot mobility. Individuals with high arches should choose a shoe with added shock absorption features to provide increased absorption of the forces imposed during running. Ultimately, shoe choice should be based on comfort and success in preventing overuse injury.

Alignment of the Knee Joint   The knee joint is the articulation between the femur and the tibia. This joint is exposed to high stress during running because of the imposed ground reaction force and the tension in muscles crossing the joint. Knee joint function is affected by its structure and anatomical factors such as the shape of the femoral condyles and the tibial plateau, as well as the presence of the medial and lateral menisci. Individual knee joint alignment characteristics are important in assessing the relative risk of injury during running because they affect the magnitude and pattern of stress imposed on the tissues of the joint and the joint's contribution to shock absorption.

The **Q angle** is a measure of alignment between the femur and the tibia. To measure the Q angle, lines representing the long axis of the tibia and femur are drawn on the front of the leg. As shown in figure 15.12, one line is drawn from the tibial tuberosity to the midpoint of the superior border of the patella to represent the tibia, and another line is drawn from the anterior superior iliac spine of the pelvis to the midpoint of the superior border of the patella to represent the femur. The Q angle is measured as the smaller angle between the two lines. A Q angle greater than 20° is considered excessive.

The alignment of the femur and tibia is important because it affects patellar tracking, or the path the patella follows as it rides between the femoral condyles during knee flexion and extension. Because the patella is the distal insertion site for the quadriceps muscles, a higher Q angle (poor alignment of the femur and tibia) means that

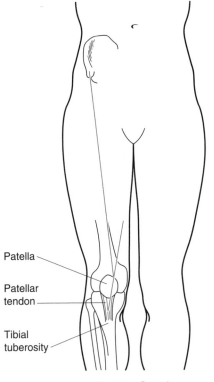

Patella

Patellar tendon

Tibial tuberosity

Measure Q angle

Figure 15.12 The Q angle measures the alignment between the femur and the tibia, a factor affecting the pattern and magnitude of stress imposed on the tissues of the joint.

the patella will not track smoothly between the condyles. This leads to excessive stress on the posterior and lateral borders of the patella, contributing to the development of pain in this region (commonly referred to as patellofemoral pain).

Despite the obvious link between the Q angle and patellar tracking, factors such as muscle tightness in the soft tissue of the lateral knee and weakness of the vastus medialis also need to be considered with regard to the development of patellofemoral pain. Tightness in the lateral soft tissues will pull the patella laterally even if the Q angle is within normal range. If the vastus medialis is weak relative to the vastus lateralis, the unequal pull on the patella causes it to glide along the lateral condyle. If a runner reports patellofemoral pain, then Q angle, lateral tightness, and muscle strength should all be evaluated to identify the cause of the probable patellar mistracking.

Psychological Considerations

Psychological factors related to overuse injury include the individual's pain tolerance level and motivation to continue with a training program. Both factors are related

to the onset and progression of an injury. Clinical studies of running-related injury frequently report on individuals who continue to run despite medical advice to rest and seek treatment. Surveys of marathon runners, both competitive and recreational, frequently list multiple running-related injuries in a single individual. Common injuries that sufferers try to "run through" include a variety of musculoskeletal ailments such as sprains, joint pain, and even fractures. Although some coaches, sport commentators, and athletes believe continued participation despite pain to be admirable, this attitude belies the fact that pain is a sign that the body needs time for healing. Continuing to participate through injury exposes the injured tissue to stress when its ability to withstand stress is already compromised. As a result, the injury can worsen, usually to the point of requiring more extensive medical intervention and a longer rest period than if it had been treated promptly.

Continuing through injury requires an adaptation in movement pattern commonly known as favoring the injured body segment. In mechanical terms, a modified movement pattern is used to reduce the stress imposed at the site of injury and eliminate the cause of the pain. However, alterations in the movement pattern necessary to reduce stress at one site mean that higher stress will be imposed somewhere else. The change in stress patterns, which often occurs in a short time if the runner continues to perform, frequently leads to a secondary injury. For example, some runners report to a clinic complaining of pain in both limbs, sometimes at the same site and other times at different sites on each leg. The athlete admits to a delay in seeking treatment until the pain became unbearable or constant. The injury history frequently reveals that the pain began unilaterally, limited to a single leg. The development of pain in the other limb results from the altered stress pattern as the athlete has either consciously or unconsciously favored the injured limb.

Prompt treatment of the cause of pain is critical to preventing more extensive injury. Conservative intervention such as rest provides time for healing to occur. It is necessary, in fact imperative, to identify the reason for the injury. Typically, this is simply a matter of "too much, too soon." When returning to running after the injury has healed, the athlete must adhere to a systematic approach of increasing distance or speed and consider using more appropriate shoes or running on a different surface. Individuals with significant skeletal malalignment may need an orthotic to align the musculoskeletal system properly or may need to incorporate specific strength and flexibility exercises to correct the malalignment. Some people may have to choose a more appropriate activity, one that reduces loading to a level that does not place excessive stress on damage-prone tissue. These individuals should seek the advice of a practitioner knowledgeable about the biomechanics of common cardiovascular training activities for help in selecting the appropriate training modality.

Summary

Mechanical stress refers to the distribution of force over the body on which it acts. The body responds to chronically low levels of stress by atrophying and to systematically applied high levels of stress by hypertrophying. However, if the stress level imposed is too high or is applied too frequently, a traumatic or overuse injury will occur. Preventing injury can be a matter of reducing either the magnitude or frequency of imposed stress or modifying the tissue's stress threshold to better withstand the stress. Examples of stress imposed during running demonstrate that although the stress–injury relationship is conceptually simple, preventing injury during skill performance is difficult.

The first three chapters in part III have provided guidelines for conducting qualitative biomechanical analyses with specific emphases on technique improvement, training, and injury prevention. Although these three types of analysis have been presented separately, and indeed are different in their approaches, they are inseparable in any thorough qualitative biomechanical analysis of an activity. The technical improvements suggested by a qualitative biomechanical analysis to improve technique should be examined to ensure that these changes in technique will not impose stresses within the pathologic overload zone of the stress continuum and thus lead to injury. Will the technique improvement increase the risk of injury? A qualitative biomechanical analysis to improve training relies on the model of the most effective technique suggested by a qualitative biomechanical analysis to improve technique. What physical parameters must be trained to allow production of the most effective technique? Practitioners must also be able to identify the anatomical structures that are at risk for injury as a result of the performance of a skill. This identification results from a qualitative biomechanical analysis to understand injury development. The results of this type of analysis may also be used to improve training to strengthen the tissues that are at risk for injury. What tissues must be strengthened to prevent injury?

Qualitative biomechanical analysis requires a sound knowledge and understanding of mechanical principles. The material presented in the first two parts of this book should have helped you gain this knowledge and understanding. Qualitative biomechanical analysis is also a complex and difficult task. The general guidelines and procedures for conducting qualitative biomechanical analyses have been presented in the last part of this book. These guidelines and procedures are an attempt to simplify the process of conducting a qualitative biomechanical analysis. In most areas of life, learning to do something well requires practice. Qualitative biomechanical analysis also requires much practice and experience to perfect. Opportunities to apply biomechanics and conduct qualitative biomechanical analyses will occur every day in your future as a human movement professional. Take advantage of these opportunities and use your knowledge of biomechanics. The payoff will be more satisfied students, athletes, or clients and an increase in your effectiveness as a teacher, coach, or other human movement professional.

KEY TERMS

extrinsic factors (p. 368)
impulsive load (p. 374)
intrinsic factors (p. 368)
overuse injury (p. 367)

physiologic loading zone (p. 365)
Q angle (p. 378)
rearfoot motion (p. 374)
remodeling (p. 365)

stress continuum (p. 365)
traumatic injury (p. 366)
Wolff's law (p. 364)

REVIEW QUESTIONS

1. Explain how the magnitude of imposed stress relates to the size of the applied force and the area over which the force is distributed. Give examples of interventions to alter stress by changing the size of the imposed force and examples of interventions to alter stress by changing the area over which the force is distributed.

2. Differentiate between tensile, compressive, and shear stress. Give an example of each type of stress imposed on the human body, identifying the source of the force and the tissue loaded.

3. Draw the stress continuum, identifying the four regions.

4. Explain the concept of the stress threshold. Relate the concept to the development of traumatic and overuse injury.

5. Differentiate between intrinsic and extrinsic factors related to injury.

6. Identify intrinsic and extrinsic factors related to overuse injury in (a) swimming, (b) long jumping, (c) pitching, and (d) an assembly task in industry.

7. What factors affect the stress threshold of a particular tissue? That is, why do some people have stronger bones than other people?

8. Explain how the concept of within-activity cross-training is similar to and different from that of cross-training. Basing your answer on the mechanics behind injury development, explain why either mode of training may be effective in reducing injury risk. Suggest ways in which the concept of within-activity cross-training can be used to avoid the types of injury sustained frequently by swimmers.

9. Jim begins a program of strength training. He is pleased by his initial rapid gain in strength and happily adds more resistance to his lifts. Although most of Jim's initial strength gain can be attributed to neural aspects, he proudly notes the hypertrophy of muscle that also contributes to his strength increase. Unfortunately, about one month into his lifting program, Jim notices annoying pain in his joints. Use the concept of the stress continuum to explain what is occurring in Jim's body, and provide suggestions to avoid the situation in beginning lifters.

10. Sally buys new running shoes and reports that they make her feel "lighter and faster." She announces that she is ready to "really push her workouts." What advice would you provide to Sally to help her avoid developing overuse injury?

11. In general, hospitalized and surgical patients are encouraged to begin weight-bearing activity much sooner than they were 25 years ago. Explain why starting weight-bearing activity is of benefit to these patients.

12. Commenting on the pattern of injury in baseball, Keith Law, the assistant general manager of the Toronto Blue Jays was quoted as saying, "What's going on now, especially when it comes to keeping pitchers healthy, is a lot of guesswork. We still don't know whether it's worse for a pitcher to throw 130 pitches in a complete game or 100 pitches over 5 innings" (Verducci 2004). Relate the quote to the conceptual basis for injury presented in this chapter.

PROBLEMS

1. During a tackle in football, the running back imposes a force of 765 N on the tackler. Calculate the magnitude of the mechanical stress if the force is
   a. applied over 9 cm^2 of the tackler's helmet or
   b. over 100 cm^2 of the tackler's shoulder pads.

2. During a fastball pitch, the tensile force in an anterior bundle of ulnar collateral ligament (UCL) reaches a maximum of 300 N. The ultimate strength of the ulnar collateral ligament is 80 MPa.
   a. If the minimum cross-sectional area of the anterior bundle of the UCL is 0.05 cm^2, what is the maximum tensile stress in the anterior bundle of the UCL as a result of the 300 N force?
   b. What percent of the ligament's ultimate strength does this stress represent?

3. During the fastball pitch in the previous question, the UCL transmits the 300 N peak tensile force to the medial epicondyle of the humerus. This force is distributed over a 0.06 cm² area of the epicondyle. The ultimate tensile strength of cortical bone is 110 MPa.

a. What is the maximal tensile stress in the medial epicondyle as a result of the 300 N force?

b. What percent of the bone's ultimate tensile strength does this stress represent?

chapter 16

Technology in Biomechanics

objectives

When you finish this chapter, you should be able to do the following:

- Define quantitative biomechanical analysis
- Discuss how measurement of biomechanical variables may influence the variables themselves
- Discuss instrumentation used for measuring kinematic parameters in biomechanics
- Discuss instrumentation used for measuring kinetic parameters in biomechanics

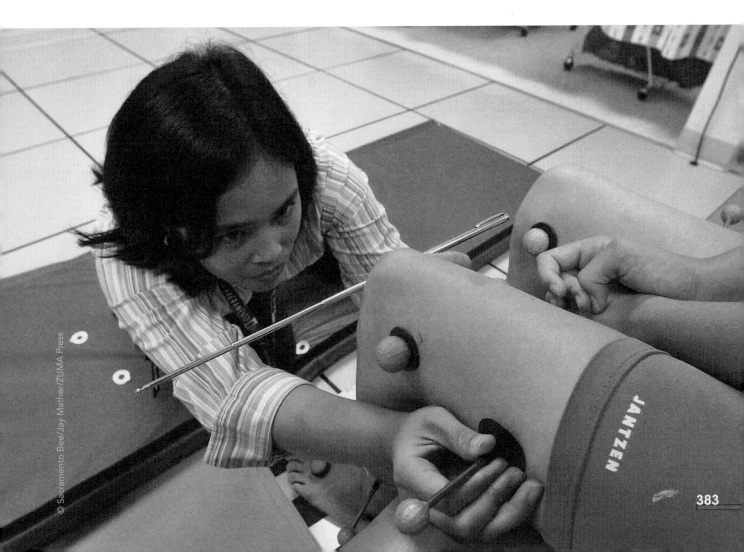

© Sacramento Bee/Jay Mather/ZUMA Press

A golfer lines up a chip shot. But the golfer looks different—he's not wearing normal golf attire. Bright dots are attached to the club and various parts of the golfer's body. A small box is strapped onto his back, and cables lead out of this box and into his shoes. More wires attach to devices on his legs. The golfer is standing on some sort of plate. Bright lights illuminate him. He's not hitting the ball on a golf course; he's in a laboratory where a variety of devices are recording his movements, muscle actions, and reaction forces under his feet. What sort of lab is this? What kinds of devices are used to measure the biomechanical variables we have discussed in this book? In the previous three chapters we learned about qualitative biomechanical analyses. This chapter is about the technology used in quantitative biomechanical analyses.

Quantitative biomechanical analysis of human movement involves actual measurement of human movement and the underlying causes of the movement. If human movement or any of its aspects is quantified or measured (described with numbers), the resulting analysis based on these measurements is a quantitative biomechanical analysis. This chapter presents an overview of the technology used to measure the biomechanical features of human movement.

Quantitative Biomechanical Analysis

Teachers, coaches, and therapists often perform qualitative biomechanical analyses, but they rarely perform any quantitative biomechanical analyses. When is a quantitative biomechanical analysis warranted? In sport, quantitative biomechanical analyses are usually done only at the elite or professional level because of the expense and time involved. These analyses may be done throughout an athlete's season or career to monitor changes in technique, to monitor critical changes in biomechanical parameters resulting from training improvements, to monitor progress in rehabilitation from injury, to provide data for biomechanical research about the specific sport skill, and so on. Ergonomists and human factors specialists may use quantitative biomechanical analyses of workers to determine the causes of overuse injuries in workplace environments and develop solutions for them. Clinical biomechanists affiliated with hospitals or other medical establishments may conduct quantitative biomechanical analyses of patients to determine effects of various medical interventions on gait, to diagnose musculoskeletal diseases or injuries, to monitor rehabilitation, and so on.

Quantitative biomechanical analyses are warranted in all of these cases because changes in the biomechanical variables being measured may be indistinguishable without special instruments. The movements occur too quickly to be readily perceived by the human eye, or the differences in position and displacement are too subtle to be noticed. In other cases, the biomechanical variable being measured may be too difficult to perceive by anyone other than the athlete (or patient or client). As an observer, how do you detect the magnitude and direction of the ground reaction forces acting on a runner? You can't see the forces—you can see only their effects. We need special instruments for measuring these variables.

Measurement Issues

Quantitative biomechanical analyses of human movement involve measurements of biomechanical variables. The variables measured may be temporal (timing), kinematic (position, displacement, velocity, acceleration), or kinetic (force, energy, work, power). In any case, some sort of instrument is used to measure the variable. The instrument itself and the setting that it is used in may affect the performance of the athlete, patient, or client. The process of measuring something influences the parameter being measured. The validity of the measured parameter is thus threatened by the measurement process. Measurement technology that minimizes the measurement effects on the performer is preferred.

⮌ The process of measuring something influences the parameter being measured.

Laboratory Data Collection

Ideally, the environment in which you measure the performance should be carefully controlled, if possible. Most data for quantitative biomechanical analyses are collected in a biomechanics laboratory, where the environment can be controlled. The drawback is that a biomechanics laboratory is not the setting in which the athlete, patient, or client normally moves or performs. The novelty of the environment may influence the movements being measured. The laboratory should be set up to duplicate as closely as possible the environment in which the movements normally occur.

The benefit of data collection in a laboratory is control of the environment. The cameras, lights, temperature, and so forth are always the same. The subject thus performs in the same conditions each time she is evaluated. Much of the instrumentation is permanently set in position so the time to prepare for data collection is minimal. In addition, sensors, markers, or data collection packs can be attached to the performer.

⟳ The benefit of data collection in a laboratory is control of the environment.

The drawback is that the environment is not the same as the real-life environment in which the athlete, patient, or client usually moves. A baseball pitch thrown in the lab may be very different from the same type of pitch thrown in a game. The lights and cameras and the technicians watching and measuring the patient's movements may make the patient self-conscious and alter the movement. The attachment of markers, sensors, or cables to the performer will have some effect on the movements being measured. In laboratory data collection, it is very important for the subject to become familiar with the equipment and the laboratory environment before data collection begins.

In-the-Field Data Collection

An actual athletic competition may be the best environment for measuring the biomechanics of an athlete's performance because it is one in which the athlete normally performs (see figure 16.1). The competition setting, however, may not be the best for the biomechanist. Most biomechanics technology is not very portable. To record ground reaction forces, force plates would have to be mounted in the competition setting. To record muscle activity, electrodes would have to be attached to the

Figure 16.1 A biomechanist operates a high-speed motion picture camera to record the strides of 400 m sprinters at the USA Track and Field Championships. Data collection in the field may not have an effect on the performance of the athletes being studied.

athlete's body and signals from these sent to a receiver for recording. Alternatively, a device for recording the signals could be attached to the athlete. The type of data regularly collected in athletic competitions is kinematic data. Technology for measuring kinematic data includes electronic timing devices, video recordings or motion picture film and their computerized analysis systems, and radar or laser velocity-measuring devices. Most of these measurement devices are relatively noninvasive. The performance of the athlete is minimally affected by their use. The directors of the athletic competitions will be more likely to grant biomechanists permission to use these non-invasive types of measurement equipment during competitions than the more invasive types of measurement equipment.

The major drawback of collecting biomechanical data during an athletic competition is lack of control of the environment. The biomechanist has no control over the performer or the factors influencing the performance, and the positions of the data collection instruments may be restricted. The film or video cameras and the radar or laser velocity-measuring devices all require direct views of the performance. The radar or laser velocity-measuring devices require setup along the line of the performer's motion. Officials, spectators, other athletes, and so on may block these views during periods of data recording. Changes in lighting may limit the use of film or video cameras. Inclement weather may also limit the use of the equipment (e.g., too cold, too hot, too wet). All of these devices require electrical power. Multiple batteries must be on hand, or an accessible power source at the athletic venue must be found. Transporting expensive and fragile electronic equipment to and from the competition venue exposes it to risks of damage or theft. The preplanning and setup time for data collection at an athletic competition is also extensive. Additionally, it is difficult to duplicate the use of the given cameras in exactly the same positions from one competition to the next. In spite of these drawbacks, sport biomechanists regularly collect biomechanical data at the Olympic Games and world and national championships for a variety of sports.

Sampling Rate

Most biomechanical parameters vary with time, so they must be measured throughout the movement. A computer is usually involved in the data collection to store and process the data. Most biomechanical variables are **analog signals**—they vary continuously with time. Before being stored and analyzed by the computer, however, the data must be converted to digital form. Digital data are numerical (for the computer, the data are represented in binary form—as 1s and 0s). To convert an analog signal to digital, the analog signal is measured at discrete intervals (the signal is sampled), and then the measured value is converted into binary form. How often the signal is sampled is referred to as the **sampling rate** or sampling frequency.

The sampling frequency of an instrument indicates how often the instrument records a measurement. The sampling frequency of some biomechanics instruments, such as a force platform, may be as high as several thousand samples per second, while that of a typical video recording is usually 30 or 60 pictures or samples per second. The sampling rate of most measuring tools in biomechanics can be adjusted to suit the motion being measured. For slow, deliberate movements, sampling rates below 100 samples per second are adequate, but for movements that involve impacts or quickly changing states, much faster sampling rates are required.

Tools for Measuring Biomechanical Variables

Tools for measuring biomechanical variables vary in sophistication and cost from simple stopwatches to highly sensitive force platforms and multi-camera motion capture systems. Measurement techniques are constantly evolving as technology improves. The review of measurement tools presented here is only a brief overview of some of the technologies available for measuring biomechanical variables. The tools have been categorized as tools for measuring kinematics and tools for measuring kinetics.

Tools for Measuring Kinematics

Kinematic variables are based on position and time or the changes in each. Popular tools for measuring kinematic variables in biomechanics include timing systems, velocity-measuring systems (based on radar or laser light), accelerometers, microelectromechanical systems (MEMS) inertial sensors, and optical imaging systems (film cameras, video cameras, and so on). Full-body motion capture (mocap) systems may use one or more of these technologies to record and quantify human motion in two or three dimensions.

Timing Devices

Time is a fundamental dimension in mechanics, so the measurement of time is important. Watches are the simplest devices for measuring time. If the duration of an event being timed is long enough, a simple stopwatch may be an appropriate timing device. If more accuracy is needed and if the duration of the event being timed is short, then an automatic timing device is more appropriate.

Most automatic timing devices use electronic clocks in a computer or other digital device. Electronic or mechanical switches start and stop the clocks. These switches may be triggered by a variety of means. For instance, pressure-sensitive mats may be used to start or stop the clock as a person steps onto or off the mat. If light is the trigger for the device, it is considered a photogate timer. Photogates may be sensitive to specific wavelengths or to a wide frequency. In any case, if a light source is shining on the sensor and this beam of light is broken by a person or a limb or an implement, the change in intensity of light on the sensor triggers the clock to start or stop.

These automatic timing devices obviously measure time, but they may also be used to measure average speed. If the triggering sensors are positioned a known distance apart, then average velocity can be computed from the distance and time measures. Adding more sensors can provide a more detailed data set for a movement. For example, multiple photogates positioned in an array along a track or runway can provide information about step rate, step length, stance time, flight time, and velocity over a number of steps.

Velocity-Measuring Systems

Timing systems are the simplest kinematic measuring tools. They are useful for measuring average velocities of humans or objects, but what about instantaneous velocity? The radar gun that troopers use to catch speeders on the highways has been adapted to capture the instantaneous velocities of objects in sport. A radar gun transmits a microwave radio signal at a specific frequency and measures the frequency of the signals that are reflected back to it. A stationary object will reflect the radio signal at the same frequency that was transmitted by the radar gun. If the object is moving, the reflected signal will experience a shift in its frequency—the Doppler effect. The velocity of the object is determined by this frequency shift.

Radar guns are limited to measurements of speed (or components of velocity) directly toward or away from the radar gun. They are most widely used to measure the speed of pitched baseballs, but radar guns are also marketed for use in golf, tennis, hockey, soccer, lacrosse, and other sports. Their use in measuring the velocity of an athlete's body is limited unless a radar-reflective marker is worn by the athlete.

A laser-based velocity-measuring device is more effective at measuring the speed of athletes, especially runners. It is similar in operation to a radar gun, but it uses a laser and the reflection of the laser light to measure velocity. Unlike radar, whose signal disperses as distance increases, the laser in a laser-based velocity-measuring device is tightly focused. So, if more than one object is moving toward the device, the laser velocity-measuring device can measure the velocity of the specific object of interest whereas a radar gun will receive multiple reflected signals. The laser velocity-measuring device is more accurate for measuring the velocity of a runner.

Optical Imaging Systems

In a qualitative biomechanical analysis, we discern most of the qualities of a performance using our vision. It seems appropriate that the technology for recording these visual images of performance is the most widely used tool in biomechanics. The most popular optical imaging systems in biomechanics are video cameras.

Video cameras provide sequential two-dimensional images of movement at specific time intervals depending on the speed of the camera. In a single recorded image, the position of the body and its parts can be measured relative to each other or to a fixed reference in the field of view. If an object of known dimensions is also recorded in the field of view and in the plane of motion, the position data can be converted to real-life units. In subsequent pictures, the changes in position or displacements can be determined. The time elapsed between the exposure of sequential frames of video can be determined from the frame rate of the camera. For example, the time between two adjacent frames of video recorded by a camera operated at 30 frames per second is 1/30 second or 0.033 s. Velocities can thus be determined from the displacement and time measures. Once velocities are computed, accelerations can be determined from the velocity and time measures.

One camera may be enough to adequately record two-dimensional or planar motion, since the resulting image is also two-dimensional. Three-dimensional coordinate data can be obtained if the motion is recorded by two or more cameras. Specialized software has been developed that computes the three-dimensional coordinates from the two-dimensional data from each camera.

How are coordinate data extracted from the images from the cameras? This process is called digitizing. This is done manually or automatically, but in either case a computerized system facilitates the digitizing. First, the points of interest on the body or the object being investigated must be identified. If possible, markers are placed on these points on the object or subject before the movement is recorded. In the manual digitizing process, a single frame of the image appears on the computer monitor, and you digitize (store the coordinate data for) each point of interest by positioning a cursor over the projection of a point on the screen. This is done for each point of interest (for a full human body model, this may require more than 20 points) and for each frame of film or video of the movement. This process is extremely tedious and time-consuming. It is also prone to human error. You manually digitized points when you completed the motion analysis exercises using the MaxTRAQ software in earlier chapters, although in the exercises only one or two points were digitized, and not in every frame.

The second method of digitizing is the automatic method. Several methods of automatic digitizing exist.

In one method, highly reflective markers are attached to the object or subject (athlete, patient, client). These markers define the points of interest. The subject is illuminated so that light is reflected off the markers and into the camera lens. The resulting video image has bright spots on the subject's image wherever a marker was present (see figure 16.2). Specialized computer software identifies these bright spots and their coordinates in every video frame of the movement. In some automatic systems, the data are processed in real time—you see the computer model of the movement on the screen as the movement is executed. A more advanced version of the MaxTRAQ software includes automatic digitizing routines that track high-contrast markers in the field of view.

Another automatic digitizing method uses active markers in contrast to passive reflective markers. In these systems, the markers are usually light-emitting diodes (LEDs) that light up in a certain sequence and at a certain frequency. Specialized cameras detect their presence, and computer software determines their coordinate locations.

One drawback of all the optical imaging systems is that they depend on line of sight. Limbs move past each other and hide markers from cameras. Electromagnetic tracking systems overcome this problem by using electromagnetic markers and specialized sensing devices for detecting the locations of the markers. These electromagnetic systems do not suffer from hidden points since body parts that may hide a marker visually do not hide the marker electromagnetically, so the body parts are invisible to the sensing device.

Data collection using automatic digitizing systems is primarily limited to laboratory settings. Cameras, which are usually fixed, record movement in only a limited volume or space. Another disadvantage is that the movements of interest must be performed with markers attached to the subject's body.

Most optical imaging motion measurement systems are quite expensive. However, several inexpensive (less than $500) motion analysis programs are available that offer basic tools for analyzing human movement in two dimensions. The MaxTRAQ software that is bundled with this book is an example of this software. You used the MaxTRAQ motion analysis software in earlier chapters in this book to analyze video clips that were provided with the software. This program can also be used to analyze any motion that you record on video if you can load the video onto your computer in .avi format.

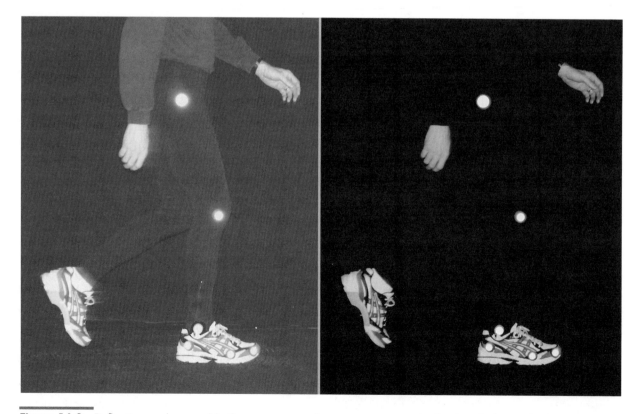

Figure 16.2 Reflective markers enable the computer and camera to automatically identify the marker locations. The photo on the left shows the markers on the subject under normal lighting conditions; the photo on the right shows the same subject with lighting and camera exposure set to highlight only the markers.

Accelerometers

Acceleration measurements can be derived from the data obtained from laser and radar velocity-measuring devices or from data obtained through optical imaging systems. These acceleration measurements may be contaminated with errors, however. The computations involved in deriving velocity from position and time data, and then acceleration from the velocity and time data, lead to error propagation. A little noise (random error) in the position data is amplified in the computation of velocity and is amplified again in the computation of acceleration. The data must be numerically filtered to eliminate this noise. In addition to this limitation, most velocity-measuring devices are limited to relatively slow sampling rates (the number of velocity measures per second is usually 60 or less). Is there a method of measuring acceleration directly so that these problems are eliminated?

An **accelerometer** is a device for measuring acceleration directly. These devices can be very light and small in size (see figure 16.3). The smallest accelerometers are smaller than the one shown, with dimensions as small as 2 mm × 2 mm × 1 mm. These tiny accelerometers are usually components of microelectromechanical systems (MEMS). Accelerometers attached to the object measure the acceleration of the object at the point of attachment.

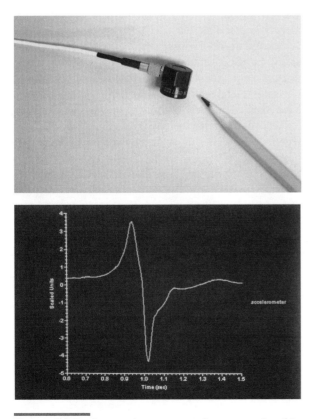

Figure 16.3 An accelerometer and an example of its output for a start–stop movement.

Accelerometers measure acceleration in a specific direction. A uniaxial or one-dimensional accelerometer measures acceleration in only one direction—along a specific axis of the unit. A triaxial or three-dimensional accelerometer measures three accelerations—along three different axes at right angles to each other. The orientation of the accelerometer determines the direction of the acceleration measured. If the accelerometer is attached to a limb that changes orientations, then the direction of the measured acceleration changes when the limb orientation changes. Accelerometers cannot be attached directly to the rigid framework of the body (bones) but instead must be attached to the skin. Because of these difficulties, accelerometers are usually not used for analyzing general whole-body movements.

Accelerometers have relatively high frequency response, so they can sample at high rates. This makes them especially well suited for analyzing impacts. In fact, accelerometers are used in automobiles as sensors to trigger air bag deployment. In biomechanics, accelerometers are used to evaluate the impact-reducing capabilities of sport safety equipment. The performance of bicycle helmets and other protective helmets is evaluated with accelerometers via measurement of the acceleration of a headform within the helmet during an impact test. The cushioning performance of materials used beneath children's playground equipment is evaluated in a similar way. Accelerometers are also well suited for measuring vibrations and their effects on the body.

Accelerometers are also the basis for physical activity monitoring devices. A uniaxial accelerometer is the basis for self-contained pocket-sized devices that can be attached to a barbell or to an athlete to measure a variety of mechanical parameters, including the power output of athletes. The technology used in simple pedometers that count steps is a crude uniaxial mechanical accelerometer. More sophisticated pedometers and activity monitors use electronic uniaxial, biaxial, or triaxial accelerometers to quantify the steps or physical activity of the wearer.

Inertial Measurement Units

MEMS inertial sensors or IMUs (inertial measurement units) use a micro-accelerometer and a gyroscope to provide kinematic measurements. These inertial sensor devices are small enough to be attached to subjects. They measure changes in position relative to an initial reference or starting position. They don't require a camera or other recording device but do require specialized software for interpretation of their output. Motion analysis systems that use IMUs are expensive relative to a basic two-dimensional video-based motion analysis system, even though the IMUs are relatively inexpensive. This difference in cost is shrinking, however, and IMU use in biomechanics is becoming more and more prevalent.

You may actually have a MEMS accelerometer or gyroscope similar to those used in IMUs. MEMS accelerometers and gyroscopes are used in numerous personal electronic devices—including most smart phones and tablets, laptops, and netbook computers—and video games such as Nintendo Wii to detect position and changes in position of the device in which they are located.

Motion Capture Systems

Motion capture (mocap) systems are used to capture in digital form the three-dimensional movements of the whole body. Components of a typical motion capture system usually include six or more video cameras, a marker system, and specialized software and hardware for reducing and organizing the data to produce the digital representation of the movement. The body is modeled as a system of rigid links connected at the joints. Marker sets consisting of two or more markers are attached to each body segment to identify the unique location and orientation of each segment in three dimensions. Some systems use mocap suits with built-in markers or inertial sensors; other systems require the markers to be placed on the subject. Typical full-body marker sets consist of more than 50 markers. Recent advances in image recognition software have led to the development of markerless optical mocap systems. These systems use advanced image recognition software to directly locate the anatomical segments and joint centers in the video image.

Three-dimensional motion capture systems are expensive, but these systems may be used in sport and exercise. They are more likely to be used in clinical gait analysis labs and research settings. The most widespread use of motion capture systems is in the entertainment industry. The Xbox Kinect game system is a crude but very inexpensive motion capture system. It uses multiple optical sensors to detect, capture, and interpret the movements of a player or players. The players can control the game with their movements and gestures. No external game controller is needed.

More complex and expensive motion capture systems are used to capture the motion of athletes and actors. The captured athletic movements of professional athletes have been used to provide realistic movement for the animated players in many popular sport-related video games and apps. The movements and gestures of actors have been captured using this technology and used to model movements of digital characters in feature films, television shows and commercials, and music videos since the late 1990s. The Na'vi characters in the 2009 film *Avatar* were created using motion capture technology. At the 77th Annual Academy Awards in 2005, Academy Awards for Technical Achievement went to Julian Morris, Michael Byrch, Paul Smyth, and Paul Tate for their development of the Vicon motion capture systems; to John O.B. Greaves, Ned Phipps, Ton J. van den Bogert, and William Hays for their development of the Motion Analysis motion capture systems; and to Nels Madsen, Vaughn Cato, Matthew Madden, and Bill Lorton for their development of the Giant Studios motion capture system. The Vicon and Motion Analysis systems were originally developed in the 1980s for use in biomechanical applications.

Tools for Measuring Kinetics

Kinetic variables are based on force—the cause of change in motion. Popular tools for measuring kinetic variables in biomechanics include force platforms, strain gauges, pressure-sensing devices, and electromyography (EMG).

Force Platforms

Force platforms are the most popular devices for measuring kinetic variables in biomechanics. **Force platforms** or force plates measure reaction forces and the point of application and direction of the resultant reaction force. Their measuring surface is rectangular and is typically about the size of a small doormat (about 40 cm by 60 cm). Force platforms are typically used to measure ground reaction forces in gait (see figure 16.4). The forces measured include the normal contact force (the vertical ground reaction force), the friction force in the anterior–posterior direction, and the friction force in the medial–lateral direction.

Force platforms are used in clinical gait laboratories to assess the effectiveness of treatments for neuromuscular diseases or to assess the progress of rehabilitation from musculoskeletal injuries: how the ground reaction forces have changed following treatment, or what changes have occurred during rehabilitation. Examining force platform records is also a way to evaluate the fit and function of prostheses.

Force platforms are used in athletics to measure the ground reaction forces exerted by shot-putters and discus throwers during their throws; by long jumpers, triple jumpers, and pole-vaulters during their takeoffs; by weightlifters during their lifts; by platform divers during their dives; and so on. The patterns revealed in the force–time histories give coaches and scientists information about technique differences that may affect performance.

Some of the larger athletic shoe manufacturers use force platforms in their biomechanics laboratories. Analysts evaluate the features of various shoe designs and the materials used in these designs by examining the ground reaction forces produced by subjects wearing the shoes.

Force Transducers

Force transducers are devices for measuring force. The force plates just described rely on multiple force transduc-

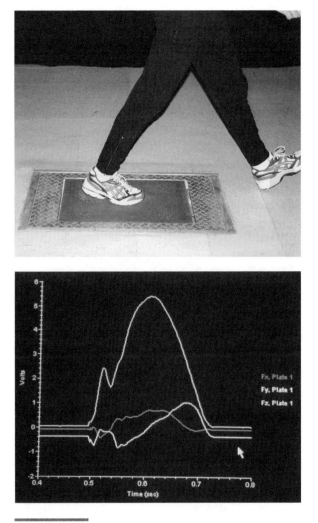

Figure 16.4 Force platforms are typically used to measure ground reaction forces. These figures show a subject walking over a force platform and the ground reaction forces produced during a walking step.

ers within them to determine their output. Another type of force transducer is a **strain gauge**. Strain gauges measure strain—change in length divided by original length. If a strain gauge is attached to a material with a known shape and elastic modulus, the stress in the material can be computed, and ultimately the external load that caused this stress and strain can be determined. Strain gauges are thus useful for measuring forces.

Force transducers have been used in a variety of sports to measure the forces applied to implements or equipment. Strain gauge force transducers have been used to measure forces on the rings and horizontal bar in gymnastics, on the wire in hammer throwing, on the oar in rowing, and on the start handle in the luge.

Clinical uses of strain gauges to measure forces have been important in improving the understanding of loads on bones and joints and even tendons and ligaments. Strain gauges have been attached to implanted artificial hips to measure the forces at the hip joint in vivo. In animals, force transducers (buckles with strain gauges attached) have been used to measure in vivo tendon forces.

Pressure Sensors

Pressure sensors are usually thin mats with arrays of force sensors imbedded in them. A force platform measures a resultant reaction force, which is really the resultant of a number of forces acting on the surface in contact with the force platform. Pressure sensors better represent the distributed nature of these forces by quantifying the pressure (force divided by area) exerted on each specified area of the pressure mat.

As with force platforms, pressure mats are most widely used in gait analysis. The regions of high pressure beneath a patient's foot may be identified when a patient walks barefoot across a pressure mat. The changes in these pressure patterns following treatment for diseases or surgical intervention and rehabilitation may be monitored.

Pressure mats may not measure the pressures exerted on the foot when a patient wears shoes, so insole pressure-measuring devices have been developed. These devices fit in the shoes between the sole of the foot and the shoe (see figure 16.5). Pressure measurements from these insole pressure-measuring devices are used by podiatrists and other medical professionals to design more effective orthotics and shoes. In sport, pressure-measuring insoles have been used in skiing to measure the pressures exerted by a skier's foot on the boot during alpine skiing.

Another clinical application of pressure-measuring devices is to measure bone-to-bone pressures in joints. Very thin pressure mats (or pressure-sensitive film) have been used to measure the pressure distributions within joints in cadavers.

Electromyography

Muscle forces produced during movement may be indirectly measured using **electromyography (EMG)**. EMG measures the electrical activity of a contracting muscle via surface electrodes placed on the skin over a superficial muscle or via indwelling electrodes that are implanted within the muscle. At the very least, EMG data indicate whether a muscle is contracting or not. With skillful processing of the EMG signal, the relative strength of the contraction may be determined. Although there is clearly a relationship between the magnitude of the muscle force and the EMG signal, quantifying this relationship and determining the muscle force from the EMG data are not yet possible. EMG is still a useful tool for clinical and sport applications.

Figure 16.5 Pressure-measuring insoles are useful tools for measuring pressure on the plantar surface of the feet during gait.

Computer Simulation and Modeling

A final tool used by biomechanists is computer simulation and modeling. This is not really a measurement tool but rather an analysis tool. In sport, computer simulation may be used to predict the outcome of a movement based on certain inputs. In clinical situations, the effects of surgical repairs or prosthetic devices may be evaluated before the intervention.

Computer simulation models are usually mathematically based, using equations derived from Newton's laws of motion. Input to the simulation usually includes inertial properties of the body and its limbs (mass, lengths, moments of inertia), initial conditions at the start of the simulation (positions and velocities of the body and limbs), and time histories of the control functions. The control functions may be relative positions of the limbs, muscle forces, resultant joint torques, and so on. The outcome of a simulation is the movement of the body that results from these inputs.

Simulations of a variety of sport skills have been developed. The limitation of computer simulation is that each simulation is specific to one individual and one set of input parameters. Results of computer simulations may be applicable only to the person whose parameters were input into the simulation. Despite this limitation, computer simulations are a promising way to investigate "what if" questions. What if the diver adducted her arms farther? What if the gymnast kept his tuck position a little longer? What if the pole-vaulter used a slightly stiffer pole?

Summary

Comprehensive quantitative biomechanical analyses are usually limited to performances by elite athletes or to clinical situations. How and where biomechanical measurements are taken may affect the parameter being measured. A number of tools are used to measure biomechanical variables. Tools for measuring kinematic variables in biomechanics include timing systems, velocity-measuring systems, optical imaging systems, accelerometers, and MEMS inertial sensors. Motion capture systems are used for capturing three-dimensional movements of the whole body. Tools for measuring kinetic variables in biomechanics include force platforms, strain gauges, pressure-sensing devices, and electromyography (EMG). Computer simulation is another useful tool for biomechanical analysis.

KEY TERMS

accelerometer (p. 389)
analog signals (p. 386)
electromyography (EMG) (p. 391)
force platform (p. 390)
sampling rate (p. 386)
strain gauge (p. 391)

REVIEW QUESTIONS

1. What are the advantages of collecting biomechanical data in a laboratory setting?
2. What are the disadvantages of collecting biomechanical data in a laboratory setting?
3. What are the advantages of collecting biomechanical data in the field?
4. What are the disadvantages of collecting biomechanical data in the field?
5. What is an analog signal?
6. What is a digital signal?
7. What is a sampling rate?
8. How can information from a video-recorded performance be used to compute velocity?
9. What does an accelerometer measure?
10. What is the difference between a uniaxial and a triaxial accelerometer?
11. What does a force platform measure?
12. How does a strain gauge measure force?
13. Which tools are best suited for measuring the mechanics of impact?
14. What tool is used to measure the electrical activity of muscles?
15. What two inertial measurement devices are part of an IMU?

Motion Analysis Exercises Using MaxTRAQ

If you haven't done so already, review the instructions for downloading and using the educational version of the MaxTRAQ motion analysis software at the beginning of this book, then download and install the software. Once this is done, you are ready to try the following two-dimensional kinematic analysis using MaxTRAQ.

1. Complete a quantitative biomechanical analysis of a planar sport skill or human movement. Before you record the movement, determine what aspects of the movement you want to analyze and what kinematic parameters you want to measure. Use a stationary video camera to record the movement. Make sure that the camera is set up so that its optical axis is perpendicular to the plane of motion and its field of view is large enough to capture the motion, but not so large that measurement accuracy is compromised. Place an object of known length in the plane of motion and use it as your reference length for scaling the video. Video record the skill. Download the video to a computer. If the video is not in .avi format, convert it to .avi format. (You can download free video conversion programs on the Internet.) Once the video clip is converted to .avi format, use MaxTRAQ to measure the kinematic parameters you previously identified.

Units of Measurement and Conversions

Table A.1 Basic Dimensions and Units in the SI System

Quantity	Symbol	SI unit	Unit abbreviation
Time	t	second	s
Length	l	meter	m
Mass	m	kilogram	kg

Table A.2 Derived Quantities and Dimensions Used in Mechanics

Quantity	Symbol	SI unit	Unit abbreviation	SI basic units
Area	A	square meter	m²	m²
Volume	V	cubic meter	m³	m³
Density	ρ	kilograms per cubic meter	kg/m³	kg/m³
Velocity	v	meters per second	m/s	m/s
Acceleration	a	meters per second per second meters per second squared	m/s/s m/s²	m/s/s m/s²
Angle	θ	radian	rad	Dimensionless
Angular velocity	ω	radians per second	rad/s	1/s
Angular acceleration	α	radians per second per second radians per second squared	rad/s/s rad/s²	1/s/s 1/s²
Momentum	L	kilogram meters per second	kg·m/s	kg·m/s
Force	F	newton	N	kg·m/s²
Weight	W	newton	N	kg·m/s²
Impulse	$F \Delta t$	newton second	N·s	kg·m/s
Pressure	P	pascal	Pa	kg/m s²
Torque	T	newton meter	Nm	kg·m²/s²
Moment of inertia	I	kilogram meter squared	kg·m²	kg·m²
Angular momentum	H	kilogram meter squared per second	kg·m²/s	kg·m²/s
Work	U	joule	J	kg·m²/s²
Energy	E	joule	J	kg·m²/s²
Power	P	watt	W	kg·m²/s³

Table A.3 Prefixes Used in the SI System

Factor of ten	Prefix name	Symbol	Name of number	Number
10^{24}	yotta	Y	1 septillion	1 000 000 000 000 000 000 000 000
10^{21}	zetta	Z	1 sextillion	1 000 000 000 000 000 000 000
10^{18}	exa	E	1 quintillion	1 000 000 000 000 000 000
10^{15}	peta	P	1 quadrillion	1 000 000 000 000 000
10^{12}	tera	T	1 trillion	1 000 000 000 000
10^{9}	giga	G	1 billion	1 000 000 000
10^{6}	mega	M	1 million	1 000 000
10^{3}	kilo	k	1 thousand	1 000
10^{2}	hecto	h	1 hundred	100
10^{1}	deka	da	1 ten	10
10^{0}			one	1
10^{-1}	deci	d	1 tenth	0.1
10^{-2}	centi	c	1 hundredth	0.01
10^{-3}	milli	m	1 thousandth	0.001
10^{-6}	micro	θ	1 millionth	0.000 001
10^{-9}	nano	n	1 billionth	0.000 000 001
10^{-12}	pico	p	1 trillionth	0.000 000 000 001
10^{-15}	femto	f	1 quadrillionth	0.000 000 000 000 001
10^{-18}	atto	a	1 quintillionth	0.000 000 000 000 000 001
10^{-21}	zepto	z	1 sextillionth	0.000 000 000 000 000 000 001
10^{-24}	yocto	y	1 septillionth	0.000 000 000 000 000 000 000 001

Tables A.4 through A.11 are used to convert from one unit of measure to another for a variety of mechanical quantities. Some instruction on reading these tables is warranted. In the leftmost column of each chart are the units you wish to convert from. The top row of the chart indicates the units you wish to convert to. First, read down the left column and find the unit you wish to convert from. Now, read across the row for that unit until you get to the column for the unit you wish to convert to. Multiply this number by your measure to get the units listed in the top row of that column. For example, in the time conversion chart, to convert 86 minutes to hours, look in the left column and find the minutes row. Read across to the right until you get to the cell in the hour column. The number in this cell is 0.016667. Multiply 86 minutes times 0.016667 hours/minute and get 1.43 hours. The conversion factors listed in boldface type are exact conversions.

Mathematically, each cell expresses the following:

1 unit of leftmost column = x units of top-row cell

For our time example,

1 minute = 0.016667 hours

Reading across the whole minute row:

1 minute = 60 seconds = 1 minute = 0.016667 hour = 0.00069444 day

Table A.4 Time Conversions

Time	Second	Minute	Hour	Day
1 second =	**1**	0.016 667	0.000 277 78	0.000 011 574
1 minute =	**60**	**1**	0.016 667	0.000 694 44
1 hour =	**3 600**	**60**	**1**	0.041 667
1 day =	**86 400**	**1 440**	**24**	**1**

Table A.5 Length Conversions

Length	Inch	Foot	Yard	Mile	Centimeter	Meter	Kilometer
1 inch =	**1**	0.083 333 33	0.027 778	0.000 015 782	**2.54**	**0.0254**	**0.000 0254**
1 foot =	**12**	**1**	0.333 33	0.000 189 39	**30.48**	**0.304 8**	**0.000 304 8**
1 yard =	**36**	**3**	**1**	0.000 568 18	**91.44**	**0.914 4**	**0.000 914 4**
1 mile =	**63 360**	**5 280**	**1 760**	**1**	160 934.4	1609.344	1.609 344
1 centimeter =	0.393 70	0.032 808	0.010 094	0.000 006 213 7	**1**	**0.01**	**0.000 010**
1 meter =	39.3708	3.280 840	1.093 613	0.000 621 37	**100**	**1**	**0.001**
1 kilometer =	39370.8	3 280.840	1 093.613	0.621 371	**100 000**	**1 000**	**1**

Table A.6 Mass Conversions

Mass	Ounce (mass)	Pound (mass)	Slug	Milligram	Gram	Kilogram
1 ounce (mass) =	**1**	**0.062 5**	0.001 942 6	**28 349.523 125**	**28.349 523 125**	**0.028 349 523 125**
1 pound (mass) =	**16**	**1**	0.031 081	**453 592.37**	**453.592 37**	**0.453 592 37**
1 slug =	514.784 79	32.174 05	**1**	14 593 903.	14 593.903	14.593 903
1 milligram =	0.000 035 273	0.000 002 204 6	0.000 000 068 521	**1**	**0.001**	**0.000 001**
1 gram =	0.035 273	0.002 204 6	0.000 685 21	**1000**	**1**	**0.001**
1 kilogram =	35.273 96	2.204 623	0.068 521	**1 000 000**	**1 000**	**1**

Table A.7 Angle Conversions

Angle	Degree	Radian	Revolution
1 degree =	**1**	0.017 453	0.002 777 8
1 radian =	57.295 78	**1**	0.159 16
1 revolution =	**360**	6.283 19	**1**

Table A.8 Velocity Conversions

Velocity	Feet per second	Miles per hour	Meters per second	Kilometers per hour
1 foot per second =	**1**	0.681 82	**0.304 8**	1.097 28
1 mile per hour =	1.466 67	**1**	**0.447 04**	**1.609 34**
1 meter per second =	3.280 84	2.236 94	**1**	**3.6**
1 kilometer per hour =	0.911 34	0.621 37	0.277 78	**1**

Table A.9 Force Conversions

Force	Pound	Newton
1 pound =	**1**	4.448 22
1 newton =	0.224 81	**1**

Table A.10 Work or Energy Conversions

Work or energy	Foot-pounds	Joules
1 foot-pound =	**1**	1.355 81
1 Joule =	0.737 56	**1**

Table A.11 Pressure or Stress Conversions

Pressure or stress	Pound per square inch	Pascal
1 pound per square inch =	**1**	6 894.757
1 pascal =	0.000 145 038	**1**

Table A.12 Torque Conversions

Torque	Inch-pound	Foot-pound	Newton meter
1 inch-pound =	**1**	0.083 333	0.112 98
1 foot-pound =	**12**	**1**	1.355 81
1 newton meter =	8.850 74	0.737 56	**1**

Answers to Selected Review Questions, Problems, and MaxTRAQ Exercises

This appendix contains answers to most of the review questions and problems listed at the end of each chapter. It also contains answers to the MaxTRAQ exercises presented at the end of chapters 2, 3, 4, and 6. Before referring to this appendix, you should make a genuine effort to answer each question—use the appendix only to check your answers.

Introduction

Review Questions

1. Kinesiology is the study of human movement. In the broad sense, it includes study of the anatomy, physiology, psychology, sociology, and mechanics of human movement. Biomechanics is the study of forces and their effects on living things. Kinesiology and biomechanics are thus two disciplines of study that overlap in the study of the mechanics of human movement.

8. Performance improvement and injury prevention

9. Improving techniques, equipment, and training

10. International Society of Biomechanics, American Society of Biomechanics, International Society of Biomechanics in Sports

11. Digital computers

12. Length, time, and mass

13. Meter, second, kilogram

Problems

1. Multiply your height in inches by 0.0254 m/in.

2. Divide your weight in pounds by 2.2 lb/kg.

3. 218 kg

4. 40 lb

5. 1609 m

6. 437 yd

7. 2.3 yd longer or 2.1 m longer

8. 8 ft 1/2 in.

9. 220 lb

10. 42.2 km

11. 62.2 mi

12. 91.4 cm

13. 91.4 m

14. 1.25 lb

15. a. 27.0-27.4 in.
    b. 0.92-0.98 lb

Chapter 1

Review Questions

1. Isaac Newton

2. Mass is a measure of linear inertia and is a scalar quantity. Weight is a measure of the force of gravity acting on an object. Weight is a vector quantity with a magnitude and direction (downward). The SI unit for mass is kg and the SI unit for weight is N.

3. In a tug-of-war, the team that weighs more usually wins. Assuming that each member of the heavier team weighs more than his counterpart on the lighter team, then larger contact forces act on the heavier team because of each team member's heavier weight. This larger normal contact force allows each member of the heavier team to produce larger frictional forces. If the heavier team

can maintain their grip on the rope, the larger friction forces under their feet will give them an advantage.

4. At the start of this "unweighting" action, as the skier drops toward the skis, the normal contact force between the skis and the snow is reduced. The smaller normal contact force results in a smaller frictional force between the skis and the snow, and the skis are easier to turn at this instant.

6. Pushing forward and upward on the object is better since it decreases the normal contact force on the object and thus reduces friction on the object. Pushing forward and downward on the object increases the normal contact force and thus increases friction on the object.

7. Pumping the brakes is better because the tires don't skid—static friction acts between the tire and the road. If the tires skid, dynamic friction acts between the tires and the road. Dynamic friction is smaller than limiting static friction.

8. The tangent of the angle that the flat surface makes with horizontal at the instant the object begins to slide equals the coefficient of static friction. The normal contact force equals the weight of the object times the cosine of the angle. The friction force equals the weight of the object times the sine of the angle. The coefficient of friction is friction force divided by normal contact force. This comes out to be the sine of the angle divided by the cosine of the angle. The tangent of an angle is equal to the sine of the angle divided by the cosine of the angle.

9. Yes. The vector sum of two forces can be equal in magnitude to the algebraic sum of the forces if the forces are collinear and in the same direction.

10. No. The largest magnitude of the vector sum of the two forces is the algebraic sum of the two forces, and this is possible only if the forces are collinear and in the same direction.

Problems

1. 1825 N
2. 4522 N
3. 50 kg
4. 640 N
5. 444 N or 222 N force from fingers and 222 N force from palm of hand
6. 938 N
7. a. 0.75
   b. 0.50
8. >432 N

9. a. 1281 N
   b. The force acts downward and backward on the blocks at an angle 51° below horizontal.
10. 1600 N
11. 300 N or 150 N at each plate
12. a. 1217 N
    b. The force acts upward and backward on Daisy at an angle of 80.5° above horizontal.
13. 1000 N force pushing anteriorly and superiorly on the patella at an angle of 30° above horizontal
14. a. 4402 N
    b. 936 N
    c. 3831 N
15. a. 42 N
    b. The force acts in a direction 12° north of west.
16. >62 N
17. The resultant force acting on the skier is 244 N directed down the hill, parallel to the 30° slope.
18. a. 7533 N
    b. 5° forward and upward above horizontal

Chapter 2

Review Questions

4. a. Tyler's
   b. Displacement was the same for both riders.
   c. Tyler
   d. Average velocity was the same for both riders.
   e. The outcome of the race was a tie.

5. Generally, the faster the speed of the ball (or puck), the smaller the size of the goal. Speed and goal size are inversely related. Actually, it is the average horizontal velocity from release to goal that is most strongly related to the inverse of area of the goal. The reaction and movement times of the goalie are directly related to the goal size. The sports that require shorter reaction and movement times of the goalie (or goaltender or goalkeeper) have goals with smaller areas.

6. This question is similar to question 5, but in this case, the faster the speed of the ball (or shuttlecock), the larger the court size. In this case, it is the average horizontal speed from release to opponent that is most strongly related to the length of the court.

7. Several factors affect the speeds of the ball and implements listed in table 2.3. These maximum reported speeds are instantaneous speeds, typically measured by radar guns. The fastest speeds are all achieved with the use of an implement

that provides an extension to an athlete's limb or limbs. Chapters 3, 4, and 6 all include information that helps explain how a longer lever (limb plus implement length) may help increase linear speed at the distal end of the implement. The other factor that may affect the speed of the balls and implements listed in the table is the mass of the implement. The golf ball and the squash ball are the two balls with the smallest masses, and these balls have the fastest and third-fastest speeds among the balls and implements listed.

8. The sprinter's fastest average horizontal velocity will occur during the flight phase of a step. When a sprinter is running at her fastest speed, her average horizontal velocity stays almost constant from step to step. During the flight phase, the sprinter is a projectile, so her horizontal velocity is constant. Her horizontal velocity at the takeoff into the flight phase (the start of the flight phase and the end of the previous support phase) is the same as her horizontal velocity at the instant just before touchdown (the end of the flight phase and the start of the support phase). During the first part of the support phase, the sprinter's horizontal velocity slows down as a braking force acts on her foot. During the second part of the support phase, her horizontal velocity speeds up as a propulsive force acts on her foot. The sprinter's fastest horizontal velocity during the support phase occurs at touchdown and again at takeoff. This fastest support phase velocity is identical to the average flight phase velocity. But the horizontal velocity during the rest of the support phase is slower than this fastest velocity, so the average horizontal velocity during the support phase is slower than that of the flight phase.

9. Yes. The runner must be accelerating since the runner must change directions to move around a curve.

10. To the left toward the center of the circle

13. Long jumpers have the whole length of the runway and plenty of time to generate horizontal velocity, but only about 0.10 s (the time the takeoff foot is on the board) to generate vertical velocity.

Problems

1. a. 21 yd
   b. 15 yd
   c. 12 yd
2. The shot from 5 m away will take 0.50 s to reach the goal, while the shot from 10 m away will take 0.25 s to reach the goal. The goalie will have a

better chance of blocking the 5 m shot because it will take longer to reach him.

3. The batter has 0.44 s to react to the pitch after the pitcher has released the ball.

4. The fastest average speed was for the 100 m world-record race at 10.44 m/s.

5. 12.7 m/s in a direction 23° east of north or 11.7 m/s north and 5 m/s east

6. a. 1.83 m/s² toward the finish line
   b. Zero—his velocity did not change.
   c. 9.09 m/s toward the finish line
   d. 0.33 m/s² toward the starting line

7. a. The puck is in the goal 0.11 m past the goal line.
   b. The Flyers win the game by 2 points.

8. 5.91 m

9. a. 9.81 m/s² downward
   b. 0.41 s
   c. 2.07 m

10. a. 15 m/s forward
    b. 4.62 m/s downward
    c. 30 m forward
    d. 10.38 m upward

11. a. 8.19 m/s
    b. 20°
    c. 7.7 m/s
    d. 5.47 m
    e. 4.17 m/s downward
    f. 1.40 m
    g. 0.52 m

12. a. 0.14 s
    b. 1.4 m

13. a. 2.96 m
    b. 0.94 s

14. a. 5.06 m
    b. 1.64 s
    c. 1.73 m
    d. Yes

15. a. 22.77 m/s
    b. 4.69 s

MaxTRAQ Exercises

There are no exact answers to the MaxTRAQ exercises since the process of digitizing is prone to error. Your answers should be close to those listed, however.

1. a. 2.73 m
   b. 1.36 strides/s
   c. 3.72 m/s
2. a. 3.20 m
   b. 1.43 strides/s
   c. 4.58 m/s

3. a. 3.71 m
    b. 1.62 step/s
    c. 6.01 m/s

4. a. 3.68 m
    b. 1.76 strides/s
    c. 6.48 m/s

5. a. 5.12 m
    b. 1.15 steps/s
    c. 5.91 m/s

6. Measured: 4.93 m; theoretical: 4.905 m

7. a. 2.85 m upward
    b. 2.83 m downward
    c. 2 cm difference; theoretically, the displacements should be the same.

8. a. Frames 5-14: 59 cm; Frames 14-23: 59 cm; Frames 23-32: 58 cm; Frames 32-41: 58 cm; Frames 41-50: 60 cm; Frames 50-59: 60 cm; Frames 59-68: 59 cm; Frames: 68-77: 59 cm; Frames 77-86: 60 cm. The horizontal displacements are all similar – between 58 and 60 cm. The displacements should be the same.
    b. Frames 5-14: 91 cm; Frames 14-23: 67 cm; Frames 23-32: 46 cm; Frames 32-41: 22 cm; Frames 41-50: 0 cm; Frames 50-59: -22 cm; Frames: 59-68: -46 cm; Frames 68-77: -68 cm; Frames 77-86: -91 cm. The vertical displacements are similar in the following way: the first upward displacement matches the last downward displacement, the second upward displacement matches the second to last downward displacement, etc. The displacements should match up if they are measured symmetrically around the time of peak height.
    c. $\Delta x = 2.64$ m; $\Delta y = 2.29$ m
    d. $\Delta x = 2.67$ m; $\Delta y = -2.27$ m
    e. Yes the answers are similar, except that the vertical displacement is downward in d and upward in c. The displacements should be similar since they are over the same time intervals before or after the peak height of the projectile's motion.

Chapter 3

Review Questions

1. Yes, if it was moving to begin with—but it cannot change its state of motion (accelerate). That is, it cannot speed up, slow down, or change direction since an external force is required to cause acceleration.

2. Yes, but only if the forces sum to zero.

3. Friction is the force that causes your change in direction. It acts on the soles of your shoes.

4. Newton's first law of motion is just a special case of Newton's second law of motion where the external forces sum to zero, so acceleration must be zero.

    $\Sigma F = ma = 0$
    $a = 0$

5. The equal but opposite force is a force (equal to your weight) exerted upward on the earth. This is so small compared to the mass of the earth that it has no effect on the earth.

6. No. A change in direction of motion is an acceleration, and an external force is required to cause acceleration.

7. The two teams are pulling with the same magnitude of force but in opposite directions. Since velocity is constant, acceleration must be zero, and thus the net force acting on the rope is zero.

8. The weight feels "heavier" at the beginning of the lift because it is accelerating upward. It is speeding up in the upward direction. To accelerate something upward, the net force acting on the object must be upward. The net force acting on the weight is the force you exert minus the weight itself. To achieve a net upward force, your upward force must be larger than the weight. Near the end of the lift, the weight feels "lighter" because it is accelerating downward. It is slowing down in the upward direction, and this is a downward acceleration. To accelerate something downward, the net force acting on the object must be downward. The net force acting on the weight is the force you exert minus the weight itself. To achieve a net downward force, your upward force must be smaller than the weight.

9. American football is a game of collisions. Players collide with one another to try to prevent or change the motion of the opposing player. These collisions are close to inelastic. If momentum is conserved in these collisions, then the heavier player's larger mass gives him an advantage. Newton's second law could also be applied here. To change the motion of a heavier, more massive player requires more force.

Problems

1. The dumbbell will accelerate 1.0 m/s² upward.

2. a. 5.0 N horizontal force in the same direction as the acceleration
    b. 9.81 N upward

3. 146 N

4. a. 1000 N in the opposite direction of the force Tonya exerts on Nancy.
   b. Tonya will accelerate 16.67 m/s^2 in the direction of the 1000 N force that Nancy exerts on her. Nancy will accelerate 20 m/s^2 in the opposite direction.
   c. 0 m/s

5. An upward force just greater than 147 N.

6. a. 8 m/s^2 backward
   b. 9.81 m/s^2 downward
   c. 200 m/s^2 forward
   d. 9.81 m/s^2 downward

7. 1.00 m/s^2 backward (south)

8. 20.0 N in the opposite direction of the ball's velocity

9. 1075 N forward

10. a. 900 N·s
    b. 1800 N upward
    c. 2781 N upward

11. 4.64 m/s^2 down the 30° slope

12. a. 0.82
    b. 0.88

13. 0.25 m

14. 64 m/s

15. a. 2.70 m/s upward
    b. 1.87 m high

MaxTRAQ Exercises

There are no exact answers to the MaxTRAQ exercises since the process of digitizing is prone to error. Your answers should be close to those listed, however.

1. a. 3.00 m/s
   b. 2.52 m/s
   c. 0.84

2. a. 42 m/s
   b. 28 m/s
   c. 48 m/s
   d. 0.42

Chapter 4

Review Questions

1. The sand does negative work on the jumper during impact. It reduces the jumper's energy as it undergoes displacement while exerting a force on the jumper. The displacement of the sand reduces the impact force on the jumper.

2. The windup or backswing puts the limb in a position to move through a greater displacement during the throw or swing. The greater displacement of the limb or implement during the throw or swing increases the total work done and thus increases the kinetic energy of the implement. More kinetic energy means more velocity—a faster throw.

3. During the follow-through, muscles do negative work on the swinging limb or implement to reduce the limb or implement's kinetic energy. The greater the displacement of the limb or implement during the follow-through, the smaller the force required by the muscles to slow down the limb or implement and the less the likelihood of a muscle injury.

4. Yes, this is an advantage since a taller, longer-limbed pitcher can move the ball through a greater displacement during the pitch, do more work on the ball, cause a greater increase in its kinetic energy, and thus throw the ball faster.

5. The helmet liner absorbs impact energy. During an impact, the head does work on the liner by deforming it—the force exerted on the head is reduced if the deformation is larger. The thicker the liner, the greater the deformation and the smaller the impact force on the head.

6. No. Stopping distance is proportional to the square of the speed. Doubling speed quadruples kinetic energy, but the negative work done by friction to reduce this energy is friction force times displacement. If the friction force stays the same, then displacement must increase by fourfold to absorb the fourfold increase in kinetic energy.

7. The tennis ball will bounce off the basketball with quite a large velocity upward because of the large mass of the basketball relative to the tennis ball (conservation of momentum) and the relatively high coefficients of restitution between the balls.

8. The air bag increases your displacement while you come to a stop—it does negative work on you over this longer displacement. Stopping over a longer displacement requires a smaller force.

9. When you run on loose sand, the sand does negative work on you and absorbs energy. When you run on asphalt, the asphalt absorbs less energy from you.

10. A child doesn't have to do as much work (or exert as much force) to swing the lighter implements or throw the lighter balls—likewise for catching the lighter balls. Catching a softer ball is also easier since the softer ball deforms when it is caught, thus reducing the impact force acting on a child's hands during the catch.

11. The prosthetic devices may be lighter. They may also able to store and return energy (like a spring) more efficiently and effectively than the live leg.

12. Possible actions:

Look to see if there is something on the ground below the window to cushion your landing. If there isn't and you have enough time, look in the room for something you can throw out the window that will cushion your landing. Use this to increase your displacement during your impact, thus reducing the impact force.

If you have time, look for a rope or other materials in the room that will allow you to lower yourself. Tie off one end of the rope or rope substitute and throw the other end out the window. Climb down the rope or rope substitute to safety.

If there is no rope or no time to use it, climb out the window and hang from the sill by your hands to reduce your potential energy as much as possible.

Let go of the sill and push off the wall to direct yourself toward any soft cushioning materials on the ground. Keep your body in a long upright position with feet down and arms overhead. Upon first contact with the ground, crumple your body by flexing your knees, hips, and trunk and allowing your arms to come down as well—try to move though as large a vertical displacement as possible and then roll. These actions will increase the displacement (d) in the work–energy equation and thus reduce the impact force. Your primary concern is reducing the forces that act on your brain and thus reducing its acceleration.

Problems

1. 400 J

2. a. The jai alai pelota has the most energy, 494 J for the 140 g pelota.
   b. The squash ball has the least kinetic energy, only 68 J for the 23 g ball or 74 J for the 25 g ball, even though the squash ball is the third-fastest ball or implement listed in the table.

3. 1742 J

4. a. Yes, you are doing positive mechanical work against your opponent since the force you are exerting and the displacement are in the same direction.
   b. Concentric contractions
   c. Yes, your opponent is doing negative work against you since the force your opponent is exerting and the displacement are in opposite directions.
   d. Eccentric contractions

5. a. 89 J
   b. 89 J
   c. 156 N

6. a. 120 J
   b. Positive
   c. 40 m/s horizontally in the direction of the force

7. a. 120 J
   b. −120 J
   c. Negative
   d. 1500 N in the opposite direction of the ball's motion

8. The 5 kg ball rolling at 4 m/s has more energy (40 J vs. 27 J).

9. a. 196.2 J
   b. −208 J
   c. 3467 N upward

10. a. 122.5 J
    b. −125.7 J
    c. 1934 N upward
    d. 3868 N upward
    e. 764 m/s^2 upward
    f. 78 g (very close to the threshold for concussion)

11. 3924 W

12. Ginger's jump was more powerful.

13. 4.26 m

14. 7416 N upward force

MaxTRAQ Exercises

There are no exact answers to the MaxTRAQ exercises since the process of digitizing is prone to error. Your answers should be close to those listed, however.

1. a. 0.0 m/s
   b. 30 cm
   c. 3.20 m/s
   d. 206 J
   e. 358 J
   f. 564 J
   g. 1880 N

Chapter 5

Review Questions

1. Leaning over to lift something creates a large moment arm between the object being lifted and the lumbar spine. When the lift begins, this moment arm creates a large torque about the lumbosacral joint, requires large forces in the back extensor muscles, and creates large stresses on the intervertebral discs in the lower back. The muscles and structures of the lower back are thus

susceptible to injury when this technique is used to lift heavy objects.

2. The bent-arm pullover is easier to do because the moment arm of the barbell around the axis of the shoulder joint is smaller. This smaller moment arm creates a smaller torque around the shoulder joint, so the torque created around the shoulder extensor muscles is smaller as well.

3. During an abdominal crunch exercise, the abdominal muscles exert their greatest force when the moment arm of the upper body about the axis of motion is greatest. During a crunch that begins on the floor, this position occurs at the start of the crunch when you first lift your upper body off the floor. When you do crunches on an incline bench, this position of greatest torque still occurs when your upper body is farthest horizontally from the axis of motion; but as the incline increases, the position occurs with increasing trunk flexion.

4. To maintain your stability when you reach forward to pick something up off the floor, you must keep your line of gravity within your base of support—the area beneath and between your shoes. As your trunk leans forward, your center of gravity moves forward as well, and the line of gravity soon gets close to your toes. To keep the center of gravity over the base of support, you normally move your hips and legs back as you lean forward. But with the wall behind you, you can't do this. So, as you keep leaning forward, your center of gravity keeps moving forward until the line of gravity moves forward of your toes and outside your base of support, and you fall forward.

5. Normally, the seat height of a chair should put you in a position in which your feet are flat on the floor as you sit in the chair with a 90° angle at your knee joint. Rising from this position into a stable standing position requires that you move your center of gravity forward over your feet. Leaning forward at the waist while extending at the knees and hips accomplishes this fairly easily. Using your arms as well will make it even easier. If the chair seat is quite low to the ground, this task is more difficult. Your feet are now resting on the floor well in front of your center of gravity. Rising out the chair with the low seat is more difficult because you have to move your center of gravity farther forward to position it over your feet. You also have to raise your center of gravity more in order to stand up.

6. The flight path of the center of gravity is unaffected by John's actions in the air. While John is in the air, his center of gravity is a projectile, and its trajectory is governed by the equations of projectile motion. The only force acting on him is

gravity. This is the only force affecting the motion of his center of gravity.

7. Place your feet slightly wider than shoulder-width apart and aligned with the rope; that is, the rope should pass above both feet. Lean back so that most of your weight is over your rear foot—your center of gravity should be over your rear foot. Flex slightly at the knees and hips to lower your center of gravity further and so that you can quickly extend and pull on the rope when the signal to begin the competition is given. This stance increases the distance from the line of gravity to the front edge of the base of support and lowers the center of gravity as well.

8. A two-handed dunk is more difficult because you have to jump higher. Your center of gravity must be higher to complete a two-handed dunk because the distance between your outstretched hands and your center of gravity is shorter than it would be if you extended only one arm.

9. Yes, it is possible for a high jumper to clear a crossbar even if her center of gravity does not go above the height of the crossbar. The jumper must drape herself over the bar in an inverted U position when she is at her peak height. In this position, her center of gravity may be outside of her body— below the peak of the inverted U—and the crossbar may be above her center of gravity but below that part of her body that forms the peak of the inverted U.

10. The larger handles increase the moment arms and thus decrease the force needed to turn the handles.

Problems

1. 15,000 N cm or 150 Nm
2. a. 3000 N cm or 30 Nm
   b. 750 N
3. a. 3000 N cm or 30 Nm extensor torque
   b. 3000 N cm or 30 Nm flexor torque
   c. 3 cm
4. Yes. The flexor muscles create a flexor torque of 40 Nm, which is sufficient to overcome the extensor torque of 37.5 Nm created by the weight.
5. a. 660 Nm tipping torque
   b. Greater than 687 Nm
   c. 625 N
6. a. 8.9 N
   b. 445 N cm or 0.445 Nm
7. The right hand exerts a 40.9 N downward force on the pole, and the left hand exerts a 65.4 N upward force on the pole.
8. 9963 N

9. a. 294 N
   b. 177 Nm abductor torque
   c. 177 Nm adductor torque
   d. 3532 N

10. The center of gravity is located 115 cm to the left of the right end of the barbell (105 cm to the right of the left end of the barbell or 5 cm to the left of the center of the bar).

Chapter 6

Review Questions

1. a. The advantages of having our muscles attach close to the joints are range of motion of the limbs and speed. Our muscles attach close to the joint, the axis of rotation, while the distal end of the limb—the part we use to throw, strike, or hit with—is far from the axis of rotation. When we rotate a limb around a joint by contracting our muscles, the attachment point of the muscle moves through only a small arc length, and the muscle length change is small. During the same movement, the distal end of the limb travels through a much larger arc length (arc length = $\Delta\theta r$). If the muscle attached farther from the joint, it might not be able to stretch or contract far enough to allow the same range of motion at the joint. An increased range of motion is thus one advantage. The other advantage is speed. The distal ends of our limbs can move with fast linear velocity because of the relationship of linear velocity to angular velocity and radius. The distal ends of our limbs are much farther from the joint axis than our muscle attachment sites; this longer radius compared to our muscles means that the distal ends of the limbs move much faster linearly than the attachment sites of the muscles ($v = \omega r$). There is a limit to the speed of shortening of a muscle, but the long length of our limbs allows a slow muscle contraction to produce a much faster linear velocity at the end of the limb.
   b. The disadvantage of having the muscle attachments close to the joints is that the moment arm of most muscles is relatively short; thus muscles must produce very large forces to create relatively modest torques about a joint.

2. If the two clubs can be swung with the same angular velocity, then the longer swing radius of the driver will cause a faster linear velocity of the club head according to the relationship $v = \omega r$.

3. Besides having the ability to do greater work due to greater displacement (see review question 4 in chapter 4), a longer-limbed individual has a longer swing radius, which may cause a faster linear velocity at the end of the limb according to the relationship $v = \omega r$.

4. The angular displacement of the leg during a step is related to the linear distance traveled by the foot during a step (which is related to the step length) as shown in the following equation:

   Arc length = θr

5. It is more difficult to run fast in lane 1 because your centripetal acceleration is greater due to the smaller radius of lane 1 compared to lane 8. This means that you must generate greater frictional forces under your feet in the radial direction to cause your centripetal acceleration.

6. Sagittal plane and transverse axis

7. Horizontal abduction (or horizontal extension) in the transverse plane around a longitudinal axis

8. Internal rotation, adduction, and extension at the shoulder joint; extension at the elbow joint

9. Adduction in the frontal plane around the AP axis.

10. In a narrow-grip pull-up, extension and flexion occur at the shoulder joint in the sagittal plane around a transverse axis. In a wide-grip pull-up, adduction and abduction occur at the shoulder joint in the frontal plane around the AP axis.

11. Extension

Problems

1. 96° or 1.68 rad of flexion

2. 3.75 rev/s; or 1350°/s; or 23.6 rad/s

3. 0.25 s

4. 20 rad/s^2 in the direction of his spin

5. 20 in.

6. a. 100 rad/s^2
   b. 40 m/s

7. a. 25.1 m/s tangent to the circular path of the hammer
   b. 526.3 m/s^2 toward the axis of rotation

8. a. The 120 rad/s angular velocity of the arm creates 42 m/s of the linear velocity of the baseball or 93% of the baseball's linear velocity.
   b. 5040 m/s^2 directed toward the internal rotation axis of the shoulder
   c. 731 N directed toward the internal rotation axis of the shoulder

9. 100 m

10. 1500 cm/s or 15 m/s

11. a. Yes; at an average angular velocity of 12 rad/s, the bat will rotate through an angular displacement of 1.65 rad to reach a hitting position above the plate by the time the ball reaches the plate.
    b. 37.5 m/s

MaxTRAQ Exercises

There are no exact answers to the MaxTRAQ exercises since the process of digitizing is prone to error. Your answers should be close to those listed, however.

1. a. 70°
   b. −89°
   c. 89°
   d. −88°
   e. 88°
   f. −156°/s
   g. −175°/s
   h. −171°/s
   i. −199°/s
   j. −167°/s
   k. −188°/s
   l. During the upswing she does work on the bar and pulls herself toward it by flexing at the hip joint and extending at the shoulder joint.
   m. −327°/s
   n. −116°/s
   o. The gymnast has large kinetic energy at the bottom of the giant swing and transforms most of this kinetic energy to potential energy at the top of the giant swing, so she has a much faster velocity at the bottom of the swing than at the top.

Chapter 7

Review Questions

1. a. Catherine
   b. Catherine, Anna, Bryn, Donna
   c. Catherine
   d. Donna

2. Flexing the recovery leg more during the swing-through is more effective since it reduces the moment of inertia of the leg about the transverse axis through the hip joint, thus speeding up the swing-through.

3. a. Dive 1
   b. Dive 3
   c. Dive 1
   d. Dive 3

4. Their smaller stature gives them smaller moments of inertia—an advantage in spinning, somersaulting, and tumbling maneuvers.

5. Her legs rotate in the opposite angular direction.

6. He should rotate his arms clockwise, in the same direction as the fall.

7. The medial–lateral friction force that acts under a runner's feet to cause him to change direction also causes a moment of force (a torque) that tends to tip him over away from the center of the curve. Leaning allows the normal contact force to produce a torque in the opposite direction and balance the torques created by the friction force.

8. The banked track allows some component of a runner's normal contact force to create the change in direction necessary to negotiate the turn, so the runner does not have to rely solely on friction force to cause the change in direction.

9. The gymnast should swing her arms in the same angular direction of her fall (in a clockwise direction as viewed from the front)—her left arm should adduct at the shoulder and her right arm should further abduct at the shoulder.

10. During the descent phase of the giant swing, the gymnast's objective is to generate enough angular momentum to carry him through the ascent phase. The torque that causes the increase in angular momentum is created by the reaction force from the bar and the moment arm of this force around his center of gravity. Maximizing the distance between the bar and his center of gravity increases the length of this moment arm. During the ascent phase, the reaction force from the bar causes a torque that acts in the opposite direction of the swing, so the gymnast must minimize this torque. He does this by shortening the distance between the bar and his center of gravity slightly, thus reducing the moment arm of this torque. This slight shortening also decreases his moment of inertia about the bar, the axis of rotation.

11. Adducting her arms during the helicopter 1080 trick reduces her moment of inertia about the longitudinal (twist) axis. This will speed up her rate of spin, since she is airborne and angular momentum is conserved. Abducting her arms just before landing increases her moment of inertia about the longitudinal (twist) axis. This will slow down her rate of spin and improve her chances of not falling when she lands.

Problems

1. a. 18.1 kg·cm² or 0.0018 kg·m²
   b. 140.0 kg·cm² or 0.014 kg·m²

2. 5 cm

3. 3150 kg·cm² or 0.3150 kg·m²

4. 0.30 kg·m²

5. a. 8.0 kg·m^2
   b. 2.5 rad/s
   c. 2.0 kg·m^2
   d. 10.0 rad/s

6. 5 rad/s

7. $694 \text{ rad/s or } 110.5 \text{ rev/s}$

8. 39.3 Nm

9. 13.2 Nm

10. a. 200 Nm
    b. 22.7 rad/s

Chapter 8

Review Questions

1. Your center of buoyancy and your center of gravity do not line up. Your center of buoyancy is superior to your center of gravity. This creates a torque about your transverse axis, and your legs begin to drop and your body rotates toward a vertical position.

2. Form drag

3. The prone position would create the least amount of drag force. However, the Union Cycliste Internationale, the international governing body for cycling, has rules that restrict the geometry and dimensions of a bicycle and thus prevent road racing cyclists from using a bicycle that would allow them to adopt a prone position.

4. As objects increase in size (volume), the surface area to mass ratio decreases (larger objects have less surface area relative to their mass). A bug's surface area to mass ratio is much larger than a human's. The effect of air resistance on a bug is therefore much larger as well, since the drag force acting on the bug is larger relative to its mass (due to the relatively larger surface area), and the effect of this drag force is greater because of the bug's smaller mass.

5. By inducing areas of turbulent flow around the ball

6. By inducing the Magnus effect

7. In a drive shot, the tennis player hits the ball hard to impart a fast horizontal velocity. The ball must have enough vertical velocity to clear the net, however. The combination of the fast horizontal velocity and enough vertical velocity to clear the net produces a trajectory that would put the ball past the service line and out of bounds if there were no air resistance. Imparting topspin to the bar induces a Magnus effect that causes the ball to veer below the trajectory it would have if there

were no air resistance so that it hits the court in bounds.

8. In a golf drive, backspin is imparted to the ball to induce a Magnus effect that causes the ball to veer above the trajectory it would have if there were no air resistance. This extra lift gives the ball more flight time and thus more distance.

9. Air resistance has the largest effect on 500 m speed skating due to the greater speeds in this event.

10. Air resistance may provide a lift force on the discus, thus prolonging its flight. The vertical component of the initial velocity could be reduced because of this. So the optimal projection angle would be smaller than if air resistance had no effect.

11. From largest to smallest: Frisbee, beach ball, Nerf ball, tennis ball, baseball, discus, basketball, shot

12. The sail acts as a wing and maximizes lift force while minimizing drag force due to the wind. The keel or centerboard of the boat also provides a lift force that prevents the boat from being pushed sideways. The sum of these forces, along with the drag forces from the water and the wind, is a resultant force that may have a component into the wind, thus providing the boat with an acceleration component into the wind.

13. To reduce total drag force; shaving makes the swimmer's surface (skin) smoother and reduces surface drag.

14. To reduce total drag forces; the helmet shape increases surface drag, since the surface area is larger, but the shape reduces form drag. The reduction in form drag is greater than the increase in surface drag, so the total drag force is decreased.

15. Drafting reduces drag forces on the trailing rider. Following the leading cyclist is a trailing wake of turbulent air that creates a pocket of lower pressure. The cyclist who is following, by moving into this pocket of lower-pressure turbulent air, reduces the drag force acting on both herself and the leading cyclist.

16. When the boat is still in the water, the only upward force acting on the boat is the buoyant force that is equal to the weight of the displaced water. If the boat is still, the buoyant force is equal in size to the weight of the boat and its contents. When the boat is moving, two upward forces act on the boat—the buoyant force and the lift force from the water. If the boat is not accelerating up and down, the sum of these two forces equals the weight of the boat and its contents. The buoyant

force is now less than the weight of the boat and its contents, so less water is displaced by the boat, and the boat rides higher in the water.

Problems

1. a. No
   b. Yes
2. 36%
3. 3 m/s north
4. 12.5 m/s
5. 33 N
6. 12 N
7. Amy
8. 24 N

Chapter 9

Review Questions

1. The lateral side of your knee undergoes compressive stress, while the medial side undergoes tensile stress.
2. The shear stress will be greatest at the outermost surfaces of the bone.
3. a. The two leg shafts have the same strength under a uniaxial tensile load.
   b. The two leg shafts have the same strength under a uniaxial compressive load.
   c. Leg shaft B has greater strength under a torsional load.
   d. Leg shaft B has greater strength under a bending load.
4. Tendon
5. Bone
6. Ligament
7. Bone
8. Live bone
9. a. The bone is strongest in resisting compressive stress.
   b. The bone is weakest in resisting shear stress.
10. Yield strength—stress at the elastic limit of the stress–strain curve

    Ultimate strength—maximum stress that a material can withstand

    Failure strength—stress achieved just before a material fails

    Failure strain—strain achieved just before a material fails

    Elastic modulus—slope of the stress–strain curve in the elastic region

Toughness—energy that a material can absorb before failure; the area under the stress–strain curve

Problems

1. 0.001 cm/cm or 0.1%
2. 200 N/mm²; or 20,000 N/cm²; or 200,000,000 Pa; or 200 MPa
3. 0.005 cm/cm or 0.5%
4. a. 48.0 MPa (48×10^6 Pa)
   b. 2.08 cm
5. The tendon will elongate 0.6 cm, that is, stretch to a length of 10.6 cm.
6. 10 GPa
7. 6,283 N
8. The anterior cruciate ligament is mechanically tougher.

Chapter 10

Review Questions

1. Support, movement, and protection
2. Mineral storage and hemopoeisis (blood cell production)
3. a. Long bone
   b. Irregular bone
   c. Short bone
   d. Long bone
   e. Sesamoid bone
4. The mineral component
5. Puberty
6. Gliding, pivot, hinge, ellipsoidal, saddle, and ball and socket
10. Ball and socket
11. Synovial fluid lubricates a synovial joint, provides nourishment to the articular cartilage, cleanses the joint cavity, and provides some shock absorption.
12. The synovial membrane and articular cartilage keep the synovial fluid in the joint cavity.
13. Articular cartilage improves bone-to-bone fit, increases joint stability, reduces bearing pressure, reduces friction, and provides some shock absorption.
14. Stability of a synovial joint is affected by the articular capsule and ligaments that cross the joint, the muscles and tendons that cross the joint, the degree of fit between the articulating surfaces, and the negative pressure gradient between the atmosphere and the joint cavity.

15. The joints of the lower extremities are generally more stable since they support the body.

Chapter 11

Review Questions

1. Movement and posture maintenance, heat production, protection, and pressure alteration

2. The sarcomere

3. The sarcolemma is the cell membrane of a muscle cell.

4. Force developed between the actin and myosin filaments is transferred to the Z-lines within a sarcomere. The Z-lines transmit the force to adjacent sarcomeres. The combined forces from the sarcomeres making up a single muscle cell are then transmitted to the endomysium, then to the endomysium of adjacent cells, and eventually to the perimysium and epimysium of the whole muscle. The force in the epimysium is then transferred to the tendon of the muscle.

5. Concentric activity occurs during the delivery phase of the throw, while eccentric activity occurs during the follow-through.

6. Eccentric activity occurs during the down (coiling or preparatory) phase, followed by concentric activity during the up or propulsive phase while the jumper is on the ground.

7. Biceps brachii and brachialis

8. Rectus femoris, pectineus, iliopsoas, tensor fascia latae, and sartorius

9. Pronator teres and the pronator quadratus

10. Muscles with longitudinal fiber arrangements usually have greater range of motion, while muscles with pennate fiber arrangements are usually stronger than similarly sized muscles with longitudinal fiber arrangements.

11. A maximal eccentric contraction is stronger.

12. A slower concentric contraction is stronger.

13. Your hip extensor muscles produce greater forces during landing because their contractions are eccentric during this phase.

14. Your shoulder muscles produce greater forces to put a shot because their contraction velocity is slower during this activity than during a baseball throw.

15. As the joint angle changes, the length of a muscle crossing that joint changes. This change in length affects the maximal contractile force of the muscle. As the joint angle changes, the moment arm of a muscle crossing that joint also changes. The torque produced by a muscle about a joint is muscle force times moment arm. Both of these variables, muscle force and moment arm, are affected by joint angle changes.

16. The advantage is range of motion and speed—the muscle does not have to shorten very far or very quickly to produce a large movement and speed of movement of the limb. The disadvantage is that the torque-producing capability of the muscle is compromised by its short moment arm about the joint.

Chapter 12

Review Questions

1. Peripheral nervous system

2. Away from the cell body

3. The central nervous system can increase the active tension in a muscle by recruiting more motor units or by increasing the rate of stimulation.

4. Muscle spindles

5. Golgi tendon organs

6. Muscle spindles

7. Golgi tendon organs

9. Stretch reflex

10. Golgi tendon reflex

11. Flexing the neck and turning the head to the opposite side of the arm doing the curl exercise

12. Pressure

13. Quickly stretching a muscle stimulates the muscle spindle and evokes the stretch reflex, which causes the muscle you are trying to stretch to contract. This detracts from the effectiveness of the stretching routine. Maximally contracting a muscle isometrically before stretching it stimulates the Golgi tendon organ and inhibits contraction of the muscle. This enhances the stretching routine since the muscle is more relaxed when you stretch it.

Chapter 13

Review Questions

1. Propose a model, examine the skill as it is performed, test or evaluate the performance, and educate the performer.

2. A quantitative biomechanical analysis involves measuring performance parameters and repre-

senting them numerically; in a qualitative biomechanical analysis, they are observed and described subjectively.

3. a. Time to finish

 b. Delivering the ball to one's teammate without interference from the opponent. This is affected by velocity and accuracy of the pass. The specific game situation may also impose constraints.

 c. The game situation will determine the performance criterion for a football punt. In all situations, however, kicking the ball quickly enough and at a high enough angle to prevent the punt from being blocked is important. In some situations, the performance criterion is to punt the ball so that it goes out of bounds as close to the opponent's goal line as possible. In other situations, the performance criterion is to punt the ball as far as possible toward the opponent's goal and with enough flight time to ensure adequate coverage by the punter's team.

 d. Measured distance of the throw: horizontal distance from the inside edge of the throwing ring to the first mark made where the discus first landed in the throwing sector

 e. The performance criterion in baseball batting depends on the game situation. It may be to hit the ball over the fence for a home run. It may be hit the ball to advance one or more base runners. It may be to hit the ball to get the hitter on base.

 f. Scoring points with the judges, usually by jumping higher and maintaining clean form during the jump and landing

 g. Scoring a goal; velocity and accuracy of the shot are two subgoals.

 h. Turning the caber so that it falls in the 12 o'clock position—with its top end closest to the thrower and the bottom end farthest from the thrower and the caber itself aligned so that it points directly back to the thrower.

4. a. To the right or left side of the runway for a sagittal view of the takeoff, flight, and landing of the jumper

 b. In front of or behind the hurdler for a frontal view

 c. To the right or left side of the swimmer for a sagittal view

 d. To the rear to view the flight of the ball toward the goal or to the side for a sagittal view of the kicker

 e. To the side (perpendicular to the arrow's path) for a frontal view of the archer

 f. To the right or left side of the weightlifter for a sagittal view

 g. Superior to the weightlifter looking along the longitudinal axis for a transverse plane view

5. Any error that endangers the athlete or exposes the athlete to increased risk of injury should be corrected first. Errors should then be grouped and ranked from major to minor in terms of effect on performance, from earliest to latest in terms of their occurrence during the performance, and finally from easiest to most difficult in terms of the time and effort required to correct them. Errors appearing in each category near the top should be attended to first. Special circumstances may also affect which errors to correct.

Chapter 14

Review Questions

1. Palpation of superficial muscles during activity; identification of sore muscles following activity; EMG; inverse dynamic analysis

2. Elbow extensors and shoulder horizontal flexors

3. Hip extensors, knee extensors, ankle plantar flexors

4. Hip extensors, knee extensors, ankle plantar flexors

5. Eccentric

6. Shoulder external rotators, flexors, horizontal abductors

7. Eccentric

Chapter 15

Review Questions

1. Stress is defined as force/area. Stress may be reduced through either a decrease in the magnitude of the applied force or an increase in the area over which the force is distributed. Using a softer material to construct a striking implement, such as a bat or hockey stick, reduces the size of the imposed force at contact because of the deformation of the softer material. The hard outer shell of a modern protective helmet is supported with a suspension system that rests on the head. At contact, the hard shell and suspension system distribute the force over the surface of the head, reducing the stress at any single point.

2. Tensile stress occurs when the forces acting on a tissue tend to pull apart the bonds between the molecules of the tissue.

Compressive stress occurs when the forces acting on a tissue tend to push the molecules of the tissue together and crush them.

Shear stress occurs when the forces acting on a tissue tend to slide adjacent molecules of the tissue past each other.

3. The four regions of the stress continuum from low stress to high stress are pathologic underload zone, physiologic loading zone, physiologic training zone, and pathologic overload zone.

4. The stress threshold refers to the level of stress that exceeds the level of stress to which a tissue has adapted. Loads above the stress threshold cause microdamage and stimulate the remodeling process that leads to a tissue's getting stronger, or hypertrophying. However, if a load above the stress threshold is imposed before the tissue has time to remodel, the cumulative effect of stressing a tissue that has not repaired itself is an overuse injury. Traumatic injury occurs when a single level of force far exceeding the stress threshold is applied, causing damage to the tissue and impairing performance.

5. Intrinsic factors related to injury are factors specifically related to the individual. Extrinsic factors related to injury are factors external to the individual (outside the body) and usually related to the task or the environment.

7. Some of the factors that affect the injury threshold and thus make the injury threshold different among people are genetics, training, and tissue adaptation (the rate of tissue remodeling and recovery).

8. Cross-training and within-activity cross-training are both intended to alter the pattern and magnitude of stress imposed on the body. Cross-training refers to performing two different activities, such as walking and swimming; within-activity cross-training refers to performing the same task with slight modifications, such as jogging on a track and jogging on a wood-chip trail. Either may be successful in preventing injury because the change in movement pattern imposes different levels of stress on tissues of the body, reducing the cumulative effect that may lead to an overuse injury.

11. Participation in weight-bearing activity maintains muscle and bone by ensuring that mechanical loads are imposed on the body (keeps the level of stress in the physiologic loading zone). On the other hand, prolonged bed rest or immobilization represents the decreased level of mechanical loading (physiologic underload zone) that leads to atrophy of tissues.

Problems

1. a. 850,000 Pa
   b. 76,500 Pa
2. a. 60 MPa
   b. 75%
3. a. 50 MPa
   b. 45%

Chapter 16

Review Questions

1. The major advantage of collecting biomechanical data in a laboratory setting is control of the environment.

2. The disadvantage of collecting biomechanical data in a laboratory setting is that the laboratory environment is not the environment through which a subject normally moves. The novelty of the lab and its instrumentation may affect the performance of the subject. The biomechanical data collected may not accurately represent the mechanics of the subject's movements in the real world.

3. The major advantage of collecting biomechanical data in the field is that the subject is moving in her natural environment. The biomechanical data collected will be a more accurate representation of the mechanics of the subject's movements in the real world.

4. The major disadvantage of collecting biomechanical data in the field is lack of control of the environment—weather, lighting, other people, and so on. Certain biomechanical measures may also be difficult to make since the measurement instruments may not be portable or suitable for use in the field.

5. An analog signal is a continuous variable that describes a physical characteristic.

6. A digital signal describes a physical characteristic that has been measured at discrete time intervals and represented numerically.

7. Sampling rate refers to the frequency with which a signal is measured or sampled.

8. If the performance was recorded with a stationary video camera positioned perpendicular to the plane of motion, and if an object of known length (a reference object) was positioned in the plane of motion and recorded, then the position of an object in a frame of video can be determined relative to some reference point in the video or on the monitor screen. With use of the known length of the reference object, the position of the object can be transformed to real-life units. The same process can be used to determine the real-life location of the object in another frame of video. The displacement of the object can then be computed from these position data. If the frame rate of the video camera that recorded the video is known, the number of frames from the object in its initial position to its final position can be used to compute time (time = number of frames/ frame rate). Displacement divided by time then gives the average velocity of the object.

9. Acceleration

10. A uniaxial accelerometer measures acceleration along only one axis of the accelerometer, while a triaxial accelerometer measures accelerations along three orthogonal axes of the accelerometer.

11. Reaction forces

12. A strain gauge measures strain in a known material with a known shape. Using the elastic modulus of the material and the strain measure, stress is computed. Using this stress measure and the cross-sectional area of the item that the strain gauge is fastened to allows computation of the force.

13. Force platforms and accelerometers

14. Electromyography (EMG)

15. A triaxial accelerometer and gyroscope

abduction—Starting from anatomical position, the joint action that occurs around an AP axis through a joint and causes limb movements in a frontal plane through the largest range of motion; the opposite of adduction; when referring to shoulder girdle movement, the movement of the scapula away from the midline of the body.

absolute angular position—Orientation of a line relative to another line or plane that is fixed relative to the earth; expressed in degrees or radians.

acceleration—Rate of change in velocity; expressed as units of length per unit of time squared or as meters per second per second (m/s^2) in SI; a vector quantity.

acceleration due to gravity—See *gravitational acceleration*.

accelerometer—A device that measures acceleration directly.

adduction—The joint action that occurs around an AP axis through a joint and causes limb movement in a frontal plane back toward anatomical position; the opposite of abduction; when referring to shoulder girdle movement, the movement of the scapula toward the midline of the body.

agonist—Role of a muscle whose torque aids the action referred to; prime mover; opposite in meaning to antagonist.

analog signal—A signal that represents a continuous measure of a time-varying signal, in contrast to a digital signal, which measures a time-varying signal at discrete intervals.

anatomical position—The position assumed by the body when it is standing erect, facing forward, with the feet aligned parallel to each other, toes forward, arms and hands hanging straight below the shoulders at the sides, fingers extended, and palms facing forward.

angular acceleration—Rate of change of angular velocity; measured in units of angular displacement divided by time squared or expressed as radians per second per second, degrees per second per second, or revolutions per second per second; a vector quantity.

angular displacement—Change in absolute angular position experienced by a rotating line with the direction of the change indicated; the angle between a line segment in its initial position and in its final position with the direction of rotation noted; measured as units of angular position or expressed as radians, degrees, or revolutions.

angular impulse—Average torque times the duration of application of the torque; causes and thus is equal to change in angular momentum; measured in units of torque times units of time or expressed as newton-meter-seconds in SI; a vector quantity.

angular inertia—Property of an object to resist changes in its angular motion; also referred to as rotary inertia.

angular momentum—Moment of inertia times angular velocity; measured in units of mass times units of velocity or expressed as kilogram-meters squared per second in SI; a vector quantity.

angular motion—Change in position that occurs when all points on a body or object move in circular paths about the same fixed axis; also referred to as rotary motion or rotation.

angular position—Orientation of a line relative to some other line or plane; expressed in degrees or radians.

angular velocity—Rate of angular displacement with direction of rotation indicated; measured in units of angular displacement divided by units of time or expressed as radians per second, degrees per second, or revolutions per second; a vector quantity.

anisotropic—Exhibiting material properties that are dependent on the direction; that is, an anisotropic material may have greater yield strength and a stiffer elastic modulus when pulled in one direction compared to another.

antagonist—Role of a muscle whose torque opposes the action referred to or the muscle referred to; opposite in meaning to agonist.

anteroposterior axis—Any one of the imaginary lines running from anterior to posterior and perpendicular to the frontal planes; abbreviated as AP axis; sagittal axis; sagittal–transverse axis.

AP axis—See *anteroposterior axis*.

aponeurosis—A flat sheet of tendon that connects a broad muscle to a line of attachment on a bone.

appendicular skeleton—That part of the skeleton composed of the bones of the appendages, including the bones of the shoulder girdle (scapula and clavicle) and the bones of the pelvic girdle (ilium, ischium, and pubis) with the exception of the sacrum.

articular capsule—Sleeve of ligamentous connective tissue surrounding a synovial joint and attached to the bones on either side of the joint; also referred to as joint capsule.

articular cartilage—Hyaline cartilage covering the articular surfaces of bones in a synovial joint.

autonomic nervous system—Part of the nervous system involved with unconscious sensations and actions; also referred to as involuntary nervous system.

average acceleration—Change in velocity divided by the time it took for the change to occur; expressed as units of length per unit of time squared or as meters per second per second in SI; a vector quantity.

average angular velocity—Angular displacement divided by the time it took for that angular displacement to occur with direction of rotation indicated; measured in units of angular displacement divided by units of time or expressed as radians per second, degrees per second, or revolutions per second; a vector quantity.

average speed—Distance traveled divided by the time it took to travel that distance; expressed as units of length per unit of time or as meters per second in SI; a scalar quantity.

average velocity—Displacement divided by the time it took for the displacement to occur; expressed as units of length per unit of time or as meters per second in SI; a vector quantity.

axial skeleton—That part of the skeleton composed of the bones of the vertebral column, ribs, and skull.

base of support—Area beneath and between the points of contact an object has with the ground.

bending load—Combination of forces that produce tensile stresses near one surface of an object, compressive stresses near the opposite surface, and shear stress throughout the object so that the object becomes a beam; as a result of a bending load, an object will deform by deflecting in a curved shape.

Bernoulli's principle—Lateral pressure in a moving fluid decreases as the velocity of the fluid increases.

biomechanics—The study of forces and their effects on living systems.

buoyant force—Upward force acting on an object in a fluid that is equal to the weight of the fluid displaced by the object.

cancellous bone—Porous, less dense bone tissue found deep to cortical bone near the ends of long bones; also referred to as spongy bone or trabecular bone.

cardinal plane—A plane that passes through the midpoint or center of gravity of the body.

Cartesian coordinate system—A system for locating the position of a point in two (or three) dimensions; the two (or three) coordinates represent displacements from a fixed point or origin in specific directions at right angles to each other.

cartilaginous joint—Joint in which fibrous cartilage or epiphyseal cartilage joins bones together, usually allowing slight movement; examples include the pubic symphysis and the epiphyseal growth plates in immature long bones.

center of gravity—Imaginary point through which the resultant force of gravity acts on an object; the point at which the entire weight of the body may be assumed to be concentrated; the point about which the torques created by the weights of the various body parts balance; the point of balance of the body.

center of pressure—The theoretical point of application of the dynamic fluid force to an object.

central nervous system—All nervous tissue lying within the skull and vertebral column; the brain and the spinal cord.

centric force—Force whose line of action passes through the center of gravity of an object.

centripetal acceleration—Linear acceleration of a point on a rotating object measured in the direction perpendicular to the circular path of the object (along a line through the axis of rotation or along a radial line); measured in units of length divided by units of time squared or expressed as meters per second per second in SI; a vector quantity.

centripetal force—An external force directed toward the axis of rotation of an object moving in a circular path.

circumduction—Flexion combined with abduction and then adduction, or extension and hyperextension combined with abduction and then adduction; the trajectory of a limb being circumducted forms a cone-shaped surface, and the end of the circumducted limb traces a circle.

coefficient of restitution—Ratio of the velocity of separation to the velocity of approach between two colliding objects; abbreviated as e.

colinear forces—Two or more forces that have the same line of action (but not necessarily the same direction along this common line of action).

combined load—Combination of forces that produce axial compression, axial tension, bending, torsion, shear loads, or any combination of these loads on an object.

compact bone—See *cortical bone*.

compression—State of an object as a result of forces pushing on it that are producing compressive stress;

compressive stress is axial stress that tends to push molecules together and squash the object.

compressive force—Pushing force whose direction and point of application would tend to shorten or squeeze an object along the dimension coinciding with the line of action of the force.

concentric contraction—Muscular activity that occurs when the muscle develops tension and its points of attachment move closer together; muscular activity that occurs when the muscle does positive work; also referred to as concentric action or concentric activity.

concurrent forces—Two or more forces whose lines of action intersect at a single point.

contact force—A force that results when two objects touch each other.

cortical bone—Solid, dense bone tissue found in the outer layer of bones; also referred to as compact bone.

creep—Material behavior characterized by a continued increase in strain under a constant stress.

curvilinear translation—Linear motion that occurs when an object maintains its orientation during a movement so that all points on the object move the same distance, in the same direction, and in the same time, but not in straight lines; also referred to as curvilinear motion.

density—Mass of an object divided by its volume; measured as units of mass divided by units of length cubed or expressed as kilograms per cubic meter in SI.

depression—Inferior movement of the scapula in a frontal plane; the opposite of elevation.

diaphysis—Central part or shaft of a long bone, which is separated from the ends or epiphyses by the epiphyseal cartilage before ossification of the epiphyseal cartilage.

displacement—Change in location of a point expressed as the length and direction of the vector from the starting position to the ending position; expressed as units of length or as meters in SI.

distance traveled—Length of the path followed by an object when moving from a starting position to an ending position; expressed as units of length or as meters in SI.

dorsiflexion—Starting from anatomical position, the ankle joint action that occurs around a transverse axis and causes the foot to move in a sagittal plane such that it moves forward and upward toward the leg; the opposite of plantar flexion.

downward rotation—Rotation of the scapula in the frontal plane such that its medial border moves superiorly and the shoulder joint moves inferiorly; the opposite of upward rotation.

drag force—Component of resultant dynamic fluid force that acts on an object in opposition to the relative motion of the object through the fluid.

dynamic fluid force—The resultant of the lift and drag forces that act on an object; the result of pressures exerted on the surfaces of the object.

dynamic friction—Frictional force that develops between two surfaces in contact that are moving or sliding relative to each other; sliding friction; kinetic friction.

dynamics—Branch of rigid-body mechanics concerned with the accelerated motion of objects.

eccentric contraction—Muscular activity that occurs when a muscle develops tension and its points of attachment move farther apart; muscular activity that occurs when the muscle does negative work; also referred to as eccentric action or eccentric activity.

eccentric force—Force whose line of action does not pass through the center of gravity of an object.

elastic—Able to return from a deformed shape to the original dimensions when the stress causing the deformation is removed.

elastic cartilage—Cartilage with more elastin than fibrous cartilage; found in the ears, the epiglottis, and part of the larynx; also referred to as yellow fibrocartilage.

elastic limit—Point on the stress–strain curve beyond which plastic deformation will occur; also referred to as proportional limit.

elastic modulus—Ratio of stress to strain; slope of the elastic region of the stress–strain curve for a material; also referred to as modulus of elasticity and Young's modulus.

electromyography (EMG)—A method of measuring the electrical activity of a contracting muscle via electrodes placed on the skin or implanted within the muscle.

elevation—Superior movement of the scapula in a frontal plane; the opposite of depression.

endolymph—Thick fluid that fills the inner ear.

endomysium—Connective tissue sheath or envelope encasing a muscle fiber.

energy—The capacity to do work; expressed as units of force times units of length or as joules in SI; a scalar quantity.

epimysium—Connective tissue sheath or envelope encasing a whole muscle.

epiphyseal cartilage—Cartilage that separates the diaphysis from the epiphyses of long bones; responsible for the growth in length of long bones before their ossification, usually during or shortly after puberty; also referred to as epiphyseal plate, epiphyseal disc, or growth plate.

epiphysis—Part of a long bone that is separated from the diaphysis or shaft of the long bone by the epiphyseal cartilage before ossification of the epiphyseal cartilage.

eversion—Starting from anatomical position, the joint action that occurs around an AP axis through the ankle when the lateral side of the sole of the foot is lifted; the opposite of inversion.

extension—The joint action that occurs around a transverse axis through a joint and causes limb movement in a sagittal plane back toward anatomical position; the opposite of flexion.

external force—A force that acts on an object as a result of its interaction with the environment surrounding it.

external rotation—Shoulder or hip joint actions that occur around longitudinal axes through these joints and cause limb movements in the transverse plane such that the knees turn away from each other or the palms of the hands turn away from the body; the opposite of internal rotation; lateral rotation; outward rotation.

exteroceptor—Sensory receptor that responds to stimuli from sources external to the body.

extrinsic factors—Factors related to injury and to characteristics of the task and the environment; they include the nature of the task being performed, the intensity and frequency of the performance, the playing surface, equipment, and level of participation.

failure strain—Strain at which a material breaks or fails.

failure strength—Stress at which a material breaks or fails.

fascicle—Bundle of muscle fibers; also referred to as a fasciculus.

fibrous cartilage—Cartilage with heavier collagen fibers than hyaline cartilage; the intervertebral discs and articular discs such as the menisci are examples of fibrous cartilage; also referred to as fibrocartilage or white fibrocartilage.

fibrous joint—Joint in which fibrous connective tissue joins bones together, usually in a rigid manner; sutures and syndesmoses are types of fibrous joints.

flexion—Starting from anatomical position, the joint action that occurs around a transverse axis through a joint and causes limb movement in a sagittal plane away from anatomical position through the largest range of motion; the opposite of extension.

force—A push or pull expressed as units of mass times units of length divided by units of time squared or as newton in SI; a vector quantity.

force couple—Torque created by a pair of oppositely directed forces about an axis; expressed as units of force times units of length or as newton meters in SI; a vector quantity.

force platform—A device used to measure kinetic variables, typically used to measure ground reaction forces and the point of application and direction of the resultant reaction force.

form drag—Drag force acting on an object within a fluid and caused by the impact forces of the fluid molecules with the object; also referred to as shape drag, profile drag, or pressure drag.

free-body diagram—A tool for analyzing forces and torques; a drawing of the object of analysis with all external forces acting on the object represented as arrows showing their points of application and directions.

friction—The component of a contact force that acts parallel to the surfaces in contact; the magnitude of friction is the product of the coefficient of friction and the normal contact force (the component of the contact force acting perpendicular to the surfaces in contact).

frontal plane—An imaginary plane running side to side and superior to inferior that divides the body into anterior and posterior parts; coronal plane; lateral plane.

general motion—A change in position that results from a combination of linear and angular motion.

Golgi tendon organ—Proprioceptor that responds to increases in muscle tension.

gravitational acceleration—The rate of change in velocity caused by the force of gravity; approximately 9.81 m/s^2 downward or 32 ft/s^2 downward; also referred to as acceleration due to gravity; often abbreviated as g.

gravitational potential energy—Energy due to the vertical position of an object; weight times height above some reference or mass times acceleration due to gravity times height above some reference; expressed as units of force times units of length or as joules in SI; a scalar quantity.

horizontal abduction—Starting from a position of hip or shoulder flexion, the shoulder or hip joint action that causes movement of the arm or thigh in the transverse plane around a longitudinal axis such that the arm or leg moves away from the midline of the body; the opposite of horizontal adduction; horizontal extension.

horizontal adduction—Starting from a position of hip or shoulder abduction, the shoulder or hip joint action that causes movement of the arm or leg in the transverse plane around a longitudinal axis such that the arm or thigh moves back toward the midline of the body; the opposite of horizontal abduction; horizontal flexion.

hyaline cartilage—Shiny, white, smooth cartilage that makes up the articular cartilage covering the articular surfaces of bones in synovial joints.

hyperextension—The joint action that occurs around a transverse axis and is a continuation of extension past anatomical position.

impulse—Average force times the duration of application of the force; causes and thus is equal to change in linear momentum; measured in units of force times units of time or expressed as newton seconds in SI; a vector quantity.

impulsive load—An applied force that reaches a relatively high magnitude in a short time.

inertia—The property of an object to resist changes in its motion.

instantaneous acceleration—Rate of change in velocity measured at an instant in time rather than over a duration of time; expressed as units of length per unit of time squared or as meters per second per second in SI; a vector quantity.

instantaneous angular velocity—The rate of change of angular displacement measured at an instant in time rather than over a duration of time with direction of rotation indicated; measured in units of angular displacement divided by units of time or expressed as radians per second, degrees per second, or revolutions per second; a vector quantity.

instantaneous speed—Rate of distance traveled measured at an instant in time rather than over a duration of time; expressed as units of length per unit of time or as meters per second in SI; a scalar quantity; a car's speedometer measures instantaneous speed.

instantaneous velocity—The rate of displacement measured at an instant in time rather than over a duration of time; expressed as units of length per unit of time or as meters per second in SI; a vector quantity.

internal force—A force that acts within the object or system whose motion is being investigated; forces between the molecules of an object that hold the object together.

internal rotation—Starting from anatomical position, shoulder or hip joint actions that occur around longitudinal axes through these joints and cause limb movements in the transverse plane such that the knees turn inward toward each other or the palms of the hands turn toward the body; the opposite of external rotation; also referred to as medial rotation or inward rotation.

interneuron—Nerve cell that transmits impulses between neurons; also referred to as connector neuron.

interoceptor—Sensory receptor that responds to stimuli from sources within the body.

intrinsic factors—Factors related to injury and to an individual's ability to cope with imposed mechanical stress; they include anthropometrics, skeletal alignment, fitness, and previous history of injury.

inversion—Starting from anatomical position, the joint action that occurs around an AP axis through the ankle when the medial side of the sole of the foot is lifted; the opposite of eversion.

isometric contraction—Muscular activity that occurs when a muscle develops tension and its points of attachment do not move relative to each other; muscular activity that occurs when a muscle develops tension and does zero work; also referred to as isometric action or isometric activity.

isotropic—Exhibiting material properties that are the same in all directions.

joint—Place where two bones join or meet and connect to one another.

joule—SI unit of measure for work and energy; equal to one newton meter. 1 J = 1 Nm.

kinematics—The branch of dynamics concerned with the description of motion.

kinesiology—The study of human movement.

kinetic energy—Energy due to the motion of an object; half of the mass times the square of the velocity of the object; expressed as joules in SI; a scalar quantity.

kinetics—The branch of dynamics concerned with the forces that cause or tend to cause motion.

laminar flow—Movement of fluid molecules such that adjacent layers of fluid flow parallel to one another and closely follow the shape of an object in the fluid.

lateral flexion—Movement of the trunk or neck to the left or right in a frontal plane around an AP axis.

lift force—Component of resultant dynamic fluid force that acts on an object in a direction perpendicular to the relative motion of the object through the fluid.

limiting friction—Peak static friction that can be produced between two surfaces for a given normal contact force; the maximum friction just before the two surfaces start to slide.

linear momentum—Mass of an object times the linear velocity of the object; measured in units of mass times units of length divided by units of time or as kilogrammeters per second in SI; a vector quantity.

linear motion—Change in position that occurs when all points on an object move the same distance, in the same direction, and at the same time; also referred to as translation.

longitudinal axis—Any of the imaginary lines running from superior to inferior and perpendicular to the transverse planes; vertical axis; frontal–sagittal axis; twist axis.

Magnus effect—An object's deviation from its normal flight path due to the effect of air resistance and the spin of the object.

mass—Measure of inertia; the quantity of matter in an object.

mechanics—The study of forces and their effects on objects.

modulus of elasticity—See *elastic modulus.*

moment arm—Perpendicular distance between the line of action of a force and the axis about which a moment of force or torque is being measured; determined by measuring the shortest distance between a line drawn along the line of action of the force and another line drawn parallel to this line but through the axis about which the torque is being measured.

moment of force—Torque created by a force about an axis; force times moment arm; expressed as units of force times units of length or as newton meters in SI; a vector quantity.

moment of inertia—Measure of angular inertia; sum of the product of the mass of each part of an object and the square of the distance each part is from the center of gravity of the object; mass times radius of gyration squared; measured in units of mass times units of length squared or expressed as kilogram-meters squared in SI; a scalar quantity.

motor neuron—Nerve cell that transmits impulses away from the central nervous system; also referred to as efferent neuron.

motor unit—A single motor neuron and all the muscle fibers with which it synapses; the fundamental unit of the neuromuscular system.

muscle fiber—A single muscle cell.

muscle spindle—Proprioceptor that responds to increases in muscle length.

net force—The vector sum of all the external forces that act on an object; the resultant of all the external forces that act on an object; the resultant force.

neutralizer—Role of a muscle whose torque cancels or eliminates the undesired effect of the torque produced by another muscle at the given joint to allow a desired movement only.

overuse injury—An injury caused by repeated applications of a stress lower than that required to cause injury in a single application; also known as repetitive motion injury, cumulative trauma injury, or predictable injury.

pacinian corpuscle—Sensory receptor for pressure.

perimysium—Connective tissue sheath or envelope encasing a bundle of muscle fibers (fascicle).

peripheral nervous system—All nervous tissue lying outside the skull and vertebral column.

physical training—Training directed at altering performance limitations due to the physical condition of the performer; involves improving the components of physical fitness, including muscular strength, flexibility, and cardiovascular fitness.

physiologic loading zone—Level of imposed stress within which tissue maintains its current status; muscle maintains the same force-generating capability, bone mineral content stays the same, and tendons and ligaments maintain their ability to withstand tensile stress.

plantar flexion—The ankle joint action that occurs around a transverse axis and causes the foot to move in a sagittal plane downward and away from the leg.

plastic—Able to maintain a deformed shape when the stress causing the deformation is removed.

Poisson's ratio—Ratio between transverse strain and strain in the axial direction for an object loaded axially.

position—Location of a point in space relative to some fixed point.

potential energy—Energy stored within an object due to its vertical position or deformation; expressed as units of force times units of length or as joules in SI; a scalar quantity.

power—Rate of doing work; work done divided by time; measured in units of work divided by units of time or expressed as watts in SI; a scalar quantity.

pressure—External force divided by the area over which this force acts; measured in units of force divided by units of length squared or expressed as newtons per square meter in SI.

principal axis—Axis about which an object's moment of inertia is largest; axis about which an object's moment of inertia is smallest; axis perpendicular to the two previously defined principal axes.

projectile—An object that has no external forces acting on it other than the force of gravity.

pronation—Starting from anatomical position, the radio-ulnar joint action that occurs around the longitudinal axis of the forearm and causes the palm to turn toward the body; the opposite of supination.

proportional limit—See *elastic limit.*

proprioceptor—Interoceptor that monitors the status of the musculoskeletal system, including joint position, change in joint position, muscle length, change in muscle length, and muscle tension.

Q angle—The smaller of the two angles formed by the intersection of the long axis of the tibia and the long axis of the femur projected onto the frontal plane; a measure of alignment between the femur and the tibia.

qualitative anatomical analysis—An analysis in which the predominant muscular activity and the active muscle groups during each phase of a movement are identified.

qualitative biomechanical analysis—A biomechanical analysis in which the mechanical characteristics of the performance are not quantified but are observed and subjectively evaluated.

quantitative biomechanical analysis—A biomechanical analysis in which the mechanical characteristics of the performance are measured and quantified.

radial deviation—The joint action that occurs around an AP axis through the wrist joint and causes movement of the hand in a frontal plane in the direction of the thumb; the opposite of ulnar deviation; also referred to as abduction or radial flexion.

radian—Ratio of the arc length that the end of a radial line travels through to the length of the radial line when the radial line rotates about one end; arc length divided by radius; 1 rad equals approximately 57.3°.

radius of gyration—The distance from an axis of rotation to where the mass of an object would have to be concentrated to create the same moment of inertia of the entire object about that axis; measured in units of length or expressed as meters in SI; a scalar quantity.

reaction force—An external contact force that results when one object touches another.

rearfoot motion—The natural sequential pattern of pronation and supination during the stance phase of running; measured for research and clinical purposes in the frontal plane as the angle between the shoe and the lower leg.

rectilinear translation—Linear motion that occurs when an object maintains its orientation during a movement so that all points on the object move the same distance, in the same direction, in the same time, in straight lines; also referred to as rectilinear motion.

relative angular position—Orientation of a line relative to another line or plane that may not be fixed; expressed in degrees or radians.

remodeling—The rebuilding of tissue, or the healing of microdamage caused by an imposed stress; the healing of the tissue leads to hypertrophy, or strengthening, of the tissue.

resultant displacement—Change in location of a point expressed as the length and direction of the vector from the starting position to the ending position.

resultant force—The vector sum of two or more forces; the force that results from the vector addition of two or more forces.

rigid-body mechanics—The branch of mechanics concerned with the effects of forces on objects that are assumed to be perfectly rigid.

rotary inertia—See *angular inertia.*

sagittal plane—An imaginary plane running anterior to posterior and superior to inferior that divides the body into right and left parts; anteroposterior plane.

sampling rate—The frequency or interval at which an analog signal is measured in order to convert it to a digital signal.

sarcolemma—Thin membrane covering a muscle cell.

sarcomere—Fundamental contractile unit of muscle, found between two adjacent Z lines in a myofibril.

sensory neuron—Nerve cell that transmits sensory impulses to the central nervous system; also referred to as afferent neuron.

sesamoid bone—Bone completely encased in connective tissue such as tendon or ligament, for example, the patella (kneecap).

shear—Force or stress acting parallel to the analysis plane or perpendicular to the long axis of the object; shear stress tends to slide molecules past each other and skew the object.

somatic nervous system—Part of the nervous system involved in conscious sensations and actions; also referred to as voluntary nervous system.

specific gravity—Density of an object divided by the density of water.

speed—See *instantaneous speed.*

spongy bone—See *cancellous bone.*

sport and exercise biomechanics—The study of forces and their effects on humans in exercise and sport.

stability—The resistance of an object to being toppled; the likelihood of an object returning to its original position after it is displaced.

stabilizer—A muscle whose torque prevents movement at a joint.

static equilibrium—The state or condition of an object that results when the object is not moving and the net force and net torque acting on the object are zero.

static friction—Frictional force that develops between two surfaces in contact that are not moving relative to each other.

statics—The branch of rigid-body mechanics concerned with the mechanics of objects at rest or moving at constant velocity.

strain energy—Energy due to the deformation of an object; for stretching or compressing, it is equal to half the stiffness constant of the material times the square of change in length of the object; expressed as units of force times units of length or as joules in SI; a scalar quantity.

strain gauge—A type of force transducer used to measure strain (change in length divided by original length).

stress—Internal force divided by the cross-sectional area of the surface upon which the internal force acts; measured in units of force divided by units of length squared; expressed as newtons per square meter or pascals in SI.

stress continuum—A graphical representation of the level of imposed stress, including zones in both pathologically high and low levels (distress) and physiological levels (eustress).

supination—The radioulnar joint action that occurs around the longitudinal axis of the forearm and causes limb movement in a transverse plane and returns the forearm and hand to anatomical position after being pronated or moves them beyond anatomical position; the opposite of pronation.

surface drag—Drag force acting on an object within a fluid and caused by friction between the fluid and the surface of the object; also referred to as skin friction or viscous drag.

synergy—Combined action of two or more muscles that cross the same joint to produce a desired result.

synovial joint—A freely movable joint characterized by a joint cavity filled with synovial fluid contained within a synovial membrane that lines a joint capsule.

synovial membrane—Thin membrane that is the interior lining of the joint capsule; it produces synovial fluid.

tangential acceleration—Linear acceleration of a point on a rotating object measured in the direction tangent to the circular path of the object; measured in units of length divided by units of time squared or expressed as meters per second per second in SI; a vector quantity.

technical training—Training devoted to improving technique; may involve performing the actual skill or performing drills that mimic specific aspects of the skill.

tendon—Cord or sheet of collagenous connective tissue that attaches muscle to bone.

tensile forces—Pulling forces whose direction and point of application would tend to lengthen or stretch an object along the dimension coinciding with the line of action of the force.

tensile stress—Axial or normal stress that occurs at an analysis plane as a result of a force or load that tends to pull apart the molecules bonding an object together at that plane.

tension—State of an object as a result of forces pulling on it and producing tensile stress; tensile stress is axial stress that tends to pull molecules apart and stretch the object.

tonic neck reflex—Rudimentary reflex associated with the position of the neck and affecting the muscles of the upper extremity.

torque—The turning effect created by a force about an axis; force times moment arm; expressed as units of force times units of length or as newton meters in SI; a vector quantity.

torsion load—A load that causes an object to twist due to torque; torques in opposite directions acting on either end and around the longitudinal axis of an object produce shear stresses in the analysis plane that become larger with increased distance from the longitudinal axis; as a result of a torsion load, an object will deform by twisting.

toughness—Ability of a material to absorb energy before failing.

trabecular bone—See *cancellous bone.*

transverse axis—Any of the imaginary lines running from left to right and perpendicular to the sagittal planes; horizontal axis; frontal axis; mediolateral axis; frontal–transverse axis.

transverse plane—An imaginary plane running from side to side and anterior to posterior that divides the body into superior and inferior parts; horizontal plane.

traumatic injury—An injury caused by a single application of a relatively high level of stress; an accidental injury.

turbulent flow—Movement of fluid molecules such that adjacent layers of fluid do not flow parallel to one another and they separate from the surface of the object in a fluid.

type I muscle fiber—Smaller-diameter muscle fiber characterized by aerobic metabolism, slow development of maximum tension, smaller maximum tension, and longer duration of tension development (endurance); first to be recruited; also referred to as slow-twitch oxidative (SO) fiber.

type IIA muscle fiber—Larger-diameter muscle fiber characterized by aerobic and anaerobic metabolism, faster development of maximum tension, larger maximum tension, and long duration of tension development (endurance); second to be recruited; also referred to as fast-twitch oxidative-glycolytic (FOG) fiber.

type IIB muscle fiber—Largest-diameter muscle fiber characterized by anaerobic metabolism, fastest development of maximum tension, largest maximum tension, and shortest duration of tension development (low endurance); last to be recruited; also referred to as fast-twitch glycolytic (FG) fiber.

ulnar deviation—Starting from anatomical position, the joint action that occurs around an AP axis through the wrist joint and causes hand movement in a frontal plane toward the little finger; the opposite of radial deviation; also referred to as adduction or ulnar flexion.

ultimate strength—Maximum stress a material can withstand.

uniform acceleration—Constant rate of change in velocity; constant, unchanging acceleration.

upward rotation—Rotation of the scapula in a frontal plane such that its medial border moves inferiorly and the shoulder joint moves superiorly; the opposite of downward rotation.

vector—Mathematical representation of any quantity that is defined by its size or magnitude (a number) and its direction (its orientation); vector quantities may be represented graphically by arrows, with the length of the arrow scaled to represent the magnitude of the vector and the shaft and head of the arrow oriented to represent the direction and sense of the vector along that direction.

velocity—See *instantaneous velocity*.

viscosity—Measure of resistance of a fluid to shear forces; measure of internal friction between layers of molecules of fluid.

watt—SI unit of measure for power; equal to 1 J/s.

weight—Measure of the force of gravity acting on an object; mass times the acceleration due to gravity; expressed as newtons in SI.

Wolff's law—A tissue adapts to the level of stress imposed on it; the level of adaptation in a tissue reflects the level of typical loading.

work—The product of force exerted on an object times the displacement of the object at the point of application of the force along the line of action of the force; expressed as units of force times units of length or as joules in SI; a scalar quantity.

yield strength—Stress above which plastic deformation will occur.

Young's modulus—See *elastic modulus*.

references and suggested readings

Abbott, A.V., and Wilson, D.G. (Eds.). (1996). *Human-powered vehicles.* Champaign, IL: Human Kinetics.

Adrian, M.J. (1980). The true meaning of biomechanics. In J.M. Cooper and B. Haven (Eds.), *Proceedings of the Biomechanics Symposium* (pp. 14-21). Indianapolis: Indiana State Board of Health.

Alexander, R.M. (1992). *The human machine.* New York: Columbia University Press.

Arend, S., and Higgins, J.R. (1976). A strategy for the classification, subjective analysis and observation of human movement. *Journal of Human Movement Studies,* 2:36-52.

Aristotle. (1912). *De motu animalium* (A.S.L. Farquharson, Trans.). In J.A. Smith and W.D. Ross (Eds.), *The works of Aristotle* (Vol. V, pp. 698-704). Oxford: Clarendon Press.

Atwater, A.E. (1980). Kinesiology/biomechanics: Perspectives and trends. *Research Quarterly for Exercise and Sport,* 51:193-218.

Bartonietz, K. and Borgtom, A. (1995). The throwing events at the World Championships in Athletics 1995, Goteborg: Techniques of the world's best athletes. Part 1: Shot put and hammer throw. *New Studies in Athletics,* 10(4):43-63.

Bennell, K.L., Malcolm, S.A., Wark, J.D., and Brukner, P.D. (1996). Models for the pathogenesis of stress fractures in athletes. *British Journal of Sports Medicine,* 30(3):200-204.

Blackwell, J.R., and Cole, K.J. (1994). Wrist kinematics differ in expert and novice tennis players performing the backhand stroke: Implications for tennis elbow. *Journal of Biomechanics,* 27(5):509-516.

Brancazio, P.J. (1984). *Sports science: Physical laws and optimum performance.* New York: Simon & Schuster.

Braun, G.L. (1941). Kinesiology: From Aristotle to the twentieth century. *Research Quarterly,* 12:163-173.

Brody, D.M. (1987). Running injuries: Prevention and management. *Clinical Symposia,* 39(3). New Jersey: Ciba-Geigy Corporation.

Brown, E.W. (1982). Visual evaluation techniques for skill analysis. *Journal of Physical Education, Recreation and Dance,* 53(1):21-26, 29.

Brown, R.M., and Councilman, J.E. (1971). The role of lift in propelling swimmers. In J.M. Cooper (Ed.), *Selected topics on biomechanics: Proceedings of the C.I.C. Symposium on Biomechanics* (pp. 179-188). Chicago: Athletic Institute.

Bunn, J. (1955). *Scientific principles of coaching.* Englewood Cliffs, NJ: Prentice-Hall.

Cavanagh, P.R. (1990). The mechanics of distance running: A historical perspective. In P.R. Cavanagh (Ed.), *Biomechanics of distance running* (pp. 1-34). Champaign, IL: Human Kinetics.

Chow, J.W., & Knudson, D.V. (2011). Use of deterministic models in sports and exercise biomechanics research. *Sports Biomechanics,* 10, 219-233.

Cureton, T.K. Jr. (1930). Mechanics and kinesiology of the crawl flutter kick. *Research Quarterly,* 1(4):93-96.

Cureton, T.K. Jr. (1939). Elementary principles and techniques of cinematographic analysis. *Research Quarterly,* 10(2):3-24.

Damask, A.C., and Damask, J.N. (1990). *Injury causation analyses: Case studies and data sources.* Charlottesville, VA: Michie Co.

Fenn, W.O. (1930). Frictional and kinetic factors in the work of sprint running. *American Journal of Physiology,* 92:583-611.

Fenn, W.O. (1931a). Work against gravity and work due to velocity changes in running. *American Journal of Physiology,* 93:433-462.

Fenn, W.O. (1931b). A cinematographic study of sprinters. *Scientific Monthly,* 32:346-354.

Frey, C. (1997). Footwear and stress fractures. *Clinics in Sports Medicine,* 16(2):249-257.

Grimston, S.K., Engsberg, J.R., Kloiber, R., and Hanley, D.A. (1991). Bone mass, external loads, and stress fractures in female runners. *International Journal of Sport Biomechanics,* 7:293-302.

Grimston, S.K., Willows, N.D., and Hanley, D.A. (1993). Mechanical loading regime and its relationship to bone mineral density in children. *Medicine and Science in Sports and Exercise,* 25(11):1203-1210.

Haapasalo, H., Sievanen, H., Kannus, P., Heinonen, A., Oja, P., and Vuori, I. (1996). Dimensions and estimated mechanical characteristics of the humerus after long-term tennis loading. *Journal of Bone and Mineral Research,* 11(6):864-872.

Hall, S.J. (2012). *Basic biomechanics* (6th ed.). New York: McGraw-Hill.

Hamill, J., and Knutzen, K.M. (2008). *Biomechanical basis of human movement* (3rd ed.). Baltimore: Williams & Wilkins.

Hatze, H. (1974). The meaning of the term 'biomechanics.' *Journal of Biomechanics,* 7:189-190.

Hay, J.G. (Winter 1982, No. 9). Biomechanics of sport—exploring or explaining (Part I). *International Society of Biomechanics Newsletter,* pp. 9-12.

Hay, J.G. (Spring 1983, No. 10). Biomechanics of sport—exploring or explaining (Part II). *International Society of Biomechanics Newsletter,* pp. 5-9.

Hay, J.G. (1984). The development of deterministic models for qualitative analysis. In R. Shapiro and J.R. Marett (Eds.), *Proceedings: Second National Symposium on Teaching Kinesiology and Biomechanics in Sports* (pp. 71-83). Colorado Springs, CO: NASPE.

Hay, J.G., and Reid, J.G. (1988). *Anatomy, mechanics, and human motion* (2nd ed.). Englewood Cliffs, NJ: Prentice-Hall.

Hill, A.V. (1928). The air resistance to a runner. *Proceedings of the Royal Society, B, 102:*43-50.

International Association of Athletics Federations. (2009). Scientific Research Project: Biomechanical analysis: 12th IAAF World Championships in Athletics Berlin, 15.23.08.2009: 100m men final: Usain Bolt. Retrieved from http://berlin.iaaf.org/mm/Document/Development/Research/05/31/54/20090817073528_httppostedfile_Analysis100mMenFinal_Bolt_13666.pdf.

International Association of Athletics Federations. (2009). Scientific Research Project: Biomechanical analysis: 12th IAAF World Championships in Athletics Berlin, 15.23.08.2009: 100 m men, Semifinal/Final. Retrieved from http://berlin.iaaf.org/mm/Document/Development/Research/05/30/83/20090817081546_httppostedfile_wch09_m100_final_13529.pdf.

James, S.L., Bates, B.T., and Osternig, L.R. (1978). Injuries to runners. *American Journal of Sports Medicine, 6*(2):40-50.

Jenkins, D.B. (1991). *Hollinshead's functional anatomy of the limbs and back* (6th ed.). Philadelphia: Saunders.

Knudson, D., and Morrison, C. (2002). *Qualitative analysis of human movement* (2nd ed.). Champaign, IL: Human Kinetics.

Kreighbaum, E.F., and Smith, M.A. (Eds.). (1995). *Sports and fitness equipment design.* Champaign, IL: Human Kinetics.

Lane, F.C. (1912). One hundred and twenty-two feet a second. *Baseball Magazine, 10*(2):22-25, 104, 106, 110.

LeVeau, B.F. (1992). *Williams & Lissner's biomechanics of human motion* (3rd ed.). Philadelphia: Saunders.

Maffulli, N., and King, J.B. (1992). Effects of physical activity on some components of the skeletal system. *Sports Medicine, 13*(6):393-407.

Marey, E.J. (1972). *Movement* (E. Pritchard, Trans.). New York: Amo. (Reprint edition; original translation published 1895 by D. Appleton Co., New York.)

McCaw, S.T. (1992) Leg length inequality: Implications for running injury prevention. *Sports Medicine, 14*(2):422-429.

McClay, I., and Manal, K. (1997). Coupling parameters in runners with normal and excessive pronation. *Journal of Applied Biomechanics, 13:*109-124.

McNitt-Gray, J. (1991). Kinematics and impulse characteristics of drop landings from three heights. *International Journal of Sport Biomechanics, 7:*201-224.

McNitt-Gray, J., Yokoi, T., and Millward, C. (1993). Landing strategy adjustments made by female gymnasts in response to drop height and mat composition. *Journal of Applied Biomechanics, 9:*173-190.

McNitt-Gray, J., Yokoi, T., and Millward, C. (1994). Landing strategies used by gymnasts on different landing surfaces. *Journal of Applied Biomechanics, 10:*237-252.

McPherson, M.N. (1988). The development, implementation, and evaluation of a program designed to promote competency in skill analysis. *Dissertation Abstracts International,* 48:3071A.

Messier, S.P., Davis, S.E., Curl, W.W., Lowery, R.B., and Pack, R.J. (1991). Etiologic factors associated with patellofemoral pain in runners. *Medicine and Science in Sports and Exercise,* 23:1008-1015.

Morris, M., Jobe, F.W., and Perry J. (1989). Electromyographic analysis of elbow function in tennis players. *American Journal of Sports Medicine,* 17:241-247.

Nelson, R.C. (1970). Biomechanics of sport: An overview. In J.M. Cooper (Ed.), *Selected topics on biomechanics: Proceedings of the C.I.C. Symposium on Biomechanics* (pp. 31-37). Chicago: Athletic Institute.

Nelson, R.C. (1980). Biomechanics: Past and present. In J.M. Cooper and B. Haven (Eds.), *Proceedings of the Biomechanics Symposium* (pp. 4-13). Indianapolis: Indiana State Board of Health.

Newton, I. (1934). *Principia* (Vol. I-II, Andrew Motte's translation revised by Florian Cajoari). Berkeley: University of California Press. (Original work published 1686, Motte's English translation 1729.)

Nigg, B.M. (Ed.). (1986). *Biomechanics of running shoes.* Champaign, IL: Human Kinetics.

Nordin, M., and Frankel, V.H. (1989). *Basic biomechanics of the musculoskeletal system* (2nd ed.). Philadelphia: Lea & Febiger.

Norman, R.W. (1977). An approach to teaching the mechanics of human motion at the undergraduate level. In C.J. Dillman and R.G. Sears (Eds.), *Proceedings: Kinesiology, a national conference on teaching* (pp. 113-123). Champaign, IL: University of Illinois.

Riek, S., Chapman, A.E., and Milner, T. (1999). A simulation of muscle force and internal kinematics of extensor carpi radialis brevis during backhand tennis stroke: Implications for injury. *Clinical Biomechanics,* 14:477-483.

Rodgers, M.M. (1993). Biomechanics of the foot during locomotion. In Grabiner, M.D. (Ed.), *Current issues in biomechanics* (pp. 33-52). Champaign, IL: Human Kinetics.

Scott, S.H., and Winter, D.M. (1990). Internal forces at chronic running injury sites. *Medicine and Science in Sports and Exercise,* 22(3):357-369.

Steindler, A. (1935). *Mechanics of normal and pathological locomotion in man.* Springfield, IL: Charles C Thomas.

Verducci, T. (2004, April 5). Out on the data frontier: where will the numbers game go in the future? Beyond hitting and pitching. *Sports Illustrated, 100* (14):64-65.

Viano, D.C., King, A.I., Melvin, J.W., and Weber, K. (1989). Injury biomechanics research: An essential element in the prevention of trauma. *Journal of Biomechanics,* 22(5):403-417.

Westfall, R. (1993). *The life of Isaac Newton.* Cambridge: Cambridge University Press.

Whitt, F.R., and Wilson, D.G. (1982). *Bicycling science* (2nd ed.). Cambridge, MA: MIT Press.

Williams, K.R. (1985). Biomechanics of running. In Terjung, R.L. (Ed.), *Exercise and Sport Science Reviews,* 13: 389-441.

Williams, K.R. (1993). Biomechanics of distance running. In Grabiner, M.D. (Ed.), *Current issues in biomechanics* (pp. 3-31). Champaign, IL: Human Kinetics.

Woodburne, R.T. (1978). *Essentials of human anatomy* (6th ed.). New York: Oxford University Press.

Yamada, H. (1970). *Strength of biological materials.* Baltimore: Williams & Wilkins.

Zatsiorsky, V.M. (1978). The present and future of the biomechanics of sports. In F. Landry and W.A.R. Orban (Eds.), *Biomechanics of sports and kinanthropometry* (pp. 11-17). Miami: Symposia Specialists, Inc.

Zebas, C., and Chapman, M. (1990). *Prevention of sports injuries: A biomechanical approach.* Dubuque, IA: Eddie Bowers.

Zernicke, R.F., Garhammer, J., and Jobe, F.W. (1977). Human patellar-tendon rupture. *Journal of Bone and Joint Surgery (American),* 59-A(2):179-183.

The following websites of professional biomechanics organizations can help you learn more about current events and research in the field.

American Society of Biomechanics
www.asbweb.org

Australian and New Zealand Society of Biomechanics
www.anzsb.asn.au

Canadian Society for Biomechanics
www.health.uottawa.ca/biomech/csb

European Society of Biomechanics
www.esbiomech.org

Gait and Clinical Movement Analysis Society
www.gcmas.org

International Society of Biomechanics
http://isbweb.org

International Society of Biomechanics in Sports
www.isbs.org

Biomechanics information and educational materials for coaches
http://coachesinfo.com/index.php

Note: The italicized *f* and *t* following page numbers refer to figures and tables, respectively.

Peter M. McGinnis, PhD, is a professor in the department of kinesiology at the State University of New York, College at Cortland, where he has taught since 1990. He is also the men's and women's pole vault coach at SUNY Cortland. Before 1990, Dr. McGinnis was an assistant professor in the department of kinesiology at the University of Northern Colorado. During that time he served as a sport biomechanist in the Sports Science Division of the U.S. Olympic Committee in Colorado Springs, where he conducted applied sport biomechanics research, tested athletes, taught biomechanics courses to coaches, and developed educational materials for coaches.

Dr. McGinnis is also the biomechanist for the pole vault event for USA Track and Field. As a member of the American Society of Testing Materials, he serves as chair of the pole vault equipment subcommittee and the task group on pole vault helmets. He has authored numerous articles and technical reports about the biomechanics of pole vaulting and has been a reviewer for *Sports Biomechanics,* the *Journal of Applied Biomechanics, Research Quarterly for Exercise and Sport,* and the *Journal of Sports Sciences.*

Dr. McGinnis is a member of numerous professional organizations, including the American College of Sports Medicine, American Society of Biomechanics, and the International Society of Biomechanics in Sport. He received a PhD in physical education from the University of Illinois in 1984 and a BS in engineering from Swarthmore College in 1976.

Quick-Reference Equations

MATHEMATICAL FORMULAS

Pythagorean theorem

$$A^2 + B^2 = C^2 \qquad (1.5)$$

Trigonometric functions

$$\sin\theta = \frac{\text{opposite side}}{\text{hypotenuse}} \qquad (1.6)$$

$$\cos\theta = \frac{\text{adjacent side}}{\text{hypotenuse}} \qquad (1.7)$$

$$\tan\theta = \frac{\text{opposite side}}{\text{adjacent side}} \qquad (1.8)$$

$$\theta = \arcsin\left(\frac{\text{opposite side}}{\text{hypotenuse}}\right) \qquad (1.9)$$

$$\theta = \arccos\left(\frac{\text{adjacent side}}{\text{hypotenuse}}\right) \qquad (1.10)$$

$$\theta = \arctan\left(\frac{\text{opposite side}}{\text{adjacent side}}\right) \qquad (1.11)$$

LINEAR KINEMATICS

Average speed

$$\bar{s} = \frac{\ell}{\Delta t} \qquad (2.5)$$

Average velocity

$$\bar{v} = \frac{d}{\Delta t} \qquad (2.6)$$

Average acceleration

$$\bar{a} = \frac{v_f - v_i}{\Delta t} \qquad (2.9)$$

PROJECTILE EQUATIONS

Vertical motion (y)

Vertical position:

$$y_f = y_i + v_i\Delta t + \frac{1}{2}g(\Delta t)^2 \qquad (2.14)$$

$$y_f = \frac{1}{2}g(\Delta t)^2 \quad \text{if } y_i = 0 \text{ and } v_i = 0 \qquad (2.16)$$

Vertical velocity:

$$v_f = v_i + g\Delta t \qquad (2.11)$$

$$v^2 = v^2 + 2g\Delta y \qquad (2.15)$$

$$v_{peak} = 0 \qquad (2.19)$$

$$v_f = g\Delta t \quad \text{if } y_i = 0 \text{ and } v_i = 0 \qquad (2.17)$$

$$v^2 = 2g\Delta y \quad \text{if } v_i = 0 \qquad (2.18)$$

Vertical acceleration:

$$a = g = -9.81 \text{ m/s}^2 \qquad (2.10)$$

Horizontal motion (x)

Horizontal position:

$$x = v\Delta t \qquad (2.26)$$

Horizontal velocity:

$$v = v_f = v_i = \text{constant} \qquad (2.22)$$

Horizontal acceleration:

$$a = 0 \qquad (2.23)$$

Other equations governing projectile motion

Time of flight:

$$\Delta t_{up} = \Delta t_{down} \quad \text{if } y_f = y_i \qquad (2.20)$$

$$\Delta t_{flight} = 2\Delta t_{up} \quad \text{if } y_f = y_i \qquad (2.21)$$

Parabolic equation:

$$y_f = y_i + v_{y_i}\left(\frac{x}{v_x}\right) + \frac{1}{2}g\left(\frac{x}{v_x}\right)^2 \qquad (2.27)$$

LINEAR KINETICS

Weight

$$W = mg \qquad (1.2)$$

Static and dynamic friction

$$F_s = \mu_s R \qquad (1.3)$$

$$F_d = \mu_d R \qquad (1.4)$$

Static equilibrium

$$\Sigma F = 0 \qquad (1.12)$$

$$\Sigma F_x = 0 \qquad (1.13)$$

$$\Sigma F_y = 0 \qquad (1.14)$$

Newton's 1st law: law of inertia

$$v = \text{constant if } \Sigma F = 0 \qquad (3.1a)$$

or

$$\Sigma F = 0 \quad \text{if } v = \text{constant} \qquad (3.1b)$$

Linear momentum

$$L = mv \qquad (3.6)$$

Conservation of momentum

$$L = \text{constant if } \Sigma F = 0 \qquad (3.7)$$

$$L_x = \text{constant if } \Sigma F_x = 0 \qquad (3.8)$$

$$L_y = \text{constant if } \Sigma F_y = 0 \qquad (3.9)$$

$$\begin{aligned} L_i = \Sigma(mu) &= m_1u_1 + m_2u_2 + m_3u_3 \\ &+ \ldots = m_1v_1 + m_2v_2 + m_3v_3 \\ &+ \ldots = \Sigma(mv) = L_f = \text{constant} \\ &\text{if } \Sigma F = 0 \end{aligned} \qquad (3.11)$$

Perfectly elastic collision of two objects

$$v_1 = \frac{2m_2u_2 + (m_1 - m_2)u_1}{m_1 + m_2} \qquad (3.17)$$

Perfectly inelastic collision of two objects

$$m_1u_1 + m_2u_2 = (m_1 + m_2)v \qquad (3.19)$$

Coefficient of restitution

$$e = \left|\frac{v_1 - v_2}{u_1 - u_2}\right| = \left|\frac{v_2 - v_1}{u_1 - u_2}\right| \qquad (3.20)$$

Newton's 2nd law: law of acceleration

$$\Sigma F = ma \qquad (3.22)$$

$$\Sigma F_x = ma_x \qquad (3.23)$$

$$\Sigma F_y = ma_y \qquad (3.24)$$

Impulse–momentum equation

$$\Sigma \bar{F} \Delta t = m(v_f - v_i) \qquad (3.29)$$

Universal law of gravitation: gravitational force

$$F = G\left(\frac{m_1 m_2}{r^2}\right) \qquad (3.30)$$

WORK, POWER, AND ENERGY

Work

$$U = \bar{F}(d) \qquad (4.2)$$

Kinetic energy

$$KE = \frac{1}{2}mv^2 \qquad (4.4)$$

Gravitational potential energy

$$PE = Wh \qquad (4.5)$$

Strain energy

$$SE = \frac{1}{2}k\,\Delta x^2 \qquad (4.7)$$

Work–energy principle

$$U = \Delta E \qquad (4.8)$$

Power

$$P = \frac{U}{\Delta t} \qquad (4.12)$$

$$P = \bar{F}\bar{v} \qquad (4.13)$$

ANGULAR KINEMATICS

Angular position measured in radians

$$\theta = \frac{\text{arc length}}{r} = \frac{\ell}{r} \qquad (6.1)$$

Angular displacement and arc length

$$\ell = \Delta\theta r \qquad (6.4)$$

Average angular velocity

$$\bar{\omega} = \frac{\Delta\theta}{\Delta t} = \frac{\theta_f - \theta_i}{\Delta t} \qquad (6.6)$$